# Computing Statistics under Interval and Fuzzy Uncertainty

## Studies in Computational Intelligence, Volume 393

**Editor-in-Chief**
Prof. Janusz Kacprzyk
Systems Research Institute
Polish Academy of Sciences
ul. Newelska 6
01-447 Warsaw
Poland
*E-mail:* kacprzyk@ibspan.waw.pl

---

Further volumes of this series can be found on our
homepage: springer.com

Vol. 371. Leonid Perlovsky, Ross Deming, and
Roman Ilin (Eds.)
*Emotional Cognitive Neural Algorithms with Engineering
Applications*, 2011
ISBN 978-3-642-22829-2

Vol. 372. António E. Ruano and
Annamária R. Várkonyi-Kóczy (Eds.)
*New Advances in Intelligent Signal Processing*, 2011
ISBN 978-3-642-11738-1

Vol. 373. Oleg Okun, Giorgio Valentini, and Matteo Re (Eds.)
*Ensembles in Machine Learning Applications*, 2011
ISBN 978-3-642-22909-1

Vol. 374. Dimitri Plemenos and Georgios Miaoulis (Eds.)
*Intelligent Computer Graphics 2011*, 2011
ISBN 978-3-642-22906-0

Vol. 375. Marenglen Biba and Fatos Xhafa (Eds.)
*Learning Structure and Schemas from Documents*, 2011
ISBN 978-3-642-22912-1

Vol. 376. Toyohide Watanabe and Lakhmi C. Jain (Eds.)
*Innovations in Intelligent Machines – 2*, 2011
ISBN 978-3-642-23189-6

Vol. 377. Roger Lee (Ed.)
*Software Engineering Research, Management
and Applications 2011*, 2011
ISBN 978-3-642-23201-5

Vol. 378. János Fodor, Ryszard Klempous, and
Carmen Paz Suárez Araujo (Eds.)
*Recent Advances in Intelligent Engineering Systems*, 2011
ISBN 978-3-642-23228-2

Vol. 379. Ferrante Neri, Carlos Cotta, and
Pablo Moscato (Eds.)
*Handbook of Memetic Algorithms*, 2011
ISBN 978-3-642-23246-6

Vol. 380. Anthony Brabazon, Michael O'Neill, and
Dietmar Maringer (Eds.)
*Natural Computing in Computational Finance*, 2011
ISBN 978-3-642-23335-7

Vol. 381. Radosław Katarzyniak, Tzu-Fu Chiu,
Chao-Fu Hong, and Ngoc Thanh Nguyen (Eds.)
*Semantic Methods for Knowledge Management and
Communication*, 2011
ISBN 978-3-642-23417-0

Vol. 382. F.M.T. Brazier, Kees Nieuwenhuis, Gregor Pavlin,
Martijn Warnier, and Costin Badica (Eds.)
*Intelligent Distributed Computing V*, 2011
ISBN 978-3-642-24012-6

Vol. 383. Takayuki Ito, Minjie Zhang, Valentin Robu,
Shaheen Fatima, and Tokuro Matsuo (Eds.)
*New Trends in Agent-Based Complex Automated
Negotiations*, 2012
ISBN 978-3-642-24695-1

Vol. 384. Daphna Weinshall, Jörn Anemüller,
and Luc van Gool (Eds.)
*Detection and Identification of Rare Audiovisual Cues*, 2012
ISBN 978-3-642-24033-1

Vol. 385. Alex Graves
*Supervised Sequence Labelling with Recurrent Neural
Networks*, 2012
ISBN 978-3-642-24796-5

Vol. 386. Marek R. Ogiela and Lakhmi C. Jain (Eds.)
*Computational Intelligence Paradigms in Advanced Pattern
Classification*, 2012
ISBN 978-3-642-24048-5

Vol. 387. David Alejandro Pelta, Natalio Krasnogor,
Dan Dumitrescu, Camelia Chira, and Rodica Lung (Eds.)
*Nature Inspired Cooperative Strategies for Optimization
(NICSO 2011)*, 2011
ISBN 978-3-642-24093-5

Vol. 388. Tiansi Dong
*Recognizing Variable Environments*, 2012
ISBN 978-3-642-24057-7

Vol. 389. Patricia Melin
*Modular Neural Networks and Type-2 Fuzzy Systems for
Pattern Recognition*, 2012
ISBN 978-3-642-24138-3

Vol. 390. Robert Bembenik, Lukasz Skonieczny,
Henryk Rybiński, and Marek Niezgódka (Eds.)
*Intelligent Tools for Building a Scientific Information
Platform*, 2012
ISBN 978-3-642-24808-5

Vol. 391. Herwig Unger, Kyandoghere Kyamaky,
and Janusz Kacprzyk (Eds.)
*Autonomous Systems: Developments and Trends*, 2012
ISBN 978-3-642-24805-4

Vol. 392. Narendra Chauhan, Machavaram Kartikeyan,
and Ankush Mittal
*Soft Computing Methods for Microwave and Millimeter-Wave
Design Problems*, 2012
ISBN 978-3-642-25562-5

Vol. 393. Hung T. Nguyen, Vladik Kreinovich, Berlin Wu,
and Gang Xiang
*Computing Statistics under Interval and Fuzzy
Uncertainty*, 2012
ISBN 978-3-642-24904-4

Hung T. Nguyen, Vladik Kreinovich, Berlin Wu,
and Gang Xiang

# Computing Statistics under Interval and Fuzzy Uncertainty

## Applications to Computer Science and Engineering

 Springer

Authors

**Prof. Hung T. Nguyen**
ChiangMai University
Department of Economics
Chiangmai 50100
Thailand
E-mail: hunguyen@nmsu.edu

**Prof. Vladik Kreinovich**
University of Texas at El Paso
Department of Computer Science
500 W. University
El Paso, TX 79968
USA
E-mail: vladik@utep.edu

**Prof. Berlin Wu**
National Chengchi University
Department of Mathematical Sciences
National Chengchi University
Taipei 116
Taiwan
E-mail: berlin@nccu.edu.tw

**Dr. Gang Xiang**
Applied Biomathematics, Inc.
100 North Country Rd.
Setauket, NY 11733
USA
E-mail: gxiang@sigmaxi.net

ISBN 978-3-642-44570-5

ISBN 978-3-642-24905-1 (eBook)

DOI 10.1007/978-3-642-24905-1

Studies in Computational Intelligence

ISSN 1860-949X



*Typeset & Cover Design:* Scientific Publishing Services Pvt. Ltd., Chennai, India.

Printed on acid-free paper

9 8 7 6 5 4 3 2 1

springer.com

# Preface

In many areas of science and engineering, we have a class ("population") of objects, and we are interested in the values of one or several quantities characterizing objects from this population. For example, we are interested in the human population in a certain region, and we are interested in their heights, weights, etc.

Different objects from a population have, in general, different values of the desired characteristics. Measuring, storing, and processing *all* these values is often not practically feasible. Instead of measuring all these values, we therefore measure the values $x_1, \ldots, x_n$ corresponding to a *sample* from the population. Based on these sample values, we need to estimate *statistical* characteristics $c$ – i.e., characteristics that describe the population as a whole, such as the mean and the variance of different quantities, the correlation between different quantities, etc.

For each desired characteristic $c$, there exist several algorithms

$$C(x_1, \ldots, x_n)$$

for estimating $c$ based on the sample values; these algorithms are known as *statistics*. For example, a typical statistic for estimating the mean is the arithmetic average, in which case

$$C(x_1, \ldots, x_n) = \frac{x_1 + \ldots + x_n}{n}.$$

In other words, to estimate, e.g., the average height $c$ of a certain population, we measure the heights $x_1, \ldots, x_n$ of several representatives from this population, and we use the arithmetic mean $\dfrac{x_1 + \ldots + x_n}{n}$ of these heights as an estimate for the population height $c$. Slightly more complex statistics are used to estimate the variance, the correlation, etc.

Of course, an estimate based on a finite sample is only approximate. To get more accurate estimates, we need to take larger samples. For most statistics used in science and engineering, there exist estimates of the accuracy with which the sample-based value $C \stackrel{\text{def}}{=} C(x_1, \ldots, x_n)$ approximates the desired statistical characteristic $c$ of the population as a whole. In other words, there are methods for estimating the difference $C - c$ between the sample-based estimate $C$ for the desired characteristic and its (unknown) actual value $c$.

These accuracy estimates are based on the assumption that for each object $i$, we know the *exact* value $x_i$ of the corresponding quantity. In practice, however, the values $x_i$ come from measurements (or from expert estimates), and measurements are never absolutely accurate: the measurement result $\widetilde{x}_i$ is, in general, different from the actual (unknown) value $x_i$ of the measured quantity. In other words, the estimation error $\Delta x_i \stackrel{\text{def}}{=} \widetilde{x}_i - x_i$ is, in general, different from 0. As a result, the statistic $\widetilde{C} \stackrel{\text{def}}{=} C(\widetilde{x}_1, \ldots, \widetilde{x}_n)$ computed based on the measurement results is, in general, different from the ideal value $C \stackrel{\text{def}}{=} C(x_1, \ldots, x_n)$.

We want to know how accurate are these estimates $\widetilde{C}$, i.e., what is the size of the quantity $\widetilde{C} - c$. Statistical analysis enables us to estimate a (slightly different quantity) $C - c$, with $C \neq \widetilde{C}$. Thus, to estimate the value of the quantity $\widetilde{C} - c$, we need to estimate the difference $(\widetilde{C} - c) - (C - c) \, (= \widetilde{C} - C)$ between the quantity $\widetilde{C} - c$ that we *want to* estimate and the quantity $C - c$ that we are already *able to* estimate. In other words, we need to estimate the difference $\widetilde{C} - C$.

Estimates for this difference $\widetilde{C} - C$ are also known in statistics – but only for the case when we know the exact probability distribution of all the measurement errors $\Delta x_i$. In practice, often, we do not know these distributions. Usually, we only know the upper bound $\Delta_i$ on the measurement error: $|\Delta x_i| \leq \Delta_i$. In this case, once we know the measurement result $\widetilde{x}_i$, we can only conclude that the actual (unknown) value of the quantity $x_i$ belongs to the *interval*

$$\boldsymbol{x}_i \stackrel{\text{def}}{=} [\widetilde{x}_i - \Delta_i, \widetilde{x}_i + \Delta_i].$$

In other cases, the values $\widetilde{x}_i$ are expert estimates, and we only have *fuzzy* information about the estimation errors $\Delta x_i$.

The main objective of this book is *computing statistics under* such *interval and fuzzy uncertainty*. In more precise terms, our objective is to estimate the difference $\widetilde{C} - C = C(\widetilde{x}_1, \ldots, \widetilde{x}_n) - C(x_1, \ldots, x_n)$ under the assumption that we only have interval or fuzzy estimates of the differences $\Delta x_i = \widetilde{x}_i - x_i$.

Part I of this book contains the formulation of the problem and a brief overview of the general techniques that can be used to solve this problem. Specifically, we formulate the general problem, we briefly describe the general techniques of interval and fuzzy computations, and we also explain which statistical characteristics are most frequently used in practice – and why.

In particular, in Part I, we explain that the algorithmic problem of computing statistics under fuzzy uncertainty can be reduced to the problem of computing statistics under interval uncertainty. Specific algorithms for computing different statistics under interval uncertainty are presented in Part II.

Algorithms presented in Part II assume that the input data is of "good quality", i.e., that the interval inputs are guaranteed to contain the actual (unknown) values – and that the fuzzy inputs are also in good accordance with the actual values. In real life, not all inputs may be reliable or accurate. Part III described how to gauge the quality of the input data.

Part IV contains examples of applications of the resulting techniques. This part contains applications ranging from computer science (optimal scheduling of different processors) and computer engineering (design of computer chips) to data processing in geosciences, radar imaging, and structural mechanics.

In Part V, we sketch possible extensions of our results beyond intervals and fuzzy sets: to ranges which are more general than interval ranges, to p-boxes and subsets of p-boxes, to interval-valued and type-2 fuzzy sets, etc.

This work was supported in part by NSF grants HRD-0734825, EAR-0225670, and EIA-0080940, by Texas Department of Transportation contract No. 0-5453, by the Japan Advanced Institute of Science and Technology (JAIST) International Joint Research Grant 2006-08, and by the Max Planck Institut für Mathematik.

The authors are thankful to all their colleagues and participants of numerous conferences for valuable suggestions.

Las Cruces, New Mexico                                           Hung T. Nguyen
El Paso, Texas                                                   Vladik Kreinovich
Taipei, Taiwan                                                           Berlin Wu
El Paso, Texas                                                         Gang Xiang

August 2011

# Contents

**Part III Towards Computing Statistics under Interval and Fuzzy
Uncertainty: Gauging the Quality of the Input Data**



**Part IV Applications**

## Part V  Beyond Interval and Fuzzy Uncertainty

# Computing Statistics under Interval and Fuzzy Uncertainty: Formulation of the Problem and an Overview of General Techniques Which Can Be Used for Solving This Problem

# 1

# Formulation of the Problem

*Need for statistical analysis.* In many areas of science and engineering, we have a class ("population") of objects, and we are interested in the values of one or several quantities characterizing objects from this population. For example, we are interested in the human population in a certain region, and we are interested in their heights, weights, etc.

Different objects from a population have, in general, different values of the desired characteristics. Measuring, storing, and processing *all* these values is often not practically feasible. Instead of measuring all these values, we therefore measure the values $x_1, \ldots, x_n$ corresponding to a *sample* from the population. Based on these sample values, we need to estimate *statistical* characteristics $c$ – i.e., characteristics that describe the population as a whole, such as the mean and the variance of different quantities, the correlation between different quantities, etc.

*Need for computing statistics.* For each desired characteristic $c$, there exist several algorithms

$$C(x_1, \ldots, x_n)$$

for estimating $c$ based on the sample values; these algorithms are known as *statistics*. For example, a typical statistic for estimating the mean is the arithmetic average, in which case

$$C(x_1, \ldots, x_n) = \frac{x_1 + \ldots + x_n}{n}.$$

In other words, to estimate, e.g., the average height $c$ of a certain population, we measure the heights $x_1, \ldots, x_n$ of several representatives from this population, and we use the arithmetic mean $\dfrac{x_1 + \ldots + x_n}{n}$ of these heights as an estimate for the population height $c$. Slightly more complex statistics are used to estimate the variance, the correlation, etc.

*How accurate are sample-based estimates?* Of course, an estimate based on a finite sample is only approximate. To get more accurate estimates, we need to take larger samples. For most statistics used in science and engineering,

H.T. Nguyen et al.: Computing Stat. under Interval & Fuzzy Uncertain., SCI 393, pp. 3–8.
springerlink.com

there exist estimates of the accuracy with which the sample-based value $C \stackrel{\text{def}}{=} C(x_1, \ldots, x_n)$ approximates the desired statistical characteristic $c$ of the population as a whole.

In other words, there are methods for estimating the difference $C - c$ between the sample-based estimate $C$ for the desired characteristic and its (unknown) actual value $c$; see, e.g., [283, 305].

*Main assumption behind the traditional statistical accuracy estimates.* The existing statistical estimates for the difference $C - c$ are based on the assumption that for each object $i$, we know the *exact* value $x_i$ of the corresponding quantity.

*Measurement and estimation errors.* In practice, the values $x_i$ come from measurements (or from expert estimates), and measurements are never absolutely accurate: the measurement result $\widetilde{x}_i$ is, in general, different from the actual (unknown) value $x_i$ of the measured quantity. In other words, the estimation error $\Delta x_i \stackrel{\text{def}}{=} \widetilde{x}_i - x_i$ is, in general, different from 0. As a result, the statistic $\widetilde{C} \stackrel{\text{def}}{=} C(\widetilde{x}_1, \ldots, \widetilde{x}_n)$ computed based on the measurement results is, in general, different from the ideal value $C \stackrel{\text{def}}{=} C(x_1, \ldots, x_n)$.

*Need to compute statistics under uncertainty.* We want to know how accurate are the estimates $\widetilde{C}$, i.e., what is the size of the quantity $\widetilde{C} - c$. Statistical analysis enables us to estimate a (slightly different quantity) $C - c$, with $C \neq \widetilde{C}$. Thus, to estimate the value of the quantity $\widetilde{C} - c$, we need to estimate the difference $(\widetilde{C} - c) - (C - c) \; (= \widetilde{C} - C)$ between the quantity $\widetilde{C} - c$ that we *want to* estimate and the quantity $C - c$ that we are already *able to* estimate. In other words, we need to estimate the difference $\widetilde{C} - C$.

*Computing statistics under uncertainty: what is known.* Estimates for the difference $\widetilde{C} - C$ are also known in statistics – for the case when we know the exact probability distribution of all the measurement errors $\Delta x_i$; see, e.g., [283, 305].

*Comment.* In most practical applications, it is assumed that the corresponding measurement errors are normally distributed with 0 mean and known standard deviation. Numerous engineering techniques are known (and widely used) for processing this uncertainty; see, e.g., [283].

*How the probability distribution of $\Delta x_i$ is determined.* In practice, we can determine the desired probabilities of different values of $\Delta x_i$ by comparing

- the result $\widetilde{x}_i$ of measuring a certain quantity with this instrument and
- the result $\widetilde{x}_i'$ of measuring the same quantity by a standard (much more accurate) measuring instrument.

Since the standard measuring instrument is much more accurate than the one we use, i.e., $|\widetilde{x}_i' - x_i| \ll |\widetilde{x}_i - x_i|$, we can assume that $\widetilde{x}_i' = x_i$, and thus, that the difference $\widetilde{x}_i - \widetilde{x}_i'$ between these two measurement results is practically equal to the measurement error $\Delta x_i = \widetilde{x}_i - x_i$. Thus, the empirical distribution of the difference $\widetilde{x}_i - \widetilde{x}_i'$ is close to the desired probability distribution for measurement error.

*Sometimes, we do not know the probabilities.* In the traditional approach to estimating accuracy of the results of data processing, we assume that we know the probabilities of different values of measurement errors $\Delta x_i$.

There are two cases when the determination of these probabilities is not done:

- First is the case of *cutting-edge* measurements, e.g., measurements in fundamental science. When the Hubble telescope detects the light from a distant galaxy, there is no "standard" (much more accurate) telescope floating nearby that we can use to calibrate the Hubble: the Hubble telescope is the best we have.
- The second case is the case of real *industrial* applications (such as measurements on the shop floor). In this case, in principle, every sensor can be thoroughly calibrated, but sensor calibration is so costly – usually costing several orders of magnitude more than the sensor itself – that manufacturers rarely do it (only if it is absolutely necessary).

In both cases, we have no information about the probabilities of $\Delta x_i$; the only information we have is the upper bound on the measurement error.

*Upper bounds on measurement errors.* The manufacturers of a measuring device usually provide us with an upper bound $\Delta_i$ for the (absolute value of) possible measurement errors, i.e., with the bound $\Delta_i$ for which we are guaranteed that $|\Delta x_i| \leq \Delta_i$.

The need for such a bound comes from the very nature of a measurement process. Indeed, if no such bound is provided, this means that the actual value $x_i$ can be as different from the "measurement result" $\widetilde{x}_i$ as possible. Such a value $\widetilde{x}_i$ – which can be very different from the actual value $x_i$ – is not a measurement, it is a wild guess.

Since the (absolute value of the) measurement error $\Delta x_i = \widetilde{x}_i - x_i$ is bounded by the given bound $\Delta_i$, we can therefore guarantee that the actual (unknown) value of the desired quantity belongs to the interval

$$\boldsymbol{x}_i \stackrel{\text{def}}{=} [\widetilde{x}_i - \Delta_i, \widetilde{x}_i + \Delta_i].$$

*Example.* If the measured value of a quantity is $\widetilde{x}_i = 1.0$, and the upper bound $\Delta_i$ on the measurement error is 0.1, this means that the (unknown) actual value of the measured quantity can be anywhere between $1 - 0.1 = 0.9$ and $1 + 0.1 = 1.1$, i.e., that it can take any value from the interval $[0.9, 1.1]$.

*Case of interval uncertainty.* In the cases when our only information about the measurement error is the upper bound, after performing a measurement and getting a measurement result $\widetilde{x}_i$, the only information that we have about the actual value $x_i$ of the measured quantity is that it belongs to the interval $\boldsymbol{x}_i = [\widetilde{x}_i - \Delta_i, \widetilde{x}_i + \Delta_i]$. In other words, we do know not the actual value $x_i$ of the $i$-th quantity. Instead, we know the *interval* $[\widetilde{x}_i - \Delta_i, \widetilde{x}_i + \Delta_i]$ that contains $x_i$.

*Important specific cases of measurement-related interval uncertainty.* To describe specific cases of measurement-related interval uncertainty, let us recall the main idea of a measurement procedure, and show how different stages of this procedure contribute to measurement uncertainty.

The main objective of a measuring instrument is to measure the value of a desired quantity, In some cases – e.g., when measuring current – this quantity characterizes a signal; in other cases, a sensor transform this quantity into a signal. For example, a weight is not a signal, but the scale embedded in the electronic scales translates the weight into a signal.

This signal comes with noise. Part of this noise comes from the outside, part is generated by the measuring instrument itself. This noise is one of the main reasons why the measurement result $\tilde{x}$ is, in general, different from the actual value $x$.

The signal needs to affect the sensor. Most sensor have inertia – in the sense that if a signal is below a certain limit, the sensor does not react at all.

For signals below the detection limit, this fact results in measurement uncertainty.

Once the sensor reacts, it gets into one of the easily distinguishable states corresponding to different measurement results. Nowadays, most sensors generate digital signals describing the measurement results; these signals are the distinguishable states. In the old days, different reading on a scale were different states.

As a result of each measurement, a measuring instrument generates a finite sequence of bits, and the number of bits $B$ that it can generate during the time allocated for a measurement is bounded. There are $\leq 2^B$ combinations of $B$ bits. Thus, each measuring instrument has a finite number $\leq 2^B$) of possible measuring results. This discretization also contributes to the measurement uncertainty.

*Case of detection limits.* For some measuring instruments, noise and discretization errors are negligible, and the detection error is the main source of measurement uncertainty.

In this case, if the actual value $x$ is below the detection limit $DL$, the sensor does not measure anything, i.e., we get $\tilde{x} = 0$. Thus, the measured value $\tilde{x} = 0$ corresponds to the interval $[0, DL]$ of possible values $x$. If a sensor, e.g., did not detect any ozone, this means that the ozone concentration is below its detection limit $DL$, i.e., in the interval $[0, DL]$.

For values above the detection limit, the sensor returns, in effect, the actual value $\tilde{x} = x$.

*Case of discretized data.* In other cases, noise and detection limits and negligible, and discretization is the main source of measurement error.

For example, if we experiment on the fish and watch it daily, and a fish is alive on Day 5 but dead on Day 6, then all we know about its lifetime is that it is in the interval $[5, 6]$.

*Interval uncertainty motivated by privacy.* In addition to measurement uncertainty, there is another sources of interval uncertainty: intervals can also come from the need to preserve privacy.

Indeed, in medicine, in social studies, etc., it is important to perform statistical data analysis. By performing such an analysis, we can find, e.g., the correlation between the age and income, between the gender and side effects of a medicine, etc. People are often willing to supply the needed confidential data for research purposes. However, many of them are worried that it may be possible to extract their confidential data from the results of statistical data processing – and indeed such privacy violations have occurred in the past.

One way to prevent such privacy violations is to replace the *exact* values of the parameters that could be used to identify a person with *ranges*. Quantities whose values can be used to identify a person are called *quasi-identifiers*; see, e.g., [322].

For example, we divide the set of all possible ages into ranges $[0, 10]$, $[10, 20]$, $[20, 30]$, etc. Then, instead of storing the actual age of 26 that would enable us to identify a person, we only store the range $[20, 30]$ which contains the actual age value. Such ranges are called *generalized ranges*, to distinguish them from the actual ranges of the quantities [322].

This approach successfully protects privacy, but it leads to a computational challenge. For example, if we want to estimate the variance of ages, we can no longer simply compute the statistic such as population variance

$$V = \frac{1}{n} \cdot \sum_{i=1}^{n} x_i^2 - \left( \frac{1}{n} \cdot \sum_{i=1}^{n} x_i \right)^2;$$

since we only know the intervals $[\underline{x}_i, \overline{x}_i]$ of possible values of $x_i$, we can only compute the range $V$ of possible values of this statistic when $x_i \in \boldsymbol{x}_i$.

*Comment about roundoff errors.* In many numerical computations, there is another important source of interval uncertainty: roundoff errors. These errors come from two sources:

- first, many physical formulas use numbers like $z = \sqrt{2}$ or $z = \pi$ which are not binary rational and thus, cannot be exactly represented in the existing computers; even many rational numbers like $z = 1/3$ or $z = 1.2$ are not binary rational; similarly, the values of many useful functions like $z = \sin(x)$, $z = \ln(x)$, etc., are usually not binary rational and thus, cannot be exactly represented;
- second, even when two numbers $x$ and $y$ are exactly represented in a computer, with exactly as many binary digits as a computer can handle, their product $z = x \cdot y$ – while still binary rational – takes twice as many digits and thus, cannot be exactly represented, it must be rounded off.

In these cases, even for the ideal rounding, all we know about the actual computational result $z$ is that it belongs to the interval $[\widetilde{z} - \varepsilon, \widetilde{z} + \varepsilon]$, where

$\widetilde{z}$ is the computation result, and $\varepsilon$ is the guaranteed upper bound on the roundoff accuracy.

In general, as it is well known in numerical computations, roundoff errors are very important. However, in our problem, measurement errors, which are usually about 10%, 1%, 0.1%, are several orders of magnitude larger than these roundoff errors of $10^{-7}$ for single precision (and even smaller for double precision). Thus, in comparison with measurement errors, roundoff errors can be safely ignored.

The additional reason why roundoff errors can be safely ignored is that our objective is to estimate the values of the statistical characteristics like mean, variance, etc., based on the sample data. The relative accuracy of determining a statistical characteristic based on a sample of size $n$ is usually of order $1/\sqrt{n}$. Even for $n = 10^3$ the resulting uncertainty is $\approx 1/\sqrt{1000} \approx 3\%$, much larger than the roundoff errors.

*Need to process fuzzy uncertainty.* In some practical situations, we only have expert estimates for the inputs $x_i$. Sometimes, experts provide guaranteed bounds on the $x_i$, and even the probabilities of different values within these bounds. However, such cases are rare. Usually, the experts' opinions about the uncertainty of their estimates are described by (imprecise, "fuzzy") words from natural language. For example, an expert can say that the value $x_i$ of the $i$-th quantity is approximately equal to 1.0, with an accuracy most probably of about 0.1.

The need to process such "fuzzy" information was first emphasized in the early 1960s by L. Zadeh who designed a special technique of *fuzzy logic* for such processing; see, e.g., [156, 252].

*Our main objective.* The main objective of this book is *computing statistics under* such *interval and fuzzy uncertainty.*

In more precise terms, our objective is to estimate the difference $\Delta C \overset{\text{def}}{=} \widetilde{C} - C = C(\widetilde{x}_1, \ldots, \widetilde{x}_n) - C(x_1, \ldots, x_n)$ under the assumption that we only have interval or fuzzy estimates of the differences $\Delta x_i = \widetilde{x}_i - x_i$.

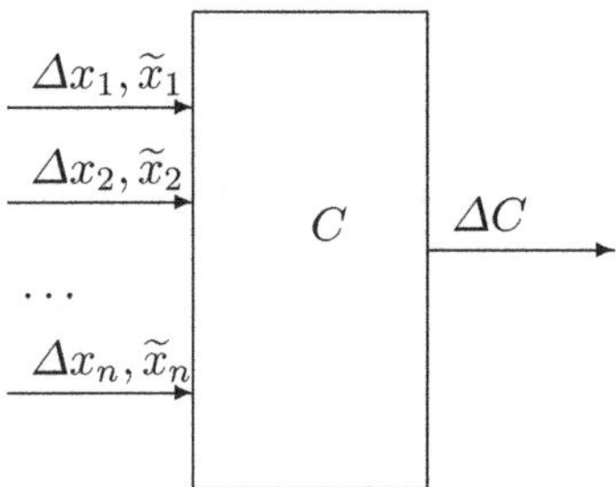

# Computing Statistics under Probabilistic and Interval Uncertainty: A Brief Description

*Computing statistics under probabilistic uncertainty.* In the case of *probabilistic uncertainty*, we know the probability distributions for measurement errors corresponding to all the inputs $x_1, \ldots, x_n$, and we want to find the probability distribution corresponding to the statistic $C(x_1, \ldots, x_n)$.

Let us formulate the corresponding problem in more precise terms. We are given:

- an integer $n$;
- $n$ measurement results $\widetilde{x}_1, \ldots, \widetilde{x}_n$;
- $n$ probability distributions corresponding to $n$ variables $\Delta x_i$ (with the assumption that these distributions are independent), and
- an algorithm $C(x_1, \ldots, x_n)$ which transforms $n$ real numbers into a real number $C = C(x_1, \ldots, x_n)$.

We need to compute the probability distribution of the difference $\Delta C = \widetilde{C} - C$, where $\widetilde{C} \stackrel{\text{def}}{=} C(\widetilde{x}_1, \ldots, \widetilde{x}_n)$, $C \stackrel{\text{def}}{=} C(x_1, \ldots, x_n)$, and $x_i \stackrel{\text{def}}{=} \widetilde{x}_i - \Delta x_i$.

*Case of interval uncertainty.* Let us consider the cases when our only information about the measurement error is the upper bound. In such cases, as we have mentioned, for each $i$, we know the interval $\boldsymbol{x}_i$ of possible values of $x_i$, and we need to find the range

$$\boldsymbol{C} \stackrel{\text{def}}{=} \{C(x_1, \ldots, x_n) : x_1 \in \boldsymbol{x}_1, \ldots, x_n \in \boldsymbol{x}_n\}$$

of the given function $C(x_1, \ldots, x_n)$ over all possible tuples $x = (x_1, \ldots, x_n)$ with $x_i \in \boldsymbol{x}_i$.

The process of computing this interval range based on the input intervals $\boldsymbol{x}_i$ is called *interval computations*; see, e.g., [142, 148].

Since the function $C(x_1, \ldots, x_n)$ is usually continuous, this range is also an interval, i.e., $\boldsymbol{y} = [\underline{y}, \overline{y}]$ for some $\underline{y}$ and $\overline{y}$. So, to find this range, it is sufficient to find the endpoints $\underline{y}$ and $\overline{y}$ of this interval.

Let us formulate the corresponding *interval computations* problem in precise terms. We are given:

H.T. Nguyen et al.: Computing Stat. under Interval & Fuzzy Uncertain., SCI 393, pp. 9–10.
springerlink.com                                           © Springer-Verlag Berlin Heidelberg 2012

- an integer $n$;
- $n$ intervals $\boldsymbol{x}_1 = [\underline{x}_1, \overline{x}_1], \ldots, \boldsymbol{x}_n = [\underline{x}_n, \overline{x}_n]$, and
- an algorithm $C(x_1, \ldots, x_n)$ which transforms $n$ real numbers into a real number $C = C(x_1, \ldots, x_n)$.

We need to compute the endpoints $\underline{y}$ and $\overline{y}$ of the interval

$$\boldsymbol{C} = \left[\underline{C}, \overline{C}\right] = \{C(x_1, \ldots, x_n) : x_1 \in [\underline{x}_1, \overline{x}_1], \ldots, [\underline{x}_n, \overline{x}_n]\}.$$

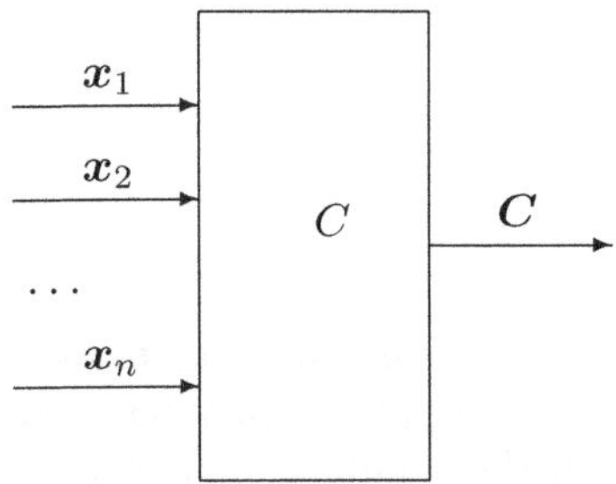

# 3

# Computing Statistics under Fuzzy Uncertainty: Formulation of the Problem

*Need to process fuzzy uncertainty.* In many practical situations, we only have expert estimates for the inputs $x_i$. Sometimes, experts provide guaranteed bounds on the $x_i$, and even the probabilities of different values within these bounds. However, such cases are rare. Usually, the experts' opinions about the uncertainty of their estimates are described by (imprecise, "fuzzy") words from natural language. For example, an expert can say that the value $x_i$ of the $i$-th quantity is approximately equal to 1.0, with an accuracy most probably of about 0.1. Based on such "fuzzy" information, what can we say about $y = f(x_1, \ldots, x_n)$?

The need to process such "fuzzy" information was first emphasized in the early 1960s by L. Zadeh who designed a special technique of *fuzzy logic* for such processing; see, e.g., [156, 252].

*Computing statistics under fuzzy uncertainty: main idea.* Intuitively, a value $y$ is a reasonable value of the statistic $C(x_1, \ldots, x_n)$ quantity if $y = C(x_1, \ldots, x_n)$ for some reasonable values $x_i$, i.e., if for some values $x_1, \ldots, x_n$,

- $x_1$ is reasonable,
- $x_2$ is reasonable,
- $\ldots$,
- $x_n$ is reasonable,
- and $y = C(x_1 \ldots, x_n)$.

Thus, to describe to what extent different values of $y$ are reasonable, we must be able:

- to describe to what extent (to what degree) different values of $x_i$ are reasonable, and
- to combine these degrees into the desired degree of belief in reasonability of $y$.

*Degrees of belief.* Let us first introduce the basic concept of degrees of belief (also known as *degree of confidence*, or *membership degree*). For example, we would like to estimate to what extent the value $x_i = 0.89$ is consistent with

H.T. Nguyen et al.: Computing Stat. under Interval & Fuzzy Uncertain., SCI 393, pp. 11–17.
springerlink.com                    © Springer-Verlag Berlin Heidelberg 2012

the statement "the value $x_i$ of the $i$-th quantity is approximately equal to 1.0, with an accuracy most probably about 0.1".

In the absence of uncertainty, every statement is either true or false. In the computer, "true" is usually represented as 1, and "false" as 0. It is therefore reasonable to use numbers between 0 and 1 to represent levels of confidence which are intermediate – intermediate between the absolute confidence that a given statement is true and the absolute confidence that a given statement is false.

How do we determine this degree of confidence? For example, we can ask several ($N$) experts whether $x_i = 0.89$ is consistent with the above statement, and if $M$ of them reply "yes", take the ratio $M/N$ as the desired degree of confidence. If we do not have access to numerous experts, we can simply ask the only available expert to describe his or her degree of confidence by marking a number on a scale from 0 to $N$ (e.g., on a scale from 0 to 5). If an expert marks his or her degree as $M$, we take the ratio $M/N$ as the desired degree of confidence.

*Membership functions.* To formally describe the original expert's statement $S$ about $x_i$, we need to know, for every real number $x_i$, the degree $\mu_S(x_i)$ to which this real number is consistent with this statement $S$.

By using the above procedure, we can determine this value $\mu_S(x_i)$ for every given real number $x_i$. This procedure includes asking questions to the expert. In practice, we can only ask finitely many questions. Thus, no matter how many questions we ask, by using the above procedure, we can only find the values $\mu_S(x_i)$ for finitely many real numbers $x_i$. To estimate the values $\mu_S(x_i)$ for all other real numbers $x_i$, we must therefore use interpolation and extrapolation. Usually, a simple piece-wise interpolation is used, but sometimes a more sophisticated procedure is applied: e.g., a piecewise quadratic interpolation.

The function $\mu_S(x_i)$ which is obtained by this approximation is called a *membership function*. This function describes, for every real number $x_i$, the degree $\mu_S(x_i)$ to which this real number is consistent with this statement $S$.

*Need for "and"- and "or"-operations: t-norms and t-conorms.* As we have mentioned earlier, we are not directly interested in the degree to which a given real number $x_i$ is consistent with the expert's knowledge $S_i$ about the $i$-th input. We are mainly interested in the degree to which $x_1$ is consistent with the knowledge about the first input *and* $x_2$ is consistent with the knowledge about the second input *and* ... *and* $x_n$ is consistent with the knowledge about the $n$-th input.

In principle, we can determine the degree of belief in such a composite statement by asking an expert, for each possible combination of values $x_1, x_2, \ldots, x_n$, what is the degree to which this combination is consistent with all the available expert knowledge. However, as we have mentioned earlier, even for a single input, we cannot realistically elicit degrees of confidence about too many values. If we consider $N$ possible values of each input, then we would need to elicit the expert's degree of confidence about $N^n \gg N$ possible combinations – which is even less realistic.

Since we cannot directly elicit the expert's degree of confidence in all composite statements, a natural idea is to estimate the degree of confidence in the composite statement based on the degrees of confidence in individual statements – such as "$x_i$ is consistent with the expert's knowledge $S_i$ about the $i$-th input."

How can we come up with such an estimate? Let us reformulate this estimation problem:

- we know the expert's degree of confidence in statements $A_1$, $A_2$, ..., $A_n$, and
- we want to estimate the expert's degree of confidence in a composite statement $A_1 \,\&\, A_2 \,\&\, \ldots \,\&\, A_n$ (i.e., "$A_1$ and $A_2$ and ... and $A_n$").

Since, e.g., $A_1 \,\&\, A_2 \,\&\, A_3$ can be represented as $(A_1 \,\&\, A_2) \,\&\, A_3$, it is sufficient to solve this estimation problem for the case of two statements. Once we have a solution for this particular case, we will then be able to solve the general problem as well:

- first, we apply the two-statement solution to the degrees of certainty in $A_1$ and $A_2$, and get an estimate for the expert's degree of certainty in $A_1 \,\&\, A_2$;
- then, we apply the same solution to the degrees of certainty in $A_1 \,\&\, A_2$ and $A_3$, and get an estimate for the expert's degree of certainty in $A_1 \,\&\, A_2 \,\&\, A_3$;
- after that, we apply the same solution to the degrees of certainty in $A_1 \,\&\, A_2 \,\&\, A_3$ and $A_4$, and get an estimate for the expert's degree of certainty in $A_1 \,\&\, A_2 \,\&\, A_3 \,\&\, A_4$;
- etc.

Eventually, we will get the degree of confidence in the desired composite statement $A_1 \,\&\, A_2 \,\&\, \ldots \,\&\, A_n$.

Thus, we need a procedure that would transform the degree of belief $d_1$ in a statement $A_1$ and the degree of belief $d_2$ in a statement $A_2$ into a (reasonable) estimate for a degree of belief in a composite statement $A_1 \,\&\, A_2$. Let us denote the estimate corresponding to given values $d_1$ and $d_2$ by $f_\&(d_1, d_2)$. The procedure $f_\&$ that maps degrees of belief $d_1$ and $d_2$ in statements $A_1$ and $A_2$ into a degree of belief $d = f_\&(d_1, d_2)$ in $A_1 \,\&\, A_2$ is called an *"and"-operation*, or, for historical reasons, a *t-norm*.

Similarly, to estimate the degree of belief in a composite statement $A_1 \vee A_2$ ("$A_1$ or $A_2$"), we need a procedure $f_\vee$ that maps degrees of belief $d_1$ and $d_2$ in statements $A_1$ and $A_2$ into a degree of belief $d = f_\vee(d_1, d_2)$ in $A_1 \vee A_2$. Such a procedure is called an *"or"-operation*. Since in logic, "or" is a kind of dual to "and", an "or"-operation can be viewed as a dual to an "and"-operation (t-norm). Because of this duality, an "or"-operation is also called a *t-conorm*.

*Properties of "and"- and "or"-operations.* From the intended meaning of the "and"- and "or"-operations, we can deduce reasonable properties of these operations.

For example, intuitively, "$A_1$ and $A_2$" means the same as "$A_2$ and $A_1$". Thus, it is reasonable to require that our estimate $f_\&(d_1, d_2)$ for the degree of confidence in "$A_1$ and $A_2$" should be the same our estimate $f_\&(d_2, d_1)$ for the degree of confidence in "$A_2$ and $A_1$". In other words, we must have $f_\&(d_1, d_2) = f_\&(d_2, d_1)$ for all possible values of $d_1$ and $d_2$. In mathematical terms, this means that the function $f_\&$ must be *commutative*.

Similarly, "$(A_1$ and $A_2)$ and $A_3$" means the same as

$$\text{"}A_1 \text{ and } (A_2 \text{ and } A_3)\text{"}$$

– because both mean the same as "$A_1$ and $A_2$ and $A_3$". For each "and"-operation $f_\&$, the expression "$(A_1$ and $A_2)$ and $A_3$" means that we:

- first estimate the degree of belief in "$A_1$ and $A_2$" as $f_\&(d_1, d_2)$, and
- then estimate the degree of belief in "$(A_1$ and $A_2)$ and $A_3$" as

$$f_\&(f_\&(d_1, d_2), d_3).$$

Similarly, the expression "$A_1$ and $(A_2$ and $A_3)$" means that we:

- first estimate the degree of belief in "$A_2$ and $A_3$" as $f_\&(d_2, d_3)$, and
- then estimate the degree of belief in "$A_1$ and $(A_2$ and $A_3)$" as

$$f_\&(d_1, f_\&(d_2, d_3)).$$

Since the expressions are equivalent, it is reasonable to require that these estimates coincide, i.e., that $f_\&(f_\&(d_1, d_2), d_3) = f_\&(d_1, f_\&(d_2, d_3))$ for all possible values of $d_1$, $d_2$, and $d_3$. In mathematical terms, this means that the function $f_\&$ must be *associative*.

There are several other reasonable properties of "and"-operations. For example, since "$A_1$ and $A_2$" implies $A_1$, our degree of belief in the composite statement "$A_1$ and $A_2$" cannot exceed our degree of belief in $A_1$. Thus, it is reasonable to require that the estimate $f_\&(d_1, d_2)$ for this degree of belief should also not exceed our degree of belief $d_1$ in the statement $A_1$. In other words, we should have $f_\&(d_1, d_2) \leq d_1$ for all possible values of $d_1$ and $d_2$.

If $A_1$ is absolutely true (i.e., $d_1 = 1$), then intuitively, the composite statement "$A_1$ and $A_2$" has exactly the same truth value as $A_2$. Thus, it is reasonable to require that $f_\&(1, d_2) = d_2$ for all possible values of $d_2$.

On the other hand, if $A_1$ is absolutely false (i.e., $d_1 = 0$), then the composite statement "$A_1$ and $A_2$" should also be absolutely false, no matter how much we may believe in $A_2$. Thus, it is reasonable to require that $f_\&(0, d_2) = 0$ for all possible values of $d_2$.

Finally, if, due to a new evidence, our degree of belief in one of the statements $A_1$ and $A_2$ increases, the resulting degree of belief in a composite statement "$A_1$ and $A_2$" will either increase or stay the same – but it cannot decrease. Thus, it is it is reasonable to require that the operation $f_\&$ be *monotonic* in the sense that if $d_1 \leq d_1'$ and $d_2 \leq d_2'$, then $f_\&(d_1, d_2) \leq f_\&(d_1', d_2')$.

All these properties are indeed required of an "and"-operation (t-norm). Similarly, it is reasonable to require that an "or"-operation (t-conorm) $f_\vee$ should be commutative, associative, monotonic, and satisfy the conditions that $d_1 \leq f_\vee(d_1, d_2)$, $f_\vee(1, d_2) = 1$, and $f_\vee(0, d_2) = d_2$ for all possible values of $d_1$ and $d_2$.

*Simplest "and"- and "or"-operations: derivation.* There exist many different "and"- and "or"-operations which satisfy the above properties; see, e.g., [156, 247, 248, 252]. In some applications such as fuzzy control, it is crucial to select appropriate operations – because we can use the corresponding additional degrees of freedom to tune the resulting control and make it an even better fit for the corresponding objective function.

However, in knowledge processing, when we are very uncertain about the inputs, it is probably more reasonable to select the simplest "and"- and "or"-operations which are consistent with the expert knowledge. To select such operations, it makes sense to consider yet another property of "and" and "or" – that for every statement $A$, "$A$ and $A$" means the same as simply $A$. Thus, it is reasonable to require that for every statement $A$ with a degree of confidence $d$, our estimate $f_\&(d, d)$ of the expert's degree of confidence in "$A$ and $A$" should be the same as the original degree of confidence $d$ in the original statement $A$. Thus, it is reasonable to require that $f_\&(d, d) = d$ for all possible values of $d$. In mathematical terms, this means that the function $f_\&$ must be *idempotent*.

Similarly, since "$A$ or $A$" means the same as simply $A$, it is reasonable to require that $f_\vee(d, d) = d$ for all possible values of $d$, i.e., that the function $f_\&$ must also be *idempotent*.

It turns out that this additional requirement leads to a unique "and"-operation $f_\&(d_1, d_2) = \min(d_1, d_2)$ and a unique "or"-operation $f_\vee(d_1, d_2) = \max(d_1, d_2)$; the proof is given at the end of this chapter.

The operations $f_\&(d_1, d_2) = \min(d_1, d_2)$ and $f_\vee(d_1, d_2) = \max(d_1, d_2)$ were actually the first designed by L. Zadeh; they are still actively used in various applications of fuzzy techniques; see, e.g., [156, 252].

*Zadeh's Extension Principle.* Let us apply the above simple operations to computing statistics under fuzzy uncertainty. In this situation:

- We know an algorithm $y = C(x_1, \ldots, x_n)$ that relates the value of the desired statistic with the sample values $x_1, \ldots, x_n$.
- We also have expert knowledge about each of the sample values $x_i$. For each $i$, this knowledge is described in terms of the corresponding membership function $\mu_i(x_i)$. For each $i$ and for each value $x_i$, the value $\mu_i(x_i)$ is the degree of confidence that this value is indeed a possible value of the $i$-th quantity.

Based on this information, we want to find the membership function $\mu(y)$ which describes, for each real number $y$, the degree of confidence that this number is a possible value of the desired statistic.

As we have mentioned earlier, $y$ is a possible value of the desired statistic if for some values $x_1, \ldots, x_n$,

- $x_1$ is a possible value of the first input quantity,
- $x_2$ is a possible value of the second input quantity,
- $\ldots$,
- $x_n$ is a possible value of the $n$-th input quantity, and
- $y = C(x_1 \ldots, x_n)$.

We know:

- that the degree of confidence that $x_1$ is a possible value of the first input quantity is equal to $\mu_1(x_1)$,
- that the degree of confidence that $x_2$ is a possible value of the second input quantity is equal to $\mu_2(x_2)$,
- etc.

The degree of confidence $d(y, x_1, \ldots, x_n)$ in an equality $y = C(x_1 \ldots, x_n)$ is, of course,

- equal to 1 if this equality holds, and
- equal to 0 if this equality does not hold.

We have already agreed to represent "and" as min. Thus, for each combination of values $x_1, \ldots, x_n$, the degree of confidence in a composite statement "$x_1$ is a possible value of the first input quantity, and $x_2$ is a possible value of the second input quantity, $\ldots$, and $x_n$ is a possible value of the $n$-th input quantity, and $y = C(x_1 \ldots, x_n)$" is equal to

$$\min(\mu_1(x_1), \mu_2(x_2), \ldots, \mu_n(x_n), d(y, x_1, \ldots, x_n)).$$

We can simplify this expression if we consider two possible cases:

- when the equality $y = C(x_1 \ldots, x_n)$ holds, and
- when this equality does not hold.

When the equality $y = C(x_1 \ldots, x_n)$ holds, we get $d(y, x_1, \ldots, x_n) = 1$, and thus, the above degree of confidence is simply equal to

$$\min(\mu_1(x_1), \mu_2(x_2), \ldots, \mu_n(x_n)).$$

When the equality $y = C(x_1 \ldots, x_n)$ does not hold, we get $d(y, x_1, \ldots, x_n) = 0$, and thus, the above degree of confidence is simply equal to 0.

We want to combine these degrees of belief into a single degree of confidence that "for some values $x_1, \ldots, x_n$, $x_1$ is a possible value of the first input quantity, and $x_2$ is a possible value of the first input quantity, $\ldots$, and $y = C(x_1 \ldots, x_n)$". The words "for some values $x_1, \ldots, x_n$" means that the following composite property holds either for one combination of real numbers $x_1, \ldots, x_n$, or for another combination – until we exhaust all (infinitely many) such combinations. We have already agreed to represent "or" as max. Thus, the desired degree of confidence $\mu(y)$ is equal to the maximum of the degrees corresponding to different combinations $x_1, \ldots, x_n$. Since we have infinitely many possible combinations, the maximum is not necessarily attained, so we should, in general, consider supremum instead of maximum:

$$\mu(y) = \sup \min(\mu_1(x_1), \mu_2(x_2), \ldots, \mu_n(x_n), d(y, x_1, \ldots, x_n)),$$

where the supremum is taken over all possible combinations.

Since we know that the maximized degree is non-zero only when $y = C(x_1 \ldots, x_n)$, it is sufficient to only take supremum over such combinations. For such combinations, we can omit the term $d(y, x_1, \ldots, x_n)$ in the maximized expression, so we arrive at the following formula:

$$\mu(y) = \sup\{\min(\mu_1(x_1), \mu_2(x_2), \ldots, \mu_n(x_n)) : y = C(x_1, \ldots, x_n)\}.$$

This formula describes a reasonable way to extend an arbitrary statistic $f(x_1, \ldots, x_n)$ from real-valued inputs to a more general case of fuzzy inputs. It was first proposed by L. Zadeh – for general functions – and is thus called *Zadeh's extension principle*.

This is the main formula that describes computing statistics (and, more generally, knowledge processing) under fuzzy uncertainty. In the following section, we will show that from the computational viewpoint, the application of this formula can be reduced to interval computations – and indeed, this is how knowledge processing under fuzzy uncertainty is usually done, by using this reduction; see, e.g., [156, 246, 252].


## Proof

*Proof that* $\min(d_1, d_2)$ *is the only idempotent t-norm, and* $\max(d_1, d_2)$ *is the only idempotent t-conorm.* Let us first show that the only idempotent "and"-operation is $f_\&(d_1, d_2) = \min(d_1, d_2)$. Without loss of generality, let us assume that $d_1 \leq d_2$. In this case, the desired equality takes the form $f_\&(d_1, d_2) = d_1$. Since the operation $f_\&$ is idempotent, we have $f_\&(d_1, d_1) = d_1$. Due to $d_1 \leq d_2$, monotonicity implies that $f_\&(d_1, d_1) \leq f_\&(d_1, d_2)$, hence $d_1 \leq f_\&(d_1, d_2)$. On the other hand, for an "and"-operation, we always have $f_\&(d_1, d_2) \leq d_1$. So, we can conclude that $f_\&(d_1, d_2) = d_1$, i.e., indeed, $f_\&(d_1, d_2) = \min(d_1, d_2)$.

Let us now prove that the only idempotent "or"-operation is $f_\vee(d_1, d_2) = \max(d_1, d_2)$. Without loss of generality, let us again assume that $d_1 \leq d_2$. In this case, the desired equality takes the form $f_\vee(d_1, d_2) = d_2$. Since the operation $f_\vee$ is idempotent, we have $f_\vee(d_2, d_2) = d_2$. Due to $d_1 \leq d_2$, monotonicity implies that $f_\vee(d_1, d_2) \leq f_\vee(d_2, d_2)$, hence $f_\vee(d_1, d_2) \leq d_2$. On the other hand, for an "or"-operation, we always have $d_2 \leq f_\vee(d_1, d_2)$. Thus, we conclude that $f_\vee(d_1, d_2) = d_2$, i.e., indeed, $f_\vee(d_1, d_2) = \max(d_1, d_2)$.

# 4

## Computing under Fuzzy Uncertainty Can Be Reduced to Computing under Interval Uncertainty

*An alternative set representation of a membership function: alpha-cuts.* To describe the desired relation between computing under fuzzy uncertainty and computing under interval uncertainty, we must first reformulate fuzzy techniques in an interval-related form.

In some situations, an expert knows exactly which values of $x_i$ are possible and which are not. In this situation, the expert's knowledge can be naturally represented by describing the *set* of all possible values.

In general, the expert's knowledge is fuzzy:

- we may still have some values about which the expert 100% believes that they are possible, and
- we may still have some values about which the expert 100% believes that they are impossible, but
- in general, the expert is not 100% confident about which values of $x_i$ are possible and which are not.

For example, a geophysicist may be confident that the density $x_i$ of some mineral can take on values ranging from 3.4 to 3.7 g/cm$^3$, and she may know that values smaller than 3.0 or larger than 4.0 are absolutely impossible, but she is not sure whether values from 3.0 to 3.4 or from 3.7 to 4.0 are indeed realistically possible.

As we have mentioned, the ultimate purpose of the measurements and estimates is to make decisions. In the geophysical example, we have measured the density at a certain depth, and we need to decide:

- whether it is possible that we have the desired mineral – in which case we should undertake more measurements, or
- whether it is not possible that we have the desired mineral – in which case we should not waste our resources on this region and move to more promising regions.

In practice, decisions are made under uncertainty. If we only have a fuzzy expert's description of possible values – in terms of the membership function

H.T. Nguyen et al.: Computing Stat. under Interval & Fuzzy Uncertain., SCI 393, pp. 19–24.
springerlink.com

$\mu_S(x_i)$ – which values $x_i$ should we then classify as possible ones and which as impossible?

Under uncertainty, a reasonable idea is to select a threshold $\alpha \in (0, 1]$. In this case,

- all the values $x_i$ for which the expert's degree of confidence is strong enough – i.e., for which $\mu_S(x_i) \geq \alpha$ – are classified as possible;
- similarly, all the values $x_i$ for which the expert's degree of confidence is not sufficiently strong – i.e., for which $\mu_S(x_i) < \alpha$ – are classified as impossible.

The resulting set of possible elements

$$\boldsymbol{x}_i(\alpha) \stackrel{\text{def}}{=} \{x_i : \mu_S(x_i) \geq \alpha\}$$

is called the $\alpha$-cut of the membership function $\mu_S(x_i)$.

The choice of a threshold $\alpha$ depends on the practical problem.

For example, if we are looking for a potentially very valuable mineral deposit, then it makes sense to continue prospecting even when our degree of confidence is not very high. In this case, it makes sense to select a reasonably small threshold $\alpha$.

On the other hand, if the potential benefit is not high and our resources are limited, it makes sense to limit our search to highly promising regions – i.e., to select a reasonably high threshold $\alpha$.

To adequately describe the expert knowledge irrespective of an application, we therefore need to know the $\alpha$-cuts corresponding to different thresholds $\alpha$. Each $\alpha$-cut $\boldsymbol{x}_i(\alpha)$ describes the set of values which are possible with degree of confidence at least $\alpha$.

By definition, $\alpha$-cuts corresponding to different $\alpha$ are *nested*: when $\alpha \leq \alpha'$, then $\mu_S(x_i) \geq \alpha'$ implies $\mu_S(x_i) \geq \alpha$ and thus,

$$\boldsymbol{x}_i(\alpha') = \{x_i : \mu_S(x_i) \geq \alpha'\} \subseteq \boldsymbol{x}_i(\alpha) = \{x_i : \mu_S(x_i) \geq \alpha\}.$$

*Comment.* It is worth mentioning that if we know the $\alpha$-cuts

$$\boldsymbol{x}_i(\alpha) = \{x_i : \mu_S(x_i) \geq \alpha\}$$

corresponding to all possible values $\alpha \in (0, 1]$, then we can uniquely reconstruct the corresponding membership function $\mu_S(x_i)$; see the proofs section for detail. Thus, we can alternatively view a membership function as a nested family of $\alpha$-cuts; see, e.g., [246].

*Fuzzy numbers and intervals.* In most practical situations, the membership function starts with 0, continuously increases until a certain value and then continuously decreases to 0. Such membership function describe usual expert's expressions such as "small", "medium", "reasonably high", "approximately equal to $a$ with an error about $\sigma$", etc. Such examples were given in the previous chapter. Since membership functions of this type are actively used in expert estimates of number-valued quantities, they are usually called *fuzzy numbers.*

For a fuzzy number $\mu_i(x_i)$, every $\alpha$-cut $\boldsymbol{x}_i(\alpha)$ is an interval. Thus, a fuzzy number can be viewed as a nested family of intervals $\boldsymbol{x}_i(\alpha)$ corresponding to different degrees of confidence.

*Simplest "and"- and "or"-operations: reformulation in terms of sets and alpha-cuts.* The main formulas for fuzzy computations (i.e., for processing fuzzy data) were derived by using the simplest "and"- and "or"-operations $f_\&(d_1, d_2) = \min(d_1, d_2)$ and $f_\vee(d_1, d_2) = \max(d_1, d_2)$. Thus, before we describe how fuzzy computations can be reduced to interval computations, let us first reformulate these "and"- and "or"-operations in terms of $\alpha$-cuts.

Specifically, let us assume that we have two properties $A$ and $B$ which are described by the membership functions $\mu_A(x)$ and $\mu_B(x)$ and, correspondingly, by the $\alpha$-cuts $\boldsymbol{x}_A(\alpha) = \{x : \mu_A(x) \geq \alpha\}$ and $\boldsymbol{x}_B(\alpha) = \{x : \mu_B(x) \geq \alpha\}$. If we use the simplest "and"-operation $f_\&(d_1, d_2) = \min(d_1, d_2)$, then the composite property $A \,\&\, B$ ("$A$ and $B$") is described by the membership function $\mu_{A \,\&\, B}(x) = \min(\mu_A(x), \mu_B(x))$. It turns out that the $\alpha$-cuts

$$\boldsymbol{x}_{A \,\&\, B}(\alpha) = \{x : \mu_{A \,\&\, B}(x) \geq \alpha\}$$

corresponding to this membership function, are the intersections of the $\alpha$-cuts corresponding to $A$ and $B$ (see the Proofs section for detail):

$$\boldsymbol{x}_{A \,\&\, B}(\alpha) = \boldsymbol{x}_A(\alpha) \cap \boldsymbol{x}_B(\alpha).$$

Therefore, to perform the simplest "and"-operation $f_\&(d_1, d_2) = \min(d_1, d_2)$, we simply take the intersection of the corresponding $\alpha$-cuts. This is a very natural operation, since, for exactly defined sets and properties, the set of all the elements which satisfy the property $A \,\&\, B$ is equal to the intersection of the set of all elements which satisfy property $A$ and the set of all elements which satisfy property $B$.

Similarly, for the simplest "or"-operation $f_\vee(d_1, d_2) = \max(d_1, d_2)$, the composite property $A \vee B$ ("$A$ or $B$") is described by the membership function $\mu_{A \vee B}(x) = \max(\mu_A(x), \mu_B(x))$. It turns out that the $\alpha$-cuts

$$\boldsymbol{x}_{A \vee B}(\alpha) = \{x : \mu_{A \vee B}(x) \geq \alpha\}$$

corresponding to this membership function are unions of the $\alpha$-cuts for $A$ and $B$:

$$\boldsymbol{x}_{A \vee B}(\alpha) = \boldsymbol{x}_A(\alpha) \cup \boldsymbol{x}_B(\alpha).$$

Therefore, to perform the simplest "or"-operation $f_\vee(d_1, d_2) = \max(d_1, d_2)$, we simply take the union of the corresponding $\alpha$-cuts. This is also a very natural operation, since, for exactly defined sets and properties, the set of all the elements which satisfy the property $A \vee B$ is equal to the union of the set of all elements which satisfy property $A$ and the set of all elements which satisfy property $B$.

*Fuzzy computations can be reduced to interval computations: derivation.* The main problem of fuzzy computation can be described as follows:

- We know an algorithm $y = C(x_1, \ldots, x_n)$ that relates the value of the desired statistic $y$ with the sample values $x_1, \ldots, x_n$.
- We also know, for every $i$ from 1 to $n$, a membership function $\mu_i(x_i)$ which describes the expert knowledge about the $i$-th sample value $x_i$.

Our objective is to compute the function

$$\mu(y) = \sup\{\min(\mu_1(x_1), \mu_2(x_2), \ldots, \mu_n(x_n)) : y = C(x_1, \ldots, x_n)\}.$$

Let us now describe this relation in terms of $\alpha$-cuts. This description will constitute the main relation between fuzzy and interval computing. This relation was first discovered and proved in [90, 245]. To describe this result in precise terms, let us first make some mathematics-related remarks.

The function $y = C(x_1, \ldots, x_n)$ describes a statistic. In statistics, such a relation is usually continuous. Thus, we will assume that the function $y = C(x_1, \ldots, x_n)$ is continuous.

We will also assume the membership functions $\mu_i(x_i)$ are continuous. If we had exact knowledge, then continuity would make no sense, since then the corresponding degree of confidence would abruptly go from 1 for possible values to 0 for impossible ones, without ever attaining any intermediate degrees. However, for fuzzy knowledge, continuity makes perfect sense. If there is some degree of confidence that a value $x_i$ is possible, then it makes sense to assume that values close to $x_i$ are possible too – with a similar degree of belief. In practice, as we mentioned earlier in this chapter and in the previous chapter, membership functions are indeed usually continuous.

It is important to mention that for continuous membership functions $\mu_i(x_i)$, $\alpha$-cuts $\{x_i : \mu_i(x_i) \geq \alpha\}$ are closed sets (i.e., sets which contain all their limit points).

Finally, we require that for every $i$ and for every $\alpha > 0$, the $\alpha$-cut is a compact set. For real numbers, since we have already assumed that the $\alpha$-cuts $\{x_i : \mu_i(x_i) \geq \alpha\}$ are closed sets, it is sufficient to require that these sets are bounded. This is true, e.g., if we assume that all the membership functions correspond to fuzzy numbers; in this case, all $\alpha$-cuts are intervals.

Suppose that we know the $\alpha$-cuts $\boldsymbol{x}_i(\alpha)$ corresponding to the inputs, and we want to find the $\alpha$-cuts $\boldsymbol{y}(\alpha)$ corresponding to the output. It turns out that the desired $\alpha$-cut $\boldsymbol{y}(\alpha)$ consists of exactly values $y = C(x_1, \ldots, x_n)$ for $x_i \in \boldsymbol{x}_i(\alpha)$:

$$\boldsymbol{y}(\alpha) = \{C(x_1, \ldots, x_n) : x_1 \in \boldsymbol{x}_1(\alpha), \ldots, x_n \in \boldsymbol{x}_n(\alpha)\}.$$

(For details, see the Proofs section.) This is exactly the range that we defined when we described interval computations, so we can rewrite this formula as

$$\boldsymbol{y}(\alpha) = C(\boldsymbol{x}_1(\alpha), \ldots, \boldsymbol{x}_n(\alpha)).$$

In particular, for fuzzy numbers, when all $\alpha$-cuts $\boldsymbol{x}_i(\alpha)$ are intervals, computing each $\alpha$-cut $\boldsymbol{y}(\alpha)$ is exactly the problem of interval computations.

*Fuzzy computations can be reduced to interval computations: conclusion.* If the inputs $\mu_i(x_i)$ are fuzzy numbers and the function $y = C(x_1, \ldots, x_n)$ is continuous, then for each $\alpha$, the $\alpha$-cut $\boldsymbol{y}(\alpha)$ of $y$ is equal to the range of possible values of $C(x_1, \ldots, x_n)$ as $x_i$ ranges over $\boldsymbol{x}_i(\alpha)$ for all $i$:

$$\boldsymbol{y}(\alpha) = C(\boldsymbol{x}_1(\alpha), \ldots, \boldsymbol{x}_n(\alpha)).$$

Thus, from the computational point of view, the problem of computing statistics under fuzzy uncertainty can be reduced to several problems of computing statistics under interval uncertainty – as many problems as there are $\alpha$-levels.

As we have mentioned, this is not just a theoretical observation: this is exactly how fuzzy data processing in general is usually performed, and this is how interval computations techniques are explained in fuzzy textbooks.


## Proofs


*Proof that a membership function can be uniquely reconstructed from $\alpha$-cuts.* The possibility for such a reconstruction follows from the fact that every real number $r$ is equal to the largest largest value $\alpha$ for which $r \geq \alpha$. In particular, for every $x_i$, the value $\mu_S(x_i)$ is equal to the largest value $\alpha$ for which $\mu_S(x_i) \geq \alpha$. By definition of the $\alpha$-cut, the inequality $\mu_S(x_i) \geq \alpha$ is equivalent to $x_i \in \boldsymbol{x}_i(\alpha)$. Thus, for every $x_i$, the value $\mu_S(x_i)$ can be reconstructed as the largest value $\alpha$ for which $x_i \in \boldsymbol{x}_i(\alpha)$.

*Proof that for the* min *t-norm, the $\alpha$-cut of $A \,\&\, B$ is the intersection of the $\alpha$-cuts corresponding to $A$ and $B$.* By definition of the minimum, the minimum of two real numbers is greater than or equal to $\alpha$ if and only if both of these numbers are greater than or equal to $\alpha$. Thus, the condition $\mu_{A \,\&\, B}(x) = \min(\mu_A(x), \mu_B(x)) \geq \alpha$ is equivalent to "$\mu_A(x) \geq \alpha$ and $\mu_B(x) \geq \alpha$". Hence, the set $\boldsymbol{x}_{A \,\&\, B}(\alpha)$ of all the values $x$ for which the condition $\mu_{A \,\&\, B}(x) = \min(\mu_A(x), \mu_B(x)) \geq \alpha$ is satisfied can be found simply as the intersection of the set of all $x$ for which $\mu_A(x) \geq \alpha$ and the set of all $x$ for which $\mu_B(x) \geq \alpha$. In other words, for every $\alpha$, we have

$$\boldsymbol{x}_{A \,\&\, B}(\alpha) = \boldsymbol{x}_A(\alpha) \cap \boldsymbol{x}_B(\alpha).$$

The statement is proven.

*Proof that for the* max *t-conorm, the $\alpha$-cut of $A \vee B$ is the union of the $\alpha$-cuts corresponding to $A$ and $B$.* To prove this statement, we can use the fact that the maximum of two real numbers is greater than or equal to $\alpha$ if and only if one of these numbers is greater than or equal to $\alpha$. Thus, the condition $\mu_{A \vee B}(x) = \max(\mu_A(x), \mu_B(x)) \geq \alpha$ is equivalent to "$\mu_A(x) \geq \alpha$ or $\mu_B(x) \geq \alpha$". Hence, the set $\boldsymbol{x}_{A \vee B}(\alpha)$ of all the values $x$ for which the condition $\mu_{A \vee B}(x) = \max(\mu_A(x), \mu_B(x)) \geq \alpha$ is satisfied can be found simply as the union of the set of all $x$ for which $\mu_A(x) \geq \alpha$ and the set of all $x$ for which $\mu_B(x) \geq \alpha$. In other words, for every $\alpha$, we have

$$\boldsymbol{x}_{A \vee B}(\alpha) = \boldsymbol{x}_A(\alpha) \cup \boldsymbol{x}_B(\alpha).$$

The statement is proven.

*Proof that the $\alpha$-cut for $C(X_1, \ldots, X_n)$ is the range of $C(x_1, \ldots, x_n)$ when each $x_i$ belongs to the corresponding $\alpha$-cut of $X_i$.* By definition of an $\alpha$-cut, $y \in \boldsymbol{y}(\alpha)$ means that $\mu(y) \geq \alpha$, i.e., that

$$\sup\{\min(\mu_1(x_1), \mu_2(x_2), \ldots, \mu_n(x_n)) : y = C(x_1, \ldots, x_n)\} \geq \alpha.$$

By definition of the supremum, this means that for every integer $k > 2/\alpha$, there exists a tuple $\left(x_1^{(k)}, x_2^{(k)}, \ldots, x_n^{(k)}\right)$ for which $y = C\left(x_1^{(k)}, \ldots, x_n^{(k)}\right)$ and

$$\min\left(\mu_1\left(x_1^{(k)}\right), \mu_2\left(x_2^{(k)}\right), \ldots, \mu_n\left(x_n^{(k)}\right)\right) \geq \alpha - \frac{1}{k}.$$

The minimum of several numbers is $\geq \alpha - 1/k$ if and only if all these numbers are $\geq \alpha - 1/k$, i.e., $\mu_i(x_i^{(k)}) \geq \alpha - 1/k$ for all $i$. Since $k > 2/\alpha$, we have $1/k < \alpha/2$ and $\alpha - 1/k > \alpha/2$. Thus, for each $i$ and all $k$, the value $x_i^{(k)}$ belongs to the compact $(\alpha/2)$-cut $\boldsymbol{x}_i(\alpha/2)$. Since the tuples $\left(x_1^{(k)}, x_2^{(k)}, \ldots, x_n^{(k)}\right)$ belong to the compact set

$$\boldsymbol{x}_1(\alpha/2) \times \boldsymbol{x}_2(\alpha/2) \times \ldots \times \boldsymbol{x}_n(\alpha/2),$$

the sequence of these tuples has a convergent subsequence converging to some tuple $(x_1, x_2, \ldots, x_n)$. Since both $C$ and $\mu_i$ are continuous, for this limit tuple, we get $y = C(x_1, \ldots, x_n)$ and $\mu_i(x_i) \geq \alpha$. In other words, every element $y \in \boldsymbol{y}(\alpha)$ can be represented as $y = C(x_1, \ldots, x_n)$ for some values $x_i \in \boldsymbol{x}_i(\alpha)$.

Conversely, if $x_i \in \boldsymbol{x}_i(\alpha)$ and $y = C(x_1, \ldots, x_n)$, then $\mu_i(x_i) \geq \alpha$ and therefore, $\min(\mu_1(x_1), \mu_2(x_2), \ldots, \mu_n(x_n)) \geq \alpha$ and hence

$$\sup\{\min(\mu_1(x_1), \mu_2(x_2), \ldots, \mu_n(x_n)) : y = C(x_1, \ldots, x_n)\} \geq \alpha,$$

i.e., $\mu(y) \geq \alpha$ and $y \in \boldsymbol{y}(\alpha)$

Thus, the desired $\alpha$-cut $\boldsymbol{y}(\alpha)$ indeed consists of exactly values $y = C(x_1, \ldots, x_n)$ for $x_i \in \boldsymbol{x}_i(\alpha)$:

$$\boldsymbol{y}(\alpha) = \{C(x_1, \ldots, x_n) : x_1 \in \boldsymbol{x}_1(\alpha), \ldots, x_n \in \boldsymbol{x}_n(\alpha)\}.$$

# 5

# Computing under Interval Uncertainty: Traditional Approach Based on Uniform Distributions

*Traditional statistical approach: main idea.* In  the case of interval uncertainty, we only know the intervals, we do not know the probability distributions on these intervals. The traditional statistical approach to situations in which we have several alternatives with unknown probabilities is to use Laplace Principle of Indifference, according to which,

- if we have several possible alternatives,
- and we have no information about the probability of different alternatives,
- we assume all these probabilities to be equal.

For example, if ten people were present at the place of a theft, and we have no information about them, it is reasonable to assume that for each of them, the probability of this person having committed a theft is exactly the same: 1/10. Similarly, in the continuous case,

- if all we know that the measurement error is somewhere within a given interval $[-\Delta, \Delta]$,
- and we do not know which values within this interval are more probable and which are less probable,
- it is reasonable to assume that all the values from this interval are equally probable, i.e., that the measurement error is uniformly distributed on this interval.

*Maximum entropy approach.* In general, a traditional statistical approach to the situation when several probability distributions are possible is to select the "most uncertain" distribution, i.e., the distribution which has the largest possible value of the entropy $S \stackrel{\text{def}}{=} -\int \rho(x) \cdot \ln(\rho(x))\, dx$ (here $\rho(x)$ denotes the probability density). For details on this Maximum Entropy approach and its relation to interval uncertainty (and Laplace's principle of indifference), see, e.g., Jaynes at al. [144].

*Maximum entropy approach leads to the uniform distribution.* One can easily check that for a single variable $x_1$, among all distributions located on a given

H.T. Nguyen et al.: Computing Stat. under Interval & Fuzzy Uncertain., SCI 393, pp. 25–27.
springerlink.com

interval, the entropy is indeed the largest when this distribution is *uniform* on this interval; see the Proof section for details.

*Case of several variables.* In the case of several variables, we can similarly conclude that the distribution with the largest value of the entropy is the one which is uniformly distributed in the corresponding box $\boldsymbol{x}_1 \times \ldots \times \boldsymbol{x}_i \times \ldots \times \boldsymbol{x}_n$, i.e., a distribution in which:

- each variable $\Delta x_i$ is uniformly distributed on the corresponding interval $[-\Delta_i, \Delta_i]$, and
- variables corresponding to different inputs are statistically independent.

*Resulting engineering approach to interval uncertainty.* This is indeed one of the main ways how interval uncertainty is treated in engineering practice: if we only know that the value of some variable is in the interval $[\underline{x}_i, \overline{x}_i]$, and we have no information about the probabilities, then we assume that the variable $x_i$ is uniformly distributed on this interval.

*Limitations of the uniform distribution approach.* To explain the limitations of this engineering approach, let us consider the simplest possible algorithm $y = f(x_1, \ldots, x_i, \ldots, x_n) = x_1 + \ldots + x_i + \ldots + x_n$. For simplicity, let us assume:

- that the measured values of all $n$ quantities are zeroes: $\widetilde{x}_1 = \ldots = \widetilde{x}_i = \ldots = \widetilde{x}_n = 0$, and
- that all $n$ measurements have the same error bound $\Delta_x$: $\Delta_1 = \ldots = \Delta x_i = \ldots = \Delta_n = \Delta_x$.

In this case, $\Delta y = \Delta x_1 + \ldots + \Delta x_i + \ldots + \Delta x_n$. Each of $n$ component measurement errors can take any value from $-\Delta_x$ to $\Delta_x$, so the largest possible value of $\Delta y$ is attained when all of the component errors attain the largest possible value $\Delta x_i = \Delta_x$. In this case, the largest possible value $\Delta$ of $\Delta y$ is equal to $\Delta = n \cdot \Delta_x$.

Let us see what the maximum entropy approach will predict in this case. According to this approach, we assume that $\Delta x_i$ are independent random variables, each of which is uniformly distributed on the interval $[-\Delta, \Delta]$. According to the Central Limit theorem (see, e.g., Sheskin [305]), when $n \to \infty$, the distribution of the sum of $n$ independent identically distributed bounded random variables tends to Gaussian. This means that for large values $n$, the distribution of $\Delta y$ is approximately normal.

A normal distribution is uniquely determined by its mean and variance. When we add several independent variables, their means and variances add up. For each uniform distribution $\Delta x_i$ on the interval $[-\Delta_x, \Delta_x]$ of width $2\Delta_x$, the probability density is equal to $\rho(x) = \dfrac{1}{2\Delta_x}$, so the mean is 0 and the variance is

$$V = \int_{-\Delta_x}^{\Delta_x} x^2 \cdot \rho(x)\, \mathrm{d}x = \frac{1}{2\Delta_x} \cdot \int_{-\Delta_x}^{\Delta_x} x^2\, \mathrm{d}x = \frac{1}{2\Delta_x} \cdot \frac{1}{3} \cdot x^3 \Big|_{-\Delta_x}^{\Delta_x} = \frac{1}{3} \cdot \Delta_x^2. \quad (5.1)$$

Thus, for the sum $\Delta y$ of $n$ such variables, the mean $E$ is 0, and the variance is equal to $(n/3) \cdot \Delta_x^2$. Hence, the standard deviation is equal to $\sigma = \sqrt{V} = \Delta_x \cdot \dfrac{\sqrt{n}}{\sqrt{3}}$.

It is known that in a normal distribution, with probability close to 1, all the values are located within the $k \cdot \sigma$ vicinity of the mean: for $k = 3$, it is true with probability 99.9%, for $k = 6$, it is true with probability $1 - 10^{-6}\%$, etc. So, practically with certainty, $\Delta y$ is located within an interval $[E - k \cdot \sigma, E + k \cdot \sigma]$ whose width grows with $n$ as $\sqrt{n}$.

For large $n$, we have $k \cdot \Delta_x \cdot \dfrac{\sqrt{n}}{\sqrt{3}} \ll \Delta_x \cdot n$, so we get a serious underestimation of the resulting measurement error. This example shows that estimates obtained by selecting a uniform distribution can be very misleading.

## Proof

*Proof that the Maximum Entropy approach leads to the uniform distribution.* A function $\rho(x) \geq 0$ is a probability density function on the given interval if $\int \rho(x)\,dx = 1$. Thus, to find the probability density function that maximizes entropy, we must maximize entropy $-\int \rho(x) \cdot \ln(\rho(x))\,dx$ under the constraint $\int \rho(x)\,dx = 1$. According to the Lagrange multiplier method, for some value $\lambda$ (Lagrange multiplier), the desired constraint optimization problem is equivalent to an unconstraint optimization problem of maximizing the expression $-\int \rho(x) \cdot \ln(\rho(x))\,dx + \lambda \cdot (\int \rho(x)\,dx - 1)$. Differentiating this expression with respect to each of the variables $\rho(x)$ and equating the derivative to 0, we conclude that $-\ln(\rho(x)) - 1 + \lambda = 0$, hence $\rho(x) = \exp(\lambda - 1)$. The probability density has the same value for all $x$ from the given interval, hence we indeed have a uniform distribution.

# 6

# Computing under Interval Uncertainty: When Measurement Errors Are Small

*Linearization: main idea.* When the measurement errors $\Delta x_i$ are relatively small, we can use a simplification called *linearization*. The main idea of linearization is as follows.

By definition of the measurement error $\Delta x_i = \widetilde{x}_i - x_i$, hence $x_i = \widetilde{x}_i - \Delta x_i$. When the measurement errors $\Delta x_i$ of direct measurements are relatively small, we can expand the expression

$$\Delta y = \widetilde{y} - y = f(\widetilde{x}_1, \ldots, \widetilde{x}_i, \ldots, \widetilde{x}_n) - f(x_1, \ldots, x_n) =$$

$$f(\widetilde{x}_1, \ldots, \widetilde{x}_i, \ldots, \widetilde{x}_n) - f(\widetilde{x}_1 - \Delta x_1, \ldots, \widetilde{x}_i - \Delta x_i, \ldots, \widetilde{x}_n - \Delta x_n) \quad (6.1)$$

in Taylor series and only keep linear terms in the resulting expansion. Since

$$y = f(\widetilde{x}_1 - \Delta x_1, \ldots, \widetilde{x}_i - \Delta x_i, \ldots, \widetilde{x}_n - \Delta x_n) \approx$$

$$f(\widetilde{x}_1, \ldots, \widetilde{x}_i, \ldots, \widetilde{x}_n) - \sum_{i=1}^{n} \frac{\partial f}{\partial x_i} \cdot \Delta x_i, \quad (6.2)$$

we conclude that $\Delta y = \widetilde{y} - y = \sum_{i=1}^{n} c_i \cdot \Delta x_i$, where $c_i = \dfrac{\partial f}{\partial x_i}$.

The dependence of $\Delta y$ on $\Delta x_i$ is linear: it is

- increasing relative to $x_i$ if $c_i \geq 0$ and
- decreasing if $c_i < 0$.

So, to find the largest possible value $\Delta$ of $\Delta y$, we must take:

- the largest possible value $\Delta x_i = \Delta_i$ when $c_i \geq 0$, and
- the smallest possible value $\Delta x_i = -\Delta_i$ when $c_i < 0$.

In both cases, the corresponding term in the sum has the form $|c_i| \cdot \Delta_i$, so we can conclude that

$$\Delta = \sum_{i=1}^{n} |c_i| \cdot \Delta_i. \quad (6.3)$$

H.T. Nguyen et al.: Computing Stat. under Interval & Fuzzy Uncertain., SCI 393, pp. 29–33.
springerlink.com        © Springer-Verlag Berlin Heidelberg 2012

Similarly, the smallest possible value of $\Delta y$ is equal to $-\Delta$. Thus, the range of possible values of $y$ is equal to $[\underline{y}, \overline{y}] = [\widetilde{y} - \Delta, \widetilde{y} + \Delta]$. So, to compute $\Delta$, it is sufficient to know the partial derivatives $c_i$.

*Case of analytical formulas.* In the simplest case when the algorithm

$$f(x_1, \ldots, x_i, \ldots, x_n) \tag{6.4}$$

consists of a simple analytical expression, we can find explicit analytical formulas for the partial derivatives and thus compute the desired bound $\Delta$.

*Techniques based on sensitivity analysis (automatic differentiation).* In the general case, a natural way to compute partial derivatives comes directly from the definition. By definition, a partial derivative is defined as a limit

$$\frac{\partial f}{\partial x_i} =$$

$$\lim_{h_i \to 0} \frac{f(\widetilde{x}_1, \ldots, \widetilde{x}_{i-1}, \widetilde{x}_i + h_i, \widetilde{x}_{i+1}, \ldots, \widetilde{x}_n) - f(\widetilde{x}_1, \ldots, \widetilde{x}_i, \ldots, \widetilde{x}_n)}{h_i}. \tag{6.5}$$

In turn, a limit, by its definition, means that when the values of $h_i$ is small, the corresponding ratio is very close to the partial derivative. Thus, we can estimate the partial derivative as the ratio

$$c_i = \frac{\partial f}{\partial x_i} \approx$$

$$\frac{f(\widetilde{x}_1, \ldots, \widetilde{x}_{i-1}, \widetilde{x}_i + h_i, \widetilde{x}_{i+1}, \ldots, \widetilde{x}_n) - f(\widetilde{x}_1, \ldots, \widetilde{x}_i, \ldots, \widetilde{x}_n)}{h_i} \tag{6.6}$$

for some small value $h_i$.

After we have computed $n$ such ratios, we can then compute the desired bound $\Delta$ on $|\Delta y|$ as $\Delta = \sum_{i=1}^{n} |c_i| \cdot \Delta_i$.

In general, this procedure takes $n$ divisions by $h_i$ and $n$ multiplications by $\Delta_i$. The procedure can be made faster if we select $h_i = \Delta_i$. In this case, we get

$$\Delta = \sum_{i=1}^{n} |f(\widetilde{x}_1, \ldots, \widetilde{x}_{i-1}, \widetilde{x}_i + \Delta_i, \widetilde{x}_{i+1}, \ldots, \widetilde{x}_n) - \widetilde{y}|. \tag{6.7}$$

*Advanced Monte-Carlo simulation techniques.* In the above algorithm, we call the data processing algorithm $n + 1$ times:

- first to compute the value $\widetilde{y} = f(\widetilde{x}_1, \ldots, \widetilde{x}_i, \ldots, \widetilde{x}_n)$, and
- then $n$ more times to compute the values

$$f(\widetilde{x}_1, \ldots, \widetilde{x}_{i-1}, \widetilde{x}_i + \Delta_i, \widetilde{x}_{i+1}, \ldots, \widetilde{x}_n) \tag{6.8}$$

and thus, the corresponding partial derivatives.

In many practical situations, the data processing algorithms are time-consuming, and we process large amounts of data, with the number $n$ of data points in thousands. In this case, the use of the above linearization algorithm would take thousands of times longer than data processing itself – which itself is already time consuming. Is it possible to estimate $\Delta$ faster?

The answer is "yes", it is possible to have a Monte-Carlo-type algorithm which estimates $\Delta$ by using only a constant number of calls to the data processing algorithm $f$; for details, see, e.g. [177, 178].

At first glance, since we know that the measurement errors are located within the intervals $[-\Delta_i, \Delta_i]$, it sounds reasonable to select distributions located on these intervals. However, it can be shown that this does not lead to the desired estimates. It turns out that it is possible to estimate the interval uncertainty if we use a distribution $d$ which is *not* located on the interval $[-\Delta_i, \Delta_i]$ – namely, a distribution related to the *basic Cauchy* distribution with the probability density function $\rho(x) = \dfrac{1}{\pi \cdot (x^2 + 1)}$. The resulting Cauchy deviate method works in the linearized case – when the function $f(x_1, \ldots, x_i, \ldots, x_n)$ is reasonably smooth and the box

$$[\underline{x}_1, \overline{x}_1] \times \ldots \times [\underline{x}_i, \overline{x}_i] \times \ldots \times [\underline{x}_n, \overline{x}_n]$$

is small enough, so that on this box, we can reasonably approximate the function $f$ by its linear terms.

If we multiply a random variable distributed according to the above basic Cauchy distribution $d$ by a value $\Delta$, then we get a Cauchy distribution with a parameter $\Delta$, i.e., a distribution described by the following density function: $\rho(x) = \dfrac{\Delta}{\pi \cdot (x^2 + \Delta^2)}$. It is known that:

- if $\xi_1, \ldots, \xi_i, \ldots, \xi_n$ are independent variables distributed according to Cauchy distributions with parameters $\Delta_i$,
- then, for every $n$ real numbers $c_1, \ldots, c_i, \ldots, c_n$, the corresponding linear combination $c_1 \cdot \xi_1 + \ldots + c_i \cdot \xi_i + \ldots + c_n \cdot \xi_n$ is also Cauchy distributed, with the parameter $\Delta$ equal to the desired value

$$\Delta = |c_1| \cdot \Delta_1 + \ldots + |c_i| \cdot \Delta_i + \ldots + |c_n| \cdot \Delta_n.$$

Thus, if for some number of iterations $N$, we simulate $\delta x_i^{(k)}$ $(1 \le k \le N)$ as Cauchy distributed with parameter $\Delta_i$, then, in the linear approximation, the corresponding differences

$$\delta y^{(k)} \stackrel{\text{def}}{=} f\left(\tilde{x}_1 + \delta x_1^{(k)}, \ldots, \tilde{x}_i + \delta x_i^{(k)}, \ldots, \tilde{x}_n + \delta x_n^{(k)}\right) - \tilde{y} \qquad (6.9)$$

are distributed according to the Cauchy distribution with the parameter $\Delta$. The resulting values $\delta y^{(1)}, \ldots, \delta y^{(k)}, \ldots, \delta y^{(N)}$ are therefore a sample from the Cauchy distribution with the desired parameter $\Delta$. Based on this sample, we can estimate the value $\Delta$.

In order to estimate $\Delta$, we can apply the Maximum Likelihood Method which leads to the following equation:

$$\frac{1}{1+\left(\frac{\delta y^{(1)}}{\Delta}\right)^2}+\ldots+\frac{1}{1+\left(\frac{\delta y^{(k)}}{\Delta}\right)^2}+\ldots+\frac{1}{1+\left(\frac{\delta y^{(N)}}{\Delta}\right)^2}=\frac{N}{2}. \quad (6.10)$$

The left-hand side of this equation is an increasing function that is equal to 0 (hence smaller than $N/2$) for $\Delta=0$ and larger than $N/2$ for $\Delta=\max\left|\delta y^{(k)}\right|$; therefore the solution to this equation can be found by applying a bisection method to the interval $\left[0,\max\left|\delta y^{(k)}\right|\right]$.

Simulation of Cauchy distribution with parameter $\Delta_i$ can be based on the functional transformation of uniformly distributed sample values:

$$\delta x_i^{(k)} = \Delta_i \cdot \tan(\pi \cdot (r_i - 0.5)), \quad (6.11)$$

where $r_i$ is uniformly distributed on the interval $[0,1]$.

As a result, we arrive at the following algorithm (see, e.g., [178, 331]):

- Apply $f$ to the midpoints: $\widetilde{y} := f(\widetilde{x}_1, \ldots, \widetilde{x}_i, \ldots, \widetilde{x}_n)$;
- For $k = 1, 2, \ldots, N$, repeat the following:
  - use the standard random number generator to compute $n$ numbers $r_i^{(k)}$, $i = 1, 2, \ldots, n$, that are uniformly distributed on the interval $[0,1]$;
  - compute Cauchy distributed values $c_i^{(k)} := \tan(\pi \cdot (r_i^{(k)} - 0.5))$;
  - compute the largest value of $|c_i^{(k)}|$ so that we will be able to normalize the simulated approximation errors and apply $f$ to the values that are within the box of possible values: $K := \max_i |c_i^{(k)}|$;
  - compute the simulated approximation errors $\delta x_i^{(k)} := \Delta_i \cdot c_i^{(k)}/K$;
  - compute the simulated "actual values" $x_i^{(k)} := \widetilde{x}_i + \delta x_i^{(k)}$;
  - apply the program $f$ to the simulated measurement results and compute the simulated approximation error for $y$:

$$\Delta y^{(k)} := K \cdot \left(f\left(x_1^{(k)}, \ldots, x_i^{(k)}, \ldots, x_n^{(k)}\right) - \widetilde{y}\right); \quad (6.12)$$

- Compute $\Delta$ by applying the bisection method to solve the equation

$$\frac{1}{1+\left(\frac{\Delta y^{(1)}}{\Delta}\right)^2}+\ldots+\frac{1}{1+\left(\frac{\Delta y^{(k)}}{\Delta}\right)^2}+\ldots+\frac{1}{1+\left(\frac{\Delta y^{(N)}}{\Delta}\right)^2}=\frac{N}{2}. \quad (6.13)$$

In [178] and [331], we found the number of iterations $N$ that would provide the desired (relative) accuracy $\varepsilon$ in estimating $\Delta$, i.e., the number of iterations that are needed to guarantee that

$$(1-\varepsilon)\cdot\widetilde{\Delta} \leq \Delta \leq (1+\varepsilon)\cdot\widetilde{\Delta} \quad (6.14)$$

with a given certainty $p_0$.

In practice, it is reasonable to get a certainty $p_0 = 95\%$ and accuracy $\varepsilon = 0.2$ (20%).

To get an accuracy $\varepsilon$ with 95% certainty, we must pick $N = 8/\varepsilon^2$. In particular, to get a 20% accuracy $(0.2 \cdot \Delta)$ with 95% certainty, i.e., to guarantee that

$$0.8 \cdot \widetilde{\Delta} \leq \Delta \leq 1.2 \cdot \widetilde{\Delta} \tag{6.15}$$

with certainty $\geq 95\%$, we need $N = 8/(0.2)^2 = 200$ runs.

In general, the number of calls to a model depends only on the desired accuracy $\varepsilon$ and not on $n$ – so for large $n$, these methods are much faster.

*Comment.* It is important to mention that we assumed that the function $f$ is reasonably linear within the box

$$[\widetilde{x}_1 - \Delta_1, \widetilde{x}_1 + \Delta_1] \times \ldots \times [\widetilde{x}_i - \Delta_i, \widetilde{x}_i + \Delta_i] \times \ldots \times [\widetilde{x}_n - \Delta_n, \widetilde{x}_n + \Delta_n]. \tag{6.16}$$

However, the simulated values $\delta_i$ may be outside the box. When we get such values, we do not use the function $f$ for them, we use a linearized function which is equal to $f$ within the box, and which is extended linearly for all other values.

# Computing under Interval Uncertainty: General Algorithms

*Need for interval computations.* In many application areas, it is sufficient to have an approximate estimate of $y$ – e.g., an estimate obtained from linearization. However, in some applications, it is important to guarantee that the (unknown) actual value $y$ of a certain quantity does not exceed a certain threshold $y_0$. The only way to guarantee this is to have an interval $Y = [\underline{Y}, \overline{Y}]$ which is guaranteed to contain $y$ (i.e., for which $y \subseteq Y$) and for which $\overline{Y} \leq y_0$.

For example, in nuclear engineering, we must make sure that the temperatures and the neutron flows do not exceed the critical values; when planning a space flight, we want to guarantee that the space ship lands on the planet and does not fly past it, etc.

The interval $Y$ which is guaranteed to contain the actual range $y$ is usually called an *enclosure* for this range. So, in such situations, we need to compute either the original range or at least an enclosure for this range. Computing such an enclosure is also one of the main tasks of interval computations.

*Traditional numerical methods are often not sufficient.* The main limitations of the traditional numerical mathematics approach to error estimation was that often, no clear distinction was made between approximate (non-guaranteed) and guaranteed (= interval) error bounds.

For example, for iterative methods, many papers on numerical mathematics consider the rate of convergence as an appropriate measure of approximation error. Clearly, if we know that the error decreases as $O(1/n)$ or as $O(a^{-n})$, we gain some information about the corresponding algorithms – and we also gain a knowledge that for large $n$, the second method is more accurate. However, in real life, we make a fixed number $n$ of iterations. If the only information we have about the approximation error is the above asymptotics, then we still have no idea how close the result of $n$-th iteration is to the actual (desired) value.

It is therefore important to emphasize the need for guaranteed methods, and to develop techniques for producing *guaranteed* estimates. Such guaranteed estimates is what interval computations are about.

H.T. Nguyen et al.: Computing Stat. under Interval & Fuzzy Uncertain., SCI 393, pp. 35–45.
springerlink.com                                            © Springer-Verlag Berlin Heidelberg 2012

*Interval computations: a brief history.* The notion of interval computations is reasonably recent, it dates back to the 1950s, but the main problem is known since Archimedes who used guaranteed two-sided bounds to compute $\pi$; see, e.g., Achimedes [14].

Since then, many useful guaranteed bounds have been developed for different numerical methods. There have also been several general descriptions of such bounds, often formulated in terms similar to what we described above. For example, in the early 20th century, the concept of a function having values which are bounded within limits was discussed by W. H. Young in [362]. The concept of operations with a set of multi-valued numbers was introduced by R. C. Young, who developed a formal algebra of multi-valued numbers [363]. The special case of closed intervals was further developed by P. S. Dwyer in [92].

Interval computations in their current form were independently invented by three researchers in three different parts of the world: by M. Warmus in Poland [342], by T. Sunaga in Japan [319], and by R. Moore in the USA [224].

The active interest in interval computations started with Moore's 1966 monograph [225]. This interest was enhanced by the fact that in addition to estimates for general numerical algorithms, Moore's monograph also described *practical* applications which have already been developed in his earlier papers and technical reports: in particular, interval computations were used to make sure that even when we take all the uncertainties into account, the trajectory of a space flight is guaranteed to reach the Moon.

Since then, interval computations have been actively used in many areas of science and engineering; see, e.g., interval website [139] and books such as [142, 229].

*Comment.* Early papers on interval computations can be found on the interval computations website [139].

*First step: interval arithmetic.* Our goal is to find the range of a given function $f(x_1, \ldots, x_i, \ldots, x_n)$ on the given intervals $\boldsymbol{x}_1 = [\underline{x}_1, \overline{x}_1], \ldots, \boldsymbol{x}_i = [\underline{x}_i, \overline{x}_i], \ldots, \boldsymbol{x}_n = [\underline{x}_n, \overline{x}_n]$.

This function $f(x_1, \ldots, x_i, \ldots, x_n)$ is given as an algorithm. In particular, we may have an explicit analytical expression for $f$, in which case this algorithm consists of simply computing this expression.

When we talk about algorithms, we usually mean an algorithm (program) written in a high-level programming language like Java or C. Such programming languages allows us to use arithmetic expressions and many other complex constructions. Most of these constructions, however, are not directly hardware supported inside a computer. Usually, only simple arithmetic operations are implemented: addition, subtraction, multiplication, and $1/x$ (plus branching). Even division $a/b$ is usually not directly supported, it is performed as a sequence of two elementary arithmetic operations:

- first, we compute $1/b$;
- then, we multiply $a$ by $1/b$.

When we input a general program into a computer, the computer *parses* it, i.e., represents it a sequence of elementary arithmetic operations.

Since a computer performs this parsing anyway, we can safely assume that the original algorithm $f(x_1, \ldots, x_i, \ldots, x_n)$ is already represented as a sequence of such elementary arithmetic operations.

Let us start our analysis of the interval computation techniques with the simplest possible case when the algorithm $f(x_1, \ldots, x_i, \ldots, x_n)$ simply consists of a single arithmetic operation: addition, subtraction, multiplication, or computing $1/x$.

Let us start by estimating the range of the addition function $f(x_1, x_2) = x_1 + x_2$ on the intervals $[\underline{x}_1, \overline{x}_1]$ and $[\underline{x}_2, \overline{x}_2]$. This function is increasing with respect to both its variables. We already know how to compute the range $[\underline{y}, \overline{y}]$ of a monotonic function. So, the range of addition is equal to $[\underline{x}_1 + \underline{x}_2, \overline{x}_1 + \overline{x}_2]$.

The desired range is usually denoted as $f(\boldsymbol{x}_1, \ldots, \boldsymbol{x}_i, \ldots, \boldsymbol{x}_n)$; in particular, for addition, this notation takes the form $\boldsymbol{x}_1 + \boldsymbol{x}_2$. Thus, we can define "addition" of two intervals as follows:

$$[\underline{x}_1, \overline{x}_1] + [\underline{x}_2, \overline{x}_2] = [\underline{x}_1 + \underline{x}_2, \overline{x}_2 + \overline{x}_2]. \tag{7.1}$$

This formula makes perfect intuitive sense: if one town has between 700 and 800 thousand people, and it merges with a nearby town whose population is between 100 and 200 thousand, then:

- the smallest possible value of the total population of the new big town is when both populations are the smallest possible, $700 + 100 = 800$, and
- the largest possible value is when both populations are the largest possible, i.e., $800 + 200 = 1000$.

The subtraction function $f(x_1, x_2) = x_1 - x_2$ is increasing with respect to $x_1$ and decreasing with respect to $x_2$, so we have

$$[\underline{x}_1, \overline{x}_1] - [\underline{x}_2, \overline{x}_2] = [\underline{x}_1 - \overline{x}_2, \overline{x}_1 - \underline{x}_2]. \tag{7.2}$$

These operations are also in full agreement with common sense. For example, if a warehouse originally had between 6.0 and 8.0 tons, and we moved between 1.0 and 2.0 tons to another location, then the smallest amount left is when we start with the smallest possible value 6.0 and move the largest possible value 2.0, resulting in $6.0 - 2.0 = 4.0$. The largest amount left is when we start with the largest possible value 8.0 and move the smallest possible value 1.0, resulting in $8.0 - 1.0 = 7.0$.

For multiplication $f(x_1, x_2) = x_1 \cdot x_2$, the direction of monotonicity depends on the actual values of $x_1$ and $x_2$: e.g.,

- when $x_2 > 0$, the product increases with $x_1$,
- otherwise it decreases with $x_1$.

So, unless we know the signs of the product beforehand, we cannot tell whether the maximum is attained at $x_1 = \underline{x}_1$ or at $x_1 = \overline{x}_1$. However, we know that it is always attained at one of these endpoints. So, to find the range of the product, it is sufficient to try all $2 \cdot 2 = 4$ combinations of these endpoints:

$$[\underline{x}_1, \overline{x}_1] \cdot [\underline{x}_2, \overline{x}_2] =$$

$$[\min(\underline{x}_1 \cdot \underline{x}_2, \underline{x}_1 \cdot \overline{x}_2, \overline{x}_1 \cdot \underline{x}_2, \overline{x}_1 \cdot \overline{x}_2), \max(\underline{x}_1 \cdot \underline{x}_2, \underline{x}_1 \cdot \overline{x}_2, \overline{x}_1 \cdot \underline{x}_2, \overline{x}_1 \cdot \overline{x}_2)]. \quad (7.3)$$

Finally, the function $f(x_1) = 1/x_1$ is decreasing wherever it is defined (when $x_1 \neq 0$), so if $0 \notin [\underline{x}_1, \overline{x}_1]$, then

$$\frac{1}{[\underline{x}_1, \overline{x}_1]} = \left[\frac{1}{\overline{x}_1}, \frac{1}{\underline{x}_1}\right]. \quad (7.4)$$

The formulas for addition, subtraction, multiplication, and reciprocal of intervals are called formulas of *interval arithmetic.*

*Straightforward ("naive") interval computations.* Historically the first method for computing the enclosure for the general case is the method which is sometimes called "straightforward" interval computations. In this method, we repeat the computations forming the program $f$ step-by-step, replacing each operation with real numbers by the corresponding operation of interval arithmetic. It is known that, as a result, we get an enclosure $\boldsymbol{Y} \supseteq \boldsymbol{y}$ for the desired range.

In some cases, this enclosure is exact. For example, straightforward interval computations lead to the exact range when $f(x_1, \ldots, x_n)$ is a *single-use expression* (SUE), i.e., when in this expression, each variable $x_i$ only occurs once; see, e.g., [131].

In more complex cases (see example below), the enclosure has excess width.

*Example.* Let us illustrate the above idea on the example of estimating the range of the function $f(x_1) = x_1 - x_1^2$ on the interval $x_1 \in [0, 0.8]$.

We start with parsing the expression for the function, i.e., describing how a computer will compute this expression; it will implement the following sequence of elementary operations:

$$r_1 := x_1 \cdot x_1; \quad r_2 := x_1 - r_1. \quad (7.5)$$

According to straightforward interval computations, we perform the same operations, but with *intervals* instead of *numbers*:

$$\boldsymbol{r}_1 := [0, 0.8] \cdot [0, 0.8] = [0, 0.64]; \quad \boldsymbol{r}_2 := [0, 0.8] - [0, 0.64] = [-0.64, 0.8]. \quad (7.6)$$

For this function, the actual range is $f(\boldsymbol{x}_1) = [0, 0.25]$; see Fig. 7.1.

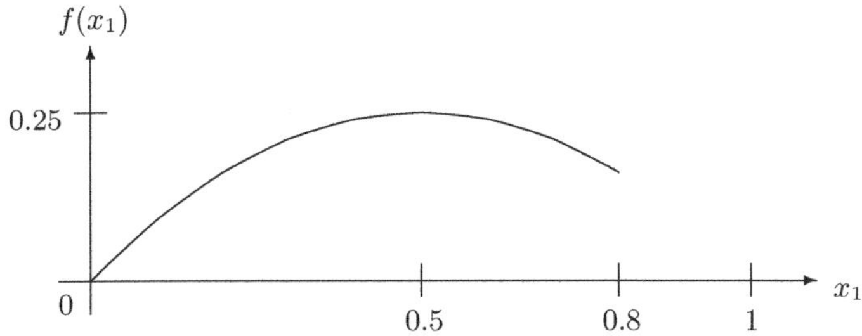

**Fig. 7.1.** Range of the function $f(x_1) = x_1 - x_1^2$ on the interval $[0, 0.8]$

*Interval computations go beyond straightforward technique.* People who are vaguely familiar with interval computations sometimes erroneously assume that the above straightforward ("naive") techniques is all there is in interval computations. In conference presentations (and even in published papers), one often encounters a statement: "I tried interval computations, and it did not work". What this statement usually means is that they tried the above straightforward approach and – not surprisingly – it did not work well.

In reality, interval computations is *not a single algorithm*, it is a *problem* for which many different techniques exist. Let us now describe some of such techniques.

*Comment.* For each of the known techniques, there are cases when we get an excess width. The reason is that the problem of computing the exact range is NP-hard even for polynomial functions $f(x_1, \ldots, x_i, \ldots, x_n)$ – actually, even for quadratic functions $f$ (see next chapter).

*Mean value form.* One of such techniques is the mean value form; see, e.g., [142]. This technique is based on the same Taylor series expansion ideas as linearization. We start by representing each interval $x_i = [\underline{x}_i, \overline{x}_i]$ in the form $[\widetilde{x}_i - \Delta_i, \widetilde{x}_i + \Delta_i]$, where $\widetilde{x}_i = (\underline{x}_i + \overline{x}_i)/2$ is the midpoint of the interval $x_i$ and $\Delta_i = (\overline{x}_i - \underline{x}_i)/2$ is the half-width of this interval.

After that, we use the Taylor expansion. In linearization, we simply ignored quadratic and higher order terms. Here, instead, we use the Taylor form with a remainder term. Specifically, the mean value form is based on the formula

$$f(x_1, \ldots, x_i, \ldots, x_n) = f(\widetilde{x}_1, \ldots, \widetilde{x}_i, \ldots, \widetilde{x}_n) +$$

$$\sum_{i=1}^{n} \frac{\partial f}{\partial x_i}(\eta_1, \ldots, \eta_i, \ldots, \eta_n) \cdot (x_i - \widetilde{x}_i), \tag{7.7}$$

where each $\eta_i$ is some value from the interval $x_i$.

Since $\eta_i \in x_i$, the value of the $i$-th derivative belongs to the interval range of this derivative on these intervals. We also know that $x_i - \widetilde{x}_i \in [-\Delta_i, \Delta_i]$. Thus, we can conclude that

$$f(x_1, \ldots, x_i, \ldots, x_n) \subseteq f(\widetilde{x}_1, \ldots, \widetilde{x}_i, \ldots, \widetilde{x}_n) +$$

$$\sum_{i=1}^{n} \frac{\partial f}{\partial x_i}(x_1, \ldots, x_i, \ldots, x_n) \cdot [-\Delta_i, \Delta_i]. \tag{7.8}$$

To compute the ranges of the partial derivatives, we can use straightforward interval computations.

*Example.* Let us illustrate this method on the above example of estimating the range of the function $f(x_1) = x_1 - x_1^2$ over the interval $[0, 0.8]$. For this interval, the midpoint is $\widetilde{x}_1 = 0.4$; at this midpoint, $f(\widetilde{x}_1) = 0.24$. The half-width is $\Delta_1 = 0.4$. The only partial derivative here is $\dfrac{\partial f}{\partial x_1} = 1 - 2x_1$, its range on $[0, 0.8]$ is equal to

$$1 - 2 \cdot [0, 0.8] = [-0.6, 1].$$

Thus, we get the following enclosure for the desired range $y$:

$$y \subseteq Y = 0.24 + [-0.6, 1] \cdot [-0.4, 0.4] = 0.24 + [-0.4, 0.4] = [-0.16, 0.64]. \tag{7.9}$$

This enclosure is narrower than the "naive" estimate $[-0.64, 0.8]$, but it still contains excess width.

*How can we get better estimates?* In the mean value form, we, in effect, ignored quadratic and higher order terms, i.e., terms of the type $\dfrac{\partial^2 f}{\partial x_i \partial x_j} \cdot \Delta x_i \cdot \Delta x_j$. When the estimate is not accurate enough, it means that this ignored term is too large. There are two ways to reduce the size of the ignored term:

- we can try to decrease this quadratic term, or
- we can try to explicitly include higher order terms in the Taylor expansion formula, so that the remainder term will be proportional to say $\Delta x_i^3$ and thus, be much smaller.

Let us describe these two ideas in detail.

*First idea: bisection.* Let us first describe the situation in which we try to minimize the second-order remainder term. In the above expression for this term, we cannot change the second derivative. The only thing we can decrease is the difference $\Delta x_i = x_i - \widetilde{x}_i$ between the actual value and the midpoint. This value is bounded by the half-width $\Delta_i$ of the box. So, to decrease this value, we can subdivide the original box into several narrower subboxes. Usually, we divide into two subboxes, so this subdivision is called *bisection.*

The range over the whole box is equal to the union of the ranges over all the subboxes. The widths of each subbox are smaller, so we get smaller $\Delta x_i$ and hopefully, more accurate estimates for ranges over each of this subbox. Then, we take the union of the ranges over subboxes.

*Example.* Let us illustrate this idea on the above $x_1 - x_1^2$ example. In this example, we divide the original interval $[0, 0.8]$ into two subintervals $[0, 0.4]$ and $[0.4, 0.8]$. For both intervals, $\Delta_1 = 0.2$.

In the first subinterval, the midpoint is $\widetilde{x}_1 = 0.2$, so $f(\widetilde{x}_1) = 0.2 - 0.04 = 0.16$. The range of the derivative is equal to $1 - 2 \cdot [0, 0.4] = 1 - [0, 0.8] = [0.2, 1]$, hence we get an enclosure $0.16 + [0.2, 1] \cdot [-0.2, 0.2] = [-0.04, 0.36]$.

For the second interval, $\widetilde{x}_1 = 0.6$, $f(0.6) = 0.24$, the range of the derivative is $1 - 2 \cdot [0.4, 0.8] = [-0.6, 0.2]$, hence we get an enclosure

$$0.24 + [-0.6, 0.2] \cdot [-0.2, 0.2] = [0.12, 0.36]. \tag{7.10}$$

The union of these two enclosures is the interval $[-0.04, 0.36]$. This enclosure is much more accurate than before.

Further bisection leads to even more accurate estimates – the smaller the subintervals, the more accurate the enclosure.

*Bisection: general comment.* The more subboxes we consider, the smaller $\Delta x_i$ and thus, the more accurate the corresponding enclosures. However, once we have more boxes, we need to spend more time processing these boxes. Thus, we have a trade-off between computation time and accuracy: the more computation time we allow, the more accurate estimates we will be able to compute.

*Additional idea: monotonicity checking.* If the function $f(x_1, \ldots, x_i, \ldots, x_n)$ is monotonic over the original box $\boldsymbol{x}_1 \times \ldots \times \boldsymbol{x}_i \ldots \times \boldsymbol{x}_n$, then we can easily compute its exact range. Since we used the mean value form for the original box, this probably means that on that box, the function is not monotonic: for example, with respect to $x_1$, it may be increasing at some points in this box, and decreasing at other points.

However, as we divide the original box into smaller subboxes, it is quite possible that at least some of these subboxes will be outside the areas where the derivatives are 0 and thus, the function $f(x_1, \ldots, x_i, \ldots, x_n)$ will be monotonic. So, after we subdivide the box into subboxes, we should first check monotonicity on each of these subboxes – and if the function is monotonic, we can easily compute its range.

In calculus terms, a function is increasing with respect to $x_i$ if its partial derivative $d_i \stackrel{\text{def}}{=} \dfrac{\partial f}{\partial x_i}$ is non-negative everywhere on this subbox. Thus, to check monotonicity, we should find the range $[\underline{d}_i, \overline{d}_i]$ of this derivative (we need to do it anyway to compute the mean value form expression):

- if $\underline{d}_i \geq 0$, this means that the derivative is everywhere non-negative and thus, the function $f$ is increasing in $x_i$;
- if $\overline{d}_i \leq 0$, this means that the derivative is everywhere non-positive and thus, the function $f$ is decreasing in $x_i$.

If $\underline{d}_i < 0 < \overline{d}_i$, then we have to use the mean value form.

If the function is monotonic (e.g., increasing) only with respect to some of the variables $x_i$, then

- to compute $\overline{y}$, it is sufficient to consider only the value $x_i = \overline{x}_i$, and
- to compute $\underline{y}$, it is sufficient to consider only the value $x_i = \underline{x}_i$.

For such subboxes, we reduce the original problem to two problems with fewer variables, problems which are thus easier to solve.

*Example.* For the example $f(x_1) = x_1 - x_1^2$, the partial derivative is equal to $1 - 2 \cdot x_1$.

On the first subbox $[0, 0.4]$, the range of this derivative is $1 - 2 \cdot [0, 0.4] = [0.2, 1]$. Thus, the derivative is always non-negative, the function is increasing on this subbox, and its range on this subbox is equal to $[f(0), f(0.4)] = [0, 0.16]$.

On the second subbox $[0.4, 0.8]$, the range of the derivative is

$$1 - 2 \cdot [0.4, 0.8] = [-0.6, 0.2].$$

Here, we do not have guaranteed monotonicity, so we can use the mean value form to get the enclosure $[0.12, 0.36]$ for the range.

The union of these two enclosures is the interval $[0, 0.36]$, which is slightly more accurate than before. Further bisection leads to even more accurate estimates.

*Comment.* We got the exact range because of the simplicity of our example, in which the extreme point 0.5 of the function $f(x_1) = x_1 - x_1^2$ is exactly in the middle of the interval $[0, 1]$. Thus, when we divided the box in two, both subboxes have the monotonicity property. In the general case, the extremal point will be inside one of the subboxes, so we will have excess width.

*General Taylor techniques.* As we have mentioned, another way to get more accurate estimates is to use so-called *Taylor techniques*, i.e., to explicitly consider second-order and higher-order terms in the Taylor expansion; see, e.g., [42, 242], and references therein.

Let us illustrate the main ideas of Taylor analysis on the case when we allow second order terms. In this case, the formula with a remainder takes the form

$$f(x_1, \ldots, x_i, \ldots, x_n) = f(\widetilde{x}_1, \ldots, \widetilde{x}_i, \ldots, \widetilde{x}_n) + \sum_{i=1}^{n} \frac{\partial f}{\partial x_i}(\widetilde{x}_1, \ldots, \widetilde{x}_n) \cdot (x_i - \widetilde{x}_i) +$$

$$\frac{1}{2} \cdot \sum_{i=1}^{n} \sum_{j=1}^{m} \frac{\partial^2 f}{\partial x_i \partial x_j}(\eta_1, \ldots, \eta_n) \cdot (x_i - \widetilde{x}_i) \cdot (x_j - \widetilde{x}_j). \qquad (7.11)$$

Thus, we get the enclosure

$$f(\boldsymbol{x}_1, \ldots, \boldsymbol{x}_i, \ldots, \boldsymbol{x}_n) \subseteq \mathbf{Y} \stackrel{\text{def}}{=} f(\widetilde{x}_1, \ldots, \widetilde{x}_i, \ldots, \widetilde{x}_n) +$$

$$\sum_{i=1}^{n} \frac{\partial f}{\partial x_i}(\widetilde{x}_1, \ldots, \widetilde{x}_i, \ldots, \widetilde{x}_n) \cdot [-\Delta_i, \Delta_i] + \qquad (7.12)$$

$$\frac{1}{2} \cdot \sum_{i=1}^{n} \sum_{j=1}^{m} \frac{\partial^2 f}{\partial x_i \partial x_j}(\boldsymbol{x}_1, \ldots, \boldsymbol{x}_n) \cdot [-\Delta_i, \Delta_i] \cdot [-\Delta_j, \Delta_j].$$

*Example.* Let us illustrate this idea on the above example of $f(x_1) = x_1 - x_1^2$. Here, $\Delta_1 = 0.4$, $\widetilde{x}_1 = 0.4$, so $f(\widetilde{x}_1) = 0.24$ and $\dfrac{\partial f}{\partial x_1}(\widetilde{x}_1) = 1 - 2 \cdot 0.4 = 0.2$. The second derivative is equal to $-2$, so the Taylor estimate takes the form

$$\boldsymbol{Y} = 0.24 + 0.2 \cdot [-0.4, 0.4] - [-0.4, 0.4]^2.$$

Strictly speaking, if we interpret $\Delta x_1^2$ as $\Delta x_1 \cdot \Delta x_1$ and use the formulas of interval multiplication, we get the interval

$$[-0.4, 0.4]^2 = [-0.4, 0.4] \cdot [-0.4, 0.4] = [-0.16, 0.16]$$

and thus, the enclosure

$$\boldsymbol{Y} = 0.24 + [-0.08, 0.08] - [-0.16, 0.16] = [0.16, 0.32] - [-0.16, 0.16] = [0, 0.48]$$

for the desired range. However, we can view $x^2$ as a special function, for which the range over $[-0.4, 0.4]$ is known to be $[0, 0.16]$. In this case, the above enclosure takes the form

$$\boldsymbol{Y} = 0.24 + [-0.08, 0.08] - [0, 0.16] = [0.16, 0.32] - [0, 0.16] = [0, 0.32]$$

which is much closer to the actual range $[0, 0.25]$.

*Taylor methods: general comment.* The more terms we consider in the Taylor expansion, the smaller the remainder term and thus, the more accurate the corresponding enclosures. However, once we have more terms, we need to spend more time computing these terms. Thus, for Taylor methods, we also have a trade-off between computation time and accuracy: the more computation time we allow, the more accurate estimates we will be able to compute.

*An alternative version of affine and Taylor arithmetic.* The main idea of Taylor methods is to approximate the given function $f(x_1, \ldots, x_i, \ldots, x_n)$ by a polynomial of a small order plus an interval remainder term.

In these terms, straightforward interval computations can be viewed as 0-th order Taylor methods in which all we have is the corresponding interval (or, equivalently, the constant term plus the remainder interval). To compute this interval, we repeated the computation of $f$ step by step, replacing operations with numbers by operations with intervals.

We can do the same for higher-order Taylor expansions as well. Let us illustrate how this can be done for the first order Taylor terms. We start with the expressions $x_i = \widetilde{x}_i - \Delta x_i$. Then, at each step, we keep a term of the type $a = \widetilde{a} + \sum_{i=1}^{n} a_i \cdot \Delta x_i + \boldsymbol{a}$. (To be more precise, the keep the coefficients $\widetilde{a}$ and $a_i$ and the interval $\boldsymbol{a}$.)

Addition and subtraction of such terms are straightforward:

$$\left(\widetilde{a} + \sum_{i=1}^{n} a_i \cdot \Delta x_i + \boldsymbol{a}\right) + \left(\widetilde{b} + \sum_{i=1}^{n} b_i \cdot \Delta x_i + \boldsymbol{b}\right) =$$

$$(\widetilde{a} + \widetilde{b}) + \sum_{i=1}^{n}(a_i + b_i) \cdot \Delta x_i + (\boldsymbol{a} + \boldsymbol{b});$$

$$\left(\widetilde{a} + \sum_{i=1}^{n} a_i \cdot \Delta x_i + \boldsymbol{a}\right) - \left(\widetilde{b} + \sum_{i=1}^{n} b_i \cdot \Delta x_i + \boldsymbol{b}\right) =$$

$$(\widetilde{a} - \widetilde{b}) + \sum_{i=1}^{n}(a_i - b_i) \cdot \Delta x_i + (\boldsymbol{a} - \boldsymbol{b}).$$

For multiplication, we add terms proportional to $\Delta x_i \cdot \Delta x_j$ to the interval part:

$$\left(\widetilde{a} + \sum_{i=1}^{n} a_i \cdot \Delta x_i + \boldsymbol{a}\right) \cdot \left(\widetilde{b} + \sum_{i=1}^{n} b_i \cdot \Delta x_i + \boldsymbol{b}\right) =$$

$$(\widetilde{a} \cdot \widetilde{b}) + \sum_{i=1}^{n}(\widetilde{a} \cdot b_i + \widetilde{b} \cdot a_i) \cdot \Delta x_i +$$

$$\left(\widetilde{a} \cdot \boldsymbol{b} + \widetilde{b} \cdot \boldsymbol{a} + \sum_{i=1}^{n} a_i \cdot b_i \cdot [0, \Delta_i^2] + \sum_{i=1}^{n}\sum_{j \neq i} a_i \cdot b_j \cdot [-\Delta_i, \Delta_i] \cdot [\Delta_j \cdot \Delta_j]\right).$$

$$(7.13)$$

At the end, we get an expression of the above type for the desired quantity $y$: $y = \widetilde{y} + \sum_{i=1}^{n} y_i \cdot \Delta x_i + \boldsymbol{y}$. We already know how to compute the range of a linear function, so we get the following enclosure for the final range: $\boldsymbol{Y} = \widetilde{y} + [-\Delta, \Delta] + \boldsymbol{y}$, where $\Delta = \sum_{i=1}^{n} |y_i| \cdot \Delta_i$.

*Example.* For $f(x_1) = x_1 - x_1^2$, we first compute $x_2 = x_1^2$ and then $y = x_1 - x_2$. We start with the interval $\boldsymbol{x}_1 = \widetilde{x}_1 - \Delta x_1 = 0.4 + (-1) \cdot \Delta_1 + [0, 0]$.

On the next step, we compute the square of this expression. This square is equal to $0.16 + (-0.8) \cdot \Delta x_1 + \Delta x_1^2$. Since $\Delta x_1 \in [-0.4, 0.4]$, we conclude that $\Delta x_1^2 \in [0, 0.16]$ and thus, that $x_2 = 0.16 + (-0.8) \cdot \Delta x_1 + [0, 0.16]$.

For $y = x_1 - x_2$, we now have

$$y = (0.4 - 0.16) + ((-1) - (-0.8)) \cdot \Delta x_1 + ([0, 0] - [0, 0.16]) =$$

$$0.24 + (-0.2) \cdot \Delta x_1 + [-0.16, 0]. \qquad (7.14)$$

Since $\Delta x_1 \in [-0.4, 0.4]$, we get the enclosure

$$\boldsymbol{Y} = 0.24 + (-0.2) \cdot [-0.4, 0.4] + [-0.16, 0] = [0, 0.32]. \qquad (7.15)$$

*Comment.* We have described several methods and several ideas. On our simple example, some ideas work better, some lead to wider enclosures. The fact that a method works better on the simple example does not mean that it always works better, it depends on the function. In large-scale practical examples, it is useful to *combine* all these methods and ideas – e.g., bisect and use mean value form and monotonicity on subboxes; see, e.g., [142].

The interval method – one of the above or their combination – has to be carefully chosen to match the function at hand. There exist several semi-empirical heuristics on which method to choose; see, e.g., [142].

# 8

# Computing under Interval Uncertainty: Computational Complexity

In this chapter, we will briefly describe the computational complexity of the range estimation problem under interval uncertainty.

*Linear case.* Let us start with the simplest case of a linear function

$$y = f(x_1, \ldots, x_n) = a_0 + \sum_{i=1}^{n} a_i \cdot x_i.$$

In this case, substituting the (approximate) measured values $\widetilde{x}_i$, we get the approximate value

$$\widetilde{y} = a_0 + \sum_{i=1}^{n} a_i \cdot \widetilde{x}_i$$

for $y$.

The approximation error $\Delta y = \widetilde{y} - y$ of this approximation can be described as

$$\Delta y = \sum_{i=1}^{n} a_i \cdot \Delta x_i,$$

where each input error $\Delta x_i$ can take any value from $-\Delta_i$ to $\Delta_i$.

The sum $\sum_{i=1}^{n} a_i \cdot \Delta x_i$ attains its largest possible value if each term $a_i \cdot \Delta x_i$ in this sum attains the largest possible value:

- If $a_i \geq 0$, then this term is a monotonically non-decreasing function of $\Delta x_i$, so it attains its largest value at the largest possible value $\Delta x_i = \Delta_i$; the corresponding largest value of this term is $a_i \cdot \Delta_i$.
- If $a_i < 0$, then this term is a decreasing function of $\Delta x_i$, so it attains its largest value at the smallest possible value $\Delta x_i = -\Delta_i$; the corresponding largest value of this term is $-a_i \cdot \Delta_i = |a_i| \cdot \Delta_i$.

In both cases, the largest possible value of this term is $|a_i| \cdot \Delta_i$, so, the largest possible value of the sum $\Delta y$ is

H.T. Nguyen et al.: Computing Stat. under Interval & Fuzzy Uncertain., SCI 393, pp. 47–50.
springerlink.com                              © Springer-Verlag Berlin Heidelberg 2012

$$\Delta = |a_1| \cdot \Delta_1 + \ldots + |a_n| \cdot \Delta_n.$$

Similarly, the smallest possible value of $\Delta y$ is $-\Delta$.

Hence, the interval of possible values of $\Delta y$ is $[-\Delta, \Delta]$, and the interval of possible values of the actual value $y$ is $[\widetilde{y} - \Delta, \widetilde{y} + \Delta]$.

The corresponding range can be computed in linear time, i.e., efficiently.

*Quadratic case.* Already for quadratic functions

$$y = f(x_1, \ldots, x_n) = a_0 + \sum_{i=1}^{n} a_i \cdot x_i + \sum_{i=1}^{n} \sum_{j=1}^{n} a_{ij} \cdot x_i \cdot x_j,$$

the problem of computing the exact range

$$\boldsymbol{y} = f(\boldsymbol{x}_1, \ldots, \boldsymbol{x}_n) =$$

$$\{f(x_1, \ldots, x_n) : x_1 \in \boldsymbol{x}_1, \ldots, x_n \in \boldsymbol{x}_n\}$$

over interval inputs $x_i \in \boldsymbol{x}_i = [\widetilde{x}_i - \Delta_i, \widetilde{x}_i + \Delta_i]$ is, in general, NP-hard; see, e.g., [182, 334].

*What is NP-hard? A brief description.* NP-hard means, crudely speaking, that no feasible (polynomial time) algorithm can compute the exact endpoints of the range $\boldsymbol{y}$ for all possible intervals $\boldsymbol{x}_1, \ldots, \boldsymbol{x}_n$. (Strictly speaking, this interpretation is only true under the widely believed but still unproven hypothesis that P$\neq$NP).

*Towards a more precise description of NP-hardness: the notion of a feasible algorithm.* The notion of NP-hardness is related to the fact that some algorithms take so much computation time that even for inputs of reasonable size, the computation time exceeds the lifetime of the Universe – and thus, cannot be practically computed. For example, if for $n$ inputs, the algorithm takes time $2^n$, then for $n \approx$ 300-400, the resulting computation time is unrealistically large. How can we separate "realistic" ("feasible") algorithms from non-feasible ones?

The running time of an algorithm depends on the size of the input. In the computer, every object is represented as a sequence of bits (0s and 1s). Thus, for every computer-represented object $x$, it is reasonable to define its *size* (or *length*) $\mathrm{len}(x)$ as the number of bits in this object's computer representation.

It is known that in most feasible algorithms, the running time on an input $x$ is bounded either by the size of the input, or by the square of the size of the input, or, more generally, by a polynomial of the size of the input. It is also known that in most non-feasible algorithms, the running time grows exponentially (or even faster) with the size, so it cannot be bounded by any polynomial. In view of this fact, in theory of computation, an algorithm is usually called feasible if its running time is bounded by a polynomial of the size of the input. This definition is not perfect: e.g., if the running time on input of size $n$ is $10^{40} \cdot n$, then this running time is bounded by a polynomial but it is clearly not feasible. However, this definition is the closest to the intuitive notion of feasible, and thus, the best we have so far.

According to this definition, an algorithm $A$ is called *polynomial time* if there exists a polynomial $P(n)$ such that on every input $x$, the running time of the algorithm $A$ does not exceed $P(\text{len}(x))$. The class of all the problems which can be solved by polynomial-time algorithms is denoted by P.

*The notion of a problem.* What do we mean by "a problem"? In most practical situations, to solve a problem means to find a solution that satisfies some (relatively) easy-to-check constraint: e.g., to design a bridge that can withstand a certain amount of load and wind, to design a spaceship and its trajectory that enables us to deliver a robotic rover to Mars, etc. In all these cases, once we have a candidate for a solution, we can check, in reasonable (polynomial) time whether this candidate is indeed a solution. In other words, once we guessed a solution, we can check its correctness in polynomial time. In theory of computation, this procedure of guess-then-compute is called *non-deterministic computation*, so the class of all problems for which solution can be checked in polynomial time is called Non-deterministic Polynomial, or NP, for short.

*The notion of NP-hardness.* Most computer scientists believe that not all problems from the class NP can be solved in polynomial time, i.e., that $\text{NP} \neq P$. However, no one has so far been able to prove that this belief is indeed true. What is known is that some problems from the class NP are the hardest in this class – in the sense that every other problem from the class NP can be reduced to such a problem.

Specifically, a general problem (not necessarily from the class NP) is called *NP-hard* if every problem from the class NP can be reduced to particular cases of this problem. If a problem from the class NP is NP-hard, we say that it is *NP-complete.*

*Propositional satisfiability: historically the first example of an NP-hard problem.* One of the best known examples of NP-complete problems is the problem of *propositional satisfiability* for formulas in 3-Conjunctive Normal Form (3-CNF). Let us describe this problem is some detail. We start with $v$ Boolean variables $z_1, \ldots, z_v$, i.e., variables which can take only values "true" or "false". A *literal* $\ell$ is defined as a variable $z_i$ or its negation $\neg z_i$. A *clause* is defined as a formula of the type $\ell_1 \vee \ell_2 \vee \ldots \vee \ell_m$. Finally, a *propositional formula in Conjunctive Normal Form (CNF)* is defined as a formula $F$ of the type $C_1 \& \ldots \& C_n$, where $C_1, \ldots, C_n$ are clauses. This formula is called a *3-CNF* formula if every clause has at most 3 literals, and a *2-CNF* formula if every clause has at most 2 literals.

The propositional satisfiability problem is as follows:

- *Given* a propositional formula $F$ (e.g., a formula in CNF);
- *Find* the values of the variables $z_1, \ldots, z_v$ which make the formula $F$ true.

*Other known NP-hard problems.* In this book, we will use several other known NP-hard problem.

One such problem is a *subset* problem: given $n$ positive integers $s_1, \ldots, s_n$, to check whether there exist signs $\eta_i \in \{-1, +1\}$ for which the signed sum

$$\sum_{i=1}^{n} \eta_i \cdot s_i \text{ equals } 0.$$

*How to prove NP-hardness of different problems.* For the propositional satisfiability problem, the proof of NP-hardness is somewhat complex. However, once this NP-hardness is proven, we can prove the NP-hardness of other problems by reducing satisfiability to these problems.

Indeed, by definition, NP-hardness of satisfiability means that every problem from the class NP can be reduced to satisfiability. If we can reduce satisfiability to some other problem, this means that by combining these two reductions, we can reduce every problem from the class NP to this new problem – and thus, that this new problem is also NP-hard.

Similarly, if we can reduce a known NP-hard problem (not necessarily propositional satisfiability) to a new problem, this means that every problem from the class NP can be reduced to this known problem and this known problem can be reduced to the new problem. Thus, we can reduce every problem from the class NP to this new problem – so the new problem is NP-hard.

For a more detailed and more formal definition of NP-hardness, see, e.g., [182, 274].

# 9

# Towards Selecting Appropriate Statistical Characteristics: The Basics of Decision Theory and the Notion of Utility

In the previous chapter, we mentioned that in general, the problem of estimating statistical characteristics under interval uncertainty is NP-hard. This means, crudely speaking, that it is not possible to design a feasible algorithm that would compute all statistics under interval uncertainty. It is therefore necessary to restrict ourselves to statistical characteristics which are practically useful.

Which statistical characteristics should we estimate? One of the main objectives of data processing is to make decisions. Thus, to find the most appropriate statistical characteristics, let us recall the traditional way of making decisions based on user's preference: the decision theory.

*How to describe preferences: general idea.* The possibility to describe preferences in precise terms comes from the fact that a decision maker can always decide which of the two alternatives is better (preferable). Thus, if we provide a continuous scale of alternatives, from a very bad to a very good one, then for each alternative in the middle, there should be an alternative on this scale which is, to this decision maker, equivalent to the given one.

*How to describe preferences: specific ideas.* Such a scale can be easy constructed as follows. We select two alternatives:

- a very negative alternative $A_0$; e.g., an alternative in which the decision maker loses all his money (and/or loses his health as well), and
- a very positive alternative $A_1$; e.g., an alternative in which the decision maker wins several million dollars.

Now, for every value $p \in [0,1]$, we can consider a lottery in which we get $A_1$ with probability $p$ and $A_0$ with the remaining probability $1 - p$. This probability will be denoted by $L(p)$.

For $p = 1$, the probability of the unfavorable outcome $A_0$ is 0, so the lottery $L(1)$ simply means the very positive alternative $A_1$. Similarly, for $p = 0$, the probability of the favorable outcome $A_1$ is 0, so the lottery $L(0)$ simply means the very negative alternative $A_0$. The larger the probability $p$,

H.T. Nguyen et al.: Computing Stat. under Interval & Fuzzy Uncertain., SCI 393, pp. 51–54.
springerlink.com        © Springer-Verlag Berlin Heidelberg 2012

the more preferable the lottery $L(p)$. Thus, the corresponding lotteries $L(p)$ form a continuous 1-D scale ranging from the very negative alternative $A_0$ to the very positive alternative $A_1$.

*The resulting notion of utility.* Practical alternatives are usually better than $L(0) = A_0$ but worse than $L(1) = A_1$: $L(0) < A < L(1)$. Thus, for each practical alternative $A$, there exists a probability $p \in (0,1)$ for which the lottery $L(p)$ is, to this decision maker, equivalent to $A$: $L(p) \sim A$. This "equivalent" probability $p$ is called the *utility* of the alternative $A$ and denoted by $u(A)$.

*How can we actually find the value of this utility $u(A)$?* We cannot just compare $A$ with different lotteries $L(p)$ and wait until we get a lottery for which $L(p) \sim A$: there are many different probability values, so such a comparison would take an impractically long time. However, there is an alternative efficient way of determining $u(A)$ which is based on the following *bisection* procedure.

The main idea of this procedure is to produce narrower and narrower intervals containing the desired value $u(A)$. In the beginning, we only know that $u(A) \in [0,1]$, i.e., we know that $u(A) \in [\underline{u}, \overline{u}]$ with $\underline{u} = 0$ and $\overline{u} = 1$. Let us assume that at some iteration of this procedure, we know that $u(A) \in [\underline{u}, \overline{u}]$, i.e., that $L(\underline{u}) \leq A \leq L(\overline{u})$. To get a narrower interval, let us take the midpoint $m \stackrel{\text{def}}{=} \dfrac{\underline{u} + \overline{u}}{2}$ of the existing interval and compare $L(m)$ with $A$.

- If $A$ is better than $L(m)$ ($L(m) \leq A$), this means that $m \leq u(A)$ and thus, that the utility $u(A)$ belongs to the upper half-interval $[m, \overline{u}]$.
- If $A$ is worse than $L(m)$ ($A \leq L(m)$), this means that $u(A) \leq m$ and thus, that the utility $u(A)$ belongs to the lower half-interval $[\underline{u}, m]$.

In both cases, we get a new interval containing $u(A)$ whose width is the half of the width of the interval $[\underline{u}, \overline{u}]$. We start with an interval of width 1. Thus, after $k$ iterations, we get an interval $[\underline{u}, \overline{u}]$ of width $2^{-k}$ that contains $u(A)$. In this case, both endpoints $\underline{u}$ and $\overline{u}$ are $2^{-k}$-approximations to $u(A)$. In particular:

- to obtain $u(A)$ with accuracy $1\% = 0.01$, it is sufficient to perform 7 iterations: since $2^{-7} = 1/128 < 0.01$;
- to obtain $u(A)$ with accuracy $0.1\% = 0.001$, it is sufficient to perform 10 iterations: since
$$2^{-10} = 1/1024 < 0.001;$$
- to obtain $u(A)$ with accuracy $10^{-4}\% = 10^{-6}$, it is sufficient to perform 20 iterations: since
$$2^{-20} = 1/(1024)^2 < 10^{-6}.$$

*The numerical value of the utility depends on the choice of extreme alternatives $A_0$ and $A_1$.* In our definition, the numerical value of the utility depends on the selection of the alternatives $A_0$ and $A_1$: e.g., $A_0$ is the alternative whose utility is 0 and $A_1$ is the alternative whose utility is 1. What if we use a different set of alternatives, e.g., $A_0' < A_0$ and $A_1' > A_1$?

Let $A$ be an arbitrary alternative between $A_0$ and $A_1$, and let $u(A)$ be its utility with respect to $A_0$ and $A_1$. In other words, we assume that $A$ is equivalent to the lottery in which we have

- $A_1$ with probability $u(A)$ and
- $A_0$ with probability $1 - p$.

In the scale defined by the new alternatives $A_0'$ and $A_1'$, let $u'(A_0)$, $u'(A_1)$, and $u'(A)$ denote the utilities of $A_0$, $A_1$, and $A$. This means, in particular, that

- $A_0$ is equivalent to the lottery in which we get $A_1'$ with probability $u'(A_0)$ and $A_0'$ with probability $1 - u'(A_0)$; and
- $A_1$ is equivalent to the lottery in which we get $A_1'$ with probability $u'(A_1)$ and $A_0'$ with probability $1 - u'(A_1)$.

Thus, the alternative $A$ is equivalent to the compound lottery, in which

- first, we select $A_1$ or $A_0$ with probabilities $u(A)$ and $1 - u(A)$, and then
- depending on the first selection, we select $A_1'$ with probability $u'(A_1)$ or $u'(A_0)$ – and $A_0'$ with the remaining probability.

As the result of this compound lottery, we get either $A_0'$ or $A_1'$. The probability $p$ of getting $A_1'$ in this compound lottery can be computed by using the formula of full probability

$$p = u(A) \cdot u'(A_1) + (1 - u(A)) \cdot u'(A_0) =$$

$$u(A) \cdot (u'(A_1) - u'(A_0)) + u'(A_0).$$

So, the alternative $A$ is equivalent to a lottery in which we get $A_1'$ with probability $p$ and $A_0'$ with the remaining probability $1 - p$. By definition of utility, this means that the utility $u'(A)$ of the alternative $A$ in the scale defined by $A_0'$ and $A_1'$ is equal to this value $p$:

$$u'(A) = u(A) \cdot (u'(A_1) - u'(A_0)) + u'(A_0).$$

So, changing the scale means a linear re-scaling of the utility values:

$$u(A) \to u'(A) = a \cdot u(A) + b$$

for some $a = u'(A_1) - u'(A_0) > 0$ and $b = u'(A_0)$.

Vice versa, for every $a > 0$ and $b$, one can find appropriate events $A_0'$ and $A_1'$ for which the re-scaling has exactly these values $a$ and $b$. In other words, utility is defined modulo an arbitrary (increasing) linear transformation.

*Utility of an action: a derivation of the expected utility formula.* What if an action leads to alternatives $a_1, \ldots, a_m$ with probabilities $p_1, \ldots, p_m$? Suppose that we know the utility $u_i = u(a_i)$ of each of the alternatives $a_1, \ldots, a_m$. By definition of the utility, this means that for each $i$, the alternative $a_i$ is equivalent to the lottery $L(u_i)$ in which we get $A_1$ with probability $u_i$ and $a_i$ with probability $1 - u_i$. Thus, the results of the action are equivalent to the "compound lottery" in which, with the probability $p_i$, we select a lottery $L(u_i)$. In this compound lottery, the results are either $A_1$ or $A_0$. The probability $p$ of getting $A_1$ in this compound lottery can be computed by using the formula for full probability:

$$p = p_1 \cdot u_1 + \ldots + p_m \cdot u_m.$$

Thus, the action is equivalent to a lottery in which we get $A_1$ with probability $p$ and $A_0$ with the remaining probability $1 - p$. By definition of utility, this means that the utility $u$ of the action in question is equal to

$$u = p_1 \cdot u_1 + \ldots + p_m \cdot u_m.$$

In statistics, the right-hand of this formula is known as the *expected value*. Thus, we can conclude that the utility of each action with different possible alternatives is equal to the expected value of the utility; see, e.g., [150, 210, 285].

# 10

# How to Select Appropriate Statistical Characteristics

*Which is the best way to describe the corresponding probabilistic uncertainty?* One of the main objectives of data processing is to make decisions. As we have seen in the previous chapter, a standard way of making a decision is to select the action $a$ for which the expected utility (gain) is the largest possible. This is where probabilities are used: in computing, for every possible action $a$, the corresponding expected utility. To be more precise, we usually know, for each action $a$ and for each actual value of the (unknown) quantity $x$, the corresponding value of the utility $u_a(x)$. We must use the probability distribution for $x$ to compute the expected value $e[u_a(x)]$ of this utility.

In view of this application, the most useful characteristics of a probability distribution would be the ones which would enable us to compute the expected value $e[u_a(x)]$ of different functions $u_a(x)$.

*Which representations are the most useful for this intended usage? General idea.* Which characteristics of a probability distribution are the most useful for computing mathematical expectations of different functions $u_a(x)$? The answer to this question depends on the type of the function, i.e., on how the utility value $u$ depends on the value $x$ of the analyzed parameter.

*Smooth utility functions naturally lead to moments.* One natural case is when the utility function $u_a(x)$ is smooth. We have already mentioned, in the previous text, that we usually know a (reasonably narrow) interval of possible values of $x$. So, to compute the expected value of $u_a(x)$, all we need to know is how the function $u_a(x)$ behaves on this narrow interval. Because the function is smooth, we can expand it into Taylor series. Because the interval is narrow, we can consider only linear and quadratic terms in this expansion and safely ignore higher-order terms: $u_a(x) \approx c_0 + c_1 \cdot (x - x_0) + c_2 \cdot (x - x_0)^2$, where $x_0$ is a point inside the interval. Thus, we can approximate the expected value of this function by the expected value of the corresponding quadratic expression: $e[u_a(x)] \approx e[c_0 + c_1 \cdot (x - x_0) + c_2 \cdot (x - x_0)^2]$, i.e., by the following expression: $e[u_a(x)] \approx c_0 + c_1 \cdot e[x - x_0] + c_2 \cdot e[(x - x_0)^2]$. So, to compute the expectations of such utility functions, it is sufficient to know the first and second moments of the probability distribution.

H.T. Nguyen et al.: Computing Stat. under Interval & Fuzzy Uncertain., SCI 393, pp. 55–58.
springerlink.com 

In particular, if we use, as the point $x_0$, the average $e[x]$, the second moment turns into the variance of the original probability distribution. So, instead of the first and the second moments, we can use the mean $E$ and the variance $V$.

*Case of several variables.* In the above text, we assumed that the situation is fully described by the value of a single random variable $x$. In practice, usually, we need several variables to describe the situation. For the case when we have several random variables $x_1, \ldots, x_n$, we can similarly expand the dependence of the smooth utility function $u_a(x_1, \ldots, x_n)$ in Taylor series and keep linear and quadratic terms in this expansion:

$$u_a(x_1, \ldots, x_n) \approx c_0 + \sum_{i=1}^{n} c_{1i} \cdot (x_i - x_{i0}) + \sum_{i=1}^{n} c_{2i} \cdot (x_i - x_{i0})^2 +$$

$$\sum_{i=1}^{n} \sum_{j \neq i} c_{2ij} \cdot (x_i - x_{i0}) \cdot (x_j - x_{j0}).$$

Thus, we can approximate the expectation of this function by the expectation of the corresponding quadratic expression:

$$e[u_a(x)] \approx e\left[ c_0 + \sum_{i=1}^{n} c_{1i} \cdot (x_i - x_{i0}) + \sum_{i=1}^{n} c_{2i} \cdot (x_i - x_{i0})^2 + \right.$$

$$\left. \sum_{i=1}^{n} \sum_{j \neq i} c_{2ij} \cdot (x_i - x_{i0}) \cdot (x_j - x_{j0}) \right],$$

i.e., by the following expression:

$$e[u_a(x)] \approx c_0 + \sum_{i=1}^{n} c_{1i} \cdot e[x_i - x_{i0}] + \sum_{i=1}^{n} c_{2i} \cdot e\left[ (x_i - x_{i0})^2 \right] +$$

$$\sum_{i=1}^{n} c_{2ij} \cdot e[(x_i - x_{i0}) \cdot (x_j - x_{j0})].$$

So, to compute the expectations of such utility functions, it is sufficient, in addition to the first and second moments of all the variables $x_i$, to also know the "mixed" moments $e[(x_i - x_{i0}) \cdot (x_j - x_{j0})]$ – corresponding, e.g., to covariance.

*In decision making, non-smooth utility functions are common.* In decision making, not all dependencies are smooth. There is often a threshold $x_0$ after which, say, a concentration of a certain chemical becomes dangerous.

This threshold sometimes comes from the detailed chemical and/or physical analysis. In this case, when we increase the value of this parameter, we see the drastic increase in effect and hence, the drastic change in utility value. Sometimes, this threshold simply comes from regulations. In this case, when we increase the value of this parameter past the threshold, there is no drastic increase in effects, but there is a drastic decrease of utility due to the necessity to pay fines, change technology, etc. In both cases, we have a utility function which experiences an abrupt decrease at a certain threshold value $x_0$.

*Non-smooth utility functions naturally lead to cumulative distribution functions (cdfs).* We want to be able to compute the expected value $e[u_a(x)]$ of a function $u_a(x)$ which

- changes smoothly until a certain value $x_0$,
- then drops it value and continues smoothly for $x > x_0$.

We usually know the (reasonably narrow) interval which contains all possible values of $x$. Because the interval is narrow and the dependence before and after the threshold is smooth, the resulting change in $u_a(x)$ before $x_0$ and after $x_0$ is much smaller than the change at $x_0$. Thus, with a reasonable accuracy, we can ignore the small changes before and after $x_0$, and assume that the function $u_a(x)$ is equal to a constant $u^+$ for $x < x_0$, and to some other constant $u^- < u^+$ for $x > x_0$.

The simplest case is when $u^+ = 1$ and $u^- = 0$. In this case, the desired expected value $e[u_a^{(0)}(x)]$ coincides with the probability that $x < x_0$, i.e., with the corresponding value $F(x_0)$ of the cumulative distribution function (cdf). A generic function $u_a(x)$ of this type, with arbitrary values $u^-$ and $u^+$, can be easily reduced to this simplest case, because, as one can easily check, $u_a(x) = u^- + (u^+ - u^-) \cdot u^{(0)}(x)$ and hence, $e[u_a(x)] = u^- + (u^+ - u^-) \cdot F(x_0)$.

Thus, to be able to easily compute the expected values of all possible non-smooth utility functions, it is sufficient to know the values of the cdf $F(x_0)$ for all possible $x_0$.

Describing the cdf is equivalent to describing the inverse *quantile* function – a function that assigns, to every possible probability $p \in [0, 1]$, the value $x = x(p)$ for which $F(x) = p$. For example, the quantile corresponding to $p = 0.5$ is the *median* of the probability distribution.

*Summarizing: which statistical characteristics we select.* Our analysis shows that the most appropriate characteristics are the moments, the covariances, and the values of the cdf (or, equivalently, the values of the quantiles). In view of this result, in the following text, we will mainly concentrate on estimating the values of these characteristics.

*How to estimate the values of the selected statistical characteristics?* We are interested in the values of the moments and other statistical characteristics of the distribution itself. For example, we are interested in the values of the mean, variance, median, etc.

If we know the probability distribution, then we can determine these characteristics. However, in practice, we usually do not know the probability distribution. Instead, we only know the measurement results $x_1, \ldots, x_n$ – which can be viewed as a *sample* from the actual (unknown) distribution. It is therefore reasonable to approximate the original (unknown) probability distribution by the "sample" distribution, in which each of the sample values occurs with the same probability $1/n$. Thus, as an estimate of a statistical characteristic (such as mean) of the actual distribution, we can take the value of this characteristic for the sample distribution.

For example, the mean of the sample distribution is equal to

$$E \overset{\text{def}}{=} \frac{x_1 + \ldots + x_n}{n};$$

the variance of the sample distribution is equal to

$$V \overset{\text{def}}{=} \frac{1}{n} \cdot \sum_{i=1}^{n} (x_i - E)^2,$$

and the median (or, more generally, any quantile) of the sample distribution is equal to the median (quantile) of the values $x_i$. These estimates are actually the most widely used in practical applications; thus, in the following text, we will mostly concentrate on how to estimate these statistical characteristics under interval uncertainty. We will also mention alternative estimates for the mean etc., and show how to estimate them as well.

# Algorithms for Computing Statistics under Interval and Fuzzy Uncertainty

# Computing under Fuzzy Uncertainty Can Be Reduced to Computing under Interval Uncertainty: Reminder

In this part, we present algorithms for computing the values of different statistical characteristics $C(x_1, \ldots, x_n)$ under interval and fuzzy uncertainty.

In Chapter 4, we have explained that the problem of computing these values under fuzzy uncertainty can be reduced to the problem of computing the values of this characteristic under interval uncertainty. Namely, for every $\alpha \in [0, 1]$, the alpha-cut $\boldsymbol{y}(\alpha)$ of the desired fuzzy value is the interval that corresponds to estimating the same characteristic $C(x_1, \ldots, x_n)$ under interval uncertainty – specifically, under the assumption that the $i$-th input belongs to the $\alpha$-cut of the corresponding fuzzy number $x_i \in \boldsymbol{x}_i(\alpha)$:

$$\boldsymbol{y}(\alpha) = \{C(x_1, \ldots, x_n) : x_1 \in \boldsymbol{x}_1(\alpha), \ldots, x_n \in \boldsymbol{x}_n(\alpha)\}.$$

Thus, from the computational point of view, the problem of computing statistics under fuzzy uncertainty can be reduced to several problems of computing statistics under interval uncertainty. Since the fuzzy degrees come from expert estimates and thus, cannot be determined with high accuracy anyway, it is sufficient to consider values $\alpha = 0, 0.1, \ldots, 0.9, 1.0$.

In view of this comment, in this part of the book, we will describe the algorithms corresponding to the case of *interval* uncertainty – with the understanding that by applying these algorithms to $\alpha$-cuts, we can also compute the values of statistical characteristics under fuzzy uncertainty as well.

H.T. Nguyen et al.: Computing Stat. under Interval & Fuzzy Uncertain., SCI 393, p. 61.
springerlink.com                                    © Springer-Verlag Berlin Heidelberg 2012

# Computing Mean under Interval Uncertainty

We have already mentioned that for the interval data $x_1 = [\underline{x}_1, \overline{x}_1], \ldots, x_n = [\underline{x}_n, \overline{x}_n]$, a reasonable estimate for the corresponding statistical characteristic $C(x_1, \ldots, x_n)$ is the range

$$C(\boldsymbol{x}_1, \ldots, \boldsymbol{x}_n) \stackrel{\text{def}}{=} \{C(x_1, \ldots, x_n) \,|\, x_1 \in \boldsymbol{x}_1, \ldots, x_n \in \boldsymbol{x}_n\}.$$

The arithmetic average $E(x_1, \ldots, x_n) = \dfrac{1}{n} \cdot \displaystyle\sum_{i=1}^{n} x_i$ is a monotonically increasing function of each of its $n$ variables $x_1, \ldots, x_n$. So:

- its smallest possible value $\underline{E}$ is attained when each value $x_i$ is the smallest possible ($x_i = \underline{x}_i$), and
- its largest possible value is attained when $x_i = \overline{x}_i$ for all $i$.

In other words, the range $\boldsymbol{E} = [\underline{E}, \overline{E}]$ of $E$ is equal to

$$[E(\underline{x}_1, \ldots, \underline{x}_n), E(\overline{x}_1, \ldots, \overline{x}_n)],$$

i.e., $\underline{E} = \dfrac{1}{n} \cdot (\underline{x}_1 + \ldots + \underline{x}_n)$ and $\overline{E} = \dfrac{1}{n} \cdot (\overline{x}_1 + \ldots + \overline{x}_n)$.

H.T. Nguyen et al.: Computing Stat. under Interval & Fuzzy Uncertain., SCI 393, p. 63.
springerlink.com 

# 13

# Computing Median (and Quantiles) under Interval Uncertainty

*Need to go beyond arithmetic average.* We have mentioned, in the Formulation of the Problem chapter, that an important source of interval uncertainty is the existence of the lower detection limits for sensors: if a sensor does not detect any signal this means that the actual value of the measured quantity is below its detection limit $DL$, i.e., in the interval $[0, DL]$.

Another practically important source of uncertainty is the fact that many sensors also have saturation values $x_{\max}$: if the sensor registers the value $\tilde{x}_i = x_{\max}$, then the only information that we know about the true value $x$ is that $x \geq x_{\max}$, i.e., that $x \in [x_{\max}, \infty)$. If one of the measurements $\tilde{x}_i$ is equal to the saturation value, then, e.g., the arithmetic average $E(x) = \frac{1}{n} \cdot (x_1 + \ldots + x_n)$ of the actual values $x_i$ can be arbitrarily large.

For such situations, we need to use different methods for estimating the expected value (mean) e$[x]$ of a random variable from the sample $x_1, \ldots, x_n$. One such method is the median.

*Estimating median under interval uncertainty.* Since the median is non-decreasing in $x_1, \ldots, x_n$, its smallest possible value is attained for $\underline{x}_1, \ldots, \underline{x}_n$, and its largest possible value is attained for $\overline{x}_1, \ldots, \overline{x}_n$.

So, to compute the exact bounds for the median, it is sufficient to apply the algorithm for computing the finite population median of $n$ numbers twice:

- first, to the values $\underline{x}_1, \ldots, \underline{x}_n$, to compute the lower endpoint for the finite population median;
- second, to the values $\overline{x}_1, \ldots, \overline{x}_n$, to compute the upper endpoint for the finite population median.

To compute each median, we can, e.g., sort the corresponding $n$ values. It is known that one can sort $n$ numbers in $O(n \cdot \log(n))$ steps; see, e.g., [73]. So, the above algorithm takes $O(n \cdot \log(n))$ steps – and is, therefore, quite feasible.

As mentioned in [73], the median of $n$ numbers can be computed in linear time $O(n)$. Since computing median under interval uncertainty means

H.T. Nguyen et al.: Computing Stat. under Interval & Fuzzy Uncertain., SCI 393, pp. 65–66.
springerlink.com 

computing two numerical medians, we can thus compute median under interval uncertainty in linear time.

*Estimating quantiles under interval uncertainty.* Similarly, to compute a given *quantile* under interval uncertainty, it is sufficient to compute the quantile of the lower endpoints and the quantile of the upper endpoints. Quantiles can also be computed in linear time; see, e.g., [73]. For example, to compute a quartile, i.e., the value that separate the first $1/4$ of the sample from the rest, it is sufficient to compute the median $m$, then select all the values $x_i$ which are smaller than $m$, and then compute the median of all these selected values. A similar "bisection" enables us to compute an arbitrary quantile in linear time.

*Beyond median.* Median is a particular case of an important class of statistical *L-estimates:* we order the values $x_i$ into a (non-strictly) increasing sequence $x_{(1)} \le x_{(2)} \le \ldots \le x_{(n)}$, and then estimate e$[x]$ as $\sum_{i=1}^{n} w_i \cdot x_{(i)}$.

Alternative methods for estimating e$[x]$ are also useful in other practical situations – e.g., if, in addition to measurement results, the values $x_i$ contain erroneously recorded values. Other widely used alternative methods for estimating e$[x]$ include [276, 337]:

- *weighted mean $E_w$* that is defined by the condition $\sum_{i=1}^{n} \dfrac{(x_i - E)^2}{\sigma_i^2} \to \min_{E}$,

  so $E_w = \sum_{i=1}^{n} p_i \cdot x_i$, where $p_i \stackrel{\text{def}}{=} \dfrac{\sigma_i^{-2}}{\sum\limits_{j=1}^{n} \sigma_j^{-2}}$;

- *M-estimates:* $\sum_{i=1}^{n} \psi(|x_i - a|) \to \max_{a}$ for some function $\psi(x)$; average is a particular case of an M-estimate, corresponding to $\psi(x) = x^2$.

They are all monotonic functions of $x_i$, so their ranges under interval uncertainty can be computed in time $O(n)$.

# Computing Variance under Interval Uncertainty: An Example of an NP-Hard Problem

## Formulation and Analysis of the Problem

*Formulation of the problem: reminder.* In many practical applications, we need to estimate the sample variance $V = \dfrac{1}{n} \cdot \sum_{i=1}^{n}(x_i - E)^2$, where $E = \dfrac{1}{n} \cdot \sum_{i=1}^{n} x_i$.

The variance can also be rewritten as $V = \dfrac{1}{n} \cdot \sum_{i=1}^{n} x_i^2 - E^2$.

As we have mentioned, in many real-life situations, we do not know the exact values $x_1, \ldots, x_n$, we only know the intervals $\boldsymbol{x}_1 = [\underline{x}_1, \overline{x}_1], \ldots, \boldsymbol{x}_n = [\underline{x}_n, \overline{x}_n]$ that contain the actual (unknown) values of $x_i$.

Intervals coming from measurements have the form $[\widetilde{x}_i - \Delta_i, \widetilde{x}_i + \Delta_i]$, where $\widetilde{x}_i$ is the measurement result and $\Delta_i$ is an upper bound on the measurement error. A general interval $[\underline{x}_i, \overline{x}_i]$ can be represented in the same way if we take the midpoint $\widetilde{x}_i = \dfrac{\underline{x}_i + \overline{x}_i}{2}$ as $\widetilde{x}_i$ and the radius (half-width) as $\Delta_i$: $\Delta_i = \dfrac{\overline{x}_i - \underline{x}_i}{2}$.

In such situations, different values $x_i \in \boldsymbol{x}_i$ lead, in general, to different values of the variance. We need to find the range

$$V(\boldsymbol{x}_1, \ldots, \boldsymbol{x}_n) \stackrel{\text{def}}{=} \{V(x_1, \ldots, x_n) \,|\, x_1 \in \boldsymbol{x}_1, \ldots, x_n \in \boldsymbol{x}_n\}$$

of possible values of the variance.

Since the function $V(x_1, \ldots, x_n)$ is continuous, its range on the box

$$[\underline{x}_1, \overline{x}_1] \times \ldots \times [\underline{x}_n, \overline{x}_n]$$

is an interval. This range interval will be denoted by $\boldsymbol{V} = [\underline{V}, \overline{V}]$.

H.T. Nguyen et al.: Computing Stat. under Interval & Fuzzy Uncertain., SCI 393, pp. 67–78.
springerlink.com　　　　　　　　　　　　　　© Springer-Verlag Berlin Heidelberg 2012

*For this problem, traditional interval methods sometimes lead to excess width.*
Let us show that for this problem, traditional interval methods sometimes
lead to excess width.

*Straightforward interval computations.* Let us first show what will happen is
we use "straightforward" interval computations.

As we have mentioned in Chapter 7, in straightforward interval computations, we repeat the computations forming the program $f$ step-by-step,
replacing each operation with real numbers by the corresponding operation
of interval arithmetic. It is known that, as a result, we get an enclosure for
the desired range.

For the problem of computing the range of finite population average, as we
have mentioned, straightforward interval computations lead to exact bounds.
The reason: in the above formula for $E$, each interval variable only occurs
once.

For the problem of computing the range of the population variance, the
situation is somewhat more difficult, because in the expression for $V$, each
variable $x_i$ occurs several times:

- explicitly, in $(x_i - E)^2$, and
- implicitly, in the expression for $E$.

In this cases, often, dependence between intermediate computation results
leads to excess width of the results of straightforward interval computations.
Not surprisingly, we do get excess width when applying straightforward interval computations to the formula for $V$.

For example, for $\boldsymbol{x}_1 = \boldsymbol{x}_2 = [0, 1]$, the actual variance is $V = (x_1 - x_2)^2/4$
and hence, its actual range is $\boldsymbol{V} = [0, 0.25]$. On the other hand, $\boldsymbol{E} = [0, 1]$,
hence

$$\frac{(\boldsymbol{x}_1 - \boldsymbol{E})^2 + (\boldsymbol{x}_2 - \boldsymbol{E})^2}{2} = [0, 1] \supset [0, 0.25].$$

It is worth mentioning that in the alternative formula

$$V = \frac{1}{n} \cdot \sum_{i=1}^{n} x_i^2 - E^2,$$

each variable $x_i$ also occurs several times, as a result of which we also get
excess width: for $\boldsymbol{x}_1 = \boldsymbol{x}_2 = [0, 1]$, we get $\boldsymbol{E} = [0, 1]$ and

$$\frac{\boldsymbol{x}_1^2 + \boldsymbol{x}_2^2}{2} - \boldsymbol{V}^2 = [-1, 1] \supset [0, 0.25].$$

Unless there is a general formula for computing the variance of a finite
population in which each interval variable only occurs once, then without
using a numerical algorithm (as contrasted with am analytical expression),
it is probably not possible to avoid excess interval width caused by dependence. The fact that we prove that the problem of computing of computing
the exact bound for the finite population variance is computationally difficult

(in precise terms, NP-hard) makes us believe that no such formula for finite population variance is possible.

*Mean value form.* As we have mentioned in Chapter 7, a better range is often provided by a *mean value form*, in which a range $f(\boldsymbol{x}_1, \ldots, \boldsymbol{x}_n)$ of a smooth function on a box $\boldsymbol{x}_1 \times \ldots \times \boldsymbol{x}_n$ is estimated as

$$f(\boldsymbol{x}_1, \ldots, \boldsymbol{x}_n) \subseteq f(\widetilde{x}_1, \ldots, \widetilde{x}_n) + \sum_{i=1}^{n} \frac{\partial f}{\partial x_i}(\boldsymbol{x}_1, \ldots, \boldsymbol{x}_n) \cdot [-\Delta_i, \Delta_i],$$

where $\widetilde{x}_i = (\underline{x}_i + \overline{x}_i)/2$ is the interval's midpoint and $\Delta_i = (\underline{x}_i - \overline{x}_i)/2$ is its half-width.

When all the intervals are the same, e.g., when $\boldsymbol{x}_i = [0, 1]$, the centered form does not lead to the desired range. Indeed, the mean value form always produced an interval centered in the point $f(\widetilde{x}_1, \ldots, \widetilde{x}_n)$. In this case, all midpoints $\widetilde{x}_i$ are the same (e.g., equal to 0.5), hence the population variance $f(\widetilde{x}_1, \ldots, \widetilde{x}_n)$ is equal to 0 on these midpoints. Thus, as a result of applying the centered form, we get an interval centered at 0, i.e., the interval whose lower endpoint is negative. In reality, $V$ is always non-negative, so negative values of $V$ are impossible.

The upper endpoint produced by the mean value form is also different from the upper endpoint of the actual range: e.g., for $\boldsymbol{x}_1 = \boldsymbol{x}_2 = [0, 1]$, we have $\frac{\partial f}{\partial x_1}(x_1, x_2) = (x_1 - x_2)/2$, hence

$$\frac{\partial f}{\partial x_1}(\boldsymbol{x}_1, \boldsymbol{x}_2) = \frac{\boldsymbol{x}_1 - \boldsymbol{x}_2}{2} = [-0.5, 0.5].$$

A similar formula holds for the derivative with respect to $x_2$. Since $\Delta_i = 0,5$, the centered form leads to:

$$f(\boldsymbol{x}_1, \ldots, \boldsymbol{x}_n) \subseteq 0 + [-0.5, 0.5] \cdot [-0.5, 0.5] + [-0.5, 0.5] \cdot [-0.5, 0.5] = [-0.5, 0.5]$$

– an excess width in comparison with the actual range $[0, 0.25]$.

*For this problem, traditional optimization methods sometimes take unreasonably long time.* A natural way to solve the problem of computing the exact range $[\underline{V}, \overline{V}]$ of the population variance is to solve it as a constrained optimization problem. Specifically, to find $\underline{V}$, we must find the minimum of the function $V$ under the conditions $\underline{x}_1 \leq x_1 \leq \overline{x}_1, \ldots, \underline{x}_n \leq x_n \leq \overline{x}_n$. Similarly, to find $\overline{V}$, we must find the maximum of the function $V$ under the same conditions.

There exist optimization techniques that lead to computing "sharp" (exact) values of $\min(f(x))$ and $\max(f(x))$. For example, there is a method described in [147] (and effectively implemented). However, the behavior of such general constrained optimization algorithms is not easily predictable, and can, in general, be exponential in $n$.

For small $n$, this is quite doable, but for large $n$, the exponential computation time grows so fast that for reasonable $n$, it becomes unrealistically large: e.g., for $n \approx 300$, it becomes larger than the lifetime of the Universe.

*We need new methods.* Summarizing: the existing methods are either not always efficient, or do not always provide us with sharp estimates for $\underline{V}$ and $\overline{V}$. So, we need new methods.

In this chapter, after proving that the problem is, in general, NP-hard, we describe several new methods for computing the variance under interval uncertainty. These algorithms were first described in [79, 102, 353]

*Estimating variance under interval uncertainty is NP-hard.* The problem of computing the exact range $\boldsymbol{V} = [\underline{V}, \overline{V}]$ for the variance $V$ over interval data $x_i \in [\widetilde{x}_i - \Delta_i, \widetilde{x}_i + \Delta_i]$ is, in general, NP-hard; this result appeared in [100, 101, 102, 188, 197].

**Theorem 14.1.** *Computing $\overline{V}$ is NP-hard.*

The very fact that computing the range of a quadratic function is NP-hard was first proven by Vavasis [334] (see also [182]). The above result shows that this difficulty happens even for the very simple quadratic function $V$ frequently used in data processing.

A natural question is: maybe the difficulty comes from the requirement that the range be computed exactly? In practice, it is often sufficient to compute, in a reasonable amount of time, a usefully accurate estimate $\widetilde{\overline{V}}$ for $\overline{V}$, i.e., an estimate $\widetilde{\overline{V}}$ which is accurate with a given accuracy $\varepsilon > 0$: $\left| \widetilde{\overline{V}} - \overline{V} \right| \leq \varepsilon$. Alas, for any $\varepsilon$, such computations are also NP-hard:

**Theorem 14.2.** *For every $\varepsilon > 0$, the problem of computing $\overline{V}$ with accuracy $\varepsilon$ is NP-hard.*

*Comment.* This result shows that the problem of computing variance under interval uncertainty is NP-hard if we want to find the range of the variable with *absolute* accuracy $\varepsilon$. It turns out that if we only need *relative* accuracy $\varepsilon$ (relative to the measured values), then this range can be actually computed in polynomial time; see Chapter 28.

It is worth mentioning that $\overline{V}$ can be computed exactly in exponential time $O(2^n)$:

**Theorem 14.3.** *There exists an algorithm that computes $\overline{V}$ in exponential time.*

*Comment.* The algorithms will be presented in the next chapter.

# Proofs

*Proof of Theorem 14.1*

$1°$. By definition, a problem is NP-hard if any problem from the class NP can be reduced to it. Therefore, to prove that a problem $\mathcal{P}$ is NP-hard, it is sufficient to reduce one of the known NP-hard problems $\mathcal{P}_0$ to $\mathcal{P}$.

In this case, since $\mathcal{P}_0$ is known to be NP-hard, this means that every problem from the class NP can be reduced to $\mathcal{P}_0$, and since $\mathcal{P}_0$ can be reduced to $\mathcal{P}$, thus, the original problem from the class NP is reducible to $\mathcal{P}$.

For our proof, as the known NP-hard problem $\mathcal{P}_0$, we take a *subset* problem (see Chapter 8): given $n$ positive integers $s_1, \ldots, s_n$, to check whether there exist signs $\eta_i \in \{-1, +1\}$ for which the signed sum $\sum_{i=1}^{n} \eta_i \cdot s_i$ equals 0.

We will show that this problem can be reduced to the problem of computing $\overline{V}$, i.e., that to every instance $(s_1, \ldots, s_n)$ of the problem $\mathcal{P}_0$, we can put into correspondence such an instance of the $\overline{V}$-computing problem that based on its solution, we can easily check whether the desired signs exist.

As this instance, we take the instance corresponding to the intervals $[\underline{x}_i, \overline{x}_i] = [-s_i, s_i]$. We want to show that for the corresponding problem, $\overline{V} = C_0$, where we denoted

$$C_0 \stackrel{\text{def}}{=} \frac{1}{n} \cdot \sum_{i=1}^{n} s_i^2, \tag{14.1}$$

if and only if there exist signs $\eta_i$ for which $\sum_{i=1}^{n} \eta_i \cdot s_i = 0$.

$2°$. Let us first show that in all cases, $\overline{V} \leq C_0$.

Indeed, it is known that the formula for the finite population variance can be reformulated in the following equivalent form:

$$V = \frac{1}{n} \cdot \sum_{i=1}^{n} x_i^2 - E^2. \tag{14.2}$$

Since $x_i \in [-s_i, s_i]$, we can conclude that $x_i^2 \leq s_i^2$ hence $\sum_{i=1}^{n} x_i^2 \leq \sum_{i=1}^{n} s_i^2$. Since $E^2 \geq 0$, we thus conclude that

$$V \leq \frac{1}{n} \cdot \sum_{i=1}^{n} s_i^2 = C_0.$$

In other words, every possible value $V$ of the population variance is smaller than or equal to $C_0$. Thus, the largest of these possible values, i.e., $\overline{V}$, also cannot exceed $C_0$, i.e., $\overline{V} \leq C_0$.

$3°$. Let us now prove that if the desired signs $\eta_i$ exist, then $\overline{V} = C_0$.

Indeed, in this case, for $x_i = \eta_i \cdot s_i$, we have $E = 0$ and $x_i^2 = s_i^2$, hence

$$V = \frac{1}{n} \cdot \sum_{i=1}^{n} (x_i - E)^2 = \frac{1}{n} \cdot \sum_{i=1}^{n} s_i^2 = C_0.$$

So, the population variance $V$ is always $\leq C_0$, and it attains the value $C_0$ for some $x_i$. Therefore, $\overline{V} = C_0$.

$4°$. To complete the proof of Theorem 14.1, we must show that, vice versa, if $\overline{V} = C_0$, then the desired signs exist.

Indeed, let $\overline{V} = C_0$. Finite population variance is a continuous function on a compact set $\boldsymbol{x}_1 \times \ldots \times \boldsymbol{x}_n$, hence its maximum on this compact set is attained for some values $x_1 \in \boldsymbol{x}_1 = [-s_1, s_1], \ldots, x_n \in \boldsymbol{x}_n = [-s_n, s_n]$. In other words, for the corresponding values of $x_i$, the variance $V$ is equal to $C_0$.

Since $x_i \in [-s_i, s_i]$, we can conclude that $x_i^2 \leq s_i^2$; since $E^2 \geq 0$, we get $V \leq C_0$. If $|x_i|^2 < s_i^2$ or $E^2 > 0$, then we would have $V < C_0$. Thus, the only way to have $V = C_0$ is to have $x_i^2 = s_i^2$ and $E = 0$. The first equality leads to $x_i = \pm s_i$, i.e., to $x_i = \eta_i \cdot s_i$ for some $\eta_i \in \{-1, +1\}$. Since $E$ is, by definition, the (arithmetic) average of the values $x_i$, the equality $E = 0$ then leads to $\sum_{i=1}^{n} \eta_i \cdot s_i = 0$. So, if $\overline{V} = C_0$, then the desired signs do exist.

The theorem is proven.

*Proof of Theorem 14.2*

$1°$. Let $\varepsilon > 0$ be fixed. We will show that the subset problem can be reduced to the problem of computing $\overline{V}$ with accuracy $\varepsilon$, i.e., that to every instance $(s_1, \ldots, s_n)$ of the subset problem $\mathcal{P}_0$, we can put into correspondence such an instance of the $\varepsilon$-approximate $\overline{V}$-computation problem that based on its solution, we can easily check whether the desired signs exist.

For this reduction, we will use two parameters. The first one – $C_0$ – is the same as in the proof of Theorem 14.1. We will also need a new real-valued parameter $k$; its value depend on $\varepsilon$ and $n$. We could produce this value right away, but we believe that the proof will be much clearer if we keep it undetermined until it becomes clear what value $k$ we need to choose for the proof to be valid.

As the desired instance, we take the instance corresponding to the intervals $[\underline{x}_i, \overline{x}_i] = [-k \cdot s_i, k \cdot s_i]$ for an appropriate value $k$. Let $\widetilde{\overline{V}}$ be a number produced, for this problem, by a $\varepsilon$-accurate computation algorithm, i.e., a number for which $\left| \widetilde{\overline{V}} - \overline{V} \right| \leq \varepsilon$. We want to to show that $\widetilde{\overline{V}} \geq k^2 \cdot C_0 - \varepsilon$ if and only if there exist signs $\eta_i$ for which $\sum_{i=1}^{n} \eta_i \cdot s_i = 0$.

$2°$. When we multiply each value $x_i$ by a constant $k$, the variance is multiplied by $k^2$. As a result, the upper bound $\overline{V}$ corresponding to $x_i \in [-k \cdot s_i, k \cdot s_i]$

is exactly $k^2$ times larger than the upper bound $\overline{v}$ corresponding to $k$ times smaller values $z_i \in [-s_i, s_i]$: $\overline{v} = \overline{V}/k^2$.

Hence, when $\widetilde{\overline{V}}$ approximates $\overline{V}$ with an accuracy $\varepsilon$, the corresponding value $\widetilde{\overline{v}} \overset{\text{def}}{=} \widetilde{\overline{V}}/k^2$ approximates $\overline{v}$ $(= \overline{V}/k^2)$ with the accuracy $\delta \overset{\text{def}}{=} \varepsilon/k^2$.

In terms of $\widetilde{\overline{v}}$, the above inequality $\widetilde{\overline{V}} \geq k^2 \cdot C_0 - \varepsilon$ takes the following equivalent form: $\widetilde{\overline{v}} \geq C_0 - \delta$.

Thus, in terms of $\widetilde{\overline{v}}$, the desired property can be formulated as follows: $\widetilde{\overline{v}} \geq C_0 - \delta$ if and only if there exist signs $\eta_i$ for which $\sum\limits_{i=1}^{n} \eta_i \cdot s_i = 0$.

$3°$. Let us first show that if the desired signs $\eta_i$ exist, then $\widetilde{\overline{v}} \geq C_0 - \delta$.

Indeed, in this case, similarly to the proof of Theorem 14.1, we can conclude that $\overline{v} = C_0$. Since $\widetilde{\overline{v}}$ is a $\delta$-approximation to the actual upper bound $\overline{v}$, we can therefore conclude that $\widetilde{\overline{v}} \geq \overline{v} - \delta = C_0 - \delta$. The statement is proven.

$4°$. Vice versa, let us assume that $\widetilde{\overline{v}} \geq C_0 - \delta$. Let us prove that in this case, the desired signs exist.

$4.1°$. Since $\widetilde{\overline{v}}$ is a $\delta$-approximation to the upper bound $\overline{v}$, we thus conclude that $\overline{v} \geq \widetilde{\overline{v}} - \delta$ and therefore, $\overline{v} \geq C_0 - 2\delta$.

Similarly to the proof of Theorem 14.1, we can conclude that the maximum is attained for some values $z_i \in [-s_i, s_i]$ and therefore, there exist values $z_i \in [-s_i, s_i]$ for which the finite population variance $v$ exceeds $C_0 - 2\delta$:

$$v \overset{\text{def}}{=} \frac{1}{n} \cdot \sum_{i=1}^{n} z_i^2 - E_z^2 \geq C_0 - 2\delta,$$

where

$$E_z \overset{\text{def}}{=} \frac{1}{n} \cdot \sum_{i=1}^{n} z_i,$$

i.e., substituting the expression (14.1) for $C_0$, that

$$\frac{1}{n} \cdot \sum_{i=1}^{n} z_i^2 - (E_z)^2 \geq \frac{1}{n} \cdot \sum_{i=1}^{n} s_i^2 - 2\delta. \tag{14.3}$$

$4.2°$. The following proof will be similar to the corresponding part of the proof of Theorem 14.1. The main difference is that we have approximate equalities instead of exact ones:

- In the proof of Theorem 14.1, we used the fact that $V = C_0$ to prove that the corresponding values $x_i$ are equal to $\pm s_i$, and that their sum is equal to 0.

- Here, $v$ is only approximately equal to $C_0$. As a result, we will only be able to show that the values $z_i$ are *close* to $\pm s_i$, and that the sum of $z_i$ is *close* to 0. From these closenesses, we will then be able to conclude (for sufficiently large $k$) that the sum of the corresponding terms $\pm s_i$ is exactly equal to 0.

$4.3^\circ$. Let us first prove that for every $i$, the value $z_i^2$ is close to $s_i^2$. Specifically, we know that $z_i^2 \leq s_i^2$; we will prove that

$$z_i^2 \geq s_i^2 - 2(n-1) \cdot \delta. \tag{14.4}$$

We will prove this inequality by reduction to a contradiction. Indeed, let us assume that for some $i_0$, this inequality is not true. This means that

$$z_{i_0}^2 < s_{i_0}^2 - 2(n-1) \cdot \delta. \tag{14.5}$$

Since $z_i \in [-s_i, s_i]$, for all $i$, in particular, for all $i \neq i_0$, we conclude, for all $i \neq i_0$, that

$$z_i^2 \leq s_i^2. \tag{14.6}$$

Adding the inequality (14.5) and $(n-1)$ inequalities (14.6) corresponding to all values $i \neq i_0$, we get

$$\sum_{i=1}^{n} z_i^2 < \sum_{i=1}^{n} s_i^2 - 2(n-1) \cdot \delta. \tag{14.7}$$

Dividing both sides of this inequality by $n-1$, we get a contradiction with (14.3). This contradiction shows that (14.4) indeed holds for every $i$.

$4.4^\circ$. The inequality (14.4) says, crudely speaking, that $z_i^2$ is close to $s_i^2$. According to our "action plan" (as outlined in Part 4.2 of this proof), we want to conclude that $z_i$ is close to $\pm s_i$, i.e., that $|z_i|$ is close to $s_i$.

To be able to make a meaningful conclusion about $z_i$ from the inequality (14.4), we must make sure that the right-hand side of the inequality (14.4) is positive: otherwise, this inequality is true simply because its left-hand side is non-negative, and the right-hand side is non-positive.

The value $s_i$ is a positive integer, so $s_i^2 \geq 1$. Therefore, to guarantee that the right-hand side of (14.4) is positive, it is sufficient to select $k$ for which, for the corresponding value $\delta = \varepsilon/k^2$, we have

$$2(n-1) \cdot \delta < 1. \tag{14.8}$$

In the following text, we will assume that this condition is indeed satisfied.

$4.5^\circ$. Let us show that under the condition (14.8), the value $|z_i|$ is indeed close to $s_i$. To be more precise, we already know that $|z_i| \leq s_i$; we are going to prove that

$$|z_i| \geq s_i - 2(n-1) \cdot \delta. \tag{14.9}$$

Indeed, since the right-hand side of the inequality (14.4) is supposed to be close to $s_i$, it makes sense to represent it as $s_i^2$ times a factor close to 1. To be more precise, we reformulate the inequality (14.4) in the following equivalent form:

$$z_i^2 \geq s_i^2 \cdot \left( 1 - \frac{2(n-1) \cdot \delta}{s_i^2} \right). \tag{14.10}$$

Since both sides of this inequality are non-negative, we can extract the square root from both sides and get the following inequality:

$$|z_i| \geq s_i \cdot \sqrt{1 - \frac{2(n-1) \cdot \delta}{s_i^2}}. \tag{14.11}$$

The square root in the right-hand side of (14.11) is of the type $\sqrt{1-t}$, with $0 \leq t \leq 1$. It is known that for such $t$, we have $\sqrt{1-t} \geq 1 - t$. Therefore, from (14.11), we can conclude that

$$|z_i| \geq s_i \cdot \sqrt{1 - \frac{2(n-1) \cdot \delta}{s_i^2}} \geq s_i \cdot \left( 1 - \frac{2(n-1) \cdot \delta}{s_i^2} \right),$$

i.e., that

$$|z_i| \geq s_i - \frac{2(n-1) \cdot \delta}{s_i}.$$

Since $s_i \geq 1$, we have

$$\frac{2(n-1) \cdot \delta}{s_i} \leq 2(n-1) \cdot \delta,$$

hence

$$|z_i| \geq s_i - \frac{2(n-1) \cdot \delta}{s_i} \geq s_i - 2(n-1) \cdot \delta.$$

So, the inequality (14.9) is proven.

$4.6^\circ$. Let us now prove that for the values $z_i$ selected on Step 4.1, the average $E_z$ is close to 0. To be more precise, we will prove that

$$(E_z)^2 \leq 2\delta. \tag{14.12}$$

Similarly to Part 4.3 of this proof, we will prove this inequality by reduction to a contradiction. Indeed, assume that this inequality is not true, i.e., that

$$(E_z)^2 > 2\delta. \tag{14.13}$$

Since $z_i^2 \leq s_i^2$, we therefore conclude that

$$\sum_{i=1}^{n} z_i^2 \leq \sum_{i=1}^{n} s_i^2,$$

hence

$$\frac{1}{n} \cdot \sum_{i=1}^{n} z_i^2 \leq \frac{1}{n} \cdot \sum_{i=1}^{n} s_i^2. \tag{14.14}$$

Adding, to both sides of the inequality (14.14), the inequality (14.13), we get an inequality

$$\frac{1}{n} \cdot \sum_{i=1}^{n} z_i^2 - (E_z)^2 < \frac{1}{n} \sum_{i=1}^{n} s_i^2 - 2\delta,$$

which contradicts to (14.3). This contradiction proves that that the inequality (14.12) is true.

$4.7^\circ$. From the fact that the average $E_z$ is close to 0, we can now conclude that the sum $\sum z_i$ is also close to 0. Specifically, we will now prove that

$$\left| \sum_{i=1}^{n} z_i \right| \leq n \cdot \sqrt{2\delta}. \tag{14.15}$$

Indeed, from (14.12), we conclude that $(E_z)^2 \leq 2\delta$, hence $|E_z| \leq \sqrt{2\delta}$. Multiplying both sides of this inequality by $n$, we get the desired inequality (14.15).

$4.8^\circ$. Let us now show that for appropriately chosen $k$, we will be able to conclude that there exist signs $\eta_i$ for which $\sum \eta_i \cdot s_i = 0$.

From the inequalities (14.9) and $|z_i| \leq s_i$, we conclude that

$$|s_i - |z_i|| \leq 2(n-1) \cdot \delta. \tag{14.16}$$

Hence, $|z_i| \leq s_i - 2(n-1) \cdot \delta$. Each value $s_i$ is a positive integer, so $s_i \geq 1$. Due to the inequality (14.8), we have $2(n-1) \cdot \delta < 1$, so $|z_i| > 1 - 1 = 0$. Therefore, $z_i \neq 0$, hence each value $z_i$ has a sign. Let us take, as $\eta_i$, the sign of the value $z_i$. Then, the inequality (14.16) takes the form

$$|\eta_i \cdot s_i - z_i| \leq 2(n-1) \cdot \delta. \tag{14.17}$$

Since the absolute value of the sum cannot exceed the sum of absolute values, we therefore conclude that

$$\left| \sum_{i=1}^{n} \eta_i \cdot s_i - \sum_{i=1}^{n} z_i \right| = \left| \sum_{i=1}^{n} (\eta_i \cdot s_i - z_i) \right| \leq \sum_{i=1}^{n} |\eta_i \cdot s_i - z_i| \leq$$

$$\sum_{i=1}^{n} 2(n-1) \cdot \delta = 2n \cdot (n-1) \cdot \delta. \tag{14.18}$$

From (14.18) and (14.15), we conclude that

$$\left| \sum_{i=1}^{n} \eta_i \cdot s_i \right| \leq \left| \sum_{i=1}^{n} z_i \right| + \left| \sum_{i=1}^{n} \eta_i \cdot s_i - \sum_{i=1}^{n} z_i \right| = n \cdot \sqrt{2\delta} + 2n \cdot (n-1) \cdot \delta. \tag{14.19}$$

All values $s_i$ are integers, hence, the sum $\sum_{i=1}^{n} \eta_i \cdot s_i$ is also an integer, and so is its absolute value $\left| \sum_{i=1}^{n} \eta_i \cdot s_i \right|$. Thus, if we select $k$ for which the right-hand side of the inequality (14.19) is less than 1, i.e., for which

$$n \cdot \sqrt{2\delta} + 2n \cdot (n-1) \cdot \delta < 1, \tag{14.20}$$

we therefore conclude that the absolute value of an integer $\sum_{i=1}^{n} \eta_i \cdot s_i$ is smaller than 1, so it must be equal to 0: $\sum_{i=1}^{n} \eta_i \cdot s_i = 0$.

Thus, to complete the proof, it is sufficient to find $k$ for which, for the corresponding value $\delta = \varepsilon/k^2$, both the inequalities (14.8) and (14.20) hold. To guarantee the inequality (14.20), it is sufficient to have

$$n \cdot \sqrt{2\delta} \leq \frac{1}{3} \tag{14.21}$$

and

$$2n \cdot (n-1) \cdot \delta \leq \frac{1}{3}. \tag{14.22}$$

The inequality (14.21) is equivalent to

$$\delta \leq \frac{1}{18n^2};$$

the inequality (14.22) is equivalent to

$$\delta \leq \frac{1}{6n \cdot (n-1)};$$

and the inequality (14.8) is equivalent to

$$\delta \leq \frac{1}{2(n-1)}.$$

Thus, to satisfy all three inequalities, we must choose $\delta$ for which $\delta = \varepsilon/k^2 = \delta_0$, where we denoted

$$\delta_0 \overset{\text{def}}{=} \min\left( \frac{1}{18n^2}, \frac{1}{6n \cdot (n-1)}, \frac{1}{2(n-1)} \right).$$

The original expression for the population variance $V$ only works for $n \geq 2$. For such $n$, $18n^2 > 6n \cdot (n-1)$ and $18n^2 > 2(n-1)$, hence the above formula can be simplified into

$$\delta_0 = \frac{1}{18n^2}.$$

To get this $\delta$ as $\delta_0 = \varepsilon/k^2$, we must take $k = \sqrt{\varepsilon/\delta_0} = 3n \cdot \sqrt{2\varepsilon}$. For this $k$, as we have shown before, the reduction holds, so the theorem is proven.

*Proof of Theorem 14.3.* The proof is straightforward: by computing the appropriate second derivatives, we can check that the function $V(x_1, \ldots, x_n)$ is convex. Therefore, it maximum is attained at one of the $2^n$ vertices of the box $\boldsymbol{x}_1 \times \ldots \times \boldsymbol{x}_n$.

For readers who are not very familiar with the ideas of convexity, we can provide a slightly longer version of this proof. Let $x_1^{(0)} \in \boldsymbol{x}_1, \ldots, x_n^{(0)} \in \boldsymbol{x}_n$ be the values for which the population variance $V$ attains maximum on the box.

Let us pick one of the $n$ variables $x_i$, and let fix the values of all the other variables $x_j$ $(j \neq i)$ at $x_j = x_j^{(0)}$. When we substitute $x_j = x_j^{(0)}$ for all $j \neq i$ into the expression for population variance, $V$ becomes a quadratic function of $x_i$.

This function of one variable should attain its maximum on the interval $\boldsymbol{x}_i$ at the value $x_i^{(0)}$.

By definition, the population variance $V$ is a sum of non-negative terms; thus, its value is always non-negative. Therefore, the corresponding quadratic function of one variable always has a global minimum. This function is decreasing before this global minimum, and increasing after it. Thus, its maximum on the interval $\boldsymbol{x}_i$ is attained at one of the endpoints of this interval.

In other words, for each variable $x_i$, the maximum is attained either for $x_i = \underline{x}_i$, or for $x_i = \overline{x}_i$. Thus, to find $\overline{V}$, it is sufficient to compute $V$ for $2^n$ possible combinations $(x_1^{\pm}, \ldots, x_n^{\pm})$, where $x_i^{-} \overset{\text{def}}{=} \underline{x}_i$ and $x_i^{+} \overset{\text{def}}{=} \overline{x}_i$, and find the largest of the resulting $2^n$ numbers.

# 15

# Types of Interval Data Sets: Towards Feasible Algorithms

## Types of Interval Data Sets: General Idea

*Need to consider specific types of interval data sets.* The main objective of this book is to compute statistics under interval uncertainty. The simplest and most widely used statistical characteristics are mean and variance. We already know that computing the mean under interval uncertainty is straightforward. However, as the previous chapter shows, computing variance $V$ under interval uncertainty is, in general, an NP-hard (computationally difficult) problem. As we will see in the following chapters, a similar problem is NP-hard for many other statistical characteristics $C$ as well.

Crudely speaking, NP-hardness means that (unless P=NP), it is not possible to have an efficient algorithm that *always* computes the desired range $C$. It is therefore desirable to describe practically meaningful cases – i.e., practically meaningful types of interval data sets – when such a computation is possible. Several such cases are described in this chapter. In the next chapter, we show how variance can be computed for some of these cases; in the following chapters, we show how other characteristics can be efficiently computed under such cases.

*How to describe different types of interval data sets.* To get a full understanding of an interval data set, we need to know two things:

- the *procedure* that was used to obtain these intervals; e.g., whether the interval uncertainty came from measurements or it was introduced artificially to maintain privacy, and
- the *results* of this procedure, i.e., the intervals themselves.

In general, the procedure affects the results – some sets of intervals can appear as the result of this procedure, some cannot. For example, intervals introduced to maintain privacy are formed by fixing a sequence of thresholds like 0, 10, 20, 30, etc., so we can have intervals $[0, 10]$, $[20, 30]$, $[10, 20]$ in which each pair either does not intersect or coincide or have a single common point, but we cannot have intervals $[0, 10]$ and $[5, 15]$.

H.T. Nguyen et al.: Computing Stat. under Interval & Fuzzy Uncertain., SCI 393, pp. 79–94.
springerlink.com                                © Springer-Verlag Berlin Heidelberg 2012

Thus, it is reasonable to expect that both the procedure and the intervals themselves affect the complexity of computing statistical characteristics based on the corresponding intervals. Because of this, we will start by classifying interval data sets in a similar way:

- by the procedure, and
- by the results.

Of course, ideally, we should take *both* the procedure and the results into account. Therefore, in addition to types corresponding only to procedure or only to the results, we will also describes types of interval data that are characterized by both.

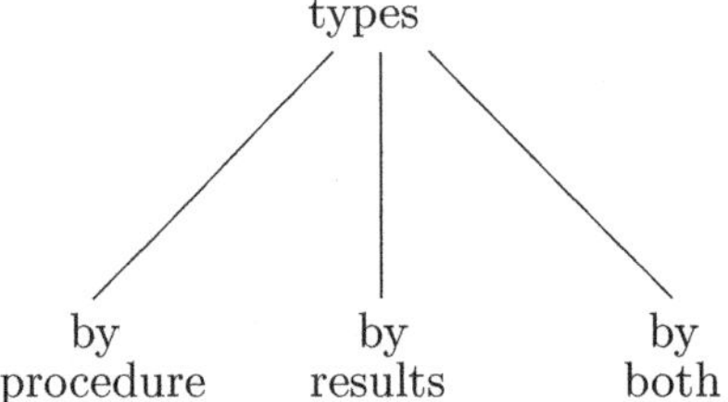


## Types of Interval Data Sets: Classification by Procedure

*Classification based on procedure.* As we have mentioned earlier, interval uncertainty can originate from several different sources:

- it can come from measurement uncertainty, and
- it can come from our desire to preserve privacy.

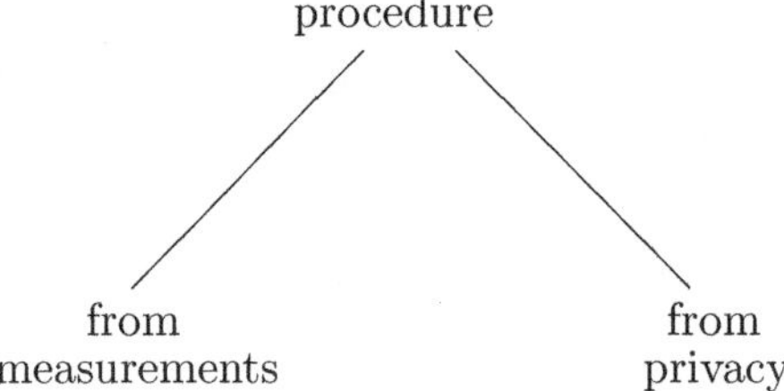


Let us analyze the interval data sets originated from these two types of procedures in detail.

*Intervals coming from measurements: general case.* First, intervals can come from *measurements*, when the only information about the actual (unknown) values $x_i$ of the corresponding quantity is that it belongs to the interval $\boldsymbol{x}_i = [\underline{x}_i, \overline{x}_i] = [\widetilde{x}_i - \Delta_i, \widetilde{x}_i + \Delta_i]$, where $\widetilde{x}_i$ is the measurement result and $\Delta_i$ is the upper bound on the measurement error $\Delta x_i = \widetilde{x}_i - x_i$.

*Single measuring instrument vs several measuring instruments.* In some cases, all the measurements are performed by a single measurement instrument (MI). In other situations, several different measuring instruments are used.

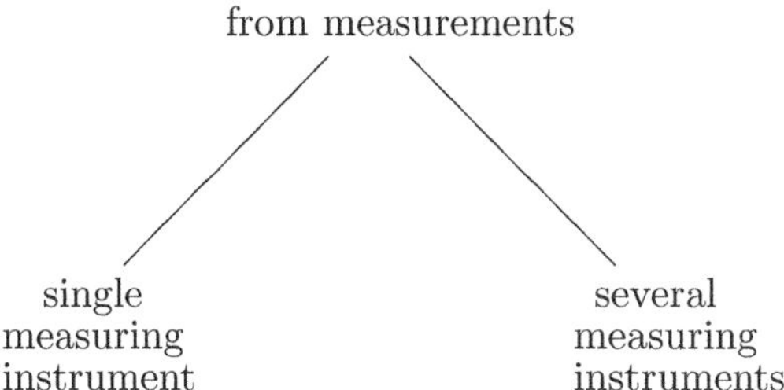


*Single measuring instrument: case of the same accuracy.* It is often reasonable to assume that the instrument with which we perform the measurements has the same accuracy over the whole range. In this case, all the values $\Delta_i$ are the same: $\Delta_1 = \Delta_2 = \ldots = \Delta_n = \Delta$ for some $\Delta > 0$.

Then, all resulting intervals $\boldsymbol{x}_i = [\widetilde{x}_i - \Delta, \widetilde{x}_i + \Delta]$ have the same width $2 \cdot \Delta$ – and are, therefore, not proper subsets of one another.

*Single measuring instrument: general case.* In other cases, the accuracy may differ across the range. However, even in this case, it is not reasonable to expect that one of the intervals is a proper subset of another one. Informally, the upper endpoint $\overline{x}_i$ is obtained from the measured value $\widetilde{x}_i$ by adding the value $\Delta_1$ corresponding to the accuracy of this measuring instrument. If $\widetilde{x}_i$ increases, i.e., if $\widetilde{x}_i < \widetilde{x}_j$, then the sum should increase (or at least non-decrease) too, i.e., we should have $\overline{x}_i \leq \overline{x}_j$. Similarly, it is reasonable to expect that the same is true for lower endpoints: $\underline{x}_i \leq \underline{x}_j$.

Thus, it is not possible to have $[\underline{x}_i, \overline{x}_i] \subseteq (\underline{x}_j, \overline{x}_j)$, because in this case, we would have $\underline{x}_j < \underline{x}_i$ but $\overline{x}_i < \overline{x}_j$.

In other words, it is quite possible to get two intervals $[5.0, 6.0]$ and $[5.4, 5.6]$ by using *two* different measuring instruments – a less accurate one and a more accurate one, but it is not realistic to expect that the *same* measuring instrument can exhibit two different accuracies within the same range of values.

Thus, formally, we can describe the general case of a single measuring instrument as the following *no-subset (no-nesting)* property: no interval is a proper subset of another one in the sense that $[\underline{x}_i, \overline{x}_i] \not\subseteq (\underline{x}_j, \overline{x}_j)$.

*Single measuring instrument: case of detection limits.* In some cases, as we have mentioned earlier, the sensors are reasonably accurate, so accurate that we can safely ignore the corresponding measurement error and assume that each measured value $\widetilde{x}_i$ is simply equal to the actual value. However, this is only true when the actual value exceeds a certain *detection limit DL* below which the sensor does not work, i.e., its reading $\widetilde{x}_i$ stays at 0.

In this case, the sensor can produce the value $\widetilde{x}_i = 0$ and all possible values $\widetilde{x}_i \geq DL$:

- the value $\widetilde{x}_i = 0$ corresponds to the interval $[0, DL]$, while
- every value $\widetilde{x}_i \geq DL$ corresponds to the degenerate interval $\boldsymbol{x}_i = [\widetilde{x}_i, \widetilde{x}_i]$.

*Single measuring instrument: case of discretized data.* As we have mentioned earlier, in some cases, discretization is the main source of measurement error. For example, we have a fixed sequence of observation times $t_0, t_1, \ldots, t_k, \ldots$, and for each change, we only know that the moment of time at which this change occurred is somewhere between $t_k$ and $t_{k+1}$; in this case, we only have intervals of the type $[t_k, t_{k+1}]$, with values $t_k$ from a pre-defined sequence.

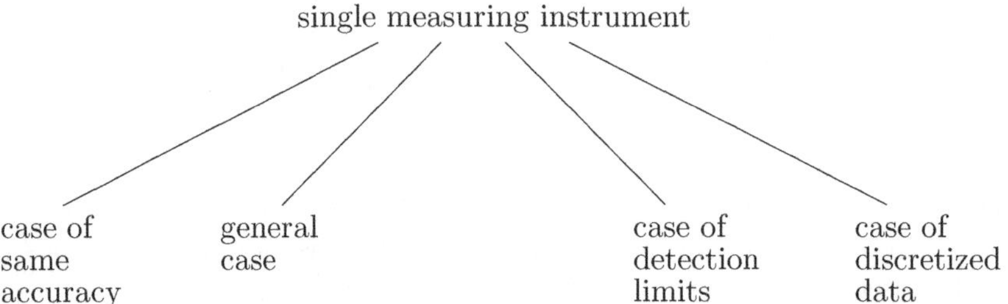


*Case of several (m) measuring instruments.* In this case, intervals can be divided into $m$ families within each of which intervals satisfy the no-subset property.

In this case, we have two different classifications by measurement results:

- similar to the case of a single measuring instrument, we can classify the intervals based on individual intervals;
- we can also distinguish between the case when we know the *provenance* of each interval, i.e., we know which interval comes from which measuring instrument, and the case when we only have the intervals, but we have not recorded which intervals was measured by which measuring instrument.

*Case of several (m) measuring instruments: classification based on individual intervals.* In this case, similarly to the case of a single measuring instrument, we can distinguish between four different types of instruments.

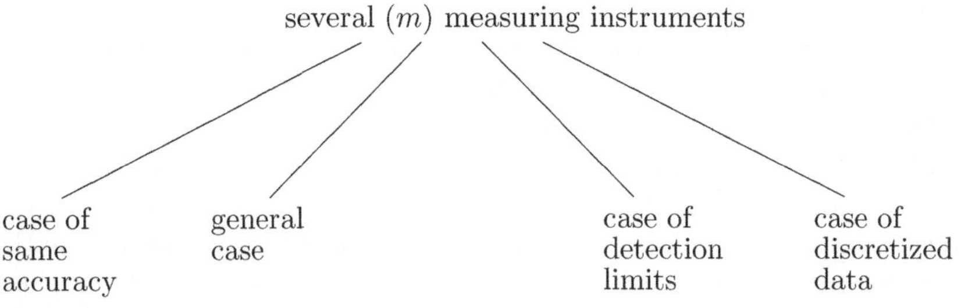

*Case of several measuring instruments: case of same accuracy.* In this case, we have $m$ families of intervals, and within each family $i$, all intervals correspond to the same accuracy, i.e., have the form $[\widetilde{x}_{ij} - \Delta_i, \widetilde{x}_{ij} + \Delta_i]$ for the same value $\Delta_i$.

*Case of several measuring instruments: general case.* In the general case, there is no restriction on intervals coming from different measuring instruments. So, in this case, the information about the number of measuring instruments does not in any way restrict the resulting collection of intervals – it is simply a general collection of intervals.

*Case of several measuring instruments: detection limits.* In the case of detection limits, we have several intervals of the type $[0, DL_i]$ for different values $DL_i$, and several degenerate intervals $[x, x]$.

*Case of several measuring instruments: discretized data.* This situation is reasonable because in data processing, we often combine data from different sources, and different sources may have different schedules of observation times. If we combine $m$ different sources, this means that the intervals can be divided into $m$ families within each of which intervals are of the type $[t_k, t_{k+1}]$ for some pre-defined sequence $t_k$ (different sequences for different families).

*Case of several (m) measuring instruments: classification based on provenance.* We consider two possibilities:

- the case when we know the *provenance* of each interval, i.e., we know which interval comes from which measuring instrument, and
- the case when we only have the intervals, but we have not recorded which intervals was measured by which measuring instrument.

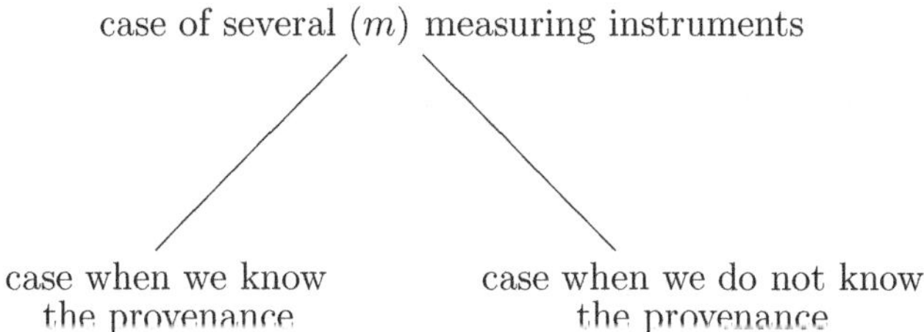


*Case when we have a limited number of different types of measuring instruments, but we do not know which measurement was made by which instrument.* In practice, often, the provenance of different measurement results is not recorded, so we may not know which measurement was made by which instrument. In this case, all we know are the intervals.

Since we know the number $m$ of different types of measuring instruments that could be used, we know that these intervals can be separated into $\leq m$ families each of which satisfies a no-subset property ($\leq m$ since it is possible that not all measuring instruments were used). It is therefore desirable to separate the intervals into $\leq m$ such families.

It should be mentioned that these families do not have to necessarily relate to the measuring instruments – as long as we have $\leq m$ families with the non-subset property, we can apply the above algorithm.

We claim that the following algorithm provides the desired separation (the justification is given in the Proofs and Justifications section).

*Algorithm for separation into families: description.* To prepare for the separation (i.e., for the assignment of intervals to families), we check, for every two intervals $[\underline{x}_i, \overline{x}_i]$ and $[\underline{x}_j, \overline{x}_j]$, whether $[\underline{x}_i, \overline{x}_i] \subseteq (\underline{x}_j, \overline{x}_j)$. If this relation holds, we denote it by $i \prec j$.

It is clear that the relation $\prec$ is transitive: if $\boldsymbol{x}_i$ is a proper subset of $\boldsymbol{x}_j$, and $\boldsymbol{x}_j$ is a proper subset of $\boldsymbol{x}_k$, then $\boldsymbol{x}_i$ is a proper subset of $\boldsymbol{x}_k$.

In the beginning, no interval is assigned to a family, so the list of all not-yet-assigned intervals consists of all $n$ original intervals $\boldsymbol{x}_i$.

On the first iteration, from the list of all not-yet-assigned intervals, we select all the intervals $\boldsymbol{x}_i$ which are not proper subsets of other intervals from this list. These non-subset intervals then form Family 1.

We then remove these intervals from the list of all not-yet-assigned intervals, and repeat the same procedure: from the list of not-yet-assigned intervals, we select all the intervals $\boldsymbol{x}_i$ which are not proper subsets of other intervals from this list. These non-subset intervals then form Family 2.

We then repeat the same procedure again and again until all the intervals from the original list are assigned to corresponding families.

*Algorithm for separation into families: computation time.* What is the computation time of this algorithm? Comparing all pairs of intervals takes time $n^2$. On each iteration, it takes linear time $O(n)$ to check whether each of $\leq n$ intervals in the not-yet-assigned list should be selected, each iteration takes $\leq O(n) \cdot n = O(n^2)$ steps. Thus, for each constant $m$, the total time for this algorithm is $m \cdot O(n^2) = O(n^2)$.

*Privacy-related intervals: case of a single database.* In addition to instruments coming from measurement inaccuracy, intervals can also come from the situations when we have the (more) exact values, but, to protect privacy, we only use intervals that contain these values.

In this case, to minimize the loss of privacy, we have a pre-defined sequence of thresholds $t_0, t_1, \ldots, t_k, \ldots$, and we replace each original value $x \in [t_k, t_{k+1}]$ (that we do not want to disclose) with the corresponding interval $[t_k, t_{k+1}]$.

*Privacy-related intervals: case of several databases.* Similarly to the case of discretized data, we may want to jointly process several different databases in which different thresholds were selected.

For example, assume that we want to compute an average salary of US and Canadian scientists by using two databases:

- the US database, in which the salary was replaced by intervals like $[70, 80]$ K based on US dollars, and

- the Canadian database, in which the salary was replaced by intervals like
[70, 80] K based on Canadian dollars.

We can easily translate the Canadian salaries into US dollars, but the resulting thresholds will be different from the thresholds used in the US salary database.

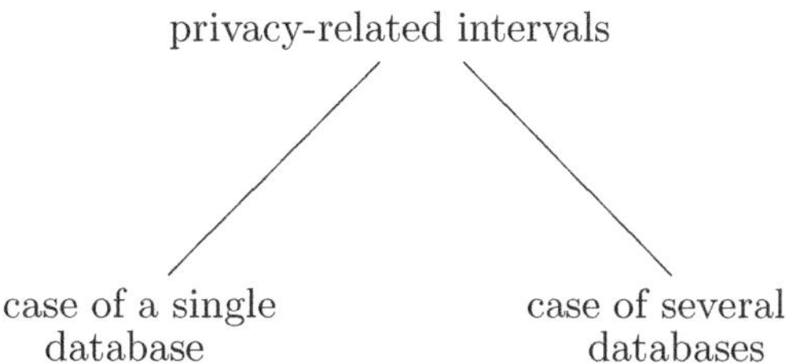


## Types of Interval Data Sets: Classification by Results

*Classification based on intervals themselves: general idea.* As we have mentioned earlier, when intervals are narrow, we can easily estimate the range by using linearization techniques. So, it is reasonable to classify the interval data sets depending on the narrowness of the corresponding intervals.

To describe this classification, let us recall that we are interested in the values of statistical characteristics. This fact implies that the actual (unknown) values $x_1, \ldots, x_n$ randomly deviate from their mean value $E$. It is also usually assumed that the random variables $x_i$ are independent. This, in turn, implies that these actual values $x_1, \ldots, x_n$ are all different – since for the usual probability distributions on the real line, the probability to get the exact same value twice is 0.

*Case of narrow intervals.* In practice, we do not observe the actual values $x_1, \ldots, x_n$, we only observe the intervals $\boldsymbol{x}_i = [\underline{x}_i, \overline{x}_i] = [\widetilde{x}_i - \Delta_i, \widetilde{x}_i + \Delta_i]$ that contain the corresponding values $x_i$. When the measurements are very accurate, the bounds $\Delta_i$ on the measurement errors are small. Thus, when they are sufficiently small, the intervals $\boldsymbol{x}_i$ do not intersect.

We can therefore use this *no-intersection* property $(\boldsymbol{x}_i \cap \boldsymbol{x}_j = \emptyset$ for $i \leq j)$ as a definition of narrow intervals.

*Next case: few intersections.* When we further increase the bounds $\Delta_i$, we will start getting intersection between the intervals.

At first, we will get intersection between intervals corresponding to the closest measurement results $\widetilde{x}_i$ and $\widetilde{x}_j$. In this case, we may have intersections between pairs of intervals, but every set of three intervals has an empty intersection.

When we increase the values $\Delta_i$ even further, we may get intersections between triples of intervals, but every of set of four intervals has an empty intersection, etc.

In general, for some integer $c_0$, we can have a property that no set of $c_0$ intervals has a common intersection. In this case, we will say that the original interval set has "few intersections of order $c_0$," or "$c_0$-few intersections", for short.

The no-intersection case of narrow intervals corresponds to $c_0 = 2$. In general, the smaller $c_0$, the more accurate the corresponding measurements.

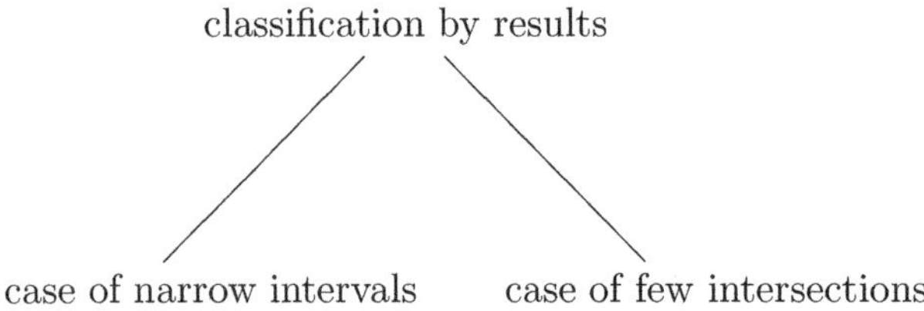


## Types of Interval Data Sets: Classification Based on Both Procedure and Results

Due to engineering progress, measurements become more and more accurate, i.e., the corresponding upper bounds $\Delta_i$ on the measurement errors decrease.

For each interval $[\widetilde{x}_i - \Delta_i, \widetilde{x}_i + \Delta_i]$, and for each factor $\lambda > 1$, we can envision a $\lambda$ times improvement in accuracy. After this improvement, the values $\Delta_i$ turn into $\dfrac{\Delta_i}{\lambda}$ and the original intervals are replaced by *narrowed* intervals

$$\left[\widetilde{x}_i - \frac{\Delta_i}{\lambda}, \widetilde{x}_i + \frac{\Delta_i}{\lambda}\right].$$

In view of this future development, if a certain property of intervals is not yet satisfied for the original intervals, it is reasonable to estimate the "degree" of this non-satisfaction by checking whether the corresponding property hold for the appropriately narrowed intervals.

If the property holds for the narrowed intervals, this means that the interval data set retains some features that make computations easier, so we expect that in this case, computations will still be sometimes easier than in the general case. In the following chapters, we will show that these expectations are indeed true. For example, it turns out that efficient algorithms for computing the range of the variance are possible not only when the original intervals satisfy a non-subset property (corresponding to a single measuring instrument), but also when the narrowed intervals $\left[\widetilde{x}_i - \dfrac{\Delta_i}{n}, \widetilde{x}_i + \dfrac{\Delta_i}{n}\right]$ satisfy this property, i.e., when

$$\left[\widetilde{x}_i - \frac{\Delta_i}{n}, \widetilde{x}_i + \frac{\Delta_i}{n}\right] \not\subset \left(\widetilde{x}_j - \frac{\Delta_j}{n}, \widetilde{x}_j + \frac{\Delta_j}{n}\right)$$

for all $i \neq j$.

Similarly, we can consider the case when the intervals can be divided into $m$ subfamilies within each of which narrowed intervals satisfy the no-subset property.

## Relation Between Different Types of Interval Data Sets

*Interval data sets of different types are related.* In the previous section, we listed a large number of different types of interval data sets, types that correspond to different origins of interval uncertainty. In this section, we analyze different types and show that from the mathematical and computational viewpoint, all these types can be reduced to a few basic ones. As a result, when designing algorithms for processing statistical data under interval uncertainty, it is sufficient to design them for a few basic ones.

Our analysis is based on a few simple observations.

*First observation: privacy is equivalent to discretized data.* Among the cases listed above, several were subcases of the case of single measuring instrument – the case that we called *no-subset* case. In particular, one important subcase was the case of discretized data.

In addition to the cases of a single and multiple measuring instruments, we also considered the privacy case. Let us show that from the mathematical and computational viewpoint, the privacy case is equivalent to the case of discretized data (and is, thus, also a subcase of the measurement cases). Specifically:

- the case of privacy-related data from single database is equivalent to the case of a single measuring instrument with discretized data, and
- the case of privacy-related data from several databases is equivalent to the case of several measuring instruments with discretized data.

*Second observation: narrow intervals are a subcase of the no-subset case.* In addition to classification by procedure, we also considered classification by result. We started with the case of narrow intervals, when no two intervals intersect. One can easily check that in this case, no interval is a subset of another one – so this case is also a subcase of the no-subset case.

Similarly, if the narrowed intervals satisfy the no-intersection property, then these narrowed intervals satisfy the no-subset property as well. So, both for original intervals and for narrowed intervals, the narrow interval property is a particular subcase of the no-subset case.

The only property which is not directly reducible to the no-subset class is the property of having $c_0$-few intersections with $c_0 > 2$. Thus, we arrive at the following conclusion.

*Conclusion: basic types of interval data sets.* The above observations show that all previously considered cases can be reduced either to the no-subset case (of a single measuring instrument), or to the case of several measuring instruments, or to the case of $c_0$-few intersections – either about the original intervals or about the narrowed intervals.

Thus, for processing each statistical characteristic under interval uncertainty, it is sufficient to design algorithms for the following basic cases:

- the *no-subset* case when for every two intervals $[\underline{x}_i, \overline{x}_i]$ and $[\underline{x}_j, \overline{x}_j]$, we have $[\underline{x}_i, \overline{x}_i] \not\subseteq (\underline{x}_j, \overline{x}_j)$;
- the case when the intervals can be divided into $m$ families each of which has a no-subset property; and
- the case when for some integer $c_0 \geq 2$, every group of $c_0$ intervals has an empty intersection.

In addition to these three classes, we also need to consider situations when *narrowed* intervals belongs to these classes.

These are the cases for which we will present algorithms.

*Auxiliary observation: properties of the original intervals vs. properties of narrowed intervals.* One can check that:

- if the original intervals satisfy the no-subset property, then the narrowed intervals also satisfy the no-subset property;
- if the original intervals do not intersect, then the corresponding narrowed intervals also do not intersect, etc.

The first implication may not be easy to observe, but it easily follows from the fact – stated and proved in the next section – that the no-subset property is equivalent to the inequalities $|\widetilde{x}_i - \widetilde{x}_j| \geq |\Delta_i - \Delta_j|$. By definition, narrowed intervals have the same midpoints $\widetilde{x}_i$ and $\widetilde{x}_j$, but $\lambda$ times smaller half-widths. Clearly, for $\lambda > 1$, the inequality $|\widetilde{x}_i - \widetilde{x}_j| \geq |\Delta_i - \Delta_j|$ (that describes the no-subset property for the original intervals) implies the inequality $|\widetilde{x}_i - \widetilde{x}_j| \geq \left| \dfrac{\Delta_i}{\lambda} - \dfrac{\Delta_j}{\lambda} \right|$, an inequality that describes a no-subset property for the narrowed intervals.

Thus, if we find algorithms that work when the narrowed intervals belong to one of the three basic classes, then these same algorithms are also applicable when the original intervals belongs to these classes – so there is no need to develop new algorithms for the original case.

## How to Detect Interval Data Sets of Different Types

How can we check whether an interval data set belongs to the given interval data type? Most above criteria can be directly detected. Let us describe the corresponding checking algorithms one by one.

*Case of measurements with the same accuracy.* By definition, this case means that $\Delta_1 = \Delta_2 = \ldots = \Delta_n$. This condition can be easily checked in time $O(n)$.

*Checking whether intervals come from the same measuring instruments.* In this case, we need to check whether the given intervals $[\underline{x}_i, \overline{x}_i]$ satisfies the no-subset property $[\underline{x}_i, \overline{x}_i] \not\subseteq (\underline{x}_j, \overline{x}_j)$ is satisfied for all $i$ and $j$. In other words, we need to check that for all $i$ and $j$, the following formula is true:

$$\neg(\underline{x}_j < \underline{x}_i \,\&\, \overline{x}_i < \overline{x}_j),$$

i.e., equivalently,

$$\underline{x}_j \geq \underline{x}_i \vee \overline{x}_i \geq \overline{x}_j.$$

To check this property for all $i$ and $j$, we need to perform $O(n^2)$ computations.

If the intervals are given in the form of lower and upper endpoints, then this is a reasonable way to check the non-subset property. If each interval is given by its midpoint $\widetilde{x}_i$ and its half-width $\Delta_i$ (which is typical in the measurement case), then we can formulate an alternative criterion in terms of these given values:

$$|\widetilde{x}_i - \widetilde{x}_j| \geq |\Delta_i - \Delta_j|;$$

(the proof of this equivalence is given in the Proofs and Justifications part of this chapter).

*Case of detection limits.* This case is easy to check: in this case, all intervals are degenerate (consist of a single point) except for intervals of type $[0, DL]$ for some value $DL > 0$ – and this value $DL$ is the same for all non-degenerate intervals and is smaller than or equal to all the other values.

Obviously, this can be done in linear time.

*Case of discretized data.* In this case, every endpoint of every interval is one of the threshold values $t_k$. Thus, to check whether the given set of intervals comes from the discretized data, we can sort all the endpoints of all the given intervals into a sequence $t_1 \leq t_2 \leq \ldots \leq t_k \leq \ldots$, and check whether each given interval has the form $[t_k, t_{k+1}]$ for some $k$. If all given intervals have this form, this means that we do have discretized data, otherwise, if one of the given intervals has the form $[t_k, t_{k+p}]$ for some $p \geq 2$, we have a more complex case.

Sorting takes time $O(n{\cdot}\log(n))$. After sorting, the above checking algorithm takes linear time $O(n)$, so the overall checking time is

$$O(n \cdot \log(n)) + O(n) = O(n \cdot \log(n)).$$

*Several (m) measuring instruments: case when we know the provenance.* When we know which interval comes from which family, the only things we need to check is whether each of these families satisfies the no-subset property.

In the case of a single measuring instrument, we check all pairs $(i, j)$ and it takes time $O(n^2)$. In the case of several measuring instruments, we only need to check the pairs that belong to the same family, so this checking takes even less time – i.e., still $O(n^2)$.

*Several (m) measuring instruments: case when we do not know the provenance.* In this case, we need to apply the above $O(n^2)$ time algorithm to divide the intervals into $m$ families.

*Case of several (m) measuring instruments: detection limits.* In this case, we have at most $m$ different non-degenerate intervals, and all these intervals should be of the type of the type $[0, DL_i]$, plus several degenerate intervals $[x, x]$ with $x \geq DL_i$ for some $i$.

This can be checked in linear time.

*Case of several measuring instruments: discretized data.* In this case, if we know that intervals come from $m$ different measuring schemes, we can divide these intervals into $m$ families by using the following algorithm.

First, we sort the intervals by their lower endpoints. As a result, we get a set of intervals for which $\underline{x}_1 \leq \underline{x}_2 \leq \ldots \leq \underline{x}_n$.

We then start forming Family 1. At each stage, we keep track of the largest value $v$ from all the intervals that have already been added to Family 1. We start with an interval $[\underline{x}_1, \overline{x}_1]$ whose lower endpoint is $\underline{x}_1$, and assign this interval $[\underline{x}_1, \overline{x}_1]$ to Family 1. As a result of this assignment, we get $v = \overline{x}_1$.

At each stage of this family forming, we select the smallest $i$ for which $\underline{x}_i \geq v$, and add the $i$-th interval $[\underline{x}_i, \overline{x}_i]$ to Family 1. As a result of this assignment, we get $v = \overline{x}_i$. Once there are no more intervals to add, Family 1 is formed.

Then, we take the remaining lower endpoints, take the smallest one, and start similarly forming Family 2, etc.

What is the computation time of this algorithm? Sorting takes time $O(n \cdot \log(n))$. At each step, finding the smallest $i$ for which $\underline{x}_i \geq v$ means a search in a sorted sequence, which takes $O(\log(n))$ steps. Thus, to find such place for all $n$ original intervals, we spend time $n \cdot O(\log(n)) = O(n \cdot \log(n))$. So, overall, this algorithm takes time $O(n \cdot \log(n)) + O(n \cdot \log(n)) = O(n \cdot \log(n))$.

The justification of this algorithm is given in the Proofs and Justifications section of this chapter.

*Case of narrow intervals.* One way to check whether $n$ given intervals $[\underline{x}_i, \overline{x}_i]$, $1 \leq i \leq n$, have a no-intersection property is to sort them in the increasing order of their lower endpoints $\underline{x}_1 \leq \underline{x}_1 \ldots \leq \underline{x}_n$, If any of these two lower endpoints coincide, the corresponding intervals have a non-empty intersection. If they are all different, i.e., if $\underline{x}_1 < \ldots < \underline{x}_n$, then it is sufficient to check whether $\overline{x}_i < \underline{x}_{i+1}$ for all $i$.

Indeed, if $\overline{x}_i < \underline{x}_{i+1}$ for all $i$, then clearly no two intervals have a common point. Vice versa, if $\overline{x}_i \geq \underline{x}_{i+1}$ for some $i$, then the intervals $[\underline{x}_i, \overline{x}_i]$ and $[\underline{x}_{i+1}, \overline{x}_{i+1}]$ both contain the point $\underline{x}_{i+1}$.

Sorting takes time $O(n \cdot \log(n))$, checking $n$ inequalities takes time $O(n)$, so overall, this checking takes time $O(n \cdot \log(n)) + O(n) = O(n \cdot \log(n))$.

*Case of few intersections.* For each given integer $c_0 \geq 2$, to check whether every subfamily of $c_0$ intervals has an empty intersection, we can sort all $2n$ endpoints $\underline{x}_i$ and $\overline{x}_i$ into a non-decreasing sequence $z_1 < z_2 < \ldots < z_m$ for some $m \leq 2n$. As a result, the real line is divided into $m+1$ zones $(-\infty, z_1]$, $[z_1, z_2], \ldots, [z_{m-1}, z_m]$, and $[z_m, +\infty)$. For each zone, we will compute the number $N$ of intervals to which elements of this zone belong. Numbers from the first zone $(-\infty, z_1)$ do not belong to any of the given intervals, so we get $N_1 = 0$. When we move from each zone $[z_{k-1}, z_k)$ to the next one $[z_k, z_{k+1})$, we add one for each $i$ for which $z_k = \underline{x}_i$ (i.e., for which we enter an interval), and check whether the resulting value is $< c_0$. If the resulting value is $\geq c_0$, we stop the algorithm and conclude that there is a point that belongs to at least $c_0$ different intervals. If the resulting value is $< c_0$, we subtract one for each $i$ for which $z_k = \overline{x}_i$ (i.e., for which we exit an interval), and continue the procedure. If after checking all the zones, we never got $N \geq c_0$, this means that the desired property is indeed satisfied.

Sorting takes time $O(n \cdot \log(n))$, computing the values $N$ takes time $O(n)$, so overall, this checking takes time $O(n \cdot \log(n)) + O(n) = O(n \cdot \log(n))$.

*Cases formulated in terms of narrowed intervals* can be handled as follows: first, we compute the narrowed intervals (which takes time $O(n)$), and then we use the above described algorithms to check that these narrowed intervals satisfy the corresponding property: e.g., that that they have a no-subset property, or that they can be divided into $m$ subfamilies with each of which narrowed intervals satisfy the no-subset property.


## Proofs and Justifications

*Proof that the algorithm for dividing intervals into families satisfying the no-subset property is correct.* To prove correctness of our algorithm, we need to prove two things:

- first, that all the families generated by the algorithm satisfy the no-subset property;
- second, that if the original intervals come from $m$ families with the no-subset property, then the algorithm will lead to $\leq m$ families.

To prove these properties, let us observe that Family 1 is formed by intervals $\boldsymbol{x}_i$ for which $i \not\prec j$ for all $j$. In other words, Family 1 is formed by intervals $i$ for which the largest number of elements $e$ in a chain (= totally ordered set) $i = i_1 \prec i_2 \prec \ldots \prec i_e$ starting with $i$ is 1.

For intervals $\boldsymbol{x}_i$ from Family 2, we may have $i \prec j$ for some $j$ – namely, for intervals $\boldsymbol{x}_j$ from Family 1. Thus, Family 2 is formed by intervals for which the largest number of elements $e$ in a chain starting with $i$ is 2.

Similarly, for every $k \leq m$, Family $k$ consists of all the intervals for which the largest number of elements $e$ in a chain starting with $i$ is $k$. Thus, if an interval $\boldsymbol{x}_i$ belongs to Family $k$, then every chain starting with $i$ must have $\leq k$ elements.

Let us first prove that each such family satisfies the no-subset property. We will prove it by contradiction. Let us assume that $i \prec j$ for some intervals $\boldsymbol{x}_i$ and $\boldsymbol{x}_j$ from Family $k$. By definition of Family $k$, this means that there is a chain with $k$ elements starting with $j$: $j_1 = j \prec j_2 \prec \ldots \prec j_k$. By adding $i$ in front of this chain, we get a chain $i \prec j = j_1 \prec j_2 \ldots \prec j_k$ with $k+1$ elements starting with $i$ – which contradicts to the fact that the interval $\boldsymbol{x}_i$ belongs to the same Family $k$ and so, every such chain must have $\leq k$ elements.

Let us now prove that when the intervals come from $m$ families with no-subset property, our algorithm will assign each of these intervals to $\leq m$ families. We will also prove this by contradiction. Let us assume that our algorithm assigns some interval to a Family with number $e > m$. By our equivalent description of the families, this means that there is a chain with $e > m$ elements. Since all these elements belong to $m$ original families, by the pigeonhole principle, at least two different elements $i \prec j$ from this chain belong to the same original family – which contradicts to the fact that each original family has a no-subset property, meaning that $i \not\prec j$ for all $i$ and $j$ from this family.

*Proof that the inequality $|\widetilde{x}_i - \widetilde{x}_j| \geq |\Delta_i - \Delta_j|$ is equivalent to the no-subset property.* The condition $|\widetilde{x}_i - \widetilde{x}_j| \geq |\Delta_i - \Delta_j|$ means that if $\widetilde{x}_i \geq \widetilde{x}_j$, then we have

$$\widetilde{x}_i - \widetilde{x}_j \geq \Delta_i - \Delta_j,$$

i.e.,

$$\widetilde{x}_i - \Delta_i \geq \widetilde{x}_j - \Delta_j$$

and also

$$\widetilde{x}_i - \widetilde{x}_j \geq \Delta_j - \Delta_i,$$

i.e.,

$$\widetilde{x}_i + \Delta_i \geq \widetilde{x}_j + \Delta_j.$$

This means that no interval is a proper subinterval of the interior of another interval.

Vice versa, if one of the intervals is a proper subinterval of another one, then the above condition is not satisfied. Thus, the above condition indeed means that two intervals are proper subintervals of each other.

*Proof that the algorithm for dividing intervals into families corresponding to discretized data is correct.* First, let us prove that, as each of the families obtained by applying this algorithm corresponds to discretized data. All the families are obtained by applying the same procedure, so it is sufficient to prove this for Family 1. We form this family step by step: we start with one interval, and then add intervals one by one. Let us denote the interval added to Family 1 on the $k$-th step by $[\underline{y}_k, \overline{y}_k]$.

We start with the interval $[\underline{y}_1, \overline{y}_1] = [\underline{x}_1, \overline{x}_1]$ with the smallest lower endpoint. In this case, we have $v = \overline{x}_1 = \overline{y}_1$. As the second element of Family 1, we select the interval $[\underline{x}_i, \overline{x}_i]$ for which the $\underline{x}_i \geq \overline{y}_1$, so we get $\underline{y}_2 \geq \overline{y}_1$. At this step, we get $v = \overline{y}_2$. At the next step, we add an interval for which

$\underline{y}_3 \geq v = \overline{y}_2$; after this step, we get $v = \overline{y}_3$, etc. As a result, we get a sequence of intervals

$$[\underline{y}_1, \overline{y}_1], [\underline{y}_2, \overline{y}_2], \ldots, [\underline{y}_k, \overline{y}_k], \ldots,$$

for which $\overline{y}_i \leq \underline{y}_{i+1}$. Thus, all the endpoints of these intervals are sorted in the natural order:

$$\underline{y}_1 \leq \overline{y}_1 \leq \underline{y}_2 \leq \overline{y}_2 \leq \cdots \leq \underline{y}_k \leq \overline{y}_k \leq \cdots$$

So, each of the intervals $[\underline{y}_k, \overline{y}_k]$ is formed by the two neighboring "thresholds", which means that we do have a discretized case here.

Let us now prove that if we start with the intervals that come from $m$ different families $F_1^{(0)}, \ldots, F_m^{(0)}$ each of which corresponds to the discretized data, then the algorithm described in this chapter results in $\leq m$ families. On each step of this algorithm, some intervals are assigned to families. We will prove, by induction over the steps, that this assignment can be extended to *all* intervals in such a way that we have $\leq m$ families each of which satisfies the discretized data property. Then, at the end of the procedure, when all intervals are assigned, we conclude that we get a subdivision into $\leq m$ families.

Indeed, at the first step, only one interval $[\underline{x}_1, \overline{x}_1]$ is assigned to a family. In this case, we simply assign each interval to one of the original families $F_j^{(0)}$.

Let us now assume that we are forming Family 1, and that the desired property is proven for the current step $k$, when we have assigned intervals $[\underline{y}_1, \overline{y}_1], \ldots, [\underline{y}_k, \overline{y}_k]$ to Family 1. This means that these $k$ intervals belong to one family with a discretized data property, a family which we will denote by $F(k)_1$, while each of the remaining intervals belongs to one of the $m$ families $F_1^{(k)}, \ldots, F(k)_m$. On this step, $v = \overline{y}_k$. On the next step, we assign, to Family 1, a new interval $[\underline{y}_{k+1}, \overline{y}_{k+1}]$ for which $\overline{y}_k \leq \underline{y}_{k+1}$ and for which no other interval has the lower endpoint $\underline{x}_i \in [\overline{y}_k, \underline{y}_{k+1})$.

If the new interval $[\underline{y}_{k+1}, \overline{y}_{k+1}]$ belongs to the same family $F_1^{(k)}$, then the desired statement for the $(k+1)$-st step holds for the same families $F_j^{(k)}$ as for the $k$-th step: $F_\ell^{(k+1)} = F_\ell^{(k)}$ for all $\ell$.

Let us now consider the case when the new interval $[\underline{y}_{k+1}, \overline{y}_{k+1}]$ belongs to a different family $F_j^{(k)} \neq F_1^{(k)}$. In this case, we can "swap" the parts of $F_1^{(k)}$ and $F_j^{(k)}$ that start with $\underline{y}_{k+1}$. Namely, we keep all the families $F_\ell^{(k)}$, $\ell \neq 1$ and $\ell \neq j$, the same $F_\ell^{(k+1)} = F_\ell^{(k)}$, but instead of the two families $F_1^{(k)}$ and $F_j^{(k)}$ we form two new families $F_1^{(k+1)}$ and $F_j^{(k+1)}$ as follows:

- to the family $F_1^{(k+1)}$, we assign all the thresholds from $F_1^{(k)}$ which are $\leq \underline{y}_{k+1}$, and all the thresholds from $F_j^{(k)}$ which are $\geq \underline{y}_{k+1}$;
- to the family $F_j^{(k+1)}$, we assign all the thresholds from $F_j^{(k)}$ which are $\leq \underline{y}_{k+1}$, and all the thresholds from $F_1^{(k)}$ which are $\geq \underline{y}_{k+1}$.

Let us show that the new families also have the discretized data property, i.e., in each of these new families, if two intervals $[\underline{x}, \overline{x}]$ and $[\underline{y}, \overline{y}]$ are different, then one of them is completely to the left of the other one, i.e., either $\overline{x} \leq \underline{y}$ or $\overline{y} \leq \underline{x}$.

Indeed, the family $F_1^{(k)}$ has an interval $[\underline{y}_k, \overline{y}_k]$ with $\overline{y}_k \leq \underline{y}_{k+1}$, and all larger intervals have lower endpoints $\geq \underline{y}_{k+1}$; thus, each of its intervals either ends before $\underline{y}_{k+1}$, or starts after $\underline{y}_{k+1}$. Due to the discretized data property, the same is true for the family $F_j'$; every interval from this family either ends before $\underline{y}_{k+1}$, or starts after $\underline{y}_{k+1}$. Thus, the intervals from both new families $F_1^{(k+1)}$ and $F_j^{(k+1)}$ also have the discretized data property.

This can be illustrated as follows. We start with the families

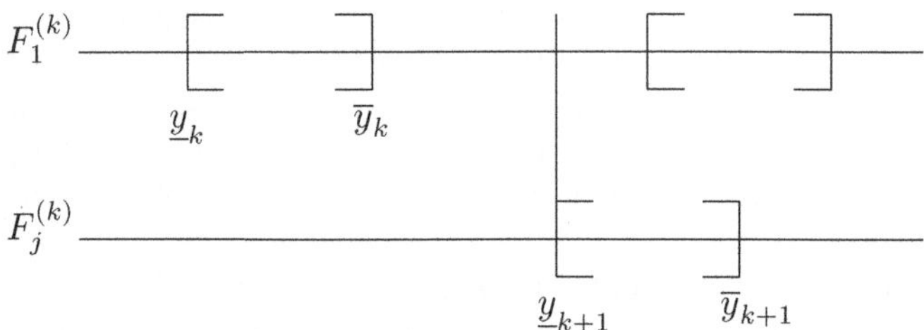

After the swap, we get the new families:

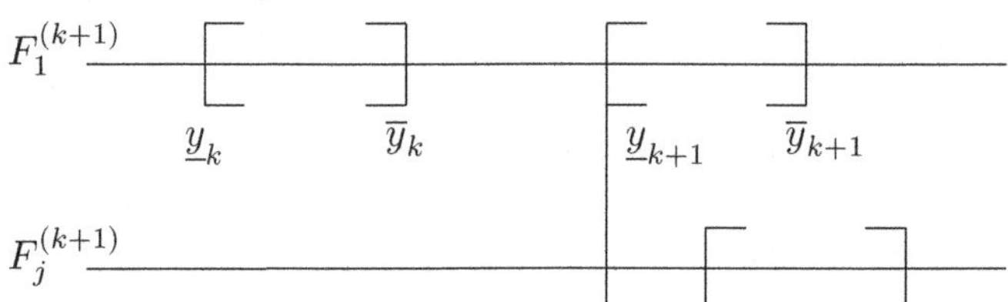

A similar swap can be described when we form Family 2, Family 3, etc. As a result, at each step of our algorithm, the assignment to families can be extended to *all* intervals in such a way that we have $\leq m$ families each of which satisfies the discretized data property (we have $\leq m$ since it may happen that after the swap, the new family $F_j^{(k+1)}$ is empty). Thus, at the end, we conclude that our algorithm divides the intervals into $\leq m$ families.

# Computing Variance under Interval Uncertainty: Efficient Algorithms

## Algorithms: General Description

We have shown that the problem of computing the upper endpoint $\overline{V}$ is, in general, NP-hard (later on, in this chapter, we will see that the lower endpoint $\underline{V}$ can be always computed in feasible (polynomial) time). Since we cannot always efficiently compute the upper endpoint $\overline{V}$, we therefore need to consider cases when such an efficient computation may be possible.

As we have shown in the previous chapter, it is reasonable to consider the following cases of interval data sets:

- case when the intervals (or, better yet, narrowed intervals) satisfy the no-subset property; this case corresponds, e.g., to the case when we process measurement results produced by the same measuring instrument;
- case when intervals can be divided into $m$ classes within each of which the no-subsect property is satisfied; this case corresponds, e.g., to the case when we process measurement results produced by $m$ different measuring instruments; and
- case when for some $c_0 \geq 2$, every group of $c_0$ intervals has an empty intersection; this case corresponds, e.g., to the case when measuring instruments are sufficiently accurate.

We will show that in all these cases, there are efficient algorithms for computing the upper endpoint $\overline{V}$. Actually, efficient algorithms for computing $\underline{V}$ will appear as natural modifications of algorithms for computing $\overline{V}$.

## Algorithms for Computing $\overline{V}$: Case When Narrowed Intervals Satisfy No-Subset Property

Let us start with the first case, of the no-subset property.

*An $O(n \cdot \log(n))$ time algorithm for computing $\overline{V}$ when narrowed intervals satisfy the no-subset property.* Let us consider the case when no two narrowed

H.T. Nguyen et al.: Computing Stat. under Interval & Fuzzy Uncertain., SCI 393, pp. 95–117.
springerlink.com                                            © Springer-Verlag Berlin Heidelberg 2012

intervals $[x_i^-, x_i^+]$ (where $x_i^- \stackrel{\text{def}}{=} \widetilde{x}_i - \dfrac{\Delta_i}{n}$ and $x_i^+ \stackrel{\text{def}}{=} \widetilde{x}_i + \dfrac{\Delta_i}{n}$) are proper subsets of one another, i.e., when $[x_i^-, x_i^+] \not\subseteq (x_j^-, x_j^+)$ for all $i$ and $j$. Here, as before, $\widetilde{x}_i$ is a midpoint of the $i$-th interval, and $\Delta_i$ its radius (= half-width).

In the previous chapter, we have shown that this condition is equivalent to the condition that the following inequality is satisfied for all $i$ and $j$:

$$|\widetilde{x}_i - \widetilde{x}_j| \geq \frac{|\Delta_i - \Delta_j|}{n}.$$

Let us show that under this condition, we can compute $\overline{V}$ in time $O(n \cdot \log(n))$. The corresponding algorithm is as follows:

- First, we sort the values $\widetilde{x}_i$ into an increasing sequence. Without losing generality, we can assume that $\widetilde{x}_1 \leq \widetilde{x}_2 \leq \ldots \leq \widetilde{x}_n$.
- Then, for every $k$ from 0 to $n$, we compute the value $V^{(k)} = M^{(k)} - (E^{(k)})^2$ of the population variance $V$ for the vector $x^{(k)} = (\underline{x}_1, \ldots, \underline{x}_k, \overline{x}_{k+1}, \ldots, \overline{x}_n)$. (For $k = 0$, $x^{(0)} = (\overline{x}_1, \ldots, \overline{x}_n)$.)
- Finally, we compute $\overline{V}$ as the largest of $n + 1$ values $V^{(0)}, \ldots, V^{(n)}$.

To compute the values $V^{(k)}$, first, we explicitly compute $M^{(0)} = \dfrac{1}{n} \cdot \sum\limits_{i=1}^{n} (\overline{x}_i)^2$,

$E^{(0)} = \dfrac{1}{n} \cdot \sum\limits_{i=1}^{n} \overline{x}_i$, and $V^{(0)} = M^{(0)} - (E^{(0)})^2$. Once we know the values $M^{(k)}$

and $E^{(k)}$, we can compute $M^{(k+1)} = M^{(k)} + \dfrac{1}{n} \cdot (\underline{x}_{k+1})^2 - \dfrac{1}{n} \cdot (\overline{x}_{k+1})^2$ and

$E^{(k+1)} = E^{(k)} + \dfrac{1}{n} \cdot \underline{x}_{k+1} - \dfrac{1}{n} \cdot \overline{x}_{k+1}$.

*Towards a faster algorithm for computing $\overline{V}$.* In the $O(n \cdot \log(n))$ algorithm, the main computation time is used on sorting: once the values $\widetilde{x}_i$ are sorted, this algorithm takes linear time.

It is possible to avoid sorting when estimating variance under interval uncertainty (see, e.g., [106, 353]), and use instead the known fact that we can compute the median of a set of $n$ elements in linear time (see, e.g., [73]). (This use of median is similar to the one from [52, 132].)

It is worth mentioning, however, that while asymptotically, the linear time algorithm for computing the median is faster than sorting, this median computing algorithm is still rather complex – so, for reasonable size $n$, sorting is faster than computing the median – and thus, sorting-based algorithms are actually faster than median-based ones.

*Linear-time algorithm for computing $\overline{V}$ for the case when narrowed intervals satisfy the no-subset property.* For simplicity, let us first consider the case when all the intervals are non-degenerate, i.e., when $\Delta_i > 0$ for all $i$.

The proposed algorithm is iterative. At each iteration of this algorithm we have three sets:

- the set $I^-$ of all the indices $i$ from 1 to $n$ for which we already know that for the optimal vector $x$, we have $x_i = \underline{x}_i$;
- the set $I^+$ of all the indices $j$ for which we already know that for the optimal vector $x$, we have $x_j = \overline{x}_j$;
- the set $I = \{1, \ldots, n\} \setminus (I^- \cup I^+)$ of the indices $i$ for which we are still undecided.

In the beginning, $I^- = I^+ = \emptyset$ and $I = \{1, \ldots, n\}$. At each iteration we also update the values of two auxiliary quantities $E^- \stackrel{\text{def}}{=} \sum_{i \in I^-} \underline{x}_i$ and $E^+ \stackrel{\text{def}}{=} \sum_{j \in I^+} \overline{x}_j$. In principle, we could compute these values by computing these sums. However, to speed up computations on each iteration, we update these two auxiliary values in a way that is faster than re-computing the corresponding two sums. Initially, since $I^- = I^+ = \emptyset$, we take $E^- = E^+ = 0$.

At each iteration we do the following:

- first, we compute the median $m$ of the set $I$ (median in terms of sorting by $\widetilde{x}_i$);
- then, by analyzing the elements of the undecided set $I$ one by one, we divide them into two subsets $P^- = \{i : \widetilde{x}_i \le \widetilde{x}_m\}$ and $P^+ = \{j : \widetilde{x}_j > \widetilde{x}_m\}$;
- we compute $e^- = E^- + \sum_{i \in P^-} \underline{x}_i$ and $e^+ = E^+ + \sum_{j \in P^+} \overline{x}_j$;
- if $n \cdot x_m^- < e^- + e^+$, then we replace $I^-$ with $I^- \cup P^-$, $E^-$ with $e^-$, and $I$ with $P^+$;
- if $n \cdot x_m^- > e^- + e^+$, then we replace $I^+$ with $I^+ \cup P^+$, $E^+$ with $e^+$, and $I$ with $P^-$;
- if $n \cdot x_m^- = e^- + e^+$, then we replace $I^-$ with $I^- \cup P^-$, $I^+$ with $I^+ \cup P^+$, and $I$ with $\emptyset$.

At each iteration the set of undecided indices is divided in half. Iterations continue until all indices are decided. After this we return, as $\overline{V}$, the value of the population variance for the vector $x$ for which $x_i = \underline{x}_i$ for $i \in I^-$ and $x_j = \overline{x}_j$ for $j \in I^+$.

*Comments*

- This same algorithm can be easily applied if one of the intervals consists of a single point only. This value is plugged in and the variable is eliminated.
- As with all asymptotic results, two natural questions arise:
  1) How practical is the new linear time $O(n)$ algorithm?
  2) For which $n$ is it better than the known $O(n \cdot \log(n))$ algorithm for computing $\overline{V}$?
  In general, the answer to these questions depends on the constants in the corresponding asymptotics. The constant for the known $O(n \cdot \log(n))$ algorithm is $\approx 1$. As one can see from the proof, for our new algorithm, the

constant is the same as for known linear time algorithm for computing the median, i.e., it is $\approx 20$ [73]; thus, the new algorithm is better when $\log_2(n) > 20$, i.e., when $n > 10^6$. We have mentioned that in many practical applications we do need to process millions of data points; in such applications, the new algorithm for computing $\overline{V}$ is indeed faster.

## Algorithms for Computing $\underline{V}$

As we have mentioned earlier, efficient algorithms for computing the lower endpoint $\underline{V}$ can be obtained by modifying the above algorithms for computing $\overline{V}$.

*An $O(n \cdot \log(n))$ algorithm for computing $\underline{V}$.* The algorithm is as follows:

- First, we sort all $2n$ values $\underline{x}_i$, $\overline{x}_i$ into a sequence $x_{(1)} \le x_{(2)} \le \ldots \le x_{(2n)}$.
- Second, we compute $\underline{\mu}$ and $\overline{\mu}$ and select all "small intervals" $[x_{(k)}, x_{(k+1)}]$ that intersect with $[\underline{\mu}, \overline{\mu}]$.
- For each of the selected small intervals $[x_{(k)}, x_{(k+1)}]$, we compute the ratio $r_k = S_k/N_k$, where

$$S_k \overset{\text{def}}{=} \sum_{i:\underline{x}_i \ge x_{(k+1)}} \underline{x}_i + \sum_{j:\overline{x}_j \le x_{(k)}} \overline{x}_j,$$

and $N_k$ is the total number of such $i$'s and $j$'s. For $k > 1$, when computing the corresponding sums and the value of $N_k$, we take into account the previously computed sums (used in computing $S_{k-1}$) and the previous value $N_{k-1}$, and only add new terms (in the second sum) or delete unnecessary terms (in the first sum).

If $r_k \in [x_{(k)}, x_{(k+1)}]$, then we compute

$$V_k' \overset{\text{def}}{=} \frac{1}{n} \cdot \left( \sum_{i:\underline{x}_i \ge x_{(k+1)}} (\underline{x}_i - r_k)^2 + \sum_{j:\overline{x}_j \le x_{(k)}} (\overline{x}_j - r_k)^2 \right).$$

These sums can also be computed based on the previous ones. For example, by explicitly performing the squaring, we conclude that the first sum has the equivalent form

$$\sum_{i:\underline{x}_i \ge x_{(k+1)}} \underline{x}_i^2 - 2 \cdot r_k \cdot \sum_{i:\underline{x}_i \ge x_{(k+1)}} \underline{x}_i + \#\{i : \underline{x}_i \ge x_{(k+1)}\} \cdot r_k^2,$$

in which both new sums can be computed by using the sums corresponding to $k - 1$.

If $N_k = 0$, we take $V_k' \overset{\text{def}}{=} 0$.
- Finally, we return the smallest of the values $V_k'$ as $\underline{V}$.

*Linear-time algorithm for computing $\underline{V}$.* The algorithm is iterative. At each iteration of this algorithm we have three sets:

- the set $J^-$ of all the endpoints $\underline{x}_i$ and $\overline{x}_j$ for which we already know that for the optimal vector $x$ we have, correspondingly, $x_i \neq \underline{x}_i$ (for $\underline{x}_i$) or $x_j = \overline{x}_j$ (for $\overline{x}_j$);
- the set $J^+$ of all the endpoints $\underline{x}_i$ and $\overline{x}_j$ for which we already know that for the optimal vector $x$ we have, correspondingly, $x_i = \underline{x}_i$ (for $\underline{x}_i$) or $x_j \neq \overline{x}_j$ (for $\overline{x}_j$);
- the set $J$ of the endpoints $\underline{x}_i$ and $\overline{x}_j$ for which we have not yet decided whether these endpoints appear in the optimal vector $x$.

In the beginning, $J^- = J^+ = \emptyset$ and $J$ is the set of all $2n$ endpoints. At each iteration we also update the values $N^- = \#(J^-)$, $N^+ = \#(J^+)$, $E^- = \sum_{\overline{x}_j \in J^-} \overline{x}_j$, and $E^+ = \sum_{\underline{x}_i \in J^+} \underline{x}_i$. Initially, $N^- = N^+ = E^- = E^+ = 0$.

At each iteration we do the following.

- First we compute the median $m$ of the set $J$.
- Then, by analyzing the elements of the undecided set $J$ one by one, we divide them into two subsets

$$Q^- = \{x \in J : x \le m\}, \quad Q^+ = \{x \in J : x > m\}.$$

We also compute $m^+ = \min\{x : x \in Q^+\}$.
- We compute $e^- = E^- + \sum_{\overline{x}_j \in Q^-} \overline{x}_j$, $e^+ = E^+ + \sum_{\underline{x}_i \in Q^+} \underline{x}_i$,

$$n^- = N^- + \#\{\overline{x}_j \in Q^-\}, \quad n^+ = N^+ + \#\{\underline{x}_i \in Q^+\},$$

and $r = \dfrac{e^- + e^+}{n^- + n^+}$.
- If $r < m$, then we replace $J^-$ with $J^- \cup Q^-$, $E^-$ with $e^-$, $J$ with $Q^+$, and $N^-$ with $n^-$.
- If $r > m^+$, then we replace $J^+$ with $J^+ \cup Q^+$, $E^+$ with $e^+$, $J$ with $P^-$, and $N^+$ with $n^+$.
- If $m \le r \le m^+$, then we replace $J^-$ with $J^- \cup Q^-$, $J^+$ with $J^+ \cup Q^+$, $J$ with $\emptyset$, $E^-$ with $e^-$, $E^+$ with $e^+$, $N^-$ with $n^-$, and $N^+$ with $n^+$.

At each iteration the set of undecided indices is divided in half. Iterations continue until all indices are decided. After this we return, as $\underline{V}$, the value of the population variance for the vector $x$ for which:

- $x_j = \overline{x}_j$ for indices $j$ for which $\overline{x}_j \in J^-$,
- $x_i = \underline{x}_i$ for indices $i$ for which $\underline{x}_i \in J^+$, and
- $x_i = r$ for all other indices $i$.

## Algorithms for Computing $\overline{V}$: Case When Narrowed Intervals Can Be Divided into $m$ Classes Each of Which Satisfies No-Subset Property

*Case when all the measurements are performed by the same measuring instrument: reminder.* For each measuring instrument, the reading of $\widetilde{x}$ means that the actual value $x$ of the measured quantity is within the interval $[\widetilde{x}-\Delta, \widetilde{x}+\Delta]$, where $\Delta$ is the measurement accuracy – i.e., the upper bound on the (absolute value of) the measurement error $\Delta x = \widetilde{x} - x$.

If we use the same measuring instrument, we do not necessarily get the same measurement accuracy $\Delta$: the accuracy may be different for different values $x$. For example, a measuring instrument is sometimes characterized not by the *absolute* accuracy $\Delta$, but by *relative* accuracy $\delta$: e.g., $\delta = 5\% = 0.05$. In this case, by definition of the relative accuracy, the absolute accuracy $\Delta$ changes with the measured value $\widetilde{x}$ as $\Delta = \delta \cdot \widetilde{x}$; see, e.g., [283].

However, for all the possible dependencies, the intervals $\boldsymbol{x}$ corresponding to different measurement results are not proper subintervals of each other. For example, if, as result of some measurement, we get the interval $[0.90, 1.10]$, then it is possible that for some other measured value, we get an interval $[0.92, 1.18]$ but we do not expect, by using the same measuring instrument, to get an interval $[0.89, 1.18]$ that would strictly contain the original interval.

We have just shown that we can effectively compute $\overline{V}$ under the condition that no two *narrowed* intervals $\left[\widetilde{x}_i - \dfrac{\Delta_i}{n}, \widetilde{x}_i + \dfrac{\Delta_i}{n}\right]$ are proper subintervals of each other. We have also shown that this no-subset condition is equivalent to the inequality $|\widetilde{x}_i - \widetilde{x}_j| \geq \dfrac{|\Delta_i - \Delta_j|}{n}$ being true for all $i$ and $j$.

Similarly, the condition that no two *original* intervals $\boldsymbol{x}_i = [\widetilde{x}_i - \Delta_i, \widetilde{x}_i + \Delta_i]$ are proper subsets of each other is equivalent to the inequality $|\widetilde{x}_i - \widetilde{x}_j| \geq |\Delta_i - \Delta_j|$ being true for all $i$ and $j$. If $|\widetilde{x}_i - \widetilde{x}_j| \geq |\Delta_i - \Delta_j|$, then, of course, $|\widetilde{x}_i - \widetilde{x}_j| \geq \dfrac{|\Delta_i - \Delta_j|}{n}$ and thus, the above efficient algorithm for computing $\overline{V}$ is applicable.

*Case when we have a limited number of different types of measuring instruments, and we know which measurement was made by which instrument.* In practice, we may have measuring instruments of different type. In this case, the interval data consists of $m$ families of intervals such that within each family, no two intervals are proper subsets of each other.

We usually know which measurement was made with which measuring instrument, so we know which interval belongs to which family.

Similarly to the justification of the above efficient algorithm for computing $\overline{V}$, we can conclude that if we sort the measured values $\widetilde{x}_i$ from each family $\alpha$ in the increasing order

$$\widetilde{x}_1 \leq \widetilde{x}_2 \leq \ldots \leq \widetilde{x}_n,$$

then, within each family, the maximum of $V$ is attained on one of the sequences $(\underline{x}_1, \ldots, \underline{x}_{k_\alpha}, \overline{x}_{k_\alpha+1}, \ldots, \overline{x}_n)$. Thus, to find the desired maximum $\overline{V}$, it is sufficient to know the value $k_\alpha \leq n$ corresponding to each of $m$ families. Overall, there are $\leq n^m$ combinations of such values. For the first combination, computing the corresponding value of the variance requires $O(n)$ steps.

Each next combination differs from the previous one by a single term, so overall, we need $O(n^m)$ steps to compute all the values of the variance – and thus, to find the largest of them, which is $\overline{V}$.

*Case when we have a limited number of different types of measuring instruments, and we do not know which measurement was made by which instrument.* The above $O(n^m)$ algorithm assumes that we know the *provenance* of each measurement, i.e., we know which measurement was made by which instrument. In practice, often, the provenance of different measurement results is not recorded, so we may not know which measurement was made by which instrument. In this case, all we know are the intervals. In the previous chapter, we showed that in this case, the intervals can be separated into $\leq m$ families each of which satisfies a no-subset property ($\leq m$ since it is possible that not all measuring instruments were used), and this separation can be performed in time $O(n^2)$.

Once the intervals are assigned to $\leq m$ families, we can apply the above $O(n^m)$ algorithm. This algorithm makes sense if we have at least two families, i.e., when $m \geq 2$; in this case, $n^2 \leq n^m$, so $O(n^2) + O(n^m) = O(n^m)$. Thus, even when we need to separate the intervals into $m$ families, the computation of $\overline{V}$ takes the same asymptotic time $O(n^m)$ as when this separation is already given.

*Case when intervals can be divided into $m$ families with no-subset property for narrowed intervals.* The same algorithm works if the given intervals can be divided into $m$ families within each of which no two *narrowed* intervals are proper subsets of one another.

## Algorithms for Computing $\overline{V}$: Case of $c_0$-Few Intersections

Finally, let us consider the case when for some integer $c_0 \geq 2$, every group of $c_0$ narrowed intervals has an empty intersection. Let us show that in this case, we can also efficiently compute the upper endpoint $\overline{V}$.

*An algorithm that takes quadratic time.* Before we start describing an $O(n \cdot \log(n))$ algorithm, let us first describe a simpler-to-describe quadratic-time algorithm for computing $\overline{V}$ in such situations:

- First, we sort all $2n$ endpoints of the narrowed intervals $\widetilde{x}_i - \Delta_i/n$ and $\widetilde{x}_i + \Delta_i/n$ into a sequence $x_{(1)} \leq x_{(2)} \leq \ldots \leq x_{(2n)}$. This enables us to divide the real line into $2n + 1$ zones $[x_{(k)}, x_{(k+1)}]$, where we denote $x_{(0)} \stackrel{\text{def}}{=} -\infty$ and $x_{(2n+1)} \stackrel{\text{def}}{=} +\infty$.

- Second, we compute $\underline{E}$ and $\overline{E}$ and pick all zones $[x_{(k)}, x_{(k+1)}]$ that intersect with $[\underline{E}, \overline{E}]$.
- For each of remaining zones $[x_{(k)}, x_{(k+1)}]$, for each $i$ from 1 to $n$, we pick the following value of $x_i$:
  - if $x_{(k+1)} \leq \widetilde{x}_i - \Delta_i/n$, then we pick $x_i = \overline{x}_i$;
  - if $\widetilde{x}_i + \Delta_i/n \leq x_{(k)}$, then we pick $x_i = \underline{x}_i$;
  - for all other $i$, we consider both possible values $x_i = \overline{x}_i$ and $x_i = \underline{x}_i$.
- As a result, we get one or several sequences of $x_i$. For each of these sequences, we check whether the average $E$ of the selected values $x_1, \ldots, x_n$ is indeed within this zone, and if it is, compute the variance by using the formula (2).
- Finally, we return the largest of the computed variances as $\overline{V}$.

*An algorithm that takes time $O(n \cdot \log(n))$.* This algorithm is, in effect, a modification of the above quadratic-time algorithm.

$1°$. First, we sort the lower endpoints $\widetilde{x}_i - \Delta_i/n$ of the narrowed intervals into an increasing sequence. Without losing generality, we can therefore assume that these lower endpoints are ordered in increasing order:

$$\widetilde{x}_1 - \Delta_1/n \leq \widetilde{x}_1 - \Delta_2/n \leq \ldots$$

$2°$. Then, we sort *all* the endpoints of the narrowed intervals into a sequence $x_{(1)} \leq x_{(2)} \leq \ldots \leq x_{(k)} \leq \ldots \leq x_{(2n)}$. Sorting means that for every $i$, we know which element $k^-(i)$ represents the lower endpoint of the $i$-th narrowed interval and which element $k^+(i)$ represents the upper endpoint of the $i$-th narrowed interval.

$3°$. On the third stage, we produce, for each of the resulting zones $[x_{(k)}, x_{(k+1)}]$, the set $S_k$ of all the indices $i$ for which the $i$-th narrowed interval

$$[\widetilde{x}_i - \Delta_i/n, \widetilde{x}_i + \Delta_i/n]$$

contains this zone.

As we have mentioned, for each $i$, we know the value $k = k^-(i)$ for which $\widetilde{x}_i - \Delta_i/n = x_{(k)}$. So, for each $i$, we place $i$ into the set $S_{k^-(i)}$ corresponding to the zone $[x_{(k^-(i))}, x_{(k^-(i)+1)}]$, into the set corresponding to the next zone, etc., until we reach the zone for which the upper endpoint is exactly $\widetilde{x}_i + \Delta_i/n$.

$4°$. On the fourth stage, for all integers $p$ from 0 to $n$, we compute the sums

$$E_p \stackrel{\text{def}}{=} \frac{1}{n} \cdot \sum_{i=1}^{p} \underline{x}_i + \frac{1}{n} \cdot \sum_{i=p+1}^{n} \overline{x}_i;$$

$$M_p \stackrel{\text{def}}{=} \frac{1}{n} \cdot \sum_{i=1}^{p} (\underline{x}_i)^2 + \frac{1}{n} \cdot \sum_{i=p+1}^{n} (\overline{x}_i)^2.$$

We compute these values sequentially. Once we know $E_p$ and $M_p$, we can compute $E_{p+1}$ and $M_{p+1}$ as $E_{p+1} = E_p + \underline{x}_{p+1} - \overline{x}_{p+1}$ and $M_{p+1} = M_p + (\underline{x}_{p+1})^2 - (\overline{x}_{p+1})^2$.

$5°$. Finally, for each zone $k$, we compute the corresponding values of the variance. For that, we first find the smallest index $i$ for which $x_{(k+1)} \leq \widetilde{x}_i - \Delta_i/n$. We will denote this value $i$ by $p(k)$.

Since the values $\widetilde{x}_i - \Delta_i/n$ are sorted, we can find this $i$ by using bisection [73].

Once $i \geq p(k)$, then $\widetilde{x}_i - \Delta_i/n \geq \widetilde{x}_{p(k)} - \Delta_{p(k)}/n \geq x_{(k+1)}$. So, we select $x_i = \overline{x}_i$, as in the sums $E_{p(k)}$ and $M_{p(k)}$.

The only values $i < p(k)$ for which we may also select $x_i = \overline{x}_i$ are the values for which the $i$-th narrowed intervals contains this zone. These values are listed in the set $S_k$ of no more than $c_0$ such intervals. So, to find all possible values of $V$, we can do the following.

We consider all subsets $s \subseteq S_k$ of the set $S_k$; there are no more than $2_0^c$ such subsets. For each subset $s$, we replace, in $E_{p(k)}$ and $M_{p(k)}$, values $\underline{x}_i$ and $(\underline{x}_i)^2$ corresponding to all $i \in s$, with, correspondingly, $\overline{x}_i$ and $(\overline{x}_i)^2$. Once we have $E$ and $V$ corresponding to the subset $s$, we can check whether $E$ belongs to the analyzed zone and, if yes, compute $V = M - E^2$.

$6°$. Finally, we find the largest of the resulting values $V$ – this will be the desired value $\overline{V}$.


## Proofs and Justifications of the Algorithms

*Justification of an $O(n \cdot \log(n))$ algorithm for computing $\overline{V}$ when narrowed intervals satisfy the no-subset property.* Let us first show that the above algorithm indeed takes $O(n \cdot \log(n))$ steps. Indeed, sorting takes $O(n \cdot \log(n))$ steps; see, e.g., [73]. Computing the initial values $M^{(0)}$, $E^{(0)}$, and $V^{(0)}$ takes linear time $O(n)$. For each $k$ from 0 to $n - 1$, we need a constant number of steps to compute the next values $M^{(k+1)}$, $E^{(k+1)}$, and $V^{(k+1)}$. Finally, finding the largest of $n + 1$ values $V^{(k)}$ also takes $O(n)$ steps. Thus, overall, we need

$$O(n \cdot \log(n)) + O(n) + O(n) + O(n) = O(n \cdot \log(n))$$

steps.

It is worth mentioning that if the measurement results $\widetilde{x}_i$ are already sorted, then we only need linear time to compute $\overline{V}$.

Let us now get to the justification of the algorithm's correctness. With respect to each variable $x_i$, the population variance is a quadratic function which is non-negative for all $x_i$. It is well known that a maximum of such a function on each interval $[\underline{x}_i, \overline{x}_i]$ is attained at one of the endpoints of this interval. Thus, the maximum $\overline{V}$ of the population variance is attained at a vector $x = (x_1, \ldots, x_n)$ in which each value $x_i$ is equal either to $\underline{x}_i$ or to $\overline{x}_i$.

We will first justify our algorithm for the case when $|\widetilde{x}_i - \widetilde{x}_j| > \dfrac{|\Delta_i - \Delta_j|}{n}$ for all $i \neq j$ and $\Delta_i > 0$ for all $i$.

To justify our algorithm, we need to prove that this maximum is attained at one of the vectors $x^{(k)}$ in which all the lower bounds $\underline{x}_i$ precede all the upper bounds $\overline{x}_i$. We will prove this by reduction to a contradiction. Indeed, let us assume that the maximum is attained at a vector $x$ in which one of the lower bounds follows one of the upper bounds. In each such vector, let $i$ be the largest upper bound index preceded by the lower bound; then, in the optimal vector $x$, we have $x_i = \overline{x}_i$ and $x_{i+1} = \underline{x}_{i+1}$.

Since the maximum is attained for $x_i = \overline{x}_i$, replacing it with $\underline{x}_i = \overline{x}_i - 2 \cdot \Delta_i$ will either decrease the value of the variance or keep it unchanged. Let us describe how variance changes under this replacement. In the sum for $M$, we replace $(\overline{x}_i)^2$ with

$$(\underline{x}_i)^2 = (\overline{x}_i - 2 \cdot \Delta_i)^2 = (\overline{x}_i)^2 - 4 \cdot \Delta_i \cdot \overline{x}_i + 4 \cdot \Delta_i^2.$$

Thus, the value $M$ changes into $M + \Delta M_i$, where

$$\Delta M_i = -\frac{4}{n} \cdot \Delta_i \cdot \overline{x}_i + \frac{4}{n} \cdot \Delta_i^2.$$

The population mean $E$ changes into $E + \Delta E_i$, where $\Delta E_i = -\dfrac{2 \cdot \Delta_i}{n}$. Thus, the value $E^2$ changes into $(E + \Delta E_i)^2 = E^2 + \Delta(E^2)_i$, where

$$\Delta(E^2)_i = 2 \cdot E \cdot \Delta E_i + \Delta E_i^2 = -\frac{4}{n} \cdot E \cdot \Delta_i + \frac{4}{n^2} \cdot \Delta_i^2.$$

So, the variance $V$ changes into $V + \Delta V_i$, where

$$\Delta V_i = \Delta M_i - \Delta(E^2)_i = -\frac{4}{n} \cdot \Delta_i \cdot \overline{x}_i + \frac{4}{n} \cdot \Delta_i^2 + \frac{4}{n} \cdot E \cdot \Delta_i - \frac{4}{n^2} \cdot \Delta_i^2 =$$

$$\frac{4}{n} \cdot \Delta_i \cdot \left( -\overline{x}_i + \Delta_i + E - \frac{\Delta_i}{n} \right).$$

By definition, $\overline{x}_i = \widetilde{x}_i + \Delta_i$, hence $-\overline{x}_i + \Delta_i = -\widetilde{x}_i$. Thus, we conclude that

$$\Delta V_i = \frac{4}{n} \cdot \Delta_i \cdot \left( -\widetilde{x}_i + E - \frac{\Delta_i}{n} \right).$$

Since $V$ attains maximum at $x$, we have $\Delta V_i \leq 0$, hence

$$E \leq \widetilde{x}_i + \frac{\Delta_i}{n}. \tag{16.1}$$

Similarly, since the maximum is attained for $x_{i+1} = \underline{x}_i$, replacing it with $\overline{x}_{i+1} = \underline{x}_{i+1} + 2 \cdot \Delta_{i+1}$ will either decrease the value of the variance or keep it unchanged. Let us describe how variance changes under this replacement. In the sum for $M$, we replace $(\underline{x}_{i+1})^2$ with

$$(\overline{x}_{i+1})^2 = (\underline{x}_{i+1} + 2 \cdot \Delta_{i+1})^2 = (\underline{x}_{i+1})^2 + 4 \cdot \Delta_{i+1} \cdot \underline{x}_{i+1} + 4 \cdot \Delta_{i+1}^2.$$

Thus, the value $M$ changes into $M + \Delta M_{i+1}$, where

$$\Delta M_{i+1} = \frac{4}{n} \cdot \Delta_{i+1} \cdot \underline{x}_{i+1} + \frac{4}{n} \cdot \Delta_{i+1}^2.$$

The population mean $E$ changes into $E + \Delta E_{i+1}$, where $\Delta E_{i+1} = \dfrac{2 \cdot \Delta_{i+1}}{n}$.
Thus, the value $E^2$ changes into $(E + \Delta E_{i+1})^2 = E^2 + \Delta(E^2)_{i+1}$, where

$$\Delta(E^2)_{i+1} = 2 \cdot E \cdot \Delta E_{i+1} + \Delta E_{i+1}^2 = \frac{4}{n} \cdot E \cdot \Delta_{i+1} + \frac{4}{n^2} \cdot \Delta_{i+1}^2.$$

So, the variance $V$ changes into $V + \Delta V_{i+1}$, where

$$\Delta V_{i+1} = \Delta M_{i+1} - \Delta(E^2)_{i+1} =$$

$$\frac{4}{n} \cdot \Delta_{i+1} \cdot \underline{x}_{i+1} + \frac{4}{n} \cdot \Delta_{i+1}^2 - \frac{4}{n} \cdot E \cdot \Delta_{i+1} - \frac{4}{n^2} \cdot \Delta_{i+1}^2 =$$

$$\frac{4}{n} \cdot \Delta_{i+1} \cdot \left( \underline{x}_{i+1} + \Delta_{i+1} - E - \frac{\Delta_{i+1}}{n} \right).$$

By definition, $\underline{x}_{i+1} = \widetilde{x}_{i+1} - \Delta_{i+1}$, hence $\underline{x}_{i+1} + \Delta_{i+1} = \widetilde{x}_{i+1}$. Thus, we conclude that

$$\Delta V_{i+1} = \frac{4}{n} \cdot \Delta_{i+1} \cdot \left( \widetilde{x}_{i+1} - E - \frac{\Delta_{i+1}}{n} \right).$$

Since $V$ attains maximum at $x$, we have $\Delta V_{i+1} \leq 0$, hence

$$E \geq \widetilde{x}_{i+1} - \frac{\Delta_{i+1}}{n}. \tag{16.2}$$

We can also change *both* $x_i$ and $x_{i+1}$ at the same time. In this case, the change $\Delta M$ in $M$ is simply the sum of the changes coming from $x_i$ and $x_{i+1}$: $\Delta M = \Delta M_i + \Delta M_{i+1}$, and the change $\Delta E$ in $E$ is also the sum of the corresponding changes: $\Delta E = \Delta E_i + \Delta E_{i+1}$. So, for

$$\Delta V = \Delta M - \Delta(E^2) = \Delta M - 2 \cdot E \cdot \Delta E - \Delta E^2,$$

we get

$$\Delta V = \Delta M_i + \Delta M_{i+1} -$$

$$2 \cdot E \cdot \Delta E_i - 2 \cdot E \cdot \Delta E_{i+1} - (\Delta E_i)^2 - (\Delta E_{i+1})^2 - 2 \cdot \Delta E_i \cdot \Delta E_{i+1}.$$

Hence,

$$\Delta V = (\Delta M_i - 2 \cdot E \cdot \Delta E_i - (\Delta E_i)^2) + (\Delta M_{i+1} - 2 \cdot E \cdot \Delta E_{i+1} - (\Delta E_{i+1})^2)$$

$$-2 \cdot \Delta E_i \cdot \Delta E_{i+1},$$

i.e.,

$$\Delta V = \Delta V_i + \Delta V_{i+1} - 2 \cdot \Delta E_i \cdot \Delta E_{i+1}.$$

We already have the expressions for $\Delta V_i$, $\Delta V_{i+1}$, $\Delta E_i = -\dfrac{2 \cdot \Delta_i}{n}$, and

$\Delta E_{i+1} = \dfrac{2 \cdot \Delta_{i+1}}{n}$, so we conclude that $\Delta V = \dfrac{4}{n} \cdot D(E)$, where

$$D(E) \stackrel{\text{def}}{=} \Delta_i \cdot \left( -\widetilde{x}_i + E - \frac{\Delta_i}{n} \right) +$$

$$\Delta_{i+1} \cdot \left( \widetilde{x}_{i+1} - E - \frac{\Delta_{i+1}}{n} \right) + \frac{2}{n} \cdot \Delta_i \cdot \Delta_{i+1}. \tag{16.3}$$

Since the function $V$ attains maximum at $x$, we have $\Delta V \leq 0$, hence $D(E) \leq 0$ (for the population mean $E$ corresponding to the optimizing vector $x$).

The expression $D(E)$ is a linear function of $E$. From (16.1) and (16.2), we know that

$$\widetilde{x}_{i+1} - \frac{\Delta_{i+1}}{n} \leq E \leq \widetilde{x}_i + \frac{\Delta_i}{n}.$$

For $E = E^- \stackrel{\text{def}}{=} \widetilde{x}_{i+1} - \dfrac{\Delta_{i+1}}{n}$, we have

$$D(E^-) = \Delta_i \cdot \left( -\widetilde{x}_i + \widetilde{x}_{i+1} - \frac{\Delta_{i+1}}{n} - \frac{\Delta_i}{n} \right) + \frac{2}{n} \cdot \Delta_i \cdot \Delta_{i+1} =$$

$$\Delta_i \cdot \left( -\widetilde{x}_i + \widetilde{x}_{i+1} + \frac{\Delta_{i+1}}{n} - \frac{\Delta_i}{n} \right).$$

We consider the case when $|\widetilde{x}_{i+1} - x_i| > \dfrac{|\Delta_i - \Delta_{i+1}|}{n}$. Since the values $\widetilde{x}_i$ are sorted in increasing order, we have $\widetilde{x}_{i+1} \geq \widetilde{x}_i$, hence

$$\widetilde{x}_{i+1} - \widetilde{x}_i = |\widetilde{x}_{i+1} - \widetilde{x}_i| > \frac{|\Delta_i - \Delta_{i+1}|}{n} \geq \frac{\Delta_i}{n} - \frac{\Delta_{i+1}}{n}.$$

So, we conclude that $D(E^-) > 0$.

For $E = E^+ \stackrel{\text{def}}{=} \widetilde{x}_i + \dfrac{\Delta_i}{n}$, we have

$$D(E^+) = \Delta_{i+1} \cdot \left( \widetilde{x}_{i+1} - \widetilde{x}_i - \frac{\Delta_i}{n} - \frac{\Delta_{i+1}}{n} \right) + \frac{2}{n} \cdot \Delta_i \cdot \Delta_{i+1} =$$

$$\Delta_{i+1} \cdot \left( -\widetilde{x}_i + \widetilde{x}_{i+1} + \frac{\Delta_i}{n} - \frac{\Delta_{i+1}}{n} \right).$$

Here, from $|\widetilde{x}_{i+1} - \widetilde{x}_i| > \dfrac{|\Delta_i - \Delta_{i+1}|}{n}$, we also conclude that $D(E^+) > 0$.

Since the linear function $D(E)$ is positive on both endpoints of the interval $[E^-, E^+]$, it must be positive for every value $E$ from this interval, which contradicts to our conclusion that $D(E) \geq 0$ for the actual population mean value $E \in [E^-, E^+]$. This contradiction shows that the maximum of the population variance $V$ is indeed attained at one of the values $x^{(k)}$, hence the algorithm is justified.

The general case when $|\tilde{x}_i - \tilde{x}_j| \geq \dfrac{|\Delta_i - \Delta_j|}{n}$ and $\Delta_i \geq 0$ can be obtained as a limit of cases when we have strict inequalities. Since the function $V$ is continuous, the value $\overline{V}$ continuously depends on the input bounds, so by tending to a limit, we can conclude that our algorithm works in the general case as well.

*Proof that the algorithm for computing $\overline{V}$ indeed takes linear time.* At each iteration, computing median takes linear time, and all other operations with $I$ take time $t$ linear in the number of elements $|I|$ of $I$: $t \leq C \cdot |I|$ for some $C$. We start with the set $I$ of size $n$. On the next iteration, we have a set of size $n/2$, then $n/4$, etc. Thus, the overall computation time is $\leq C \cdot (n + n/2 + n/4 + \ldots) \leq C \cdot 2n$, i.e. linear in $n$.

*Proof that under the no-subset property for narrowed intervals, the linear-time algorithm always computes $\overline{V}$.* In our justification of an $O(n \cdot \log(n))$ algorithm, we have shown that when we sort the intervals by their midpoints $\tilde{x}_i$, then in this sorting the value $\overline{V}$ is attained at one of the vectors $x^{(k)} = (\underline{x}_1, \ldots, \underline{x}_k, \overline{x}_{k+1}, \ldots, \overline{x}_n)$, i.e. that $V = V(x^{(k)})$ for some $k$.

We have also analyzed the change in $V(x^{(k)})$ when we replace $x^{(k)}$ with $x^{(k-1)}$, i.e. when we replace $\underline{x}_k$ with $\overline{x}_k = \underline{x}_k + 2\Delta_k$. We have shown that
$$V_{k-1} - V_k = \frac{4\Delta_k}{n} \cdot (x_k^- - E_k), \text{ where } E_k \stackrel{\text{def}}{=} E(x^{(k)}).$$

Hence $V_{k-1} < V_k$ if and only if $x_k^- < E_k$. Multiplying both sides of this inequality by $n$ we get an equivalent inequality $x_k^- < n \cdot E_k$, where $n \cdot E_k = \sum\limits_{i=1}^{k} \underline{x}_i + \sum\limits_{j=k+1}^{n} \overline{x}_j$. Similarly $V_{k-1} > V_k$ if and only if $x_k^- > E_k$, and $V_{k-1} = V_k$ if and only if $x_k^- = E_k$.

When we go from $k$ to $k+1$, we replace the larger value $\overline{x}_{k+1}$ in the sum $n \cdot E_k$ by a smaller value $\underline{x}_k$. Thus, the sequence $n \cdot E_k$ is strictly decreasing with $k$, while $x_k^-$ is (maybe non-strictly) increasing with $k$. Therefore, once we have $n \cdot x_k^- < E_k$, i.e., $V_{k-1} < V_k$, these inequalities will hold for smaller $k$ as well. Similarly, once we have $n \cdot x_k^- > E_k$, i.e., $V_{k-1} > V_k$, these inequalities will hold for larger $k$ as well.

Once we have $n \cdot x_k^- = E_k$, i.e., $V_{k-1} = V_k$, then we will have $V_k > V_{k+1} > \ldots$ and $V_k = V_{k-1} > V_{k-2} > \ldots$, i.e. $V_k = V_{k-1}$ will be the largest value of $V$.

In other words, the sequence $V_k$ first increases ($V_k > V_{k-1}$ for $k = 1, 2, \ldots$) and then starts decreasing ($V_k < V_{k-1}$ for larger $k$), with one or two top values.

For each $m$, if $V_{m-1} < V_m$ (i.e. if $n \cdot x_m^- < E_m$) this means that the value $k_{\max}$ corresponding to the maximum of $V$ is $\leq m$. Hence for all the indices $i \leq m$ we already know that in the optimal vector $x$ we have $x_i = \underline{x}_i$. Thus these indices can be added to the set $I^-$.

If $V_m > V_{m-1}$ (i.e. if $n \cdot x_m^- > E_m$) this means that the value $k_{\max}$ corresponding to the maximum of $V$ is $> m$. Hence for all the indices $i > m$ we already know that in the optimal vector $x$ we have $x_i = \overline{x}_i$. Thus these indices can be added to the set $I^+$.

Finally, if $V_m = V_{m-1}$ (i.e. if $n \cdot x_m^- = E_m$) then this $m$ is where the maximum is attained.

The algorithm has been justified.

*Proof that the algorithm for computing $\underline{V}$ is correct.* Let us prove that the above algorithm for computing $\underline{V}$ is indeed correct.

1°. Indeed, let $x_1^{(0)} \in \boldsymbol{x}_1, \ldots, x_n^{(0)} \in \boldsymbol{x}_n$ be the values for which the sample variance $V$ attains minimum on the box $\boldsymbol{x}_1 \times \ldots \times \boldsymbol{x}_n$.

Let us pick one of the $n$ variables $x_i$, and let fix the values of all the other variables $x_j$ ($j \neq i$) at $x_j = x_j^{(0)}$. When we substitute $x_j = x_j^{(0)}$ for all $j \neq i$ into the expression for sample variance, $V$ becomes a quadratic function of $x_i$.

This function of one variable should attain its minimum on the interval $\boldsymbol{x}_i$ at the value $x_i^{(0)}$.

2°. Let us start with the analysis of the quadratic function of one variable we described in Part 1 of this proof.

By definition, the sample variance $V$ is a sum of non-negative terms; thus, its value is always non-negative. Therefore, the corresponding quadratic function of one variable always has a global minimum. This function is decreasing before this global minimum, and increasing after it.

3°. Where is the global minimum of the quadratic function of one variable described in Part 1?

It is attained when $\partial V / \partial x_i = 0$. Differentiating the formula for $V$ with respect to $x_i$, we conclude that

$$\frac{\partial V}{\partial x_i} = \frac{1}{n} \cdot \left( 2(x_i - E) + \sum_{j=1}^{n} 2(E - x_j) \cdot \frac{\partial E}{\partial x_j} \right). \tag{16.4}$$

Since $\partial E / \partial x_i = 1/n$, we conclude that

$$\frac{\partial V}{\partial x_i} = \frac{2}{n} \cdot \left( (x_i - E) + \sum_{j=1}^{n} (E - x_j) \cdot \frac{1}{n} \right). \tag{16.5}$$

Here,

$$\sum_{j=1}^{n} (E - x_j) = n \cdot E - \sum_{j=1}^{n} x_j. \tag{16.6}$$

By definition of the average $E$, this difference is 0, hence the formula (16.5) takes the form

$$\frac{\partial V}{\partial x_i} = \frac{2}{n} \cdot (x_i - E).$$

So, this function attains the minimum when $x_i - E = 0$, i.e., when $x_i = E$. Since

$$E = \frac{x_i}{n} + \frac{\displaystyle\sum_{j:j\neq i} x_j}{n},$$

the equality $x_i = E$ means that

$$x_i = \frac{x_i}{n} + \frac{\displaystyle\sum_{j:j\neq i} x_j^{(0)}}{n}.$$

Moving terms containing $x_i$ into the left-hand side and dividing by the coefficient at $x_i$, we conclude that the minimum is attained when

$$x_i = E_i' \stackrel{\text{def}}{=} \frac{\displaystyle\sum_{j:j\neq i} x_j^{(0)}}{n-1},$$

i.e., when $x_i$ is equal to the arithmetic average $E_i'$ of all other elements.

$4^\circ$. Let us now use the knowledge of a global minimum to describe where the desired function attains its minimum on the interval $\boldsymbol{x}_i$.

In our general description of non-negative quadratic functions of one variable, we mentioned that each such function is decreasing before the global minimum and increasing after it. Thus, for $x_i < E_i'$, the function $V$ is decreasing, for $x_i > E_i'$, this function in increasing. Therefore:

- If $E_i' \in \boldsymbol{x}_i$, the global minimum of the function $V$ of one variable is attained within the interval $\boldsymbol{x}_i$, hence the minimum on the interval $\boldsymbol{x}_i$ is attained for $x_i = E_i'$.
- If $E_i' < \underline{x}_i$, the function $V$ is increasing on the interval $\boldsymbol{x}_i$ and therefore, its minimum on this interval is attained when $x_i = \underline{x}_i$.
- Finally, if $E_i' > \overline{x}_i$, the function $V$ is decreasing on the interval $\boldsymbol{x}_i$ and therefore, its minimum on this interval is attained when $x_i = \overline{x}_i$.

$5^\circ$. Let us reformulate the above conditions in terms of the average

$$E = \frac{1}{n} \cdot x_i + \frac{n-1}{n} \cdot E_i'.$$

- In the first case, when $x_i = E_i'$, we have $x_i = E = E_i'$, so $E \in \boldsymbol{x}_i$.
- In the second case, we have $E_i' < \underline{x}_i$ and $x_i = \underline{x}_i$. Therefore, in this case, $E < \underline{x}_i$.
- In the third case, we have $E_i' > \overline{x}_i$ and $x_i = \overline{x}_i$. Therefore, in this case, $E > \overline{x}_i$.

Thus:

- If $E \in \mathbf{x}_i$, then we cannot be in the second or third cases. Thus, we are in the first case, hence $x_i = E$.
- If $E < \underline{x}_i$, then we cannot be in the first or the third cases. Thus, we are the second case, hence $x_i = \underline{x}_i$.
- If $E > \overline{x}_i$, then we cannot be in the first or the second cases. Thus, we are in the third case, hence $x_i = \overline{x}_i$.

$6°$. So, as soon as we determine the position of $E$ with respect to all the bounds $\underline{x}_i$ and $\overline{x}_i$, we will have a pretty good understanding of all the values $x_i$ at which the minimum is attained.

Hence, to find the minimum, we will analyze how the endpoints $\underline{x}_i$ and $\overline{x}_i$ divide the real line, and consider all the resulting sub-intervals.

Let the corresponding subinterval $[x_{(k)}, x_{(k+1)}]$ by fixed. For the $i$'s for which $E \notin \mathbf{x}_i$, the values $x_i$ that correspond to the minimal sample variance are uniquely determined by the above formulas.

For the $i$'s for which $E \in \mathbf{x}_i$ the selected value $x_i$ should be equal to $E$. To determine this $E$, we can use the fact that $E$ is equal to the average of all thus selected values $x_i$, in other words, that we should have

$$E = \frac{1}{n} \cdot \left( \sum_{i:\underline{x}_i \geq x_{(k+1)}} \underline{x}_i + (n - N_k) \cdot E + \sum_{j:\overline{x}_j \leq x_{(k)}} \overline{x}_j \right), \tag{16.7}$$

where $(n - N_k) \cdot E$ combines all the points for which $E \in \mathbf{x}_i$. Multiplying both sides of (16.7) by $n$ and subtracting $n \cdot E$ from both sides, we conclude that (in notations of the algorithm), we have $E = S_k / N_k$ – what we denoted, in the algorithm's description, by $r_k$. If thus defined $r_k$ does not belong to the subinterval $[x_{(k)}, x_{(k+1)}]$, this contradiction with our initial assumption shows that there cannot be any minimum in this subinterval, so this subinterval can be easily dismissed.

The corresponding sample variance is denoted by $V'_k$. If $N_k = 0$, this means that $E$ belongs to all the intervals $\mathbf{x}_i$ and therefore, that the lower endpoint $\underline{V}$ is exactly 0 – so we assign $V'_k = 0$.

The correctness is proven.

*Proof that the algorithm for computing $\underline{V}$ takes time $O(n \cdot \log(n))$.* Indeed, sorting takes $O(n \cdot \log(n))$ steps (see, e.g., [73]). For $k = 1$, the algorithm takes linear time $(O(n))$ to compute all the original sums and then a constant $(O(1))$ number of steps to perform all the computations.

On each following iteration, we re-calculate the corresponding sums. In each of these sums, each endpoint is processed only once, so totally, all these re-calculations take $O(n)$ steps. Thus, the total computation time is indeed $O(n \cdot \log(n))) + O(n) = O(n \cdot \log(n))$.

*Proof that the new algorithm for computing $\underline{V}$ takes linear time.* At each iteration, computing median takes linear time. All other operations with $J$ take time $t$ linear in the number of elements $|J|$ of $J$. We start with the set $J$ of size $n$. On the next iteration, we have a set of size $n/2$, then $n/4$, etc. Thus, the overall computation time is $\leq C \cdot (n + n/2 + n/4 + \ldots) \leq C \cdot 2n$, i.e. linear in $n$.

*Proof that the new algorithm always computes $\underline{V}$.* In the above justification of the $O(n \cdot \log(n))$ algorithm, we proved that if we sort all $2n$ endpoints into a sequence $x_{(1)} \leq x_{(2)} \leq \ldots \leq x_{(2n)}$, then for some $k = k_{\min}$ the minimum $\underline{V}$ is attained for the vector $x$ for which the following holds:

- For all indices $j$ for which $\overline{x}_j \leq x_{(k)}$ we have $x_j = \overline{x}_j$.
- For all indices $i$ for which $\underline{x}_i \geq x_{(k+1)}$ we have $x_i = \underline{x}_i$.

- For all other indices $i$ we have $x_i = r_k \overset{\text{def}}{=} \dfrac{E_k}{N_k}$, where

$$E_k = \sum_{j:\overline{x}_j \leq x_{(k)}} \overline{x}_j + \sum_{i:\underline{x}_i \geq x_{(k+1)}} \underline{x}_i; \quad N_k = \#\{j : \overline{x}_j \leq x_{(k)}\} + \#\{i : \underline{x}_i \geq x_{(k+1)}\}.$$

It has also been proven that for the optimal $k$ we have $r_k \in [x_{(k)}, x_{(k+1)}]$.

In general, the condition $x_{(k)} \leq r_k = \dfrac{E_k}{N_k}$ is equivalent to

$$N_k \cdot x_{(k)} \leq E_k = \sum_{j:\overline{x}_j \leq x_{(k)}} \overline{x}_j + \sum_{i:\underline{x}_i \geq x_{(k+1)}} \underline{x}_i.$$

Subtracting $x_{(k)}$ from each of $N_k$ terms in the right-hand side (RHS) and moving the sum of the resulting non-positive differences into the left-hand side (LHS), we conclude that

$$\sum_{j:\overline{x}_j \leq x_{(k)}} (x_{(k)} - \overline{x}_j) \leq \sum_{i:\underline{x}_i \geq x_{(k+1)}} (\underline{x}_i - x_{(k)}). \tag{16.8}$$

When we increase $k$, we get (in general) more terms in the LHS and fewer in the RHS. Hence LHS (non-strictly) increases while the RHS non-strictly decreases. So if the inequality (16.8) holds for some $k$, it holds for all smaller values of $k$ as well. Thus this inequality holds for all $k$ until a certain value $k_0$.

Similarly, the condition $x_{(k+1)} \geq r_k = \dfrac{E_k}{N_k}$ is equivalent to

$$N_k \cdot r_{k+1} \geq \sum_{j:\overline{x}_j \leq x_{(k)}} \overline{x}_j + \sum_{i:\underline{x}_i \geq x_{(k+1)}} \underline{x}_i.$$

Subtracting $x_{(k+1)}$ from each of $N_k$ terms in RHS and moving the sum of the resulting non-positive differences into LHS, we conclude that

$$\sum_{j:\overline{x}_j \leq x_{(k)}} (x_{(k+1)} - \overline{x}_j) \geq \sum_{i:\underline{x}_i \geq x_{(k+1)}} (\underline{x}_i - x_{(k+1)}). \tag{16.9}$$

When we increase $k$, the LHS (non-strictly) increases, while the RHS non-strictly decreases. So if the inequality (16.9) holds for some $k$, it holds for all larger values of $k$ as well. Thus this inequality holds for all $k$ after a certain value $l_0$.

So both conditions (16.8) and (16.9) are satisfied (which is equivalent to the condition $r_k \in [x_{(k)}, x_{(k+1)}]$) either for a single value $k_{\min}$ or for several sequential values $l_0, l_0+1, \ldots, k_0$. Let us show that if this condition is satisfied for several sequential values, this simply means that the same minimum $\underline{V}$ is attained for all these values. For that it is sufficient to show that if both conditions (16.8) and (16.9) holds for $k$ and for $k+1$ then the variance $V$ has the same value for both $k$ and $k+1$. Indeed, since (16.8) is true for $k+1$, we have

$$\sum_{j:\overline{x}_j \leq x_{(k+1)}} (x_{(k+1)} - \overline{x}_j) \leq \sum_{i:\underline{x}_i \geq x_{(k+2)}} (\underline{x}_i - x_{(k+1)}).$$

The LHS of this new inequality is smaller than or equal to the LHS of the inequality (16.9), and its RHS is larger than or equal to the RHS of the inequality (16.9). Thus the only way for both inequalities to hold is when both sides are equal, i.e. when replacing $x_{(k)}$ with $x_{(k+1)}$ and replacing $x_{(k+1)}$ with $x_{(k+2)}$ does not change which endpoints are in $I^-$ and which are in $I^+$ – and thus, does not change the corresponding value of the variance.

So:

- for $k < k_{\min}$, we have $r_k > x_{(k+1)}$,
- for $k > k_{\min}$, we have $r_k < x_{(k)}$, and
- for $k = k_{\min}$ (or, to be more precise, for $l_0 \leq k \leq k_0$), we have $x_{(k)} \leq r_k \leq x_{(k+1)}$.

Hence:

- if $r_k < x_{(k)}$, then we cannot have $k < k_{\min}$ and $k = k_{\min}$, hence $k > k_{\min}$;
- if $r_k > x_{(k+1)}$, then we cannot have $k > k_{\min}$ and $k = k_{\min}$, hence $k < k_{\min}$;
- if $x_{(k)} \leq r_k \leq x_{(k+1)}$, then we cannot have $k < k_{\min}$ and $k > k_{\min}$, hence $k = k_{\min}$.

Thus the above algorithm finds the correct value of $k_{\min}$ and thence, the correct value of $\underline{V}$.

*Justification of an $O(n^2)$ algorithm for computing $\overline{V}$ for the case of $c_0$-few intersections.*

$1^\circ$. In order to find the maximum of $V$, it turns out to be useful to first find where the variance $V$ attains its minimum. Let $x_1^{(0)} \in \mathbf{x}_1, \ldots, x_n^{(0)} \in \boldsymbol{x}_n$ be the values for which the finite population variance $V$ attains minimum on the box $\boldsymbol{x}_1 \times \ldots \times \boldsymbol{x}_n$.

Let us pick one of the $n$ variables $x_i$, and let fix the values of all the other variables $x_j$ $(j \neq i)$ at $x_j = x_j^{(0)}$. When we substitute $x_j = x_j^{(0)}$ for all $j \neq i$ into the expression for finite population variance, $V$ becomes a quadratic function of $x_i$.

This function of one variable should attain its minimum on the interval $\mathbf{x}_i$ at the value $x_i^{(0)}$.

$2°$. By definition, the variance $V$ is a sum of non-negative terms; thus, its value is always non-negative. Therefore, the corresponding quadratic function of one variable always has a global minimum. This function is decreasing before this global minimum, and increasing after it.

$3°$. Where is the global minimum of the quadratic function of one variable described in Part 1 of this proof?

It is attained when $\dfrac{\partial V}{\partial x_i} = 0$. Differentiating the formula for the variance with respect to $x_i$, we conclude that

$$\frac{\partial V}{\partial x_i} = \frac{1}{n} \cdot \left( 2(x_i - E) + \sum_{j=1}^{n} 2(E - x_j) \cdot \frac{\partial E}{\partial x_j} \right).$$

Since $\dfrac{\partial E}{\partial x_i} = 1/n$, we conclude that

$$\frac{\partial V}{\partial x_i} = \frac{2}{n} \cdot \left( (x_i - E) + \sum_{j=1}^{n} (E - x_j) \cdot \frac{1}{n} \right).$$

Here,

$$\sum_{j=1}^{n} (E - x_j) = n \cdot E - \sum_{j=1}^{n} x_j.$$

By definition of the average $E$, this difference is 0, hence the above formula takes the form

$$\frac{\partial V}{\partial x_i} = \frac{2}{n} \cdot (x_i - E).$$

So, this function attains the minimum when $x_i - E = 0$, i.e., when $x_i = E$.

Since

$$E = \frac{x_i}{n} + \frac{\displaystyle\sum_{j:j\neq i} x_j}{n},$$

the equality $x_i = E$ means that

$$x_i = \frac{x_i}{n} + \frac{\displaystyle\sum_{j:j\neq i} x_j^{(0)}}{n}.$$

Moving terms containing $x_i$ into the left-hand side and dividing by the coefficient at $x_i$, we conclude that the minimum is attained when

$$x_i = E_i' \overset{\text{def}}{=} \frac{\displaystyle\sum_{j:j\neq i} x_j^{(0)}}{n-1},$$

i.e., when $x_i$ is equal to the arithmetic average $E_i'$ of all other elements.

$4°$. Let now $x_1, \ldots, x_n$ be the values at which the finite population variance attain its *maximum* on the box $\boldsymbol{x}_1 \times \ldots \times \boldsymbol{x}_n$. If we fix the values of all the variables but one $x_i$, then $V$ becomes a quadratic function of $x_i$. When the function $V$ attains maximum over $x_1 \in \boldsymbol{x}_1, \ldots, x_n \in \boldsymbol{x}_n$, then this quadratic function of one variable will attain its maximum on the interval $\boldsymbol{x}_i$ at the point $x_i$.

We have already shown that this quadratic function has a (global) minimum at $x_i = E'_i$, where $E'_i$ is the average of all the values $x_1, \ldots, x_n$ except for $x_i$. Since this quadratic function of one variable is always non-negative, it cannot have a global maximum. Therefore, its maximum on the interval $\boldsymbol{x}_i = [\underline{x}_i, \overline{x}_i]$ is attained at one of the endpoints of this interval.

An arbitrary quadratic function of one variable is symmetric with respect to the location of its global minimum, so its maximum on any interval is attained at the point which is the farthest from the minimum. There is exactly one point which is equally close to both endpoints of the interval $\boldsymbol{x}_i$: its midpoint $\widetilde{x}_i$. Depending on whether the global minimum is to the left, to the right, or exactly at the midpoint, we get the following three possible cases:

1. If the global minimum $E'_i$ is to the left of the midpoint $\widetilde{x}_i$, i.e., if $E'_i < \widetilde{x}_i$, then the upper endpoint is the farthest from $E'_i$. In this case, the maximum of the quadratic function is attained at its upper endpoint, i.e., $x_i = \overline{x}_i$.
2. Similarly, if the global minimum $E'_i$ is to the right of the midpoint $\widetilde{x}_i$, i.e., if $E'_i > \widetilde{x}_i$, then the lower endpoint is the farthest from $E'_i$. In this case, the maximum of the quadratic function is attained at its lower endpoint, i.e., $x_i = \underline{x}_i$.
3. If $E'_i = \widetilde{x}_i$, then the maximum of $V$ is attained at both endpoints of the interval $\boldsymbol{x}_i = [\underline{x}_i, \overline{x}_i]$.

$5°$. In the third case, we have either $x_i = \underline{x}_i$ or $x_i = \overline{x}_i$. Depending on whether $x_i$ is equal to the lower or to the upper endpoints, we can "combine" the corresponding situations with Cases 1 and 2. As a result, we arrive at the conclusion that one of the following two situations happen:

1. either $E'_i \leq \widetilde{x}_i$ and $x_i = \overline{x}_i$;
2. either $E'_i \geq \widetilde{x}_i$ and $x_i = \underline{x}_i$.

$6°$. Let us reformulate these conclusions in terms of the average $E$ of the maximizing values $x_1, \ldots, x_n$.

The average $E'_i$ can be described as

$$\frac{\sum\limits_{j:j \neq i} x_j}{n-1}.$$

By definition, $\sum\limits_{j:j \neq i} x_j = \sum\limits_{j=1}^{n} x_j - x_i$. By definition of $E$, we have

$$E = \frac{\sum\limits_{j=1}^{n} x_j}{n},$$

hence $\sum\limits_{j=1}^{n} x_j = n \cdot \mu$. Therefore,

$$E'_i = \frac{n \cdot E - x_i}{n-1}.$$

Let us apply this formula to the above three cases.

6.1°. In the first case, we have $\widetilde{x}_i \geq E'_i$. So, in terms of $E$, we get the inequality

$$\widetilde{x}_i \geq \frac{n \cdot E - x_i}{n-1}.$$

Multiplying both sides of this inequality by $n-1$, and using the fact that in this case, $x_i = \overline{x}_i = \widetilde{x}_i + \Delta_i$, we conclude that

$$(n-1) \cdot \widetilde{x}_i \geq n \cdot E - \widetilde{x}_i - \Delta_i.$$

Moving all the terms but $n \cdot E$ to the left-hand side and dividing by $n$, we get the following inequality:

$$E \leq \widetilde{x}_i + \frac{\Delta_i}{n}.$$

6.2°. In the second case, we have $\widetilde{x}_i \leq E'_i$. So, in terms of $E$, we get the inequality

$$\widetilde{x}_i \leq \frac{n \cdot E - x_i}{n-1}.$$

Multiplying both sides of this inequality by $n-1$, and using the fact that in this case, $x_i = \underline{x}_i = \widetilde{x}_i - \Delta_i$, we conclude that

$$(n-1) \cdot \widetilde{x}_i \leq n \cdot E - \widetilde{x}_i + \Delta_i.$$

Moving all the terms but $n \cdot E$ to the left-hand side and dividing by $n$, we get the following inequality:

$$E \geq \widetilde{x}_i - \frac{\Delta_i}{n}.$$

7°. Parts 6.1 and 6.2 of this proof can be summarized as follows:

- In Case 1, we have $E \leq \widetilde{x}_i + \Delta_i/n$ and $x_i = \overline{x}_i$.
- In Case 2, we have $E \geq \widetilde{x}_i - \Delta_i/n$ and $x_i = \underline{x}_i$.

Therefore:

- If $E < \widetilde{x}_i - \Delta_i/n$, this means that we cannot be in Case 2. So we must be in Case 1 and therefore, we must have $x_i = \overline{x}_i$.

- If $E > \widetilde{x}_i + \Delta_i/n$, this means that we cannot be in Case 1. So, we must be in Case 2 and therefore, we must have $x_i = \underline{x}_i$.

The only case when we do not know which endpoint for $x_i$ we should choose is the case when $\mu$ belongs to the narrowed interval $[\widetilde{x}_i - \Delta/n, \widetilde{x}_i + \Delta_i]$.

$8°$. Hence, once we know where $E$ is with respect to the endpoints of all narrowed intervals, we can determine the values of all optimal $x_i$ – except for those that are within this narrowed interval. Since we consider the case when no more than $c_0$ narrowed intervals can have a common point, we have no more than $c_0$ undecided values $x_i$. Trying all possible combinations of lower and upper endpoints for these $\leq c_0$ values requires $\leq 2^{c_0}$ steps.

Thus, the overall number of steps is $O(2^{c_0} \cdot n^2)$. Since $c_0$ is a constant, the overall number of steps is thus $O(n^2)$.

*Justification of a $O(n{\cdot}\log(n))$ algorithm for computing $\overline{V}$ for the the case when intervals have $c_0$-few intersections.* To prove that the algorithm computes the exact upper endpoint in time $o(n \cot \log(n))$, let us follow this algorithm stage-by-stage.

$1°$. First, we sort the lower endpoints $\widetilde{x}_i - \Delta_i/n$ of the narrowed intervals into an increasing sequence. It is well known that sorting requires time $O(n \cdot \log(n))$.

$2°$. Then, we sort *all* the endpoints of the narrowed intervals into a sequence $x_{(1)} \leq x_{(2)} \leq \ldots \leq x_{(k)} \leq \ldots \leq x_{(2n)}$. This sorting also requires $O(n \cdot \log(n))$ steps.

$3°$. On the third stage, we produce, for each of the resulting zones $[x_{(k)}, x_{(k+1)}]$, the set $S_k$ of all the indices $i$ for which the $i$-th narrowed interval

$$[\widetilde{x}_i - \Delta_i/n, \widetilde{x}_i + \Delta_i/n]$$

contains this zone.

As we have mentioned, for each $i$, we know the value $k = k^-(i)$ for which $\widetilde{x}_i - \Delta_i/n = x_{(k)}$. So, for each $i$, we place $i$ into the set $S_{k^-(i)}$ corresponding to the zone $[x_{(k^-(i))}, x_{(k^-(i)+1)}]$, into the set corresponding to the next zone, etc., until we reach the zone for which the upper endpoint is exactly $\widetilde{x}_i + \Delta_i/n$.

Here, we need one computational step for each new entry of $i$ into the set corresponding to a new zone. Therefore, filling in all these sets requires as many steps as there are items in all these sets. For each of $2n + 1$ zones, as we have mentioned, there are no more than $c$ items in the corresponding set; therefore, overall, all the sets contain no more than $c \cdot (2n + 1) = O(n)$ steps. Thus, this stage requires $O(n)$ time.

$4°$. On the fourth stage, for all integers $p$ from 0 to $n$, we compute the sums

$$E_p \stackrel{\text{def}}{=} \frac{1}{n} \cdot \sum_{i=1}^{p} \underline{x}_i + \frac{1}{n} \cdot \sum_{i=p+1}^{n} \overline{x}_i;$$

$$M_p \overset{\text{def}}{=} \frac{1}{n} \cdot \sum_{i=1}^{p} (\underline{x}_i)^2 + \frac{1}{n} \cdot \sum_{i=p+1}^{n} (\overline{x}_i)^2.$$

We compute these values sequentially. Once we know $E_p$ and $M_p$, we can compute $E_{p+1}$ and $M_{p+1}$ as $E_{p+1} = E_p + \underline{x}_{p+1} - \overline{x}_{p+1}$ and $M_{p+1} = M_p + (\underline{x}_{p+1})^2 - (\overline{x}_{p+1})^2$.

Transition from $E_p$ and $M_p$ to $E_{p+1}$ and $M_{p+1}$ requires a constant number of computational steps; so overall, we need $O(n)$ steps to compute all the values $E_p$ and $M_p$.

$5°$. Finally, for each zone $k$, we compute the corresponding values of the variance. For that, we first find the smallest index $i$ for which $x_{(k+1)} \leq \widetilde{x}_i - \Delta_i/n$. We will denote this value $i$ by $p(k)$.

Since the values $\widetilde{x}_i - \Delta_i/n$ are sorted, we can find this $i$ by using bisection. It is known that bisection requires $O(\log(n))$ steps, so finding such $p(k)$ for all $2n + 1$ zones requires $O(n \cdot \log(n))$ steps.

Once $i \geq p(k)$, then $\widetilde{x}_i - \Delta_i/n \geq \widetilde{x}_{p(k)} - \Delta_{p(k)}/n \geq x_{(k+1)}$. So, in accordance with the above justification for the quadratic-time algorithm, we should select $x_i = \overline{x}_i$, as in the sums $E_{p(k)}$ and $M_{p(k)}$.

In accordance with the same justification, the only values $i < p(k)$ for which we may also select $x_i = \overline{x}_i$ are the values for which the $i$-th narrowed intervals contains this zone. These values are listed in the set $S_k$ of no more than $c_0$ such intervals. So, to find all possible values of $V$, we can do the following.

We then consider all subsets $s \subseteq S_k$ of the set $S_k$; there are no more than $2^c$ such subsets. For each subset $s$, we replace, in $E_{p(k)}$ and $M_{p(k)}$, values $\underline{x}_i$ and $(\underline{x}_i)^2$ corresponding to all $i \in s$, with, correspondingly, $\overline{x}_i$ and $(\overline{x}_i)^2$.

Each replacement means subtracting no more than $c_0$ terms and then adding no more than $c_0$ terms, so each computation requires no more than $2c_0$ steps. Once we have $E$ and $V$ corresponding to the subset $s$, we can check whether $E$ belongs to the analyzed zone and, if yes, compute $V = M - E^2$.

For each subset, we need no more than $2c_0 + 2$ computations, so for all no more than $2^{c_0}$ subsets, we need no more than $(2c_0 + 2) \cdot 2^{c_0}$ computations. For a fixed $c_0$, this value does not depend on $n$; in other words, for each zone, we need $O(1)$ steps.

To perform this computation for all $2n + 1$ zones, we need $(2n + 1) \cdot O(1) = O(n)$ steps.

$6°$. Finally, we find the largest of the resulting values $V$ – this will be the desired value $\overline{V}$.

Finding the largest of $O(n)$ values requires $O(n)$ steps.

Overall, we need

$$O(n \cdot \log(n)) + O(n \cdot \log(n)) + O(n) + O(n) + O(n \cdot \log(n)) + O(n) = O(n \cdot \log(n))$$

steps. Thus, we have proven that our algorithm computes $\overline{V}$ in $O(n \cdot \log(n))$ steps.

# Computing Variance under Hierarchical Privacy-Related Interval Uncertainty

## Formulation and Analysis of the Problem and the Resulting Algorithms

*Need for hierarchical statistical analysis.* In the above text, we assumed that we have all the data in one large database, and we process this large statistical database to estimate the desired statistical characteristics.

To prevent privacy violations, we replace the original values of the quasi-identifier variables with ranges. For example, we divide the set of all possible ages into ranges $[0, 10]$, $[10, 20]$, $[20, 30]$, etc. Then, instead of storing the actual age of 26, we only store the range $[20, 30]$ which contains the actual age value.

In reality, the data is often stored hierarchically. For example, it makes sense to store the census results by states, get averages and standard deviations per state, and then combine these characteristics to get nation-wide statistics. In many real-life situations, several research groups independently perform statistical analysis of different data sets. The more data we use for statistical analysis, the better the estimates. So, it is desirable to get estimates based on the data from all the data sets. In principle, we can combine the data sets and re-process the combined data. However, this would take a large amount of time for data processing. It is known that for many statistics (e.g., for population variance), we can avoid these lengthy computations: the statistic for the combined data can be computed based on the results of processing individual data sets.

In this chapter, we show that a similar computational simplification is possible when instead of processing the exact values, we process privacy-related interval ranges for these values.

Its main results first appeared in [209] and [356].

*Formulas behind hierarchical statistical analysis.* Let the data values $x_1 \ldots, x_n$ be divided into $m < n$ groups $I_1, \ldots, I_m$. For each group $j$, we know the frequency $p_j$ of this group (i.e., the number $n_j$ of elements of this group divided

H.T. Nguyen et al.: Computing Stat. under Interval & Fuzzy Uncertain., SCI 393, pp. 119–127.
springerlink.com 

by the overall number of records), the average $E_j$ over this group, and the population variance $V_j$ within $j$-th group.

In this case, the overall average $E$ can be described as

$$E = \frac{1}{n} \cdot \sum_{i=1}^{n} x_i = \frac{1}{n} \cdot \sum_{j=1}^{m} \sum_{i \in I_j} x_i.$$

By definition of the group average $E_j$, we have $E_j = \dfrac{1}{n_j} \cdot \sum_{i \in I_j} x_i$, where $n_j = p_j \cdot n$ denotes the overall number of elements in the $j$-th group. Thus, $\sum_{i \in I_j} x_i = n_j \cdot E_j = p_j \cdot n \cdot E_j$, hence

$$E = \sum_{j=1}^{m} p_j \cdot E_j. \tag{17.1}$$

Similarly, the overall variance $V$ can be described as

$$V = \frac{1}{n} \cdot \sum_{i=1}^{n} x_i^2 - E^2 = \frac{1}{n} \cdot \sum_{j=1}^{m} \sum_{i \in I_j} x_i^2 - E^2.$$

For each $j$ and for each $i \in I_j$, we have $x_i = (x_i - E_j) + E_j$, hence

$$x_i^2 = (x_i - E_j)^2 + E_j^2 + 2(x_i - E_j) \cdot E_j.$$

Therefore,

$$\sum_{i \in I_j} x_i^2 = \sum_{i \in I_j} (x_i - E_j)^2 + n_j \cdot E_j^2 + 2E_j \cdot \sum_{i \in I_j} (x_i - E_j).$$

The first sum, by definition of population variance $V_j$, is equal to $n_j \cdot V^2$; the third sum, by definition of the population mean, is equal to 0. Thus, $\sum_{i \in I_j} x_i^2 = n_j \cdot (V_j + E_j^2)$, where $n_j = p_j \cdot n$, and thus,

$$V = V_E + V_\sigma, \tag{17.2}$$

where $V_E \stackrel{\text{def}}{=} \sum_{j=1}^{m} p_j \cdot E_j^2 - E^2$ and $V_\sigma \stackrel{\text{def}}{=} \sum_{j=1}^{m} p_j \cdot V_j$.

*Hierarchical case: situation with interval uncertainty.* When we start with values $x_i$ which are only known with interval uncertainty, we end up knowing $E_j$ and $V_j$ also with interval uncertainty. In other words, we only know the intervals $\mathbf{E}_j = [\underline{E}_j, \overline{E}_j]$ and $[\underline{V}_j, \overline{V}_j]$ that contain the actual (unknown) values of $E_j$ and $V_j$. In such situations, we must find the ranges of the possible values for the population mean $E$ and for the population variance $V$.

*Analysis of the interval problem.* The formula that describes the dependence of $E$ on $E_j$ is monotonic in $E_j$. Thus, we get an explicit formula for the range $[\underline{E}, \overline{E}]$ of the population mean $E$: $\underline{E} = \sum\limits_{j-1}^{m} p_j \cdot \underline{E}_j$ and $\overline{E} = \sum\limits_{j-1}^{m} p_j \cdot \overline{E}_j$.

Since the terms $V_E$ and $V_\sigma$ in the expression for $V$ depend on different variables, the range $[\underline{V}, \overline{V}]$ of the population variance $V$ is equal to the sum of the ranges $[\underline{V}_E, \overline{V}_E]$ and $[\underline{V}_\sigma, \overline{V}_\sigma]$ of the corresponding terms: $\underline{V} = \underline{V}_E + \underline{V}_\sigma$ and $\overline{V} = \overline{V}_E + \overline{V}_\sigma$. Due to similar monotonicity, we can find an explicit expression for the range $[\underline{V}_\sigma, \overline{V}_\sigma]$ for $V_\sigma$: $\underline{V}_\sigma = \sum\limits_{j=1}^{m} p_j \cdot \underline{V}_j$ and $\overline{V}_\sigma = \sum\limits_{j=1}^{m} p_j \cdot \overline{V}_j$.

Thus, to find the range of the population variance $V$, it is sufficient to find the range of the term $V_E$. So, we arrive at the following problem:

*Formulation of the problem in precise terms*

GIVEN: an integer $m \geq 1$, $m$ numbers $p_j > 0$ for which $\sum\limits_{j=1}^{m} p_j = 1$, and $m$ intervals $\boldsymbol{E}_j = [\underline{E}_j, \overline{E}_j]$.

COMPUTE the range $\boldsymbol{V}_E = \{V_E(E_1, \ldots, E_m) \,|\, E_1 \in \boldsymbol{E}_1, \ldots, E_m \in \boldsymbol{E}_m\}$, where

$$V_E \stackrel{\mathrm{def}}{=} \sum_{j=1}^{m} p_j \cdot E_j^2 - E^2; \quad E \stackrel{\mathrm{def}}{=} \sum_{j=1}^{m} p_j \cdot E_j.$$

*Main result.* Since the function $V_E$ is convex, we can compute its minimum $\underline{V}_E$ on the box $\boldsymbol{E}_1 \times \ldots \times \boldsymbol{E}_m$ by using known polynomial-time algorithms for minimizing convex functions over interval domains; see, e.g., [334].

For computing maximum $\overline{V}_E$, even the particular case when all the values $p_j$ are equal $p_1 = \ldots = p_m = 1/m$, is known to be NP-hard. Thus, the more general problem of computing $\overline{V}_E$ is also NP-hard. Let us show that in a reasonable class of cases, there exists a feasible algorithm for computing $\overline{V}_E$.

For each interval $\boldsymbol{E}_j$, let us denote its midpoint by $\widetilde{E}_j \stackrel{\mathrm{def}}{=} \dfrac{\underline{E}_j + \overline{E}_j}{2}$, and its half-width by $\Delta_j \stackrel{\mathrm{def}}{=} \dfrac{\overline{E}_j - \underline{E}_j}{2}$. In these terms, the $j$-th interval $\boldsymbol{E}_j$ takes the form $[\widetilde{E}_j - \Delta_j, \widetilde{E}_j + \Delta_j]$.

In this chapter, we consider "narrowed" intervals $[E_j^-, E_j^+]$, where $E_j^- \stackrel{\mathrm{def}}{=} \widetilde{E}_j - p_j \cdot \Delta_j$ and $E_j^+ \stackrel{\mathrm{def}}{=} \widetilde{E}_j + p_j \cdot \Delta_j$. We show that there exists an efficient $O(m \cdot \log(m))$ algorithm for computing $\overline{V}_E$ for the case when no two narrowed intervals are proper subsets of each other, i.e., when $[E_j^-, E_j^+] \not\subseteq (E_k^-, E_k^+)$ for all $j$ and $k$.

*Algorithm*

- First, we sort the midpoints $\widetilde{E}_1, \ldots, \widetilde{E}_m$ into an increasing sequence. Without losing generality, we can assume that $\widetilde{E}_1 \leq \widetilde{E}_2 \leq \ldots \leq \widetilde{E}_m$.

- Then, for every $k$ from 0 to $m$, we compute the value $V_E^{(k)} = M^{(k)} - (E^{(k)})^2$ of the quantity $V_E$ for the vector $E^{(k)} = (\underline{E}_1, \ldots, \underline{E}_k, \overline{E}_{k+1}, \ldots, \overline{E}_m)$.
- Finally, we compute $\overline{V}_E$ as the largest of $m + 1$ values $V_E^{(0)}, \ldots, V_E^{(m)}$.

To compute the values $V_E^{(k)}$, first, we explicitly compute $M^{(0)}$, $E^{(0)}$, and $V_E^{(0)} = M^{(0)} - E^{(0)}$. Once we computed the values $M^{(k)}$ and $E^{(k)}$, we can compute

$$M^{(k+1)} = M^{(k)} + p_{k+1} \cdot (\underline{E}_{k+1})^2 - p_{k+1} \cdot (\overline{E}_{k+1})^2 \text{ and}$$

$$E^{(k+1)} = E^{(k)} + p_{k+1} \cdot \underline{E}_{k+1} - p_{k+1} \cdot \overline{E}_{k+1}.$$

*Number of computation steps.*

- It is well known that sorting takes $O(m \cdot \log(m))$ steps.
- Computing the initial values $M^{(0)}$, $E^{(0)}$, and $V_E^{(0)}$ takes linear time $O(m)$.
- For each $k$ from 0 to $m - 1$, we need a constant number $O(1)$ of steps to compute the next values $M^{(k+1)}$, $E^{(k+1)}$, and $V_E^{(k+1)}$.
- Finally, finding the largest of $m + 1$ values $V_E^{(k)}$ also takes $O(m)$ steps.

Thus, overall, we perform

$$O(m \cdot \log(m)) + O(m) + m \cdot O(1) + O(m) = O(m \cdot \log(m)) \text{ steps.}$$

*Auxiliary result: what if the frequencies are also only known with interval uncertainty?* In the previous text, we assumed that we know the exact values of the frequencies $p_j$. In practice, we usually only know the frequencies with uncertainty. Let us show how to take this uncertainty into account.

*Reminder: hierarchical statistical data processing.* If we know the frequency of the group $j$, the mean $E_j$ of the group $j$, and its second moment $M_j = V_j + E_j^2 = \dfrac{1}{p_j \cdot n} \cdot \sum_{i \in I_j} x_i^2$, then we can compute the overall mean $E$ and the overall variance as $E = \sum\limits_{j=1}^{m} p_j \cdot E_j$ and $V = \sum\limits_{j=1}^{m} p_j \cdot M_j - E^2$.

*Reminder: hierarchical statistical data processing under interval uncertainty.* In the above text, we considered the case when the statistical characteristics $E_j$ and $V_j$ corresponding to different groups are known with interval uncertainty, but the frequencies $p_j$ are known exactly.

*New situation.* In practice, the frequencies $p_j$ may also only be known with interval uncertainty. This may happen, e.g., if instead of the full census we extrapolate data – or if we have a full census and try to take into account that no matter how thorough the census, a certain portion of the population will be missed.

In practice, the values $x_i$ (age, salary, etc.) are usually non-negative. In this case, $E_j \geq 0$. In this section, we will only consider this non-negative case. Thus, we arrive at the new formulation of the problem:

GIVEN: an integer $m \geq 1$, and for every $j$ from 1 to $m$, intervals $[\underline{p}_j, \overline{p}_j]$, $[\underline{E}_j, \overline{E}_j]$, and $[\underline{M}_j, \overline{M}_j]$ for which $\underline{p}_j \geq 0$, $\underline{E}_j \geq 0$, and $\underline{M}_j \geq 0$.

COMPUTE the range $\mathbf{E} = [\underline{E}, \overline{E}]$ of $E = \sum\limits_{j=1}^{m} p_j \cdot E_j$ and the range $\mathbf{M} = [\underline{M}, \overline{M}]$ of $M = \sum\limits_{j=1}^{m} p_j \cdot M_j - E^2$ under the conditions that $p_j \in [\underline{p}_j, \overline{p}_j]$, $E_j \in [\underline{E}_j, \overline{E}_j]$, $M_j \in [\underline{M}_j, \overline{M}_j]$, and $\sum\limits_{j=1}^{m} p_j = 1$.

*Derivation of an algorithm for computing $\underline{E}$.* When the frequencies $p_j$ are known, we can easily compute the bounds for $E$. In the case when $p_j$ are also known with interval uncertainty, it is no longer easy to compute these bounds.

Since $E$ monotonically depends on $E_j$, the smallest value $\underline{E}$ of $E$ is attained when $E_j = \underline{E}_j$ for all $j$, so the only problem is to find the corresponding probabilities $p_j$. Suppose that $p_1, \ldots, p_n$ are minimizing probabilities, and for two indices $j$ and $k$, we change $p_j$ to $p_j + \Delta p$ (for some small $\Delta p$) and $p_k$ to $p_k - \Delta p$. In this manner, the condition $\sum\limits_{j=1}^{m} p_j$ is preserved. After this change, $E$ changes to $E + \Delta E$, where $\Delta E = \Delta p \cdot (\underline{E}_j - \underline{E}_k)$.

Since we start with the values at which $E$ attains its minimum, we must have $\Delta E \geq 0$ for all $\Delta p$. If both $p_j$ and $p_k$ are strictly inside the corresponding intervals, then we can have $\Delta p$ of all signs hence we should have $\underline{E}_j = \underline{E}_k$. So, excluding this degenerate case, we should have at most one value $p_i$ strictly inside – others are at one of the endpoints.

If $p_j = \underline{p}_j$ and $p_k = \overline{p}_k$, then we can have $\Delta p > 0$, so $\Delta E \geq 0$ implies $\underline{E}_j \geq \underline{E}_k$. So, the values $\underline{E}_j$ for all $j$ for which $p_j = \underline{p}_j$ should be $\leq$ than all the values $\underline{E}_k$ for which $p_k = \overline{p}_k$. This conclusion can be reformulated as follows: if we sort the groups in the increasing order of $\underline{E}_j$, we should get first $\overline{p}_j$ then all $\underline{p}_k$. Thus, it is sufficient to consider only such arrangements of probabilities for which we have $\overline{p}_1, \ldots \overline{p}_{l_0-1}, p_{l_o}, \underline{p}_{l_0+1}, \ldots \underline{p}_m$. The value $l_0$ can be uniquely determined from the condition that $\sum\limits_{j=1}^{m} p_j = 1$. Thus, we arrive at the following algorithm:

*Algorithm for computing $\underline{E}$.* To compute $\underline{E}$, we first sort the values $\underline{E}_j$ in increasing order. Let us assume that the groups are already sorted in this order, i.e., that $\underline{E}_1 \leq \underline{E}_2 \leq \ldots \leq \underline{E}_m$.

For every $l$ from 0 to $k$, we then compute $P_l = \overline{p}_1 + \ldots + \overline{p}_l + \underline{p}_{l+1} + \ldots + \underline{p}_n$ as follows: we explicitly compute the sum $P_0$, and then consequently compute $P_{l+1}$ as $P_l + (\overline{p}_{l+1} - \underline{p}_{l+1})$. This sequence is increasing. Then, we find the value $l_0$ for which $P_{l_0} \leq 1 \leq P_{l_0+1}$, and take $\underline{E} = \sum\limits_{j=1}^{l_0-1} \overline{p}_j \cdot \underline{E}_j + p_{l_0} \cdot \underline{E}_{l_0} + \sum\limits_{j=l_0+1}^{m} \underline{p}_j \cdot \underline{E}_j$, where $p_{l_0} = 1 - \sum\limits_{j=1}^{l_0-1} \overline{p}_j - \sum\limits_{j=l_0+1}^{m} \underline{p}_j$.

*Computation time.* We need $O(m \cdot \log(m))$ time to sort, $O(m)$ time to compute $P_0$, $O(m)$ time to compute all $P_l$ and hence, to find $l_0$, and $O(m)$ time to compute $\underline{E}$ – to the total of $O(m \cdot \log(m))$.

*Algorithm for computing $\overline{E}$.* Similarly, we can compute $\overline{E}$ in time

$$O(m \cdot \log(m)).$$

We first sort the values $\overline{E}_j$ in increasing order. Let us assume that the groups are already sorted in this order, i.e., that $\overline{E}_1 \leq \overline{E}_2 \leq \ldots \leq \overline{E}_m$.

For every $l$ from 0 to $k$, we then compute $P_l = \underline{p}_1 + \ldots + \underline{p}_l + \overline{p}_{l+1} + \ldots + \overline{p}_n$ as follows: we explicitly compute the sum $P_0$, and then consequently compute $P_{l+1}$ as $P_l - (\overline{p}_{l+1} - \underline{p}_{l+1})$. This sequence is decreasing. Then, we find the value $l_0$ for which $P_{l_0} \geq 1 \geq P_{l_0+1}$, and take $\overline{E} = \sum_{j=1}^{l_0-1} \underline{p}_j \cdot \overline{E}_j + p_{l_0} \cdot \overline{E}_{l_0} + \sum_{j=l_0+1}^{m} \overline{p}_j \cdot \overline{E}_j,$

where $p_{l_0} = 1 - \sum_{j=1}^{l_0-1} \underline{p}_j - \sum_{j=l_0+1}^{m} \overline{p}_j.$

*Derivation of an algorithm for computing $\underline{M}$.* First, we notice that the minimum is attained when $M_j$ are the smallest ($M_j = \underline{M}_j$) and $E_j$ are the largest ($E_j = \overline{E}_j$). So, the only problem is to find the optimal values of $p_j$.

Similarly to the case of $\underline{E}$, we add $\Delta p$ to $p_j$ and subtract $\Delta p$ from $p_k$. Since we started with the values at which the minimum is attained we must have $\Delta M \leq 0$, i.e., $\Delta \cdot [\underline{M}_j - \underline{M}_k - 2E \cdot (\overline{E}_j - \overline{E}_k)] \leq 0$. So, at most one value $p_j$ is strictly inside, and if $p_j = \underline{p}_j$ and $p_k = \overline{p}_k$, we must have

$$\underline{M}_j - \underline{M}_k - 2E \cdot (\overline{E}_j - \overline{E}_k) \leq 0,$$

i.e., $\underline{M}_j - 2E \cdot \overline{E}_j \leq \underline{M}_k - 2E \cdot \overline{E}_j.$

Once we know $E$, we can similarly sort and get the optimal $p_j$. The problem is that we do not know the value $E$, and for different values $E$, we have different orders. The solution to this problem comes from the fact that the above inequality is equivalent to comparing $2E$ with the ratio $\dfrac{\underline{M}_j - \underline{M}_k}{\overline{E}_j - \overline{E}_k}.$

Thus, if we compute all $n^2$ such ratios, sort them, then within each zone between the consequent values, the sorting will be the same. Thus, we arrive at the following algorithm.

*Algorithm for computing $\underline{M}$.* To compute $\underline{M}$, we first compute all the ratios $\dfrac{\underline{M}_j - \underline{M}_k}{\overline{E}_j - \overline{E}_k}$, sort them, and take $E$s between two consecutive values in this sorting.

For each such $E$, we sort the groups according to the value of the expression $\underline{M}_j - 2E \cdot \overline{E}_j$. In this sorting, we select the values

$$p_j = (\overline{p}_1, \ldots, \overline{p}_{l_0-1}, p_{l_0}, \underline{p}_{l_0+1}, \ldots, \underline{p}_m)$$

and pick $l_0$ in the same manner as when we computed $\underline{E}$. For the resulting $p_j$, we then compute $\underline{M} = \sum\limits_{j=1}^{m} p_j \cdot \underline{M}_j - \left( \sum\limits_{j=1}^{m} p_j \cdot \overline{E}_j \right)^2$.

*Computation time.* We need $O(m \cdot \log(m))$ steps for each of $m^2$ zones, to the (still polynomial) total time $O(m^3 \cdot \log(m)))$.

*Algorithm for computing $\overline{M}$.* A similar polynomial-time algorithm can be used to compute $\overline{M}$. We first compute all the ratios $\dfrac{\overline{M}_j - \overline{M}_k}{\underline{E}_j - \underline{E}_k}$, sort them, and take $E$s between two consecutive values in this sorting.

For each such $E$, we sort the groups according to the value of the expression $\overline{M}_j - 2E \cdot \underline{E}_j$. In this sorting, we select the values

$$p_j = (\underline{p}_1, \ldots, \underline{p}_{l_0-1}, p_{l_0}, \overline{p}_{l_0+1}, \ldots, \overline{p}_m)$$

and pick $l_0$ in the same manner as when we computed $\overline{E}$. For the resulting $p_j$, we then compute $\overline{M} = \sum\limits_{j=1}^{m} p_j \cdot \overline{M}_j - \left( \sum\limits_{j=1}^{m} p_j \cdot \underline{E}_j \right)^2$.


## Justifications of the Algorithm

*Proof of the algorithm's correctness.* The function $V_E$ is convex. Thus, its maximum $\overline{V}_E$ on the box $E_1 \times \ldots \times E_m$ is attained at one of the vertices of this box, i.e., at a vector $(E_1, \ldots, E_m)$ in which each value $E_j$ is equal to either $\underline{E}_j$ or to $\overline{E}_j$.

To justify our algorithm, we need to prove that this maximum is attained at one of the vectors $E^{(k)}$ in which all the lower bounds $\underline{E}_j$ precede all the upper bounds $\overline{E}_j$. We will prove this by reduction to a contradiction. Indeed, let us assume that the maximum is attained at a vector in which one of the lower bounds follows one of the upper bounds. In each such vector, let $i$ be the largest upper bound index followed by the lower bound; then, in the optimal vector $(E_1, \ldots, E_m)$, we have $E_i = \overline{E}_i$ and $E_{i+1} = \underline{E}_{i+1}$.

Since the maximum is attained for $E_i = \overline{E}_i$, replacing it with $\underline{E}_i = E_i - 2\Delta_i$ will either decrease the value of $V_E$ or keep it unchanged. Let us describe how $V_E$ changes under this replacement. Since $V_E$ is defined in terms of $M$ and $E$, let us first describe how $E$ and $M$ change under this replacement. In the sum for $M$, we replace $(\overline{E}_i)^2$ with

$$(\underline{E}_i)^2 = (\overline{E}_i - 2\Delta_i)^2 = (\overline{E}_i)^2 - 4 \cdot \Delta_i \cdot \overline{E}_i + 4 \cdot \Delta_i^2.$$

Thus, the value $M$ changes into $M + \Delta_i M$, where

$$\Delta_i M = -4 \cdot p_i \cdot \Delta_i \cdot \overline{E}_i + 4 \cdot p_i \cdot \Delta_i^2.$$

The population mean $E$ changes into $E + \Delta_i E$, where $\Delta_i E = -2 \cdot p_i \cdot \Delta_i$. Thus, the value $E^2$ changes into $(E + \Delta_i E)^2 = E^2 + \Delta_i(E^2)$, where

$$\Delta_i(E^2) = 2 \cdot E \cdot \Delta_i E + (\Delta_i E)^2 = -4 \cdot p_i \cdot E \cdot \Delta_i + 4 \cdot p_i^2 \cdot \Delta_i^2.$$

So, the variance $V$ changes into $V + \Delta_i V$, where

$$\Delta_i V = \Delta_i M - \Delta_i(E^2) = -4 \cdot p_i \cdot \Delta_i \cdot \overline{E}_i + 4 \cdot p_i \cdot \Delta_i^2 + 4 \cdot p_i \cdot E \cdot \Delta_i - 4 \cdot p_i^2 \cdot \Delta_i^2 =$$

$$4 \cdot p_i \cdot \Delta_i \cdot (-\overline{E}_i + \Delta_i + E - p_i \cdot \Delta_i).$$

By definition, $\overline{E}_i = \widetilde{E}_i + \Delta_i$, hence $-\overline{E}_i + \Delta_i = -\widetilde{E}_i$. Thus, we conclude that $\Delta_i V = 4 \cdot p_i \cdot \Delta_i \cdot (-\widetilde{E}_i + E - p_i \cdot \Delta_i)$. So, the fact that $\Delta_i V \leq 0$ means that $E \leq \widetilde{E}_i + p_i \cdot \Delta_i = E_i^+$. Similarly, since the maximum of $V_E$ is attained for $E_{i+1} = \underline{E}_{i+1}$, replacing it with $\overline{E}_{i+1} = \underline{E}_{i+1} + 2\Delta_{i+1}$ will either decrease the value of $V_E$ or keep it unchanged. In the sum for $M$, we replace $(\underline{E}_{i+1})^2$ with

$$(\overline{E}_{i+1})^2 = (\underline{E}_{i+1} + 2\Delta_{i+1})^2 = (\underline{E}_{i+1})^2 + 4 \cdot \Delta_{i+1} \cdot \underline{E}_{i+1} + 4 \cdot \Delta_{i+1}^2.$$

Thus, the value $M$ changes into $M + \Delta_{i+1} M$, where

$$\Delta_{i+1} M = 4 \cdot p_{i+1} \cdot \Delta_{i+1} \cdot \underline{E}_{i+1} + 4 \cdot p_{i+1} \cdot \Delta_{i+1}^2.$$

The population mean $E$ changes into $E + \Delta_{i+1} E$, where $\Delta_{i+1} E = 2 \cdot p_{i+1} \cdot \Delta_{i+1}$. Thus, the value $E^2$ changes into $E^2 + \Delta_{i+1}(E^2)$, where

$$\Delta_{i+1}(E^2) = 2 \cdot E \cdot \Delta_{i+1} E + (\Delta_{i+1} E)^2 = 4 \cdot p_{i+1} \cdot E \cdot \Delta_{i+1} + 4 \cdot p_{i+1}^2 \cdot \Delta_{i+1}^2.$$

So, the term $V_E$ changes into $V_E + \Delta_{i+1} V$, where

$$\Delta_{i+1} V = \Delta_{i+1} M - \Delta_{i+1}(E^2) =$$

$$4 \cdot p_{i+1} \cdot \Delta_{i+1} \cdot \underline{E}_{i+1} + 4 \cdot p_{i+1} \cdot \Delta_{i+1}^2 - 4 \cdot p_{i+1} \cdot E \cdot \Delta_{i+1} - 4 \cdot p_{i+1}^2 \cdot \Delta_{i+1}^2 =$$

$$4 \cdot p_{i+1} \cdot \Delta_{i+1} \cdot (\underline{E}_{i+1} + \Delta_{i+1} - E - p_{i+1} \cdot \Delta_{i+1}).$$

By definition, $\underline{E}_{i+1} = \widetilde{E}_{i+1} - \Delta_{i+1}$, hence $\underline{E}_{i+1} + \Delta_{i+1} = \widetilde{E}_{i+1}$. Thus, we conclude that

$$\Delta_{i+1} V = 4 \cdot p_{i+1} \cdot (\widetilde{E}_{i+1} - E - p_{i+1} \cdot \Delta_{i+1}).$$

Since $V_E$ attains maximum at $(E_1, \ldots, E_i, E_{i+1}, \ldots, E_m)$, we have $\Delta_{i+1} V \leq 0$, hence $E \geq \widetilde{E}_{i+1} - p_{i+1} \cdot \Delta_{i+1} = E_{i+1}^-$.

We can also change both $E_i$ and $E_{i+1}$ at the same time. In this case, from the fact that $V_E$ attains maximum, we conclude that $\Delta V_E \leq 0$.

Here, the change $\Delta M$ in $M$ is simply the sum of the changes coming from $E_i$ and $E_{i+1}$: $\Delta M = \Delta_i M + \Delta_{i+1} M$, and the change in $E$ is also the sum of the corresponding changes: $\Delta E = \Delta_i E + \Delta_{i+1} E$. So, for $\Delta V = \Delta M - \Delta(E^2)$, we get

$$\Delta V = \Delta_i M + \Delta_{i+1} M - 2 \cdot E \cdot \Delta_i E - 2 \cdot E \cdot \Delta_{i+1} E - (\Delta_i E)^2 - (\Delta_{i+1} E)^2 -$$

$$2 \cdot \Delta_i E \cdot \Delta_{i+1} E.$$

Hence,

$$\Delta V = (\Delta_i M - 2 \cdot E_i \cdot \Delta_i E - (\Delta_i E)^2) + (\Delta_{i+1} M - 2 \cdot E_{i+1} \cdot \Delta_{i+1} E - (\Delta_{i+1} E)^2) -$$

$$2 \cdot \Delta E_i \cdot \Delta E_{i+1},$$

i.e., $\Delta V = \Delta_i V + \Delta_{i+1} V - 2 \cdot \Delta_i E \cdot \Delta_{i+1} E$.

We already have expressions for $\Delta_i V$, $\Delta_{i+1} V$, $\Delta_i E$, and $\Delta_{i+1} E$, and we already know that $E_{i+1}^- \leq E \leq E_i^+$. Thus, we have $D(E) \leq 0$ for some $E \in [E_{i+1}^-, E_i^+]$, where

$$D(E) \overset{\text{def}}{=} 4 \cdot p_i \cdot \Delta_i \cdot (-E_i^+ + E) + 4 \cdot p_{i+1} \cdot \Delta_{i+1} \cdot (E_{i+1}^- - E) + 8 \cdot p_i \cdot \Delta_i \cdot p_{i+1} \cdot \Delta_{i+1}.$$

Since the narrowed intervals are not subsets of each other, we can sort them in lexicographic order; in which order, midpoints are sorted, left endpoints are sorted, and right endpoints are sorted, hence $E_i^- \leq E_{i+1}^-$ and $E_i^+ \leq E_{i+1}^+$.

For $E = E_{i+1}^-$, we get

$$D(E_{i+1}^-) = 4 \cdot p_i \cdot \Delta_i \cdot (-E_i^+ + E_{i+1}^-) + 8 \cdot p_i \cdot \Delta_i \cdot p_{i+1} \cdot \Delta_{i+1} =$$

$$4 \cdot p_i \cdot \Delta_i \cdot (-E_i^+ + E_{i+1}^- + 2 \cdot p_{i+1} \cdot \Delta_{i+1}).$$

By definition, $E_{i+1}^- = E_{i+1} - p_{i+1} \cdot \Delta_{i+1}$, hence $E_{i+1}^- + 2 \cdot p_{i+1} \cdot \Delta_{i+1} = E_{i+1}^+$, and $D(E_{i+1}^-) = 4 \cdot p_i \cdot \Delta_i \cdot (E_{i+1}^+ - E_i^+) \geq 0$. Similarly,

$$D(E_i^+) = 4 \cdot p_{i+1} \cdot \Delta_{i+1} \cdot (E_{i+1}^- - E_i^+) \geq 0.$$

The only possibility for both values to be 0 is when interval coincide; in this case, we can easily swap them. In all other cases, all intermediate values $D(E)$ are positive, which contradicts to our conclusion that $D(E) \leq 0$. The statement is proven.

# Computing Outlier Thresholds under Interval Uncertainty

In many application areas, it is important to detect outliers. The traditional engineering approach to outlier detection is that we start with some "normal" values $x_1, \ldots, x_n$, compute the sample average $E$, the sample standard variation $\sigma$, and then mark a value $x$ as an outlier if $x$ is outside the $k_0$-sigma interval $[E - k_0 \cdot \sigma, E + k_0 \cdot \sigma]$ (for some pre-selected parameter $k_0$). In real life, we often have only interval ranges $[\underline{x}_i, \overline{x}_i]$ for the normal values $x_1, \ldots, x_n$. In this case, we only have intervals of possible values for the "outlier threshold" – bounds $E - k_0 \cdot \sigma$ and $E + k_0 \cdot \sigma$. We can therefore identify outliers as values that are outside all $k_0$-sigma intervals.

Once we identify a value as an outlier for a fixed $k_0$, it is also desirable to find out to what degree this value is an outlier, i.e., what is the largest value $k_0$ for which this value is an outlier.

In this chapter, we analyze the problem of computing outlier thresholds under interval uncertainty. The main results of this chapter first appeared in [80, 186, 187, 191].

## Formulation of the Problem

*Outlier detection is important.* In many application areas, it is important to detect *outliers*, i.e., unusual, abnormal values. In medicine, unusual values may indicate disease (see, e.g., [164, 348, 349]); in geophysics, abnormal values may indicate a mineral deposit or an erroneous measurement result (see, e.g., [121, 214, 295, 344]); in structural integrity testing, abnormal values may indicate faults in a structure (see, e.g., [95, 123, 164, 271, 272, 348, 349, 350]), etc.

The traditional engineering approach to outlier detection (see, e.g., [88, 283, 337]) is as follows:

- first, we collect measurement results $x_1, \ldots, x_n$ corresponding to normal situations;

H.T. Nguyen et al.: Computing Stat. under Interval & Fuzzy Uncertain., SCI 393, pp. 129–151.
 

- then, we compute the sample average $E = \dfrac{x_1 + \ldots + x_n}{n}$ of these normal values and the (sample) standard deviation $\sigma = \sqrt{V}$, where $V = \dfrac{(x_1 - E)^2 + \ldots + (x_n - E)^2}{n}$;

- finally, a new measurement result $x$ is classified as an outlier if it is outside the interval $[L, U]$ (i.e., if either $x < L$ or $x > U$), where $L \stackrel{\text{def}}{=} E - k_0 \cdot \sigma$, $U \stackrel{\text{def}}{=} E + k_0 \cdot \sigma$, and $k_0 > 1$ is some pre-selected value (most frequently, $k_0 = 2$, 3, or 6).

*Outlier detection under interval uncertainty.* As we have mentioned in Part I of this book, in some practical situations, we only have intervals $\boldsymbol{x}_i = [\underline{x}_i, \overline{x}_i]$ of possible values of $x_i$. For different values $x_i \in \boldsymbol{x}_i$, we get different bounds $L$ and $U$. Possible values of $L$ form an interval – we will denote it by $\boldsymbol{L} \stackrel{\text{def}}{=} [\underline{L}, \overline{L}]$; possible values of $U$ form an interval $\boldsymbol{U} = [\underline{U}, \overline{U}]$.

How do we now detect outliers? There are two possible approaches to this question: we can detect *possible* outliers and we can detect *guaranteed* outliers:

- a value $x$ is a possible outlier if it is located outside one of the possible $k_0$-sigma intervals $[L, U]$ (but is may be inside some other possible interval $[L, U]$);
- a value $x$ is a guaranteed outlier if it is located outside all possible $k_0$-sigma intervals $[L, U]$.

Which approach is more reasonable depends on a possible situation:

- if our main objective is not to miss an outlier, e.g., in structural integrity tests, when we do not want to risk launching a spaceship with a faulty part, it is reasonable to look for possible outliers;
- if we want to make sure that the value $x$ is an outlier, e.g., if we are planning a surgery and we want to make sure that there is a micro-calcification before we start cutting the patient, then we would rather look for guaranteed outliers.

The two approaches can be described in terms of the endpoints of the intervals $\boldsymbol{L}$ and $\boldsymbol{U}$:

A value $x$ guaranteed to be normal – i.e., it is not a possible outlier – if $x$ belongs to the *intersection* of all possible intervals $[L, U]$; the intersection corresponds to the case when $L$ is the largest and $U$ is the smallest, i.e., this intersection is the interval $[\overline{L}, \underline{U}]$. So, if $x > \underline{U}$ or $x < \overline{L}$, then $x$ is a possible outlier, else it is guaranteed to be a normal value.

If a value $x$ is inside *one* of the possible intervals $[L, U]$, then it can still be normal; the only case when we are sure that the value $x$ is an outlier is when $x$ is outside *all* possible intervals $[L, U]$, i.e., is the value $x$ does not belong to the *union* of all possible intervals $[L, U]$ of normal values; this union is equal to the interval $[\underline{L}, \overline{U}]$. So, if $x > \overline{U}$ or $x < \underline{L}$, then $x$ is a guaranteed outlier, else it can be a normal value.

In real life, the situation may be slightly more complicated because, as we have mentioned, measurements often come with interval inaccuracy; so, instead of the exact value $x$ of the measured quantity, we get an interval $\mathbf{x} = [\underline{x}, \overline{x}]$ of possible values of this quantity.

In this case, we have a slightly more complex criterion for outlier detection:

- the actual (unknown) value of the measured quantity is a possible outlier if some value $x$ from the interval $[\underline{x}, \overline{x}]$ is a possible outlier, i.e., is outside the intersection $[\overline{L}, \underline{U}]$; thus, the value is a possible outlier if one of the two inequalities hold: $\underline{x} < \overline{L}$ or $\underline{U} < \overline{x}$.
- the actual (unknown) value of the measured quantity is guaranteed to be an outlier if all possible values $x$ from the interval $[\underline{x}, \overline{x}]$ are guaranteed to be outliers (i.e., are outside the union $[\underline{L}, \overline{U}]$); thus, the value is a guaranteed outlier if one of the two inequalities hold: $\overline{x} < \underline{L}$ or $\overline{U} < \underline{x}$.

Thus:

- to detect possible outliers, we must be able to compute the values $\overline{L}$ and $\underline{U}$;
- to detect guaranteed outliers, we must be able to compute the values $\underline{L}$ and $\overline{U}$.

In this chapter, we consider the problem of computing these bounds.

Once we identify a value as an outlier for a fixed $k_0$, it is also desirable to find out to what degree this value is an outlier, i.e., what is the largest value $k_0$ for which this value is an outlier. In this chapter, we analyze the algorithmic solvability and computational complexity of this problem as well.

*Computing degree of outlier-ness.* Once we identify a value $x$ as an outlier for a fixed $k_0$, it is also desirable to find out to what degree this value is an outlier, i.e., what is the largest value $k_0$ for which this value $x$ is outside the corresponding $k_0$-sigma interval $[E - k_0 \cdot \sigma, E + k_0 \cdot \sigma]$.

If we know the exact values of the measurement results $x_1, \ldots, x_n$, then we can compute the exact values of $E$ and $\sigma$ and thus, determine this "degree of outlier-ness" as the ratio $r \overset{\text{def}}{=} |x - E|/\sigma$. If we only know the intervals $\mathbf{x}_i$ of possible values of $x_i$, then different values $x_i \in \mathbf{x}_i$ may lead to different values of this ratio. In this situation, it is desirable to know the *interval* of possible values of $r$.

*It is not enough to compute the ranges of $E$ and $\sigma$.* To detect outliers under interval uncertainty, we must be able to compute the range $\mathbf{L} = [\underline{L}, \overline{L}]$ of possible values of $L = E - k_0 \cdot \sigma$ and the range $\mathbf{U} = [\underline{U}, \overline{U}]$ of possible values of $U = E + k_0 \cdot \sigma$.

In the previous chapters, we have shown how to compute the intervals $\mathbf{E} = [\underline{E}, \overline{E}]$ and $[\underline{\sigma}, \overline{\sigma}]$ of possible values for $E$ and $\sigma$. In principle, we can use the general ideas of interval computations to combine these intervals and conclude, e.g., that $U$ always belongs to the interval $\mathbf{E} + k_0 \cdot [\underline{\sigma}, \overline{\sigma}]$. However, as often happens in interval computations, the resulting interval for $L$ is *wider* than the actual range – wider because the values $E$ and $\sigma$ are computed based on the same inputs $x_1, \ldots, x_n$ and cannot, therefore, change independently.

As an example that we may lose precision by combining intervals for $E$ and $\sigma$, let us consider the case when $\boldsymbol{x}_1 = \boldsymbol{x}_2 = [0, 1]$ and $k_0 = 2$. In this case, the range $\boldsymbol{E}$ of $E = (x_1 + x_2)/2$ is equal to $[0, 1]$, where the largest value 1 is attained only if $x_1 = x_2 = 1$. For the variance, we have $V = ((x_1 - E)^2 + (x_2 - E)^2)/2 = (x_1 - x_2)^2/4$; so, the range $\boldsymbol{V}$ of $V$ is $[0, 0.25]$ and, correspondingly, the range for $\sigma = \sqrt{V}$ is $[0, 0.5]$. The largest value $\sigma = 0.5$ is only attained in two cases: when $x_1 = 0$ and $x_2 = 1$, and when $x_1 = 1$ and $x_2 = 0$. When we simply combine the intervals, we conclude that $U \in [0, 1] + 2 \cdot [0, 0.5] = [0, 2]$. However, it is easy to see that $U$ cannot be equal to 2:

- The only way for $U$ to be equal to 2 is when both $E$ and $\sigma$ attain their largest values: $E = 1$ and $\sigma = 0.5$.
- However, the only pair on which the mean $E$ attains its largest value 1 is $x_1 = x_2 = 1$, and for this pair, $\sigma = 0$.

So, the actual range of $U$ must be narrower than the result $[0, 2]$ of combining intervals for $E$ and $\sigma$.

We mark a value $x$ as an outlier if it is outside the interval $[L, U]$. Thus, if, instead of the actual ranges for $L$ and $U$, we use wider intervals, we may miss some outliers. It is therefore important to compute the *exact* ranges for $L$ and $U$. In this chapter, we show how to compute these exact ranges.

## Detecting Possible Outliers: Algorithms

To find possible outliers, we must know the values $\underline{U}$ and $\overline{L}$. In this section, we design *feasible* algorithms for computing the exact lower bound $\underline{U}$ of the function $U$ and the exact upper bound $\overline{L}$ of the function $L$. Specifically, our algorithms take $O(n \cdot \log(n))$ computational steps (arithmetic operations or comparisons) for $n$ interval data points $\boldsymbol{x}_i = [\underline{x}_i, \overline{x}_i]$.

The algorithms $\underline{\mathcal{A}}_U$ for computing $\underline{U}$ and $\overline{\mathcal{A}}_L$ for computing $\overline{L}$ are as follows:

- In both algorithms, first, we sort all $2n$ values $\underline{x}_i$, $\overline{x}_i$ into a sequence $x_{(1)} \leq x_{(2)} \leq \ldots \leq x_{(2n)}$; take $x_{(0)} = -\infty$ and $x_{(2n+1)} = +\infty$. Thus, the real line is divided into $2n + 1$ zones $(x_{(0)}, x_{(1)}], [x_{(1)}, x_{(2)}], \ldots, [x_{(2n-1)}, x_{(2n)}], [x_{(2n)}, x_{(2n+1)})$.
- For each of these zones $[x_{(k)}, x_{(k+1)}]$, $k = 0, 1, \ldots, 2n$, we compute the values

$$e_k \overset{\text{def}}{=} \sum_{i:\underline{x}_i \geq x_{(k+1)}} \underline{x}_i + \sum_{j:\overline{x}_j \leq x_{(k)}} \overline{x}_j,$$

$$m_k \overset{\text{def}}{=} \sum_{i:\underline{x}_i \geq x_{(k+1)}} (\underline{x}_i)^2 + \sum_{j:\overline{x}_j \leq x_{(k)}} (\overline{x}_j)^2,$$

and $n_k$ = the total number of such $i$'s and $j$'s. Then, we solve the quadratic equation

$$A_k - B_k \cdot \mu + C_k \cdot \mu^2 = 0,$$

where

$$A_k \stackrel{\text{def}}{=} e_k^2 \cdot (1 + \alpha^2) - \alpha^2 \cdot m_k \cdot n; \quad \alpha \stackrel{\text{def}}{=} 1/k_0,$$

$$B_k \stackrel{\text{def}}{=} 2 \cdot e_k \cdot \left((1 + \alpha^2) \cdot n_k - \alpha^2 \cdot n\right); \quad C_k \stackrel{\text{def}}{=} n_k \cdot \left((1 + \alpha^2) \cdot n_k - \alpha^2 \cdot n\right).$$

For computing $\underline{U}$, we select only those solutions for which $\mu \cdot n_k \leq e_k$ and $\mu \in [x_{(k)}, x_{(k+1)}]$; for computing $\overline{L}$, we select only those solutions for which $\mu \cdot n_k \geq e_k$ and $\mu \in [x_{(k)}, x_{(k+1)}]$. For each selected solution, we compute the values of

$$E_k = \frac{e_k}{n} + \frac{n - n_k}{n} \cdot \mu, \quad M_k = \frac{m_k}{n} + \frac{n - n_k}{n} \cdot \mu^2,$$

$$U_k = E_k + k_0 \cdot \sqrt{M_k - (E_k)^2} \text{ or } L_k = E_k - k_0 \cdot \sqrt{M_k - (E_k)^2}.$$

- Finally, if we are computing $\underline{U}$, we return the smallest of the values $U_k$; if we are computing $\overline{L}$, we return the smallest of the values $L_k$.

**Theorem 18.1.** *The algorithms $\underline{\mathcal{A}}_U$ and $\overline{\mathcal{A}}_L$ always compute $\underline{U}$ and $\overline{L}$ in time $O(n \cdot \log(n))$.*

## In General, Detecting Guaranteed Outliers Is NP-Hard

As we have mentioned in the beginning of this chapter, to be able to detect guaranteed outliers, we must be able to compute the values $\underline{L}$ and $\overline{U}$. In general, this is an NP-hard problem:

**Theorem 18.2.** *For every $k_0 > 1$, computing the upper endpoint $\overline{U}$ of the interval $[\underline{U}, \overline{U}]$ of possible values of $U = E + k_0 \cdot \sigma$ is NP-hard.*

**Theorem 18.3.** *For every $k_0 > 1$, computing the lower endpoint $\underline{L}$ of the interval $[\underline{L}, \overline{L}]$ of possible values of $L = E - k_0 \cdot \sigma$ is NP-hard.*

*Comment.* The proofs of Theorems 18.2 and 18.3 show that the decision problems related to the computation of $\underline{L}$ and $\overline{U}$ are NP-complete. Therefore, NP-hardness of the computational problems does not mean that the problems are located somewhere higher in the polynomial hierarchy.

## Detecting Guaranteed Outliers: Results and Algorithms

How can we actually compute these values? First, we will show that if $1 + (1/k_0)^2 < n$ (which is true, e.g., if $k_0 > 1$ and $n \geq 2$), then the maximum of $U$ (correspondingly, the minimum of $L$) is always attained at some combination of endpoints of the intervals $\boldsymbol{x}_i$; thus, in principle, to determine the values $\overline{U}$ and $\underline{L}$, it is sufficient to try all $2^n$ combinations of values $\underline{x}_i$ and $\overline{x}_i$:

**Theorem 18.4.** *If $1 + (1/k_0)^2 < n$, then the maximum of the function $U$ and the minimum of the function $L$ on the box $\boldsymbol{x}_1 \times \ldots \times \boldsymbol{x}_n$ are attained at its vertices, i.e., when for every $i$, either $x_i = \underline{x}_i$ or $x_i = \overline{x}_i$.*

NP-hard means, crudely speaking, that there are no general ways for solving all particular cases of this problem (i.e., computing $\overline{U}$ and $\underline{L}$) in reasonable time.

However, we show that there are algorithms for computing $\overline{U}$ and $\underline{L}$ for many reasonable situations. Namely, we propose efficient algorithms that compute $\overline{U}$ and $\underline{L}$ for the case when all the interval midpoints ("measured values") $\widetilde{x}_i \overset{\text{def}}{=} (\underline{x}_i + \overline{x}_i)/2$ are definitely different from each other, in the sense that the "narrowed" intervals

$$\left[ \widetilde{x}_i - \frac{1 + \alpha^2}{n} \cdot \Delta_i, \widetilde{x}_i + \frac{1 + \alpha^2}{n} \cdot \Delta_i \right]$$

– where $\alpha = 1/k_0$ and $\Delta_i \overset{\text{def}}{=} (\underline{x}_i - \overline{x}_i)/2$ is the interval's half-width – do not intersect with each other.

The algorithms $\overline{\mathcal{A}}_U$ and $\underline{\mathcal{A}}_L$ are as follows:

- In both algorithms, first, we sort all $2n$ endpoints of the narrowed intervals $\widetilde{x}_i - \frac{1 + \alpha^2}{n} \cdot \Delta_i$ and $\widetilde{x}_i + \frac{1 + \alpha^2}{n} \cdot \Delta_i$ into a sequence $x_{(1)} \leq x_{(2)} \leq \ldots \leq x_{(2n)}$. This enables us to divide the real line into $2n + 1$ zones $[x_{(i)}, x_{(i+1)}]$, where we denoted $x_{(0)} \overset{\text{def}}{=} -\infty$ and $x_{(2n+1)} \overset{\text{def}}{=} +\infty$.
- For each of zones $[x_{(i)}, x_{(i+1)}]$, we do the following: for each $j$ from 1 to $n$, we pick the following value of $x_j$:
  - if $x_{(i+1)} < \widetilde{x}_j - \frac{1 + \alpha^2}{n} \cdot \Delta_j$, then we pick $x_j = \overline{x}_j$;
  - if $x_{(i+1)} > \widetilde{x}_j + \frac{1 + \alpha^2}{n} \cdot \Delta_j$, then we pick $x_j = \underline{x}_j$;
  - for all other $j$, we consider both possible values $x_j = \overline{x}_j$ and $x_j = \underline{x}_j$.
  As a result, we get one or several sequences of $x_j$ for each zone.
- To compute $\overline{U}$, for each of the sequences $x_j$, we check whether, for the selected values $x_1, \ldots, x_n$, the value of $E - \alpha \cdot \sigma$ is indeed within the corresponding zone, and if it is, compute the value $U = E + k_0 \cdot \sigma$. Finally, we return the largest of the computed values $U$ as $\overline{U}$.
- To compute $\underline{L}$, for each of the sequences $x_j$, we check whether, for the selected values $x_1, \ldots, x_n$, the value of $E + \alpha \cdot \sigma$ is indeed within the corresponding zone, and if it is, compute the value $L = E - k_0 \cdot \sigma$. Finally, we return the smallest of the computed values $L$ as $\underline{L}$.

**Theorem 18.5.** *Let* $1/n + 1/k_0^2 < 1$. *The algorithms* $\overline{\mathcal{A}}_U$ *and* $\underline{\mathcal{A}}_L$ *compute* $\overline{U}$ *and* $\underline{L}$ *in time* $O(n \cdot \log(n))$ *for all the cases in which the "narrowed" intervals do not intersect with each other.*

These algorithms also work when, for some fixed $c_0$, no more than $c_0$ "narrowed" intervals can have a common point:

**Theorem 18.6.** *Let* $1 + (1/k_0)^2 < n$. *For every positive integer* $c_0$, *the algorithms* $\overline{\mathcal{A}}_U$ *and* $\underline{\mathcal{A}}_L$ *compute* $\overline{U}$ *and* $\underline{L}$ *in time* $O(n \cdot \log(n))$ *for all the cases in which no more than* $c_0$ *"narrowed" intervals can have a common point.*

For each zone, we can determine the values of all optimal $x_i$ – except for the case when the zone intersects with the corresponding narrowed interval. Since we consider the case when no more than $c_0$ narrowed intervals can have a common point, we have no more than $c_0$ undecided values $x_i$. Trying all possible combinations of lower and upper endpoints for $c_0$ different values $i$ takes $2^{c_0}$ steps. Thus, the corresponding computation times are feasible (polynomial) in $n$ but grow exponentially with $c_0$. So, when $c_0$ grows, this algorithm takes more and more computation time. It is worth mentioning that the examples on which we prove NP-hardness (see proof of Theorem 18.2) correspond to the case when $n/2$ out of $n$ narrowed intervals have a common point.

Another possible generalization is to a more general case of the no-subset property, when no two narrowed intervals are proper subsets of one another – in the sense that one of them is a subset of the interior of the second one. This is a more general case because if they do not intersect, them, of course, they cannot be proper subsets of one another.

**Theorem 18.7.** *There exist algorithms that compute* $\overline{U}$ *and* $\underline{L}$ *in time* $O(n \cdot \log(n))$ *for all the cases in which no two "narrowed" intervals are proper subsets of one another.*

Let us first describe the algorithms themselves. Without losing generality, we can describe an algorithm for computing $\overline{U}$.

- First, we sort of the values $\widetilde{x}_i$ into an increasing sequence. Without losing generality, we can assume that $\widetilde{x}_1 \leq \widetilde{x}_2 \leq \ldots \leq \widetilde{x}_n$.
- Then, for every $k$ from 0 to $n$, we compute the value $V^{(k)} = M^{(k)} - (E^{(k)})^2$ of the population variance $V$ for the vector $x^{(k)} = (\underline{x}_1, \ldots, \underline{x}_k, \overline{x}_{k+1}, \ldots, \overline{x}_n)$, and we compute $U^{(k)} = E^{(k)} + k_0 \cdot \sqrt{V^{(k)}}$.
- Finally, we compute $\overline{U}$ as the largest of $n+1$ values $U^{(0)}, \ldots, U^{(n)}$.

To compute the values $V^{(k)}$, first, we explicitly compute $M^{(0)}$, $E^{(0)}$, and $V^{(0)} = M^{(0)} - (E^{(0)})^2$. Once we know the values $M^{(k)}$ and $E^{(k)}$, we can compute

$$M^{(k+1)} = M^{(k)} + \frac{1}{n} \cdot (\underline{x}_{k+1})^2 - \frac{1}{n} \cdot (\overline{x}_{k+1})^2$$

and $E^{(k+1)} = E^{(k)} + \dfrac{1}{n} \cdot \underline{x}_{k+1} - \dfrac{1}{n} \cdot \overline{x}_{k+1}.$

*Comment.* It is worth mentioning that if the measurement results $\widetilde{x}_i$ are already sorted, then we only need linear time to compute $\overline{U}$.

## Computing Degree of Outlier-Ness: Algorithms

*Simplification of the problem.* In order to compute the interval of possible values of the degree of outlier-ness, let us first reduce the problem of computing this interval to a simpler problem. This reduction will be done in three steps.

First, it turns out that the value of $r$ does not change if, instead of the original variables $x_i$ with values from intervals $\boldsymbol{x}_i$, we consider new variables $x_i' \stackrel{\text{def}}{=} x_i - x$ and a new value $x' = 0$. Indeed, in this case, $E' = E - x$ hence $E' - x' = E - x$, and the standard deviation $\sigma$ does not change if we simply shift all the values $x_i$. Thus, without losing generality, we can assume that $x = 0$, and we are therefore interested in the ratio $|E|/\sigma$.

Second, the lower bound of the ratio $r$ is attained when the reverse ratio $1/r = \sigma/|E|$ is the largest, and vice versa. Thus, to find the interval of possible values for $|E|/\sigma$, it is necessary and sufficient to find the interval of possible values of $\sigma/|E|$. Computing this interval is, in its turn, equivalent to computing the interval for the square $V/E^2$ of the reverse ratio $1/r$.

Finally, since $V = M - E^2$, where $M \stackrel{\text{def}}{=} \dfrac{x_1^2 + \ldots + x_n^2}{n}$ is the second moment, we have $V/E^2 = M/E^2 - 1$, so computing the sharp bounds for $V/E^2$ is equivalent to computing the sharp bounds for the ratio $R \stackrel{\text{def}}{=} M/E^2$.

In this section, we will describe how to compute the sharp bounds $\underline{R}$ and $\overline{R}$ for the ratio $R$; based on these sharp bounds, we can compute the desired sharp bounds on $k_0$.

*Computing $\underline{R}$: algorithm.* The algorithm $\underline{\mathcal{A}}_R$ for computing $\underline{R}$ is as follows. If all the original intervals have a common point, then we take $\underline{R} \stackrel{\text{def}}{=} 1$. Otherwise, we do the following:

- First, we sort all $2n$ values $\underline{x}_i$, $\overline{x}_i$ into a sequence $x_{(1)} \leq x_{(2)} \leq \ldots \leq x_{(2n)}$; take $x_{(0)} = -\infty$ and $x_{(2n+1)} = +\infty$. Thus, the real line is divided into $2n + 1$ zones $(x_{(0)}, x_{(1)}], [x_{(1)}, x_{(2)}], \ldots, [x_{(2n-1)}, x_{(2n)}], [x_{(2n)}, x_{(2n+1)})$.
- For each of these zones $[x_{(k)}, x_{(k+1)}]$, $k = 0, 1, \ldots, 2n$, we compute the values

$$e_k \stackrel{\text{def}}{=} \sum_{i:\underline{x}_i \geq x_{(k+1)}} \underline{x}_i + \sum_{j:\overline{x}_j \leq x_{(k)}} \overline{x}_j,$$

$$m_k \stackrel{\text{def}}{=} \sum_{i:\underline{x}_i \geq x_{(k+1)}} (\underline{x}_i)^2 + \sum_{j:\overline{x}_j \leq x_{(k)}} (\overline{x}_j)^2,$$

and $n_k$ = the total number of such $i$'s and $j$'s. Then, we find $\lambda_k \stackrel{\text{def}}{=} m_k/e_k$. If $\lambda_k \in [x_{(k)}, x_{(k+1)}]$, then we compute

$$E_k = \frac{e_k}{n} + \frac{n - n_k}{n} \cdot \lambda_k, \quad M_k = \frac{m_k}{n} + \frac{n - n_k}{n} \cdot \lambda_k^2,$$

and $R_k \stackrel{\text{def}}{=} M_k / E_k^2$.
- Finally, we return the smallest of the values $R_k$ as $\underline{R}$.

**Theorem 18.8.** *The algorithm $\underline{\mathcal{A}}_R$ always computes $\underline{R}$ in time $O(n \cdot \log(n))$.*

*Computing $\overline{R}$.* In principle, we can have $\overline{R} = +\infty$ – e.g., if $0 \in [\underline{E}, \overline{E}]$. If $0 \notin [\underline{E}, \overline{E}]$ – e.g., if $\underline{E} > 0$ – then we can guarantee that $\overline{R} < +\infty$. In this case, we can bound $\overline{R}$ by the ratio $\overline{M} / \underline{E}^2$.

When $\overline{R} < n$, the maximum $\overline{R}$ is always attained at the endpoints:

**Theorem 18.9.** *When $\overline{R} < n$, the maximum $\overline{R}$ of the function $R = M/E^2$ on the box $\boldsymbol{x}_1 \times \ldots \times \boldsymbol{x}_n$ is attained at one of its vertices, i.e., when for every $i$, either $x_i = \underline{x}_i$ or $x_i = \overline{x}_i$.*

In this case, we are able to efficiently compute $\overline{R}$ if the "narrowed" intervals $[x_i^-, x_i^+]$ have few intersections, where:

$$x_i^- \stackrel{\text{def}}{=} \frac{\widetilde{x}_i}{1 + \dfrac{\Delta_i}{\underline{E} \cdot n}}; \quad x_i^+ \stackrel{\text{def}}{=} \frac{\widetilde{x}_i}{1 - \dfrac{\Delta_i}{\underline{E} \cdot n}}, \tag{18.1}$$

and $\underline{E} \stackrel{\text{def}}{=} \dfrac{\underline{x}_1 + \ldots + \underline{x}_n}{n}$ where $\widetilde{x}_i \stackrel{\text{def}}{=} (\underline{x}_i + \overline{x}_i)/2$ and $\Delta_i \stackrel{\text{def}}{=} (\underline{x}_i - \overline{x}_i)/2$.

The corresponding algorithm $\overline{\mathcal{A}}_R$ is as follows:

- First, we sort all $2n$ values $\underline{x}_i$, $\overline{x}_i$ into a sequence $x_{(1)} \leq x_{(2)} \leq \ldots \leq x_{(2n)}$, take $x_{(0)} = -\infty$ and $x_{(2n+1)} = +\infty$, and thus divide the real line into $2n + 1$ zones $(x_{(0)}, x_{(1)}], [x_{(1)}, x_{(2)}], \ldots, [x_{(2n-1)}, x_{(2n)}], [x_{(2n)}, x_{(2n+1)})$.
- For each of these zones $[x_{(k)}, x_{(k+1)}]$, $k = 0, 1, \ldots, 2n$, and for each variable $x_i$, we take:
  - $x_i = \underline{x}_i$ if $x_i^+ \leq x_{(k)}$;
  - $x_i = \overline{x}_i$ if $x_i^- \geq x_{(k+1)}$;
  - both values $x_i = \underline{x}_i$ and $x_i = \overline{x}_i$ otherwise.

  For each of these combinations, we compute $E$, $M$, and $\lambda = M/E$, and check if $\lambda$ is within the zone; if it is, we compute $R_k = M/E^2$.

The largest of these computed values $R_k$ is the desired upper endpoint $\overline{R}$.

**Theorem 18.10.** *For every positive integer $c_0$, the algorithm $\overline{\mathcal{A}}_R$ computes $\overline{R}$ in time $O(n \cdot \log(n))$ for all the cases in which $\overline{R} < n$ and no more than $c_0$ "narrowed" intervals can have a common point.*

## Summary

In many application areas, it is important to detect outliers. Traditional engineering approach to outlier detection is that we start with some "normal" values $x_1, \ldots, x_n$, compute the sample average $E$, the sample standard variation $\sigma$, and then mark a value $x$ as an outlier if $x$ is outside the $k_0$-sigma interval $[E - k_0 \cdot \sigma, E + k_0 \cdot \sigma]$ (for some pre-selected parameter $k_0$).

In real life, we often have only interval ranges $\boldsymbol{x}_i = [\underline{x}_i, \overline{x}_i]$ for the normal values $x_1, \ldots, x_n$. For different values $x_i \in \boldsymbol{x}_i$, we get different values of $L \stackrel{\text{def}}{=} E - k_0 \cdot \sigma$ and $U \stackrel{\text{def}}{=} E + k_0 \cdot \sigma$ – and thus, different $k_0$-sigma intervals $[L, U]$. We can therefore identify *guaranteed* outliers as values that are outside *all* $k_0$-sigma intervals, and *possible* outliers as values that are outside *some* $k_0$-sigma intervals. To detect guaranteed and possible outliers, we must therefore be able to compute the *range* $\boldsymbol{L} = [\underline{L}, \overline{L}]$ of possible values of $L$ and the range $\boldsymbol{U} = [\underline{U}, \overline{U}]$ of possible values of $U$.

In the previous chapters, we have shown how to compute the intervals $\boldsymbol{E} = [\underline{E}, \overline{E}]$ and $[\underline{\sigma}, \overline{\sigma}]$ of possible values for $E$ and $\sigma$. In principle, we can combine these intervals and conclude, e.g., that $L$ always belongs to the interval $\boldsymbol{E} - k_0 \cdot [\underline{\sigma}, \overline{\sigma}]$. However, the resulting interval for $L$ is *wider* than the actual range – wider because the values $E$ and $\sigma$ are computed based on the same inputs $x_1, \ldots, x_n$ and are, therefore, not independent from each other.

If, instead of the actual ranges for $L$ and $U$, we use wider intervals, we may miss some outliers. It is therefore important to compute the *exact* ranges for $L$ and $U$.

In this chapter, we showed that computing these ranges is, in general, NP-hard, and we provided efficient algorithms that compute these ranges under reasonable conditions.

Once a value is identified as an outlier for a fixed $k_0$, we also show how to find out to what degree this value is an outlier, i.e., what is the largest value $k_0$ for which this value is an outlier.


## Proofs

*Proof of Theorem 18.1.* We will only prove the result for $\underline{U}$; for $\overline{L}$, the proof is practically identical.

Our proof is based on the fact that the minimum of a differentiable function of $x_i$ on an interval $[\underline{x}_i, \overline{x}_i]$ is attained either inside this interval or at one of the endpoints. If the minimum is attained inside, the derivative $\dfrac{\partial U}{\partial x_i}$ is equal to 0; if it is attained at $x_i = \underline{x}_i$, then $\dfrac{\partial U}{\partial x_i} \geq 0$; finally, if it is attained at $x_i = \overline{x}_i$, then $\dfrac{\partial U}{\partial x_i} \leq 0$. For our function,

$$\frac{\partial U}{\partial x_i} = \frac{1}{n} + k_0 \cdot \frac{x_i - E}{\sigma \cdot n};$$

thus, $\dfrac{\partial U}{\partial x_i} = 0$ if and only if $x_i = \mu \stackrel{\text{def}}{=} E - \alpha \cdot \sigma$; similarly, the non-positiveness and non-negativeness of the derivative can be described by comparing $x_i$ with $\mu$. Thus:

- either $x_i \in (\underline{x}_i, \overline{x}_i)$ and $x_i = \mu$,
- or $x_i = \underline{x}_i$ and $x_i = \underline{x}_i \geq \mu$,
- or $x_i = \overline{x}_i$ and $x_i = \overline{x}_i \leq \mu$.

Hence, if we know how the value $\mu$ is located with respect to all the intervals $[\underline{x}_i, \overline{x}_i]$, we can find the optimal values of $x_i$:

- if $\overline{x}_i \leq \mu$, then minimum cannot be attained inside or at the lower endpoint, so it is attained when $x_i = \overline{x}_i$;
- if $\mu \leq \underline{x}_i$, then, similarly, the minimum is attained when $x_i = \underline{x}_i$;
- if $\underline{x}_i < \mu < \overline{x}_i$, then the minimum is attained when $x_i = \mu$.

Hence, to find the minimum, we will analyze how the endpoints $\underline{x}_i$ and $\overline{x}_i$ divide the real line, and consider all the resulting zones.

Let the corresponding zone $[x_{(k)}, x_{(k+1)}]$ be fixed. For the $i$'s for which $\mu \notin (\underline{x}_i, \overline{x}_i)$, the values $x_i$ that correspond to the minimal sample variance are uniquely determined by the above formulas.

For the $i$'s for which $\mu \in (\underline{x}_i, \overline{x}_i)$, the selected value $x_i$ should be equal to the same value $\mu$. To determine this $\mu$, we will use the fact that, by definition, $\mu = E - \alpha \cdot \sigma$, where $E$ and $\sigma$ are computed by using the same value of $\mu$. This equation is equivalent to $E - \mu \geq 0$ and $\alpha^2 \cdot \sigma^2 = (\mu - E)^2$. Substituting the above values of $x_i$ into the formula for the mean $E$ and for the standard deviation $\sigma$, we get the quadratic equation for $\mu$ which is described in the algorithm. So, for each zone, we can uniquely determine the values $x_i$ that may correspond to a minimum of $U$.

For the actual minimum, the value $\mu$ is inside one of these zone, so the smallest of the values $U_k$ is indeed the desired minimum.

In this algorithm, sorting takes $O(n \cdot \log(n))$ steps (see, e.g., [73]). The rest of the algorithm takes linear time ($O(n)$) for the first zone, and then (similarly to algorithm for computing variance) each value $x_i$ is updated once when we go from one zone to another, so we need the total linear time. Thus, overall, we need time $O(n) + O(n \cdot \log(n)) = O(n \cdot \log(n))$. The theorem is proven.

*Proof of Theorem 18.2.* Since $U = E + k_0 \cdot \sigma = k_0 \cdot J$, where $J \stackrel{\text{def}}{=} \sigma + \alpha \cdot E$ and $\alpha = 1/k_0$, we have $\overline{U} = k_0 \cdot \overline{J}$, where $\overline{J}$ is the upper endpoint of the interval of possible values of $J$. Thus, to prove that computing $\overline{U}$ is NP-hard, it is sufficient to prove that computing $\overline{J}$ is NP-hard.

To prove that the problem of computing $\overline{J}$ is NP-hard, we will prove that the known NP-hard *subset* problem $\mathcal{P}_0$ can be reduced to it in polynomial time. In the subset problem, given $m$ positive integers $s_1, \ldots, s_m$, we must check whether there exist signs $\eta_i \in \{-1, +1\}$ for which the signed sum $\displaystyle\sum_{i=1}^{m} \eta_i \cdot s_i$ equals 0.

We will show that this problem can be reduced to the problem of computing $\overline{J}$ in polynomial time, i.e., that to every instance $(s_1, \ldots, s_m)$ of the problem $\mathcal{P}_0$, we can put into correspondence such an instance of the $\overline{J}$-computing problem that based on its solution, we can easily check whether the desired signs exist.

For that, we compute three auxiliary values

$$S \stackrel{\text{def}}{=} \frac{1}{m} \cdot \sum_{i=1}^{m} s_i^2; \qquad N \stackrel{\text{def}}{=} \alpha \cdot \sqrt{\frac{2S}{1 - \alpha^2}}; \qquad J_0 \stackrel{\text{def}}{=} (1 + \alpha^2) \cdot \sqrt{\frac{S}{2 \cdot (1 - \alpha^2)}};$$

since $k_0 > 1$, we have $\alpha < 1$, so these definitions make sense. Then, we take $n = 2 \cdot m$, $[\underline{x}_i, \overline{x}_i] = [-s_i, s_i]$ for $i = 1, 2, \ldots, m$, and $[\underline{x}_i, \overline{x}_i] = [N, N]$ for $i = m + 1, \ldots, 2 \cdot m$. We want to show that for the corresponding problem, we always have $\overline{J} \leq J_0$, and $\overline{J} = J_0$ if and only if there exist signs $\eta_i$ for which $\sum \eta_i \cdot s_i = 0$.

Let us first prove that $\overline{J} \leq J_0$. Since $\overline{J}$ is the upper endpoint of the interval of possible values of $J$, this inequality is equivalent to proving that $J \leq J_0$ for all possible values $J$ – i.e., for the values $J$ corresponding to all possible values $x_i \in \boldsymbol{x}_i$.

Indeed, it is known that $V = M - E^2$, where $M \stackrel{\text{def}}{=} (1/n) \cdot \sum_{i=1}^{n} x_i^2$ is the sample second moment; therefore, $J = \sqrt{M - E^2} + \alpha \cdot E$. This expression for $J$ can be viewed as a scalar (dot) product $\mathbf{a} \cdot \mathbf{b}$ of two 2-D vectors $\mathbf{a} \stackrel{\text{def}}{=} (1, \alpha)$ and $\mathbf{b} \stackrel{\text{def}}{=} (\sqrt{M - E^2}, E)$. It is well known that for arbitrary vectors $\mathbf{a}$ and $\mathbf{b}$, we have $\mathbf{a} \cdot \mathbf{b} \leq \|\mathbf{a}\| \cdot \|\mathbf{b}\|$. In our case, $\|\mathbf{a}\| = \sqrt{1 + \alpha^2}$ and $\|\mathbf{b}\| = \sqrt{M}$, hence $J \leq \sqrt{1 + \alpha^2} \cdot \sqrt{M}$.

Since $|x_i| \leq s_i$ for $i \leq m$ and $x_i = N$ for $i > m$, we conclude that

$$M \leq \frac{1}{2 \cdot m} \cdot \sum_{i=1}^{m} x_i^2 + \frac{1}{2 \cdot m} \cdot \sum_{i=m+1}^{2 \cdot m} x_i^2 = \frac{1}{2} \cdot S + \frac{1}{2} \cdot N^2;$$

therefore, $J \leq \sqrt{1 + \alpha^2} \cdot \sqrt{(S + N^2)/2}$. Substituting the expression that defines $N$ into this formula, we conclude that $J \leq J_0$.

To complete our proof, we will show that if $J = J_0$, then $x_i = \eta_i \cdot s_i$ for $i \leq m$, and $\sum_{i=1}^{m} x_i = \sum_{i=1}^{m} \eta_i \cdot s_i = 0$. Let us first prove that $x_i = \pm s_i$. Indeed:

- we know that $J = J_0$ and that $J_0 = \sqrt{1 + \alpha^2} \cdot \sqrt{(S + N^2)/2}$, so $J = \sqrt{1 + \alpha^2} \cdot \sqrt{(S + N^2)/2}$;
- we have proved that in general, $J \leq \sqrt{1 + \alpha^2} \cdot \sqrt{M} \leq \sqrt{1 + \alpha^2} \cdot \sqrt{(S + N^2)/2}$.

Therefore, $J = \sqrt{1 + \alpha^2} \cdot \sqrt{(S + N^2)/2} = \sqrt{1 + \alpha^2} \cdot \sqrt{M}$, hence $M = (S + N^2)/2$. If $|x_j| < s_j$ for some $j \leq m$, then, from the fact that $|x_i| \leq s_i$ for all $i \leq m$ and $x_i = N$ for all $i > m$, we conclude that $M < (S + N^2)/2$.

Thus, for every $i$ from 1 to $m$, we have $|x_i| = s_i$, hence $x_i = \eta_i \cdot s_i$ for some $\eta_i \in \{-1, 1\}$.

Let us now show that $a \stackrel{\text{def}}{=} \dfrac{1}{m} \cdot \sum_{i=1}^{m} x_i = 0$. Indeed, since $x_i = N$ for $i > m$, we have

$$E = \frac{1}{2 \cdot m} \cdot \sum_{i=1}^{m} x_i + \frac{1}{2 \cdot m} \cdot \sum_{i=m+1}^{2 \cdot m} x_i = \frac{1}{2} \cdot a + \frac{1}{2} \cdot N;$$

therefore, to prove that $a = 0$, it is sufficient to prove that $E = N/2$. The value of $E$ can deduced from the following:

- we have just shown that in our case, $J = \sqrt{1 + \alpha^2} \cdot \sqrt{M}$, where $M = (S + N^2)/2$, and
- we know that in general, $J = \mathbf{a} \cdot \mathbf{b} \leq \|\mathbf{a}\| \cdot \|\mathbf{b}\| = \sqrt{1 + \alpha^2} \cdot \sqrt{M}$, where the vectors $\mathbf{a}$ and $\mathbf{b}$ are defined above.

Therefore, in this case, $\mathbf{a} \cdot \mathbf{b} = \|\mathbf{a}\| \cdot \|\mathbf{b}\|$, and hence, the vectors $\mathbf{a} = (1, \alpha)$ and $\mathbf{b} = (\sqrt{M - E^2}, E)$ are parallel (proportional) to each other, i.e., $\sqrt{M - E^2}/1 = E/\alpha$ hence $E = \alpha \cdot \sqrt{M - E^2}$. From this equality, we conclude that $E > 0$ and, squaring both sides, that $E^2 = \alpha^2 \cdot (M - E^2)$ hence $(1 + \alpha^2) \cdot E^2 = \alpha^2 \cdot M = \alpha^2 \cdot (S + N^2)/2$ and $E^2 = \alpha^2 \cdot (S + N^2)/(2 \cdot (1 + \alpha^2))$. Substituting the expression that defines $N$ into this formula, we conclude that $E^2 = N^2/4$, so, since $E > 0$, we conclude that $E = N/2$ – and therefore, that $a = 0$. The theorem is proven.

*Proof of Theorem 18.3.* This proof is similar to the proof of Theorem 18.2, with the only difference that we consider $J = \sigma - \alpha \cdot E$ and we take $x_i = -N$ for $i > m$.

*Proof of Theorem 18.4.* We will only prove the result for $U$; for $L$, the proof is practically identical.

When a function $U$ attains its largest possible value at the value $x_i$ inside the interval $[\underline{x}_i, \overline{x}_i]$, then at this inside point, $\dfrac{\partial U}{\partial x_i} = 0$ and $\dfrac{\partial^2 U}{\partial x_i^2} \leq 0$. For our function $U$, we have

$$\frac{\partial U}{\partial x_i} = \frac{1}{n} + k_0 \cdot \frac{x_i - E}{\sigma \cdot n};$$

$$\frac{\partial^2 U}{\partial x_i^2} = \frac{k_0}{\sigma^3 \cdot n} \cdot \left( \left( 1 - \frac{1}{n} \right) \cdot \sigma^2 - \frac{1}{n} \cdot (x_i - E)^2 \right).$$

Since $\dfrac{\partial U}{\partial x_i} = 0$, we have $x_i - E = -\alpha \cdot \sigma$, hence

$$\frac{\partial^2 U_i}{\partial x_i^2} = \frac{k_0}{\sigma^3 \cdot n} \cdot \left( \left( 1 - \frac{1}{n} \right) - \frac{\alpha^2}{n} \right) \cdot \sigma^2.$$

Since we assumed that $1 + (1/k_0)^2 = 1 + \alpha^2 < n$, we conclude that $1 - (1/n) - (\alpha^2/n) > 0$, so the second derivative is positive and therefore, we cannot have a maximum in an internal point. The theorem is proven.

*Proof of Theorems 18.5 and 18.6.* Similarly to the case of the previous two theorems, we will only provide the result for $U$; for $L$, the proof is, in effect, the same.

Let us first prove that the algorithm described in the main text is indeed correct. Since $1 + (1/k_0)^2 < n$, we can use Theorem 18.4 and conclude that the maximum of the function $U$ is attained when for every $i$, either $x_i = \underline{x}_i$ or $x_i = \overline{x}_i$. For each $i$, we will consider both these cases.

If the maximum is attained for $x_i = \overline{x}_i$, this means, in particular, that if we keep all the other values $x_j$ the same $(x'_j = x_j)$ but replace $x_i$ by $x'_i = \underline{x}_i = x_i - 2 \cdot \Delta_i$, then the value $U$ will decrease. We will denote the values of $E$, $U$, etc., that correspond to $(x_1, \ldots, x_{i-1}, x'_i, x_{i+1}, \ldots, x_n)$, by $E'$, $U'$, etc. In these terms, the desired inequality takes the form $U \geq U'$, where $U = E + k_0 \cdot \sigma$ and $U' = E' + k_0 \cdot \sigma'$. We can represent this inequality as $k_0 \cdot \sigma \geq (E' - E) + k_0 \cdot \sigma'$, hence either $(E' - E) + k_0 \cdot \sigma' \leq 0$, or

$$k_0^2 \cdot \sigma^2 \geq (E' - E)^2 + k_0^2 \cdot (\sigma')^2 + 2(E - E') \cdot k_0 \cdot \sigma'.$$

In the second case, we move the terms linear in $\sigma'$ to one side of the inequality and square both sides again. As a result, we get an inequality that only contains variances $V = \sigma^2 = M - E^2$ (where $M$ is the sample second moment) and $V' = (\sigma')^2 = M' - (E')^2$ and no longer contains square roots.

For our choice of $x'_i$, we have $E' = E - (2 \cdot \Delta_i)/n$ and

$$M' = M - \frac{4 \cdot \Delta_i \cdot x_i}{n} + \frac{4 \cdot \Delta_i^2}{n}.$$

Substituting these expressions into the above-described inequality and simplifying the resulting algebraic expression, we conclude that

$$\widetilde{x}_i + \Delta_i \cdot \frac{1 + \alpha^2}{n} \geq E - \alpha \cdot \sigma.$$

Similarly, if the maximum is attained for $x_i = \overline{x}_i$, this means, in particular, that if we keep all the other values $x_j$ the same but replace $x_i$ by $x'_i = \overline{x}_i = x_i + 2 \cdot \Delta_i$, then the value $U$ will decrease. This property leads to the inequality

$$\widetilde{x}_i - \Delta_i \cdot \frac{1 + \alpha^2}{n} \leq E - \alpha \cdot \sigma.$$

So:

- if $x_i = \overline{x}_i$, then $E - \alpha \cdot \sigma \leq \widetilde{x}_i + \Delta_i \cdot \frac{1 + \alpha^2}{n}$;
- if $x_i = \underline{x}_i$, then $E - \alpha \cdot \sigma \geq \widetilde{x}_i - \Delta_i \cdot \frac{1 + \alpha^2}{n}$.

Therefore, if we know the value of $E - \alpha \cdot \sigma$, then:

- if $\widetilde{x}_i + \Delta_i \cdot \frac{1 + \alpha^2}{n} < E - \alpha \cdot \sigma$, then we cannot have $x_i = \overline{x}_i$ hence $x_i = \underline{x}_i$;
- similarly, if $\widetilde{x}_i - \Delta_i \cdot \frac{1 + \alpha^2}{n} > E - \alpha \cdot \sigma$, then we cannot have $x_i = \underline{x}_i$ hence $x_i = \overline{x}_i$.

The only case when we do not know what value to choose is the case when

$$\widetilde{x}_i - \Delta_i \cdot \frac{1 + \alpha^2}{n} \leq E - \alpha \cdot \sigma \leq \widetilde{x}_i + \Delta_i \cdot \frac{1 + \alpha^2}{n},$$

i.e., when the value $E - \alpha \cdot \sigma$ belongs to the $i$-th narrowed interval; in this case, we can, in principle, have both $x_i = \underline{x}_i$ and $x_i = \overline{x}_i$. Thus, the algorithm is indeed correct.

Let us prove that this algorithm takes time $O(n \cdot \log(n))$. Indeed, once we know where $E$ is with respect to the endpoints of all narrowed intervals, we can determine the values of all optimal $x_i$ – except for those that are within this narrowed interval. Since we consider the case when no more than $c_0$ narrowed intervals can have a common point, we have no more than $c_0$ undecided values $x_i$. Trying all possible combinations of lower and upper endpoints for these $\leq c_0$ values takes $\leq 2^{c_0}$ steps. For each zone and for each of these combinations, we need a linear time ($O(n)$) to compute $U$. Thus, for each zone, we need $O(2^{c_0} \cdot n)$ computational steps. Since $c_0$ is a constant, the overall number of steps is thus $O(n)$.

For the first zone, we need all this time; for every other zone, each value $x_i$ is updated once. Thus, the overall number of steps after sorting is $O(n)$, and the total number of steps is $O(n \cdot \log(n))$. The theorem is proven.

*Proof of Theorem 18.7.* To find the values $x_i$ which maximize $U$, we reduce the interval computation problem to the constraint satisfaction problem with the following constraints:

- for every $i$, if in the maximizing assignment we have $x_i = \underline{x}_i$, then replacing this value with $x_i = \overline{x}_i$ will either decrease $U$ or leave $U$ unchanged;
- similarly, for every $i$, if in the maximizing assignment we have $x_i = \overline{x}_i$, then replacing this value with $x_i = \underline{x}_i$ will either decrease $U$ or leave $U$ unchanged;
- finally, for every $i$ and $j$, replacing both values $x_i$ and $x_j$ with the opposite ends of the corresponding intervals $\mathbf{x}_i$ and $\mathbf{x}_j$ will either decrease $U$ or leave $U$ unchanged.

We will show that the solution to the resulting constraint satisfaction problem indeed leads to the above efficient algorithm for computing $\overline{U}$.

Let us first show that the above algorithm indeed takes $O(n \cdot \log(n))$ computation steps. It is well known that sorting takes $O(n \cdot \log(n))$ steps (see, e.g., [73]). Computing the initial values $M^{(0)}$, $E^{(0)}$, and $V^{(0)}$ takes linear time $O(n)$. For each $k$ from 0 to $n - 1$, we need a constant number of steps to compute the next values $M^{(k+1)}$, $E^{(k+1)}$, and $V^{(k+1)}$. Computing $U^{(k+1)}$ also takes a constant number of steps. Finally, finding the largest of $n + 1$ values $U^{(k)}$ also takes $O(n)$ steps. Thus, overall, we need

$$O(n \cdot \log(n)) + O(n) + O(n) + O(n) = O(n \cdot \log(n)) \text{ steps.}$$

It is worth mentioning that if the measurement results $\widetilde{x}_i$ are already sorted, then we only need linear time to compute $\overline{U}$.

Now, we need to justify our algorithm. We have already proven that the maximum $\overline{U}$ of the function $U$ is attained at a vector $x = (x_1, \ldots, x_n)$ in which each value $x_i$ is equal either to $\underline{x}_i$ or to $\overline{x}_i$.

To justify our algorithm, we need to prove that this maximum is attained at one of the vectors $x^{(k)}$ in which all the lower bounds $\underline{x}_i$ precede all the upper bounds $\overline{x}_i$. We will prove this by reduction to a contradiction. Indeed, let us assume that the maximum is attained at a vector $x$ in which one of the lower bounds follows one of the upper bounds. In each such vector, let $i$ be the largest upper bound index followed by the lower bound; then, in the optimal vector $x$, we have $x_i = \overline{x}_i$ and $x_{i+1} = \underline{x}_{i+1}$.

Since the maximum is attained for $x_i = \overline{x}_i$, replacing it with $\underline{x}_i = \overline{x}_i - 2 \cdot \Delta_i$ will either decrease the value of $U$ or keep it unchanged. Let us describe how $U$ changes under this replacement. Since $U$ is defined in terms of $E$, $M$, and $V$, let us first describe how $E$, $M$, and $V$ change under this replacement. In the sum for $M$, we replace $(\overline{x}_i)^2$ with

$$(\underline{x}_i)^2 = (\overline{x}_i - 2 \cdot \Delta_i)^2 = (\overline{x}_i)^2 - 4 \cdot \Delta_i \cdot \overline{x}_i + 4 \cdot \Delta_i^2.$$

Thus, the value $M$ changes into $M + \Delta M_i$, where

$$\Delta M_i = -\frac{4}{n} \cdot \Delta_i \cdot \overline{x}_i + \frac{4}{n} \cdot \Delta_i^2. \tag{18.2}$$

The population mean $E$ changes into $E + \Delta E_i$, where

$$\Delta E_i = -\frac{2 \cdot \Delta_i}{n}. \tag{18.3}$$

Thus, the value $E^2$ changes into $(E + \Delta E_i)^2 = E^2 + \Delta(E^2)_i$, where

$$\Delta(E^2)_i = 2 \cdot E \cdot \Delta E_i + \Delta E_i^2 = -\frac{4}{n} \cdot E \cdot \Delta_i + \frac{4}{n^2} \cdot \Delta_i^2. \tag{18.4}$$

So, the variance $V$ changes into $V + \Delta V_i$, where

$$\Delta V_i = \Delta M_i - \Delta(E^2)_i = -\frac{4}{n} \cdot \Delta_i \cdot \overline{x}_i + \frac{4}{n} \cdot \Delta_i^2 + \frac{4}{n} \cdot E \cdot \Delta_i - \frac{4}{n^2} \cdot \Delta_i^2 =$$

$$\frac{4}{n} \cdot \Delta_i \cdot \left( -\overline{x}_i + \Delta_i + E - \frac{\Delta_i}{n} \right).$$

By definition, $\overline{x}_i = \widetilde{x}_i + \Delta_i$, hence $-\overline{x}_i + \Delta_i = -\widetilde{x}_i$. Thus, we conclude that

$$\Delta V_i = \frac{4}{n} \cdot \Delta_i \cdot \left( -\widetilde{x}_i + E - \frac{\Delta_i}{n} \right). \tag{18.5}$$

The function $U = E + k_0 \cdot \sigma$ attains its maximum if and only if the function $u \stackrel{\text{def}}{=} \alpha \cdot U = \alpha \cdot E + \sigma$ attains its maximum. After the change, the value $u$ changes into

$$u + \Delta u_i = \alpha \cdot (E + \Delta E_i) + \sqrt{V + \Delta V_i},$$

so the condition $u + \Delta u_i \leq u$ leads to

$$\alpha \cdot (E + \Delta E_i) + \sqrt{V + \Delta V_i} \leq \alpha \cdot E + \sigma.$$

By moving the term proportional to $\alpha$ to the right-hand side, we conclude that $\sqrt{V + \Delta V_i} \leq \sigma - \alpha \cdot \Delta E_i$. In the new inequality, the left-hand side is the new value of the standard deviation, so it is a non-negative number, hence the right-hand side is also non-negative, so we can square both sides of the inequality and conclude that

$$V + \Delta V_i \leq \sigma^2 - 2 \cdot \alpha \cdot \sigma \cdot \Delta E_i + \alpha^2 \cdot (\Delta E_i)^2.$$

Moving all the terms to the left-hand side and using the fact that $V = \sigma^2$, we conclude that

$$z_i \stackrel{\text{def}}{=} \Delta V_i + 2 \cdot \alpha \cdot \sigma \cdot \Delta E_i - \alpha^2 \cdot (\Delta E_i)^2 \leq 0. \tag{18.6}$$

Substituting the known values of $\Delta V_i$ and $\Delta E_i$, we get:

$$z_i = \frac{4}{n} \cdot \Delta_i \cdot e_i, \tag{18.7}$$

where

$$e_i = -\widetilde{x}_i + E - \frac{\Delta_i}{n} - \alpha \cdot \sigma - \alpha^2 \cdot \frac{\Delta_i}{n},$$

i.e.,

$$e_i = (E - \alpha \cdot \sigma) - \left( \widetilde{x}_i + \frac{1 + \alpha^2}{n} \cdot \Delta_i \right). \tag{18.8}$$

Thus, from $z_i \leq 0$, we conclude that

$$E - \alpha \cdot \sigma \leq \widetilde{x}_i + \frac{1 + \alpha^2}{n} \cdot \Delta_i. \tag{18.9}$$

Similarly, since the maximum of $u$ is attained for $x_{i+1} = \underline{x}_{i+1}$, replacing it with $\overline{x}_{i+1} = \underline{x}_{i+1} + 2 \cdot \Delta_{i+1}$ will either decrease the value of $u$ or keep it unchanged. Let us describe how variance changes under this replacement. In the sum for $M$, we replace $(\underline{x}_{i+1})^2$ with

$$(\overline{x}_{i+1})^2 = (\underline{x}_{i+1} + 2 \cdot \Delta_{i+1})^2 = (\underline{x}_{i+1})^2 + 4 \cdot \Delta_{i+1} \cdot \underline{x}_{i+1} + 4 \cdot \Delta_{i+1}^2.$$

Thus, the value $M$ changes into $M + \Delta M_{i+1}$, where

$$\Delta M_{i+1} = \frac{4}{n} \cdot \Delta_{i+1} \cdot \underline{x}_{i+1} + \frac{4}{n} \cdot \Delta_{i+1}^2. \tag{18.10}$$

The population mean $E$ changes into $E + \Delta E_{i+1}$, where

$$\Delta E_{i+1} = \frac{2 \cdot \Delta_{i+1}}{n}. \tag{18.11}$$

Thus, the value $E^2$ changes into

$$(E + \Delta E_{i+1})^2 = E^2 + \Delta(E^2)_{i+1},$$

where

$$\Delta(E^2)_{i+1} = 2 \cdot E \cdot \Delta E_{i+1} + \Delta E_{i+1}^2 = \frac{4}{n} \cdot E \cdot \Delta_{i+1} + \frac{4}{n^2} \cdot \Delta_{i+1}^2. \tag{18.12}$$

So, the variance $V$ changes into $V + \Delta V_{i+1}$, where

$$\Delta V_{i+1} = \Delta M_{i+1} - \Delta(E^2)_{i+1} = \frac{4}{n} \cdot \Delta_{i+1} \cdot \underline{x}_{i+1} + \frac{4}{n} \cdot \Delta_{i+1}^2 - \frac{4}{n} \cdot E \cdot \Delta_{i+1} - \frac{4}{n^2} \cdot \Delta_{i+1}^2$$

$$= \frac{4}{n} \cdot \Delta_{i+1} \cdot \left( \underline{x}_{i+1} + \Delta_{i+1} - E - \frac{\Delta_{i+1}}{n} \right).$$

By definition, $\underline{x}_{i+1} = \widetilde{x}_{i+1} - \Delta_{i+1}$, hence $\underline{x}_{i+1} + \Delta_{i+1} = \widetilde{x}_{i+1}$. Thus, we conclude that

$$\Delta V_{i+1} = \frac{4}{n} \cdot \Delta_{i+1} \cdot \left( \widetilde{x}_{i+1} - E - \frac{\Delta_{i+1}}{n} \right). \tag{18.13}$$

Since $u$ attains maximum at $x$, we have $\Delta u_{i+1} \leq 0$, i.e., $z_{i+1} \leq 0$, where

$$z_{i+1} \overset{\text{def}}{=} \Delta V_{i+1} + 2 \cdot \alpha \cdot \sigma \cdot \Delta E_{i+1} - \alpha^2 \cdot (\Delta E_{i+1})^2. \tag{18.14}$$

Substituting the expressions (18.13) for $\Delta V_{i+1}$ and (18.11) for $\Delta E_{i+1}$ into this formula, we conclude that

$$z_{i+1} = \frac{4}{n} \cdot \Delta_{i+1} \cdot e_{i+1}, \tag{18.15}$$

where

$$e_{i+1} \overset{\text{def}}{=} -(E - \alpha \cdot \sigma) + \left( \widetilde{x}_{i+1} - \frac{1 + \alpha^2}{n} \cdot \Delta_{i+1} \right) \tag{18.16}$$

and

$$E - \alpha \cdot \sigma \geq \widetilde{x}_{i+1} - \frac{1 + \alpha^2}{n} \cdot \Delta_{i+1}. \tag{18.17}$$

We can also change *both* $x_i$ and $x_{i+1}$ at the same time. In this case, from the fact that $u$ attains the maximum at $x$, we conclude that $u + \Delta u \leq u$, i.e., that

$$z \overset{\text{def}}{=} \Delta V + 2 \cdot \alpha \cdot \sigma \cdot \Delta E - \alpha^2 \cdot (\Delta E)^2. \tag{18.18}$$

Here, the change $\Delta M$ in $M$ is simply the sum of the changes coming from $x_i$ and $x_{i+1}$:

$$\Delta M = \Delta M_i + \Delta M_{i+1}, \tag{18.19}$$

and the change $\Delta E$ in $E$ is also the sum of the corresponding changes:

$$\Delta E = \Delta E_i + \Delta E_{i+1}. \tag{18.20}$$

So, for

$$\Delta V = \Delta M - \Delta(E^2) = \Delta M - 2 \cdot E \cdot \Delta E - \Delta E^2,$$

we get

$$\Delta V = \Delta M_i + \Delta M_{i+1} - 2 \cdot E \cdot \Delta E_i - 2 \cdot E \cdot \Delta E_{i+1} - (\Delta E_i)^2 - (\Delta E_{i+1})^2 -$$

$$2 \cdot \Delta E_i \cdot \Delta E_{i+1}.$$

Hence,

$$\Delta V = (\Delta M_i - 2 \cdot E \cdot \Delta E_i - (\Delta E_i)^2) + (\Delta M_{i+1} - 2 \cdot E \cdot \Delta E_{i+1} - (\Delta E_{i+1})^2) -$$

$$2 \cdot \Delta E_i \cdot \Delta E_{i+1},$$

i.e.,

$$\Delta V = \Delta V_i + \Delta V_{i+1} - 2 \cdot \Delta E_i \cdot \Delta E_{i+1}. \tag{18.21}$$

Substituting expressions (18.19), (18.20), and (18.21) into the formula (18.18) for $z$, we conclude that

$$z = \Delta V + 2 \cdot \alpha \cdot \sigma \cdot \Delta E - \alpha^2 \cdot (\Delta E)^2 = \Delta V_i + \Delta V_{i+1} - 2 \cdot \Delta E_i \cdot \Delta E_{i+1} +$$

$$2\alpha \cdot \sigma \cdot \Delta E_i + 2\alpha \cdot \sigma \cdot \Delta E_{i+1} -$$

$$\alpha^2 \cdot (\Delta E_i)^2 - \alpha^2 \cdot (\Delta E_{i+1})^2 - 2 \cdot \alpha^2 \cdot \Delta E_i \cdot \Delta E_{i+1}.$$

Hence,

$$z = (\Delta V_i + 2 \cdot \alpha \cdot \sigma \cdot \Delta E_i - \alpha^2 \cdot (\Delta E_i)^2) + (\Delta V_{i+1} + 2 \cdot \alpha \cdot \sigma \cdot \Delta E_{i+1} - \alpha^2 \cdot (\Delta E_{i+1})^2) -$$

$$2 \cdot (1 + \alpha^2) \cdot \Delta E_i \cdot \Delta E_{i+1}.$$

From the formulas (18.6) and (18.14), we know that the first expression is $z_i$ and that the second expression is $z_{i+1}$, so

$$z = z_i + z_{i+1} - 2 \cdot (1 + \alpha^2) \cdot \Delta E_i \cdot \Delta E_{i+1}.$$

We already have the expressions (18.7), (18.8), (18.15), (18.16), (18.3), and (18.11) for, correspondingly, $z_i$, $z_{i+1}$, $\Delta E_i$, and $\Delta E_{i+1}$, so we conclude that $z = \dfrac{4}{n} \cdot D(E')$, where $E' \stackrel{\text{def}}{=} E - \alpha \cdot \sigma$ and

$$D(E') \stackrel{\text{def}}{=} \Delta_i \cdot \left( E' - \left( \widetilde{x}_i + \frac{1+\alpha^2}{n} \cdot \Delta_i \right) \right) +$$

$$\Delta_{i+1} \cdot \left( -E' + \left( \widetilde{x}_{i+1} - \frac{1+\alpha^2}{n} \cdot \Delta_{i+1} \right) \right) + 2 \cdot (1+\alpha^2) \cdot \frac{\Delta_i \cdot \Delta_{i+1}}{n}. \tag{18.22}$$

Since $z \leq 0$, we have $D(E') \leq 0$ (for the value $E' = E - \alpha \cdot \sigma$ corresponding to the optimizing vector $x$).

The expression $D(E')$ is a linear function of $E'$. From (18.9) and (18.17), we know that

$$\widetilde{x}_{i+1} - \frac{1+\alpha^2}{n} \cdot \Delta_{i+1} \leq E' \leq \widetilde{x}_i + \frac{1+\alpha^2}{n} \cdot \Delta_i.$$

For $E' = E^- \stackrel{\text{def}}{=} \widetilde{x}_{i+1} - \dfrac{1+\alpha^2}{n} \cdot \Delta_{i+1}$, we have

$$D(E^-) = \Delta_i \cdot f_i + \frac{2 \cdot (1+\alpha^2)}{n} \cdot \Delta_i \cdot \Delta_{i+1},$$

where

$$f_i \stackrel{\text{def}}{=} -\widetilde{x}_i + \widetilde{x}_{i+1} - \frac{1+\alpha^2}{n} \cdot \Delta_{i+1} - \frac{1+\alpha^2}{n} \cdot \Delta_i,$$

hence $D(E^-) = \Delta_i \cdot g_i$, where

$$g_i \stackrel{\text{def}}{=} -\widetilde{x}_i + \widetilde{x}_{i+1} + \frac{1+\alpha^2}{n} \cdot \Delta_{i+1} - \frac{1+\alpha^2}{n} \cdot \Delta_i.$$

We assumed that no narrowed interval is a proper subset of any other. How can we describe this condition in algebraic terms? Let us denote $\delta_i \stackrel{\text{def}}{=} \dfrac{1+\alpha^2}{n} \cdot \Delta_i$; then, the $i$-th narrowed interval has the form $[\widetilde{x}_i - \delta_i, \widetilde{x}_i + \delta_i]$. If $[\widetilde{x}_i - \delta_i, \widetilde{x}_i + \delta_i]$ is a proper subinterval of $[\widetilde{x}_j - \delta_j, \widetilde{x}_j + \delta_j]$, this means that $\widetilde{x}_i - \delta_i > \widetilde{x}_j - \delta_j$ and $\widetilde{x}_i + \delta_i < \widetilde{x}_j + \delta_j$, i.e., equivalently, that

$$\delta_i - \delta_j < \widetilde{x}_i - \widetilde{x}_j < \delta_j - \delta_i.$$

This inequality is equivalent to $\delta_j > \delta_i$ and $|\widetilde{x}_i - \widetilde{x}_j| < \delta_j - \delta_i$. Similarly, the condition that the $j$-th narrowed interval is a proper subinterval of the $i$-th is equivalent to $\delta_j < \delta_i$ and $|\widetilde{x}_i - \widetilde{x}_j| < \delta_i - \delta_j$. Both cases can be described by a single inequality $|\widetilde{x}_i - \widetilde{x}_j| < |\delta_i - \delta_j|$. Thus, the condition that no narrowed interval can be a proper subinterval of any other narrowed interval can be described as

$$|\widetilde{x}_i - \widetilde{x}_j| \geq |\delta_i - \delta_j|. \tag{18.23}$$

In particular, we have $|\widetilde{x}_i - \widetilde{x}_{i+1}| \geq |\delta_i - \delta_{i+1}|$.

Let us first consider the case when

$$|\widetilde{x}_{i+1} - x_i| > |\delta_i - \delta_{i+1}|.$$

Since the values $\widetilde{x}_i$ are sorted in increasing order, we have $\widetilde{x}_{i+1} \geq \widetilde{x}_i$, hence

$$\widetilde{x}_{i+1} - \widetilde{x}_i = |\widetilde{x}_{i+1} - \widetilde{x}_i| > |\delta_i - \delta_{i+1}| \geq \delta_i - \delta_{i+1}.$$

So, we conclude that $D(E^-) > 0$.

For $E = E^+ \stackrel{\text{def}}{=} \widetilde{x}_i + \dfrac{1 + \alpha^2}{n} \cdot \Delta_i$, we have

$$D(E^+) = \Delta_{i+1} \cdot f_{i+1} + \frac{2 \cdot (1 + \alpha^2)}{n} \cdot \Delta_i \cdot \Delta_{i+1},$$

where

$$f_{i+1} \stackrel{\text{def}}{=} -\widetilde{x}_i + \widetilde{x}_{i+1} - \frac{1 + \alpha^2}{n} \cdot \Delta_{i+1} - \frac{1 + \alpha^2}{n} \cdot \Delta_i,$$

hence $D(E^+) = \Delta_{i+1} \cdot g_{i+1}$, where

$$g_{i+1} \stackrel{\text{def}}{=} -\widetilde{x}_i + \widetilde{x}_{i+1} + \frac{1 + \alpha^2}{n} \cdot \Delta_i - \frac{1 + \alpha^2}{n} \cdot \Delta_{i+1}.$$

Here, from $|\widetilde{x}_{i+1} - \widetilde{x}_i| > |\delta_i - \delta_{i+1}|$, we also conclude that $D(E^+) > 0$.

Since the linear function $D(E')$ is positive on both endpoints of the interval $[E^-, E^+]$, it must be positive for every value $E'$ from this interval, which contradicts to our conclusion that $D(E') \leq 0$ for the actual value $E' = E - \alpha \cdot \sigma \in [E^-, E^+]$. This contradiction shows that the maximum of $U$ is indeed attained at one of the values $x^{(k)}$, hence the algorithm is justified.

The general case when $|\widetilde{x}_i - \widetilde{x}_j| \geq |\delta_i - \delta_j|$ can be obtained as a limit of cases when we have strict inequality. Since the function $U$ is continuous, the value $\overline{U}$ continuously depends on the input bounds, so by tending to a limit, we can conclude that our algorithm works in the general case as well.

*Proof of Theorem 18.8.* Let us first consider the case when all the intervals intersect. We know that the variance $V = M - E^2$ is always non-negative; therefore, $M \geq E^2$ and $R \geq 1$; hence $\underline{R} \geq 1$. If all the intervals have a common point, it is possible that all the values $x_i$ are equal to this common point; in this case, $V = 0$ hence $R = 1$. Thus, in this case, $\underline{R} = 1$.

Let us now consider the case when the intersection of $n$ intervals is empty. For this case, the proof is similar to the proof of Theorem 18.1. Indeed, the minimum of a differentiable function of $x_i$ on an interval $[\underline{x}_i, \overline{x}_i]$ is attained either inside this interval or at one of the endpoints. If the minimum is attained inside, the derivative $\dfrac{\partial R}{\partial x_i}$ is equal to 0; if it is attained at $x_i = \underline{x}_i$, then $\dfrac{\partial R}{\partial x_i} \geq 0$; finally, if it is attained at $x_i = \overline{x}_i$, then $\dfrac{\partial R}{\partial x_i} \leq 0$. For our function,

$$\frac{\partial R}{\partial x_i} = \frac{2}{n \cdot E^3} \cdot (E \cdot x_i - M);$$

thus, $\dfrac{\partial R}{\partial x_i} = 0$ if and only if $x_i = \lambda \stackrel{\text{def}}{=} M/E$; similarly, the non-positiveness and non-negativeness of the derivative can be described by comparing $x_i$ with $\lambda$. Thus:

- either $x_i \in (\underline{x}_i, \overline{x}_i)$ and $x_i = \lambda$,
- or $x_i = \underline{x}_i$ and $x_i = \underline{x}_i \geq \lambda$,
- or $x_i = \overline{x}_i$ and $x_i = \overline{x}_i \leq \lambda$.

The proof continues just like for Theorem 18.1.

*Proof of Theorem 18.9.* This proof is similar to the proof of Theorem 18.4. When a function $R = M/E^2$ attains its largest possible value $\overline{R}$ at the value $x_i$ inside the interval $[\underline{x}_i, \overline{x}_i]$, then at this inside point, $\dfrac{\partial R}{\partial x_i} = 0$ and $\dfrac{\partial^2 R}{\partial x_i^2} \leq 0$. For our function $R$, we have

$$\frac{\partial R}{\partial x_i} = \frac{2}{n \cdot E^3} \cdot (E \cdot x_i - M);$$

$$\frac{\partial^2 R}{\partial x_i^2} = \frac{2}{n \cdot E^4} \cdot \left[ \left( E - \frac{x_i}{n} \right) \cdot E - 2(E \cdot x_i - M) \cdot \frac{1}{n} \right].$$

Since $\dfrac{\partial R}{\partial x_i} = 0$, we have $x_i = M/E$, hence

$$\frac{\partial^2 R_i}{\partial x_i^2} = \frac{2}{n \cdot E^2} \left( 1 - \frac{x_i}{n \cdot E} \right) = \frac{2}{n \cdot E^2} \left( 1 - \frac{M}{n \cdot E^2} \right) = \frac{2}{n \cdot E^2} \left( 1 - \frac{R}{n} \right).$$

Since we assumed that $\overline{R} < n$, we conclude that the second derivative is positive and therefore, we cannot have a maximum in an internal point. The theorem is proven.

*Proof of Theorem 18.10.* This proof is similar to the proof of Theorems 18.5 and 18.6. Let us first prove that the algorithm described in the main text is indeed correct. Since $\overline{R}$, we can use Theorem 18.9 and conclude that the maximum of the function $R$ is attained when for every $i$, either $x_i = \underline{x}_i$ or $x_i = \overline{x}_i$. For each $i$, we will consider both these cases.

If the maximum is attained for $x_i = \overline{x}_i$, this means, in particular, that if we keep all the other values $x_j$ the same ($x'_j = x_j$) but replace $x_i$ by $x'_i = \underline{x}_i = x_i - 2 \cdot \Delta_i$, then the value $R = M/E^2$ will decrease. We will denote the values of $E$ and $M$ that correspond to $(x_1, \ldots, x_{i-1}, x'_i, x_{i+1}, \ldots, x_n)$, by $E'$, and $M'$. In these terms, the desired inequality takes the form $M/E^2 \geq M'/(E')^2$, i.e., equivalently, $M \cdot (E')^2 \geq M' \cdot E^2$.

In the proof of Theorems 18.5 and 18.6, we had expressions for $E'$ and $M'$. Substituting these expressions into the above inequality and simplifying the resulting algebraic expression, we conclude that

$$\widetilde{x}_i \leq \lambda \cdot \left( 1 + \frac{\Delta_i}{E \cdot n} \right),$$

where $\lambda \overset{\text{def}}{=} M/E$.

Similarly, if the maximum is attained for $x_i = \underline{x}_i$, we have

$$\widetilde{x}_i \geq \lambda \cdot \left( 1 - \frac{\Delta_i}{E \cdot n} \right).$$

Therefore, if we know the value of $\lambda = M/E$, then:

- if $\dfrac{\widetilde{x}_i}{1 + \dfrac{\Delta_i}{E \cdot n}} > \lambda$, then we cannot have $x_i = \underline{x}_i$ hence $x_i = \overline{x}_i$;
- if $\dfrac{\widetilde{x}_i}{1 - \dfrac{\Delta_i}{E \cdot n}} < \lambda$, then we cannot have $x_i = \overline{x}_i$ hence $x_i = \underline{x}_i$.

Similarly to the proof of Theorems 18.5 and 18.6, we can now conclude that the algorithm described in the main text is correct and that this algorithm takes time $O(n \cdot \log(n))$.

# Computing Higher Moments under Interval Uncertainty

Higher central moments $M_h = \dfrac{1}{n} \cdot \sum\limits_{i=1}^{n} (x_i - E)^h$ are very useful in statistical analysis: the third moment $M_3$ characterizes asymmetry of the corresponding probability distribution, the fourth moment $M_4$ describes the size of the distribution's tails, etc. To be more precise, *skewness* $M_3/\sigma^3$ is used to characterize asymmetry, and *kurtosis* $M_4/\sigma^4 - 3$ is used to characterize the size of the tails (3 is subtracted so that kurtosis is 0 for the practically frequent case of a normal distribution).

In addition to central moments $M_h$, sometimes, non-central sample moments are also used: $M'_h = \dfrac{x_1^h + \ldots + x_n^h}{n}$.

When we know the exact values $x_1, \ldots, x_n$, we can use known formulas for computing the corresponding sample central moments. In many practical situations, however, we only know intervals $\mathbf{x}_1, \ldots, \mathbf{x}_n$ of possible values of $x_i$; in such situations, we want to know the range of possible values of $M_h$. In this chapter, we propose algorithms that compute such ranges.

## Algorithms

*Case of non-central moments.* Due to monotonicity, we can easily compute the exact bounds for *non-central* moments $M'_h$ for odd $h$:

$$\underline{M'_h} = \frac{(\underline{x}_1)^h + \ldots + (\underline{x}_n)^h}{n}; \quad \overline{M'_h} = \frac{(\overline{x}_1)^h + \ldots + (\overline{x}_n)^h}{n}.$$

For example, for $h = 3$ and $\mathbf{x}_1 = \mathbf{x}_2 = \mathbf{x}_3 = [-1, 1]$, we get $\mathbf{M'_3} = [-1, 1]$.

For even $h$, it is known (see, e.g., [142]) that the range of $x^h$ when $x \in [\underline{x}, \overline{x}]$ is equal to $[(\min(|\underline{x}|, |\overline{x}|))^h, (\max(|\underline{x}|, |\overline{x}|))^h]$ when $0 \notin [\underline{x}, \overline{x}]$ and to $[0, (\max(|\underline{x}|, |\overline{x}|))^h]$ otherwise. Thus,

H.T. Nguyen et al.: Computing Stat. under Interval & Fuzzy Uncertain., SCI 393, pp. 153–166.
springerlink.com © Springer-Verlag Berlin Heidelberg 2012

$$\underline{M}'_h = \frac{F^h(\underline{x}_1, \overline{x}_1) + \ldots + F^h(\underline{x}_n, \overline{x}_n)}{n},$$

where $F(a, b) = 0$ if $a \leq 0 \leq b$ and $\min(|a|, |b|)$ otherwise, and

$$\overline{M}'_h = \frac{(\max(|\underline{x}_1|, |\overline{x}_1|))^h + \ldots + (\max(|\underline{x}_n|, |\overline{x}_n|))^h}{n}.$$

For example, for $[-1, 1]$ and $h = 4$, we have $[-1, 1]^4 = [0, 1]$ hence $\boldsymbol{M}'_4 = [0, 1]$ – which again makes perfect sense: this moment can be 0 if all the values $x_i$ are equal to 0; it can be equal to 1 if all the values are equal to 1; and it cannot be larger than 1 because it is the average of three values $x_i^4$, each of which is $\leq 1$.

**Theorem 19.1.** *For every even $h$, there exists an algorithm that computes $\underline{M}_h$ in time $O(n \cdot \log(n))$.*

This algorithm is as follows.

- First, we sort all $2n$ values $\underline{x}_i, \overline{x}_i$ into a sequence $x_{(1)} \leq x_{(2)} \leq \ldots \leq x_{(2n)}$. This sequence divides the real line into $2n+1$ segments $[x_{(k)}, x_{(k+1)}]$, where $k = 0, \ldots, 2n$, $x_{(0)} \stackrel{\text{def}}{=} -\infty$, and $x_{(2n+1)} \stackrel{\text{def}}{=} +\infty$.
- For each segment $[x_{(k)}, x_{(k+1)}]$, we do the following:
  - First, define $n$ values $x_1, \ldots, x_n$ as follows:
    - if $\underline{x}_i \geq x_{(k+1)}$, we take $x_i = \underline{x}_i$;
    - if $\overline{x}_i \leq x_{(k)}$, we take $x_i = \overline{x}_i$;
    - in all other cases, let $x_i = \alpha$, where $\alpha$ is a new auxiliary variable.
  - Based on these expressions for the $x_i$, we find each as a linear function of $\alpha$: $x_i = x_i(\alpha)$; combining these expressions, we get the expression for the sample average $E$ as a linear function of $\alpha$:

$$E(\alpha) = \frac{x_1(\alpha) + \ldots + x_n(\alpha)}{n}.$$

  - Then, we substitute these expressions for $x_i$ and $E$ into the equation

$$\frac{1}{n} \cdot \sum_{i=1}^{n} (x_i(\alpha) - E(\alpha))^{h-1} = (\alpha - E(\alpha))^{h-1}.$$

    This equation is a polynomial equation of order $\leq h - 1$ in terms of the unknown $\alpha$, so it has $\leq h - 1$ solutions. We compute these solutions.
  - For each of the solutions that is inside the segment $[x_{(k)}, x_{(k+1)}]$, we assign it to $\alpha$ in the formulas for $x_i$ and $E$, giving values for $x_i$ and $E$. Based on these values, we compute

$$M_h = \frac{1}{n} \cdot \sum_{i=1}^{n} (x_i - E)^h.$$

- The smallest of the computed values for $M_h$ for each $k$ and for each case of $\alpha$ in $[x_{(k)}, x_{(k+1)}]$ is the desired lower bound $\underline{M}_h$.

*Example.* For example, when $x_1 = x_2 = x_3 = [-1, 1]$ and $h = 4$, this algorithm produces the correct bound $\underline{M}_4 = 0$. That bound is attainable if, e.g., $x_1 = x_2 = x_3 = 0$.

*Computing* $\overline{M}_h$ *for even* $h$ *in time* $2^n$. We have already mentioned that for $h = 2$, the problem of computing $\overline{M}_h$ is NP-hard, but it is possible to compute this bound in time $2^n$. This computation is practical when the number of observations $n$ is small. A similar algorithm is possible for all even $h$:

**Theorem 19.2.** *For every even* $h$, *there exists an algorithm that computes* $\overline{M}_h$ *in time* $O(2^n)$.

The algorithm is as follows: for each $i$, we select either $x_i = \underline{x}_i$ or $x_i = \overline{x}_i$. For each $i$ from 1 to $n$, there are two options, so totally, we have $2^n$ combinations to try. For each of these combinations, we compute $M_h$; the largest of the resulting $2^n$ values is the desired upper bound $\underline{M}_h$.

In particular, for $\boldsymbol{x}_1 = \boldsymbol{x}_2 = \boldsymbol{x}_3 = [-1, 1]$ and $h = 4$, this algorithm produces the bound $\overline{M}_4 = 32/27 \approx 1.19$ – that is attainable if, e.g., $x_1 = x_2 = 1$ and $x_3 = -1$. Once can check that for all other combinations $x_i \in [-1, 1]$, we get smaller (or equal) value of the 4th central moment $M_4$.

*Third result: computing* $\overline{M}_h$ *for even* $h$ *in quadratic time (case when intervals have* $c_0$*-few intersections).* Sets $S_1, \ldots, S_n$ are called *pairwise disjoint* if every pair has an empty intersection, i.e., if $S_i \cap S_j = \emptyset$ for all $i \neq j$. We can generalize this definition from pairs to tuples of arbitrary size $C$:

**Definition 19.1.** *Let* $c_0 \geq 2$ *be an integer. We say that a sequence of sets* $S_1, \ldots, S_n$ *has* $c_0$-few intersections *if for every* $c_0$ *different indices* $i_1, \ldots, i_{c_0}$, *we have* $S_{i_1} \cap \ldots \cap S_{i_{c_0}} = \emptyset$.

**Theorem 19.3.** *For every even* $h$ *and for every* $c_0 \geq 2$, *there exists an algorithm that computes* $\overline{M}_h$ *in time* $O(n \cdot \log(n))$ *when the input intervals* $\boldsymbol{x}_1, \ldots, \boldsymbol{x}_n$ *have* $c_0$-few intersections.

This algorithm is as follows:

- First, we sort all $2n$ values $\underline{x}_i$, $\overline{x}_i$ into a sequence $x_{(1)} \leq x_{(2)} \leq \ldots \leq x_{(2n)}$. This sequence divides the real line into $2n+1$ segments $[x_{(k)}, x_{(k+1)}]$, where $k = 0, \ldots, 2n$, $x_{(0)} \stackrel{\text{def}}{=} -\infty$, and $x_{(2n+1)} \stackrel{\text{def}}{=} +\infty$.
- For each of these segments $[x_{(k)}, x_{(k+1)}]$, we do the following:
  - First, we describe several combinations $(x_1, \ldots, x_n)$ as follows:
    - if $\overline{x}_i < x_{(k)}$, we take $x_i = \underline{x}_i$;
    - if $\underline{x}_i > x_{(k+1)}$, we take $x_i = \overline{x}_i$;
    - for all other indices $i$ (there are $\leq c_0$ of them), we consider all possible combinations of $x_i = \underline{x}_i$ and $x_i = \overline{x}_i$.

    As a result, we get $\leq 2^{c_0}$ different combinations.
  - For each resulting combination $(x_1, \ldots, x_n)$, we compute $E$ as the average of all the values $x_i$, then we compute

$$M_{h-1} = \frac{1}{n} \cdot \sum_{i=1}^{n} (x_i - E)^{h-1},$$

and $\alpha = E + M_{h-1}^{1/(h-1)}$.

- For each combination for which the resulting value $\alpha$ is within the segment $[x_{(k)}, x_{(k+1)}]$, we compute

$$M_h = \frac{1}{n} \cdot \sum_{i=1}^{n} (x_i - E)^h.$$

- The largest of thus computed values $M_h$ is the desired upper bound $\overline{M}_h$.

In particular, for $\boldsymbol{x}_1 = \boldsymbol{x}_2 = \boldsymbol{x}_3 = [-1,1]$, $h = 4$, and $c_0 = 4$, we get the correct bound $\overline{M}_4 = 32/27 \approx 1.19$.

*Fourth result: computing $\underline{M}_h$ and $\overline{M}_h$ for odd $h$ in quadratic time (case when intervals have $c_0$-few intersections).*

**Theorem 19.4.** *For every odd $h$ and for every $c_0 \geq 2$, there exists an algorithm that computes $\underline{M}_h$ in quadratic time when the input intervals $\boldsymbol{x}_1, \ldots, \boldsymbol{x}_n$ have $c_0$-few intersections.*

The algorithm for computing $\underline{M}_h$ is as follows:

- First, we sort all $2n$ values $\underline{x}_i$, $\overline{x}_i$ into a sequence $x_{(1)} \leq x_{(2)} \leq \ldots \leq x_{(2n)}$. This sequence divides the real line into $2n+1$ segments $[x_{(k)}, x_{(k+1)}]$, where $k = 0, \ldots, 2n$, $x_{(0)} \overset{\text{def}}{=} -\infty$, and $x_{(2n+1)} \overset{\text{def}}{=} +\infty$.
- For each pair of segments $[x_{(k)}, x_{(k+1)}]$ and $[x_{(l)}, x_{(l+1)}]$, $k \leq l$, we do the following:
  - First, we describe several combinations $(x_1, \ldots, x_n)$ as linear functions of $\alpha^-$ and $\alpha^+$ as follows:
    · if $\overline{x}_i < x_{(k)}$, we take $x_i = \underline{x}_i$;
    · if $\underline{x}_i > x_{(k+1)}$, we take $x_i = \overline{x}_i$;
    · for all other indices $i$ (there are $\leq 2c_0$ of them), we consider all possible combinations of $x_i = \underline{x}_i$, $x_i = \overline{x}_i$, $x_i = \alpha^-$ (if $[x_{(k)}, x_{(k+1)}] \subseteq \boldsymbol{x}_i$) and $x_i = \alpha^+$ (if $[x_{(l)}, x_{(l+1)}] \subseteq \boldsymbol{x}_i$).
    As a result, we get $\leq 4^{2c_0}$ different combinations.
  - For each resulting combination $(x_1, \ldots, x_n)$, we find the expression for $E$ as the average of all the values $x_i$. Then, we substitute the expressions for $x_i$ and $E$ into the system of equations

$$\frac{1}{n} \cdot \sum_{i=1}^{n} (x_i - E)^{h-1} = (E - \alpha^-)^{h-1};$$

$$\frac{1}{n} \cdot \sum_{i=1}^{n} (x_i - E)^{h-1} = (\alpha^+ - E)^{h-1}.$$

We compute all solutions of this system of polynomial equations with unknowns $\alpha^-$ and $\alpha^+$.

- For each of the solutions for which $\alpha^- \in [x_{(k)}, x_{(k+1)}]$ and $\alpha^+ \in [x_{(l)}, x_{(l+1)}]$, we substitute the corresponding values $\alpha^-$ and $\alpha^+$ into the formulas for $x_i$ and $E$, thus, we compute $x_i$ and $E$; based on these values, we compute

$$M_h = \frac{1}{n} \cdot \sum_{i=1}^{n} (x_i - E)^h.$$

- The smallest of thus computed values $M_h$ is the desired lower bound $\underline{M}_h$.

**Theorem 19.5.** *For every odd $h$ and for every $c_0 \geq 2$, there exists an algorithm that computes $\overline{M}_h$ in quadratic time when the input intervals $\boldsymbol{x}_1, \ldots, \boldsymbol{x}_n$ have $c_0$-few intersections.*

Since $h$ is odd, the $h$-th central moment of the values $x_1, \ldots, x_n$ is equal to minus the $h$-th moment of the values $-x_1, \ldots, -x_n$. Turning to $-x_i$ changes largest and smallest values and vice versa. Thus, to compute $\overline{M}_h$ for the intervals $\boldsymbol{x}_i = [\underline{x}_i, \overline{x}_i]$, it is sufficient to compute the lower bound $\underline{M}_h$ for the intervals $-\boldsymbol{x}_i = [-\overline{x}_i, -\underline{x}_i]$, and then change the sign of the resulting bound. Since we can use the above cubic-time algorithm to compute $\underline{M}_h$, we thus get a cubic-time algorithm for computing $\overline{M}_h$.

In particular, for $\boldsymbol{x}_1 = \boldsymbol{x}_2 = \boldsymbol{x}_3 = [-1, 1]$, $h = 3$, and $c_0 = 4$, we get the bounds $\overline{M}_3 = 80/81 \approx 0.988$ and $\overline{M}_3 = -80/81$. The largest value $\overline{M}_3$ is attained when $x_1 = x_2 = 1$ and $x_3 = -1$; the smallest value $\underline{M}_3$ is attained, e.g., when $x_1 = x_2 = -1$ and $x_3 = 1$.

*Fifth result: computing the bounds $\underline{M}_3$ and $\overline{M}_3$ in quadratic time (case when intervals satisfy the no-subset property).* We say that intervals $[\underline{x}_i, \overline{x}_i]$ satisfy a *no-subset property* if $[\underline{x}_i, \overline{x}_i] \not\subseteq (\underline{x}_j, \overline{x}_j)$ for all $i$ and $j$.

*Comment.* As we have mentioned when we discussed computing variance under interval uncertainty, this property holds, e.g., when we perform all the measurements with the same measuring instrument. Another case when this property is satisfied is when we have a database in which, for privacy purposes, we select thresholds $t_1 < t_2 < \ldots < t_k$, and each value $x$ is replaced by the thresholds interval $[t_j, t_{j+1}]$ that contains this value.

**Theorem 19.6.** *There exists an algorithm that computes $\underline{M}_3$ and $\overline{M}_3$ in quadratic time when the intervals satisfy the no-subset property.*

When we have a limited number $m$ of measuring instruments, then the resulting intervals can be divided into $C$ families corresponding to different intervals. In this case, for each $m$, it is also possible to have a polynomial-time algorithm for computing the third central moment.

**Theorem 19.7.** *There exists an algorithm that computes $\underline{M}_3$ and $\overline{M}_3$ in time $O(n^{2m})$ when the intervals can be divided into $m$ families each of which satisfies the no-subset property.*

## Proofs of Theoretical Results and Justifications of Algorithms

*Proof of Theorem 19.1.* The central moment $M_h$ is a continuous function of $n$ variables; thus, its smallest possible value on a compact box $\boldsymbol{x}_1 \times \ldots \times \boldsymbol{x}_n$ is attained at some point $x^{(0)} = (x_1^{(0)}, \ldots, x_n^{(0)})$. Since the function $M_h$ is also smooth, for each variable $i$ for which the interval $\boldsymbol{x}_i$ is non-degenerate (i.e., of finite width), the minimum is attained either when $x_i^{(0)}$ is inside the corresponding interval $(\underline{x}_i, \overline{x}_i)$ and value $M_{m,i}$ of the derivative $\partial M_h / \partial x_i$ at $x_i = x_i^{(0)}$ is equal to 0, or when $x_i^{(0)}$ coincides with one of the endpoints of this interval. To be more precise, we must have one of the following three situations:

- either $x_i^{(0)} \in (\underline{x}_i, \overline{x}_i)$ and $\partial M_h / \partial x_i = 0$ at $x_i = x_i^{(0)}$;
- or $x_i^{(0)} = \underline{x}_i$ and $\partial M_h / \partial x_i \geq 0$ at $x_i = x^{(0)}$;
- or $x_i^{(0)} = \overline{x}_i$ and $\partial M_h / \partial x_i \leq 0$ at $x_i = x^{(0)}$.

Differentiating $M_h$ w.r.t. $x_i$, and taking into consideration that $\partial E / \partial x_i = 1/n$, we conclude that

$$\frac{\partial M_h}{\partial x_i} = \frac{1}{n} \cdot h \cdot (x_i - E)^{h-1} + \frac{1}{n} \cdot \sum_{j=1}^{n} h \cdot (x_j - E)^{h-1} \cdot \left(-\frac{1}{n}\right).$$

The summation in this formula is proportional to the $(h-1)$-st central moment $M_{h-1}$, and allows the above formula can be simplified into:

$$\frac{\partial M_h}{\partial x_i} = \frac{m}{n} \cdot ((x_i - E)^{h-1} - M_{h-1}). \tag{19.1}$$

Therefore,

$$M_{m,i} = \frac{\partial M_h}{\partial x_i}(x_1^{(0)}, \ldots, x_n^{(0)}) = \frac{h}{n} \cdot ((x_i^{(0)} - E^{(0)})^{h-1} - M_{h-1}^{(0)}), \tag{19.2}$$

where

$$E^{(0)} \stackrel{\text{def}}{=} \frac{x_1^{(0)} + \ldots + x_n^{(0)}}{n}$$

and

$$M_{h-1}^{(0)} = \frac{(x_1^{(0)} - E^{(0)})^{h-1} + \ldots + (x_n^{(0)} - E^{(0)})^{h-1}}{n}.$$

Due to this formula:

- if $\dfrac{\partial M_h}{\partial x_i} = 0$ at $x_i = x_i^{(0)}$, then $(x_i^{(0)} - E^{(0)})^{h-1} = M_{h-1}^{(0)}$, hence $x_i^{(0)} = E^{(0)} + (M_{h-1}^{(0)})^{1/(h-1)} \overset{\text{def}}{=} \alpha$;

- if $\dfrac{\partial M_h}{\partial x_i} \geq 0$ at $x_i = x_i^{(0)}$, then $(x_i^{(0)} - E^{(0)})^{h-1} \geq M_{h-1}^{(0)}$, hence (since the function $z \to z^{1/(h-1)}$ is increasing for even $h$) $x_i^{(0)} \geq \alpha$;

- if $\dfrac{\partial M_h}{\partial x_i} \leq 0$ at $x_i = x_i^{(0)}$, then $(x_i^{(0)} - E^{(0)})^{h-1} \leq M_{h-1}^{(0)}$, hence $x_i^{(0)} \leq \alpha$.

Therefore, the above conditions on $x_i^{(0)}$ can be reformulated as follows:

- either $x_i^{(0)} \in (\underline{x}_i, \overline{x}_i)$ and $x_i^{(0)} = \alpha$; in this case, $\underline{x}_i < \alpha < \overline{x}_i$;
- or $x_i^{(0)} = \underline{x}_i$ and $x_i^{(0)} \geq \alpha$;
- or $x_i^{(0)} = \overline{x}_i$ and $x_i^{(0)} \leq \alpha$.

Hence, once we know $\alpha$, we can determine all $n$ values $x_i^{(0)}$ as follows:

- if $\overline{x}_i \leq \alpha$, then we cannot have the first case (when $\alpha < \overline{x}_i$) or the second case (when $\alpha \leq \underline{x}_i$ hence $\alpha < \overline{x}_i$); therefore, we can only have the third case, when $x_i^{(0)} = \overline{x}_i$;
- similarly, if $\alpha \leq \underline{x}_i$, then we must have $x_i^{(0)} = \underline{x}_i$;
- finally, when $\underline{x}_i < \alpha < \overline{x}_i$, then we must have $x_i^{(0)} = \alpha$.

The only thing that remains is to find $\alpha$. Once we know to which of the $2n + 1$ segments $[x_{(k)}, x_{(k+1)}]$ the value $\alpha$ belongs (and if $\alpha$ is an endpoint, it can belong to two segments), we then can tell for which of the original intervals $[\underline{x}_i, \overline{x}_i] = \boldsymbol{x}_i$ the value leading to the minimal value of $\underline{M}_h$ is not an endpoint (i.e., not $\underline{x}_i$ or $\overline{x}_i$). We can uniquely describe all the values $x_i$ as linear functions of $\alpha$, and then define $\alpha$ from the condition that $\alpha = E + M_{h-1}^{1/(h-1)}$, i.e., equivalently, that $M_{h-1} = (\alpha - E)^{h-1}$. This is exactly what our algorithm does.

This proves that our algorithm is correct. To complete the proof, we must also show that this algorithm takes time $O(n \cdot \log(n))$.

Indeed, sorting takes $O(n \cdot \log(n))$ steps, and the rest of the algorithm takes linear time ($O(n)$) for each of $2n$ segments. Computations corresponding to the first segment take linear time $O(n)$. Then, we can use the some corresponding to each segment to compute the sums corresponding to the next segment. For each of the input intervals, the corresponding value in each of the sums changes only once. Thus, overall, all these changes in sums take linear time $O(n)$. So, the total computation time is

$$O(n \cdot \log(n)) + O(n) + O(n) = O(n \cdot \log(n)).$$

The theorem is proven.

*Proof of Theorem 19.2.* Similarly to the proof of Theorem 19.1, we an conclude that for every $i$, the maximum of $M_h$ over the interval $\boldsymbol{x}_i$ is attained either inside the interval (when the partial derivative is 0) or at one of the endpoint of this interval. Thus, to prove that our algorithm is correct, we must show that the maximum of $M_h$ cannot be attained for $x_i \in (\underline{x}_i, \overline{x}_i)$, when $\partial M_h/\partial x_i = 0$. Indeed, in the maximum point, the second derivative $\partial^2 M_h/\partial x_i^2$ must be non-positive. In the proof of Theorem 19.1, we have already derived an explicit formula (19.2) for $\partial M_h/\partial x_i$. The formula (19.2) describes this derivative in terms of $M_{h-1}$, so when we differentiate both sides of the formula (19.2), we can use the same expression for the derivative of $M_{h-1}$. As a result, we get the following:

$$\frac{\partial^2 M_h}{\partial x_i^2} = \frac{h}{n} \cdot T,$$

where

$$T = (h-1) \cdot (x_i - E)^{h-2} - (h-1) \cdot (x_i - E)^{h-2} \cdot \frac{1}{n} -$$

$$\frac{h-1}{n} \cdot (x_i - E)^{h-2} + \frac{h-1}{n} \cdot M_{h-2} =$$

$$\frac{h-1}{n} \cdot \left((n-2) \cdot (x_i - E)^{h-2} + M_{h-2}\right).$$

In the trivial case of $n = 1$, all central moments are 0. When $n \geq 2$, both terms are non-negative, so the second derivative is non-negative. We know that the second derivative must be non-positive, so it must be equal to 0. Since the sum of two non-negative numbers is equal to 0, both numbers are equal to 0, in particular,

$$M_{h-2} = \frac{1}{n} \cdot \sum_{i=1}^{n} (x_i - E)^{h-2} = 0.$$

Therefore, all the values $x_i$ are identically equal to $E$. In this case, $M_h = 0$, so this cannot be where the largest possible value of $M_h$ is attained. This contradiction shows that the maximum cannot be attained inside the interval $\boldsymbol{x}_i$, hence it attained at the endpoints. The theorem is proven.

*Proof of Theorem 19.3.* We have already proven, in Theorem 19.2, that maximum can only be attained at one of the endpoints of the interval $[\underline{x}_i, \overline{x}_i]$, i.e., when $x_i^{(0)} = \underline{x}_i$ or $x_i^{(0)} = \overline{x}_i$. Hence, for each $i$, we have one of the following two situations:

- either $x_i^{(0)} = \underline{x}_i$ and $\partial M_h/\partial x_i \leq 0$;
- or $x_i^{(0)} = \overline{x}_i$ and $\partial M_h/\partial x_i \geq 0$.

We already know, from the proof of Theorem 1, that the condition $\partial M_h/\partial x_i \leq 0$ is equivalent to $x_i \leq \alpha$, and the condition $\partial M_h/\partial x_i \geq 0$ is equivalent to $x_i \geq \alpha$. Thus, the above two situations can be reformulated a follows:

- either $x_i^{(0)} = \underline{x}_i$ and $x_i^{(0)} = \underline{x}_i \leq \alpha$;
- or $x_i^{(0)} = \overline{x}_i$ and $x_i^{(0)} = \overline{x}_i \geq \alpha$.

Hence:

- if $\overline{x}_i < \alpha$, then we cannot have the second case (when $\overline{x}_i \geq \alpha$) and therefore, we can only have the first case, when $x_i^{(0)} = \underline{x}_i$;
- similarly, if $\alpha < \underline{x}_i$, then we must have $x_i^{(0)} = \overline{x}_i$.

The only case when the knowledge of $\alpha$ does not help us determine $x_i$ is the case when $\underline{x}_i \leq \alpha \leq \overline{x}_i$, i.e., when $\alpha \in \boldsymbol{x}_i$.

Since intervals have $c_0$-few intersections, for each $\alpha$, there can be no more than $c_0$ such intervals, so we can try all $2^{c_0}$ possible assignments for each segment. In other words, the time increases by a constant ($\leq 2^{c_0}$) over the running time of the algorithm described in Theorem 19.1. This justifies the algorithm and proves that it runs in time $O(n \cdot \log(n))$.

*Proof of Theorem 19.4.* Similarly to the proof of Theorem 19.1, we conclude that for the point where the function $M_h$ attains its minimum, we have:

- either $x_i^{(0)} \in (\underline{x}_i, \overline{x}_i)$ and $\partial M_h / \partial x_i = 0$;
- or $x_i^{(0)} = \underline{x}_i$ and $\partial M_h / \partial x_i \geq 0$;
- or $x_i^{(0)} = \overline{x}_i$ and $\partial M_h / \partial x_i \leq 0$.

Here, the derivative $\partial M_h / \partial x_i$ is described by the same formula (19.2) as in the proof of Theorem 19.1. The difference is that $h$ is now odd, so:

- if $\partial M_h / \partial x_i = 0$, then $(x_i - E)^{h-1} = M_{h-1}$, hence $|x_i - E| = M_{h-1}^{1/(h-1)}$, so either $x_i$ is equal to $\alpha^- \overset{\text{def}}{=} E - M_{h-1}^{1/(h-1)}$, or $x_i$ is equal to $\alpha^+ \overset{\text{def}}{=} E + M_{h-1}^{1/(h-1)}$;
- if $\partial M_h / \partial x_i \geq 0$, then $(x_i - E)^{h-1} \geq M_{h-1}$, hence $|x_i - E| \geq M_{h-1}^{1/(h-1)}$, so $x_i \leq \alpha^-$ or $x_i \geq \alpha^+$;
- if $\partial M_h / \partial x_i \leq 0$, then $(x_i - E)^{h-1} \leq M_{h-1}$, hence $|x_i - E| \leq M_{h-1}^{1/(h-1)}$, so $\alpha^- \leq x_i \leq \alpha^+$.

Therefore, the above conditions on $x_i^{(0)}$ can be reformulated as follows:

- in the first case, $x_i^{(0)} \in (\underline{x}_i, \overline{x}_i)$ and either $x_i^{(0)} = \alpha^-$ or $x_i^{(0)} = \alpha^+$; in this case, either $\underline{x}_i < \alpha^- < \overline{x}_i$ or $\underline{x}_i < \alpha^+ < \overline{x}_i$;
- in the second case, $x_i^{(0)} = \underline{x}_i$ and either $x_i^{(0)} = \underline{x}_i \leq \alpha^-$ or $x_i^{(0)} = \underline{x}_i \geq \alpha^+$;
- in the third case, $x_i^{(0)} = \overline{x}_i$ and $\alpha^- \leq x_i^{(0)} = \overline{x}_i \leq \alpha^+$.

Hence, once we know $\alpha^-$ and $\alpha^+$, we can determine (at least some) some values $x_i^{(0)}$ as follows:

- if $\alpha^- < \underline{x}_i < \overline{x}_i < \alpha^+$, then we cannot have the first case (when $\alpha^- > \underline{x}_i$ or $\alpha^+ < \overline{x}_i$), and we cannot have the second case (when $\underline{x}_i \leq \alpha^-$ or $\underline{x}_i \geq \alpha^+$); therefore, we can only have the third case, when $x_i^{(0)} = \overline{x}_i$;

- if $\alpha^+ \le \underline{x}_i$, then we cannot have the first case (when either $\alpha^+ > \underline{x}_i$, or $\alpha^- > \underline{x}_i$ hence $\alpha^+ > \underline{x}_i$), and we cannot have the third case (when $\overline{x}_i \le \alpha^+$ hence $\underline{x}_i < \alpha^+$); therefore, we can only have the second case, when $x_i^{(0)} = \underline{x}_i$;
- similarly, if $\overline{x}_i < \alpha^-$, then we cannot have the first case (when either $\alpha^- < \underline{x}_i$, or $\alpha^+ < \overline{x}_i$ hence $\alpha^- < \overline{x}_i$), and we cannot have the third case (when $\alpha^- \le \overline{x}_i$); therefore, we can only have the second case, when $x_i^{(0)} = \underline{x}_i$.

We have described all the cases in which neither of the two auxiliary values $\alpha^-$ and $\alpha^+$ is in the interval $\boldsymbol{x}_i$; in all these cases, we can uniquely determine the value $x_i^{(o)}$. The only cases when we cannot uniquely determine the value $x_i^{(0)}$ are the cases when either $\alpha^-$ or $\alpha^+$ is within the interval $\boldsymbol{x}_i$.

Once we choose segments that contain $\alpha^-$ and $\alpha^+$, we have no more than $c_0$ intervals $\boldsymbol{x}_i$ that contain $\alpha^-$ and no more than $c_0$ intervals that contain $\alpha^+$. Thus, for the remaining $\ge n - 2c_0$ indices $i$, we can uniquely determine $x_i$. For the $2c_0$ indices, we try all possible combinations. This is exactly what we do in our algorithm. Thus, the algorithm is indeed correct.

The algorithm takes linear time $O(n)$ for each pair of segments. There are $O(n^2)$ pairs of segments, and for each segment, we can modify the sums from the previous pair, so all this computation takes $O(n^2)$ time. Thus, overall, we take $O(n \cdot \log(n))$ time for sorting, $O(n)$ for the first pair, and quadratic time for computing the values for all the pairs. The resulting time is thus

$$O(n \cdot \log(n)) + O(n) + O(n^2) = O(n^2),$$

hence the algorithm takes quadratic time. The theorem is proven.

*Proof of Theorem 19.6.*

$1^\circ$. Let us denote $M_3$ by $S$. We have already mentioned that for odd $h$ (including our case $h = 3$), $\overline{S}$, then we can use the fact that $S$ is an odd function $S(-x_1, \ldots, -x_n) = -S(x_1, \ldots, x_n)$ to compute $\underline{S}$ as well: since $\boldsymbol{S}(-\boldsymbol{x}_1, \ldots, -\boldsymbol{x}_n) = -\boldsymbol{S}(\boldsymbol{x}_1, \ldots, \boldsymbol{x}_n)$, we conclude that $\overline{S}(-\boldsymbol{x}_1, \ldots, -\boldsymbol{x}_n) = -\underline{S}(\boldsymbol{x}_1, \ldots, \boldsymbol{x}_n)$ and thus, $\underline{S}(\boldsymbol{x}_1, \ldots, \boldsymbol{x}_n) = -\overline{S}(-\boldsymbol{x}_1, \ldots, -\boldsymbol{x}_n)$

In view of this comment, in the remaining part of the proof, we will only consider an algorithm for computing $\overline{S}$.

$2^\circ$. As we have shown while proving the results about variance, since the interval data satisfies the subset property, after we sort these elements in lexicographic order, both the lower endpoints $\underline{x}_i$ and the upper endpoints $\overline{x}_i$ are sorted in non-decreasing order: $\underline{x}_i \le \underline{x}_{i+1}$ and $\overline{x}_i \le \overline{x}_{i+1}$.

$3^\circ$. The maximum of a differentiable function $S(x_1, \ldots, x_n)$ on an interval $[\underline{x}_i, \overline{x}_i]$ can be attained either in an internal point of this interval, or at one of the endpoints.

If the maximum is attained at an internal point, then the first derivative is $0$ $\left( \dfrac{\partial S}{\partial x_i} = 0 \right)$ and the second derivative should be non-positive $\left( \dfrac{\partial^2 S}{\partial x_i^2} \le 0 \right)$.

If the maximum is attained at the left endpoint, the function $S$ cannot be increasing at this point, so we must have $\dfrac{\partial S}{\partial x_i} \leq 0$. Similarly, if the maximum is attained at the right endpoint, the function $S$ cannot be decreasing at this point, so we must have $\dfrac{\partial S}{\partial x_i} \geq 0$.

For skewness,

$$\frac{\partial S}{\partial x_i} = \frac{3}{n} \cdot (x_i - E)^2 - \frac{3}{n} \cdot \sum_{j=1}^{n} (x_j - E)^2 \cdot \frac{\partial E}{\partial x_i}.$$

Since $\dfrac{\partial E}{\partial x_i} = \dfrac{1}{n}$, we thus get $\dfrac{\partial S}{\partial x_i} = \dfrac{3}{n} \cdot ((x_i - E)^2 - V)$. So, the first derivative of $S$ has the same sign as the expression $(x_i - E)^2 - V$.

To compute the second derivative of $S$, we must take into account that $\dfrac{\partial V}{\partial x_i} = \dfrac{2}{n} \cdot (x_i - E)$, hence

$$\frac{\partial^2 S}{\partial x_i^2} = \frac{3}{n} \cdot \left( 2(x_i - E) - 2(x_i - E) \cdot \frac{1}{n} - \frac{2}{n} \cdot (x_i - E) + \frac{2}{n} \cdot \sum_{j=1}^{n} (x_j - E) \cdot \frac{1}{n} \right).$$

Since $\displaystyle\sum_{j=1}^{n} (x_j - E) = 0$, we conclude that

$$\frac{\partial^2 S}{\partial x_i^2} = \frac{3}{n} \cdot 2 \cdot \left( 1 - \frac{2}{n} \right) \cdot (x_i - E).$$

We have already mentioned that the problem of computing skewness only makes sense for $n > 2$, because for $n \leq 2$, the skewness is identically 0. For $n > 2$, the second derivative $\dfrac{\partial^2 S}{\partial x_i^2}$ has the same sign as the expression $x_i - E$.

Thus, for skewness, we value $x_i$ at which the maximum is attained satisfies one of the following three conditions:

- either $\underline{x}_i < x_i < \overline{x}_i$, $(x_i - E)^2 - V = 0$, and $x_i - E \leq 0$,
- or $x_i = \underline{x}_i$ and $(x_i - E)^2 - V \leq 0$,
- or $x_i = \overline{x}_i$ and $(x_i - E)^2 - V \geq 0$.

In the first case, $(x_i - E)^2 = V = \sigma^2$, hence $x_i - E = \pm\sigma$. Since $x_i - E \leq 0$, we cannot have $x_i - E = \sigma$, so in this case, $x_i = E - \sigma$. In the second case, $(x_i - E)^2 \leq V = \sigma^2$, hence $E - \sigma \leq x_i \leq E + \sigma$. In the third case, $(x_i - E)^2 \geq V = \sigma^2$, so either $x_i \leq E - \sigma$ or $x_i \geq E + \sigma$. So:

- either $\underline{x}_i < x_i = E - \sigma < \overline{x}_i$,
- or $x_i = \underline{x}_i$ and $E - \sigma \leq \underline{x}_i \leq E + \sigma$,
- or $x_i = \overline{x}_i$ and either $\overline{x}_i \leq E - \sigma$ or $\overline{x}_i \geq E + \sigma$.

In all three cases, the desired maximum of the skewness $S$ is attained when $x_i$ is either at one of the endpoints of the corresponding interval $\mathbf{x}_i$, or has the value $\mu \overset{\text{def}}{=} E - \sigma$.

$4°$. Let us now deduce a more specific information about the values $x_i$ at which the maximum is attained.

Based on the above description of possible cases, once we know how the intervals are located in relation to $E - \sigma$ and $E + \sigma$, we can sometimes uniquely determine the value $x_i$ at which the maximum is attained. Namely,

- If $\overline{x}_i \leq E - \sigma$, then the maximum cannot be attained at an internal point and it cannot be attained at the value $\underline{x}_i$, so it is attained when $x_i = \overline{x}_i$.
- If $\underline{x}_i \leq E - \sigma \leq \overline{x}_i \leq E + \sigma$, then the maximum can only be attained when $x_i = E - \sigma$.
- If $E - \sigma \leq \underline{x}_i \leq E + \sigma$, then the maximum is attained at $x_i = \underline{x}_i$.
- Finally, if $E + \sigma \leq \underline{x}_i$, then the maximum is attained at $x_i = \overline{x}_i$.

These conclusions can be described in the following graphical manner, in which the arrows indicate the direction towards the corresponding maximum:

The only case when we cannot exactly determine the optimal value $x_i$ is when the interval $\mathbf{x}_i$ contains the value $E + \sigma$: it this case, we may have $x_i = \overline{x}_i$, and we may also have $x_i = \max(E - \sigma, \underline{x}_i)$.

$5°$. Let us show that the maximum of skewness is always attained at a vector $x = (x_1, \ldots, x_n)$ which can be divided into three consequent fragments (some of which may be empty):

- first, we have values $\overline{x}_i$ which are smaller than $E - \sigma$;
- then, we have the values $\max(E - \sigma, \underline{x}_i)$;
- finally, we have the values $\overline{x}_i$ which are larger than $E + \sigma$.

All the intervals $\mathbf{x}_i$ that do not contain $E + \sigma$ inside naturally fall into this scheme. The only intervals that we do need to consider to prove this result are the intervals that do contain $E + \sigma$. For each of these intervals, the corresponding values $x_i$ are either $\max(E - \sigma, \underline{x}_i)$ or $\overline{x}_i$. What we claim is that after we sort the intervals in lexicographic order, we will first have the values equal to $\max(E - \sigma, \underline{x}_i)$, and then the values equal to $\overline{x}_i$. In other words, once we have a value $x_i = \overline{x}_i$, all the following values will also be of the same type.

We will show that if there is an optimizing vector at which this condition is not satisfied, then we can rearrange it into a new vector with the same optimal value of $S$ for which this condition holds.

Indeed, let us start with a vector for which, for some $i$, for two consequent intervals $\mathbf{x}_i$ and $\mathbf{x}_{i+1}$, a value $x_i = \overline{x}_i \geq E + \sigma$ is followed by a value $x_{i+1} = \max(E - \sigma, \underline{x}_{i+1}) \leq E + \sigma$. If there are several such indices $i$, we take the smallest $i$ with this property.

According to Part 2 of this proof, we have $\underline{x}_i \leq \underline{x}_{i+1} \leq E + \sigma$ and $E + \sigma \leq \overline{x}_i \leq \overline{x}_{i+1}$. Thus, $x_i = \overline{x}_i \leq \overline{x}_{i+1}$ and $x_i = \overline{x}_i \geq E + \sigma \geq \underline{x}_{i+1}$; hence, $x_i \in \mathbf{x}_{i+1}$. Similarly, $x_{i+1} \in \mathbf{x}_i$. Thus, we can "swap" the values $x_i$ and $x_{i+1}$: as a new value of $x_i$, we take the old value of $x_{i+1}$, and vice versa. The swap does not change the average $E$ and does not change the sample skewness $S$, so the function $S$ attains the maximum at the new values as well.

As a result of this swap, if there is now a value $i'$ for which $\overline{x}_{i'}$ is followed by $\max(E - \sigma, \underline{x}_{i'+1})$, this value $i'$ has to be equal to at least $i + 1$. If there still is such an index $i'$, we apply a new swap again and thus again increase the smallest problematic value $i$. After $\leq n$ such swaps, there will be no problematic cases anymore, so we will get a sequence which has the desired property.

$6°$. To determine the optimal vector $x$, we must thus select a zone $[x_{(p)}, x_{(p+1)}]$ that contains $\mu = E - \sigma$, and an index $k$ at which the optimal value $x_i$ switches from $\max(\mu, \underline{x}_i)$ to $\overline{x}_i$.

Once $p$ and $k$ are fixed, we can uniquely determine each of the optimal values $x_i$ – some as known numbers, some as equal to the (unknown) value $\mu$:

- when $\overline{x}_i \leq x_{(p)}$, we have $x_i = \overline{x}_i$;
- when $\underline{x}_i < x_{(p)} < x_{(p+1)} \leq \overline{x}_i$ and $i < k$, we have $x_i = \mu$;
- when $x_{(p+1)} \leq \underline{x}_i$ and $i < k$, we have $x_i = \underline{x}_i$;
- finally, when $i \geq k$, we have $x_i = \overline{x}_i$.

To find $\mu$, we must use the fact that $\mu = E - \sigma$. Specifically, the average $E$ can be determined as

$$\frac{1}{n} \cdot \sum_{i \in N'} x_i + \frac{n - n'}{n} \cdot (E - \sigma) = E,$$

where the sum is taken over the set $N'$ of all the indices for which $x_i$ is known, and $n'$ is the total number of such indices. Similarly, the sample second moment $E^2 + \sigma^2$ can be determined as

$$\frac{1}{n} \cdot \sum_{i \in N'} x_i^2 + \frac{n - n'}{n} \cdot (E - \sigma)^2 = E^2 + \sigma^2.$$

From the first of these equations, we can determine $\sigma$ as a linear function of $E$. Substituting this expression into the second equation, we get a quadratic equation with the only unknown $\sigma$, from which we can determine $\sigma$. Then, we can use the first equation to find $E$ – and hence find $\mu = E - \sigma$.

If the resulting value of $\mu$ is indeed within the zone $[x_{(p)}, x_{(p+1)}]$, then we compute the sample skewness for the corresponding values $x_i$. Specifically, the skewness can be computed as

$$\frac{1}{n} \cdot \sum_{i=1}^{n} (x_i - E)^3 = \frac{1}{n} \cdot \sum_{i=1}^{n} x_i^3 - \frac{3 \cdot E}{n} \cdot \sum_{i=1}^{n} x_i^2 + \frac{3 \cdot E^2}{n} \cdot \sum_{i=1}^{n} x_i - E^3 =$$

$$\frac{1}{n} \cdot \sum_{i=1}^{n} x_i^3 - \frac{3 \cdot E}{n} \cdot \sum_{i=1}^{n} x_i^2 + 2 \cdot E^3 =$$

$$\frac{1}{n} \cdot \sum_{i \in N'} x_i^3 + \frac{n - n'}{n} \cdot \mu^3 - \frac{3 \cdot E}{n} \cdot \left( \sum_{i \in N'} x_i^2 + (n - n') \cdot \mu^2 \right) + 2 \cdot E^3.$$

The largest of these skewnesses is the desired value $\overline{S}$.

$7°$. How much times does this algorithm take? Sorting takes time $O(n \cdot \log(n))$.

For $n$ interval data points, we have $2n$ possible zone and $n$ possible indices $k$ – totally, $O(n^2)$ possible pairs $(p, k)$. For the first pair, computing the corresponding values $n'$, $\sum_{i \in N'} x_i$, $\sum_{i \in N'} x_i^2$, and $\sum_{i \in N'} x_i^3$ takes linear time. For each next pair, we, in general, change one value in comparison with the previous pair, so each new computation takes a constant number of steps. Thus, for $O(n^2)$ pairs, we take $O(n^2)$ time. (In some cases, we change more than one value, but still, each value changes only once, so we still take $O(n^2)$ times).

So, overall, we take time $O(n \cdot \log(n)) + O(n) + O(n^2) = O(n^2)$.

The theorem is proven.

*Proof of Theorem 19.7.* In this case, the interval data consists of $C$ families of intervals such that within each family, no two intervals are proper subsets of each other.

Similarly to the proof of the previous Theorem 19.6, we can conclude that if we sort each family in lexicographic order, then, within each family, the maximum of $S$ is attained on one of the sequences described in Part $6°$ of proof of the previous theorem. Thus, to find the desired maximum $\overline{S}$, it is sufficient to know the values $p_\alpha \leq n$ and $k_\alpha \leq n$ corresponding to each of $m$ families.

For each family $\alpha$, there are $\leq n^2$ such combinations, so overall, there are $\leq (n^2)^m = n^{2m}$ combinations of such values. For the first combination, computing the corresponding value of $S$ takes $O(n)$ steps. Each next combination differs from the previous one by a single term, so overall, we take $O(n^{2m})$ steps to compute all the values of $S$ – and thus, to find the largest of them, which is $\overline{S}$.

# Computing Mean, Variance, Higher Moments, and Their Linear Combinations under Interval Uncertainty: A Brief Summary

In the previous chapters, we described several results and algorithms for computing:

- the mean $E$,

- the variance $V = \sigma^2 = \dfrac{1}{n} \cdot \sum_{i=1}^{n} (x_i - E)^2$,

- more generally, higher central moments $M_h = \dfrac{1}{n} \cdot \sum_{i=1}^{n} (x_i - E)^h$, and

- statistically useful linear combinations of these characteristics – such as the lower and upper endpoints of the confidence interval $L = E - k_0 \cdot \sigma$ and $U = E + k_0 \cdot \sigma$, where the parameter $k_0$ is usually taken as $k_0 = 2$, $k_0 = 3$, and $k_0 = 6$.

For most of these problems, we considered the following situations:

- case when the intervals (or, better yet, narrowed intervals) satisfy the no-subset property; this case corresponds, e.g., to the case when we process measurement results produced by the same measuring instrument (MI); the privacy case is another case when intervals satisfy the no-subset property;

- case when intervals can be divided into $m$ classes within each of which the no-subsect property is satisfied; this case corresponds, e.g., to the case when we process measurement results produced by $m$ different measuring instruments (MI); and

- case when for some $c_0 \geq 2$, every group of $c_0$ intervals has an empty intersection; this case corresponds to narrow intervals, e.g., to the case when measuring instruments are sufficiently accurate.

The results of these chapters can be summarized in the following table; for comparison, we also add the complexity of the general case, when we have an arbitrary collection of intervals.

H.T. Nguyen et al.: Computing Stat. under Interval & Fuzzy Uncertain., SCI 393, pp. 167–168.
springerlink.com 

| Case | $E$ | $V$ | $L, U$ | $M_3$ | $M_h$ |
|---|---|---|---|---|---|
| Single MI | $O(n)$ | $O(n)$ | $O(n \cdot \log(n))$ | $O(n^2)$ | ? |
| Several $(m)$ MI | $O(n)$ | $O(n^m)$ | $O(n^m)$ | $O(n^{2m})$ | ? |
| Narrow intervals | $O(n)$ | $O(n)$ | $O(n \cdot \log(n))$ | $O(n^2)$ | $O(n^2)$ |
| General | $O(n)$ | NP-hard | NP-hard | ? | NP-hard |

# Computing Covariance under Interval Uncertainty

## Formulation of the Problem

When we have two sets of data $x_1, \ldots, x_n$ and $y_1, \ldots, y_n$, we normally compute *finite population covariance*

$$C_{xy} = \frac{1}{n} \sum_{i=1}^{n} (x_i - E_x) \cdot (y_i - E_y),$$

where

$$E_x = \frac{1}{n} \sum_{i=1}^{n} x_i; \quad E_y = \frac{1}{n} \sum_{i=1}^{n} y_i.$$

Finite population covariance is used to describe the correlation between $x_i$ and $y_i$.

If we take interval uncertainty into consideration, then, after each measurement, we do not get the exact values of $x_1, \ldots, x_n$, $y_1, \ldots, y_n$; instead, we only have *intervals* $[\underline{x}_1, \overline{x}_1], \ldots, [\underline{x}_n, \overline{x}_n]$, $[\underline{y}_1, \overline{y}_1], \ldots, [\underline{y}_n, \overline{y}_n]$. Depending on what are the actual values of $x_1, \ldots, x_n$, $y_1, \ldots, y_n$ within these intervals, we get different values of finite population covariance. To take the interval uncertainty into consideration, we need to be able to describe the interval $[\underline{C}_{xy}, \overline{C}_{xy}]$ of possible values of the finite population covariance $C_{xy}$.

So, we arrive at the following problems: given the intervals $[\underline{x}_i, \overline{x}_i]$, $[\underline{y}_i, \overline{y}_i]$, compute the lower and upper bounds $\underline{C}_{xy}$ and $\overline{C}_{xy}$ for the interval of possible values of finite population covariance.

## Results

It turns out that, similarly to the problem of estimating variance under interval uncertainty, the problems problems of estimating covariance are also NP-hard:

H.T. Nguyen et al.: Computing Stat. under Interval & Fuzzy Uncertain., SCI 393, pp. 169–171.
springerlink.com

**Theorem 21.1.** *The problem of computing $\overline{C}_{xy}$ from the interval inputs $[\underline{x}_i, \overline{x}_i]$, $[\underline{y}_i, \overline{y}_i]$ is NP-hard.*

**Theorem 21.2.** *The problem of computing $\underline{C}_{xy}$ from the interval inputs $[\underline{x}_i, \overline{x}_i]$, $[\underline{y}_i, \overline{y}_i]$ is NP-hard.*

*Comment.* These results first appeared in [102] and [271].


## Proofs

$1°$. Similarly to the proof of Theorem 14.1, we reduce a subset problem to the problem of computing $\overline{C}_{xy}$.

Each instance of the subset problem is as follows: given $n$ positive integers $s_1, \ldots, s_n$, to check whether there exist signs $\eta_i \in \{-1, +1\}$ for which the signed sum $\sum_{i=1}^{n} \eta_i \cdot s_i$ equals 0.

We will show that this problem can be reduced to the problem of computing $\overline{C}_{xy}$, i.e., that to every instance $(s_1, \ldots, s_n)$ of the subset problem $\mathcal{P}_0$, we can put into correspondence such an instance of the $\overline{C}_{xy}$-computing problem that based on its solution, we can easily check whether the desired signs exist.

As this instance, we take the instance corresponding to the intervals $[\underline{x}_i, \overline{x}_i] = [\underline{y}_i, \overline{y}_i] = [-s_i, s_i]$. We want to to show that for the corresponding problem, $\overline{C}_{xy} = C_0$ (where $C_0$ is the same as in the proof of Theorem 14.1) if and only if there exist signs $\eta_i$ for which $\sum_{i=1}^{n} \eta_i \cdot s_i = 0$.

$2°$. Let us first show that in all cases, $\overline{C}_{xy} \le C_0$.

Indeed, it is known that the finite population covariance $C_{xy}$ is bounded by the product $\sigma_x \cdot \sigma_y$ of finite population standard deviations $\sigma_x = \sqrt{V_x}$ and $\sigma_y = \sqrt{V_y}$ of $x$ and $y$. In the proof of Theorem 14.1, we have already proven that the finite population variance $V_x$ of the values $x_1, \ldots, x_n$ satisfies the inequality $V_x \le C_0$; similarly, the finite population variance $V_y$ of the values $y_1, \ldots, y_n$ satisfies the inequality $V_y \le C_0$. Hence, $C_{xy} \le \sigma_x \cdot \sigma_y \le \sqrt{C_0} \cdot \sqrt{C_0} = C_0$. In other words, every possible value $C_{xy}$ of the finite population covariance is smaller than or equal to $C_0$. Thus, the largest of these possible values, i.e., $\overline{C}_{xy}$, also cannot exceed $C_0$, i.e., $\overline{C}_{xy} \le C_0$.

$3°$. Let us now show that if $\overline{C}_{xy} = C_0$, then the desired signs exist.

Indeed, if $\overline{C}_{xy} = C$, this means that for the corresponding values of $x_i$ and $y_i$, the finite population covariance $C_{xy}$ is equal to $C_0$, i.e.,

$$C_{xy} = C_0 = \frac{1}{n} \cdot \sum_{i=1}^{n} s_i^2.$$

On the other hand, we have shown that in all cases (and in this case in particular), $C_{xy} \le \sigma_x \cdot \sigma_y \le \sqrt{C_0} \cdot \sqrt{C_0} = C_0$. If $\sigma_x < \sqrt{C_0}$, then we would have $C_{xy} < C_0$. So, if $C_{xy} = C_0$, we have $\sigma_x = \sigma_y = \sqrt{C_0}$, i.e., $V_x = V_y = C_0$. We have already shown, in the proof of Theorem 14.1, that in this case the desired signs exist.

$4°$. To complete the proof of Theorem 21.1, we must show that, vice versa, if the desired signs $\eta_i$ exist, then $\overline{C}_{xy} = C_0$.

Indeed, in this case, for $x_i = y_i = \eta_i \cdot s_i$, we have $\mu_x = \mu_y = 0$ and $x_i \cdot y_i = s_i^2$, hence

$$C_{xy} = \frac{1}{n} \cdot \sum_{i=1}^{n} (x_i - E_x) \cdot (y_i - E_y) = \frac{1}{n} \cdot \sum_{i=1}^{n} s_i^2 = C_0.$$

The theorem is proven.

*Proof of Theorem 21.2.* This proof is similar to the proof of Theorem 21.1, with the only difference that in this case, we use the other part of the inequality $|C_{xy}| \le \sigma_x \cdot \sigma_y$, namely, that $C_{xy} \ge -\sigma_x \cdot \sigma_y$, and in the last part of the proof, we take $y_i = -x_i$.

# Computing Correlation under Interval Uncertainty

## Formulation of the Problem

As we have mentioned in Chapter 21, finite population covariance $C$ between the data sets $x_1, \ldots, x_n$ and $y_1, \ldots, y_n$ is often used to compute finite population correlation

$$\rho = \frac{C}{\sigma_x \cdot \sigma_y}, \tag{22.1}$$

where $\sigma_x = \sqrt{V_x}$ is the finite population standard deviation of the values $x_1, \ldots, x_n$, and $\sigma_y = \sqrt{V_y}$ is the finite population standard deviation of the values $y_1, \ldots, y_n$.

When we only have *intervals* $[\underline{x}_1, \overline{x}_1], \ldots, [\underline{x}_n, \overline{x}_n]$, $[\underline{y}_1, \overline{y}_1], \ldots, [\underline{y}_n, \overline{y}_n]$, we have an interval $[\underline{\rho}, \overline{\rho}]$ of possible value of correlation.

## Results

It turns out that, similar to finite population covariance, computation of the endpoints of this interval problems is also an NP-hard problem:

**Theorem 22.1.** *The problem of computing $\overline{\rho}$ from the interval inputs $[\underline{x}_i, \overline{x}_i]$, $[\underline{y}_i, \overline{y}_i]$ is NP-hard.*

**Theorem 22.2.** *The problem of computing $\underline{\rho}$ from the interval inputs $[\underline{x}_i, \overline{x}_i]$, $[\underline{y}_i, \overline{y}_i]$ is NP-hard.*

## Proofs

*Proof of Theorem 22.1.*

$1°$. Similarly to the proof of Theorems 14.1 and 21.1, we reduce a subset problem to the problem of computing $\overline{\rho}$.

H.T. Nguyen et al.: Computing Stat. under Interval & Fuzzy Uncertain., SCI 393, pp. 173–176.
springerlink.com 

Each instance of the subset problem is as follows: given $m$ positive integers $s_1, \ldots, s_m$, to check whether there exist signs $\eta_i \in \{-1, +1\}$ for which the signed sum $\sum_{i=1}^{m} \eta_i \cdot s_i$ equals 0.

We will show that this problem can be reduced to the problem of computing $\overline{\rho}$, i.e., that to every instance $(s_1, \ldots, s_m)$ of the subset problem $\mathcal{P}_0$, we can put into correspondence such an instance of the $\overline{\rho}$-computing problem that based on its solution, we can easily check whether the desired signs exist.

As this instance, we take the instance corresponding to the following intervals:

- $n = m + 2$ (note the difference between this reduction and reductions from the proofs of Theorems 14.1 and 21.1, where we have $n = m$);
- $[\underline{x}_i, \overline{x}_i] = [-s_i, s_i]$ and $\mathbf{y}_i = [0, 0]$ for $i = 1, \ldots, m$;
- $\mathbf{x}_{m+1} = \mathbf{y}_{m+2} = [1, 1]$; $\mathbf{x}_{m+2} = \mathbf{y}_{m+1} = [-1, -1]$.

Like in the proof of Theorem 14.1, we define $C_1$ as

$$C_1 = \sum_{i=1}^{m} s_i^2. \tag{22.2}$$

We will prove that for the corresponding problem, $\overline{\rho} = -\sqrt{\dfrac{2}{C_1 + 2}}$ if and only if there exist signs $\eta_i$ for which $\sum_{i=1}^{n} \eta_i \cdot s_i = 0$.

$2°$. The correlation coefficient is defined as $\rho = C/\sqrt{V_x} \cdot \sqrt{V_y}$. To find the range for $\rho$, it is therefore reasonable to first find ranges for $C$, $V_x$, and $V_y$.

$3°$. Of these three, the variance $V_y$ is the easiest to compute because there is no interval uncertainty in $y_i$ at all. For $y_i$, we have $E_y = 0$ and therefore,

$$E_y = \frac{1}{n} \cdot \sum_{i=1}^{n} y_i^2 - (E_y)^2 = \frac{2}{n} = \frac{2}{m+2}. \tag{22.3}$$

$4°$. To find the range for the covariance, we will use the known equivalent formula

$$C = \frac{1}{n} \cdot \sum_{i=1}^{n} x_i \cdot y_i - E_x \cdot E_y. \tag{22.4}$$

Since $E_y = 0$, the second sum in this formula is 0, so $C$ is equal to the first sum. In this first sum, the first $m$ terms are 0's because for $i = 1, \ldots, m$, we have $y_i = 0$. The only non-zero terms correspond to $i = m + 1$ and $i = m + 2$, so

$$C = -\frac{2}{n} = -\frac{2}{m+2}. \tag{22.5}$$

5°. Substituting the formulas (22.3) and (22.5) into the definition of the correlation, we conclude that

$$\rho = -\frac{\dfrac{2}{m+2}}{\sqrt{\dfrac{2}{m+2}} \cdot \sqrt{V_x}} = -\sqrt{\frac{2}{(m+2) \cdot V_x}}. \tag{22.6}$$

Therefore, the finite population correlation $\rho$ attains its maximum $\overline{\rho}$ if and only if the finite population variance $V_x$ takes the largest possible value $\overline{V}_x$:

$$\overline{\rho} = -\sqrt{\frac{2}{(m+2) \cdot \overline{V}_x}}. \tag{22.7}$$

Thus, if we can know $\overline{\rho}$, we can reconstruct $\overline{V}_x$ as

$$\overline{V}_x = \frac{2}{(m+2) \cdot (\overline{\rho})^2}. \tag{22.8}$$

In particular, the desired value $\overline{\rho} = -\sqrt{\dfrac{2}{C_1+2}}$ corresponds to $\overline{V}_x = \dfrac{C_1+2}{m+2}$. Therefore, to complete our proof, we must show that $\overline{V}_x = \dfrac{C_1+2}{m+2}$ if and only if there exist signs $\eta_i$ for which $\sum\limits_{i=1}^{n} \eta_i \cdot s_i = 0$.

6°. Similarly to the proof of Theorem 14.1, we will use the equivalent expression for the finite population variance $V_x$; we will slightly reformulate this expression by substituting the definition of $E_x$ into it:

$$V_x = \frac{1}{n} \cdot \sum_{i=1}^{n} x_i^2 - \left(\sum_{i=1}^{n} x_i\right)^2. \tag{22.9}$$

We can (somewhat) simplify this expression by substituting the values $n = m+2$, $x_{m+1} = 1$, and $x_{m+2} = -1$. We have

$$\sum_{i=1}^{n} x_i = \sum_{i=1}^{m} x_i + x_{m+1} + x_{m+2} = \sum_{i=1}^{m} x_i$$

and

$$\sum_{i=1}^{n} x_i^2 = \sum_{i=1}^{m} x_i^2 + x_{m+1}^2 + x_{m+2}^2 = \sum_{i=1}^{m} x_i + 2.$$

Therefore,

$$V_x = \frac{1}{m+2} \cdot \sum_{i=1}^{m} x_i^2 + \frac{2}{m+2} - \frac{1}{(m+2)^2} \cdot \left(\sum_{i=1}^{m} x_i\right)^2. \tag{22.10}$$

Similarly to the proof of Theorem 14.1, we can show that always $V_x \leq \dfrac{C_1+2}{m+2}$, and that $\overline{V}_x = \dfrac{C_1+2}{m+2}$ if and only if there exist the signs $\eta_i$ for which

$$\sum_{i=1}^{n} \eta_i \cdot s_i = 0.$$

The theorem is proven.

*Proof of Theorem 22.2.* This proof is similar to the proof of Theorem 22.1, with the only difference that we take $y_{m+1} = 1$ and $y_{m+2} = -1$. In this case,

$$C = \frac{2}{m+2},$$

hence

$$\rho = \sqrt{\frac{2}{(m+2) \cdot V_x}},$$

and so the largest possible value of $V_x$ corresponds to the smallest possible value of $\rho$.

# Computing Expected Value under Interval Uncertainty

## Formulation of the Problem

In statistics, the values of the statistical characteristics are estimated either *directly* from the sample $x_1, \ldots, x_n$, or *indirectly*: based on the values of other statistical characteristics that have already been estimated based on the sample.

In most of the previous chapters, we considered the situations in which:

- we want to compute the statistical characteristics of a probability distribution based on a sample $x_1, \ldots, x_n$, and
- we know the values $x_i$ with interval uncertainty.

The only exception so far was a chapter on hierarchical statistical analysis, where the desired characteristics – the overall mean and variance – were estimated based on the means and variances of several subsamples. In this hierarchical case, due to the interval uncertainty with which we know the sample values, the subsample means and variances are also only known with interval uncertainty.

Another natural case of an indirect estimation is when:

- we know the probabilities $p_1, \ldots, p_n$ of different situations,
- we know the values $a_1, \ldots, a_n$ of a certain quantity in these situation, and
- we want to find the expected value $a = \sum_{i=1}^{n} a_i \cdot p_i$ of this quantity.

Since we assume that the probabilities $p_i$ are estimated based on the sample, and this sample is known with interval uncertainty, the probabilities are also known with interval uncertainty, i.e., we only know the intervals $[\underline{p}_i, \overline{p}_i]$ that contain the actual (unknown) probabilities $p_i$.

By taking the midpoints $\widetilde{p}_i \stackrel{\text{def}}{=} \dfrac{\underline{p}_i + \overline{p}_i}{2}$ and the radii $\Delta_i \stackrel{\text{def}}{=} \dfrac{\overline{p}_i - \underline{p}_i}{2}$, we can represent these intervals in the form $[\widetilde{p}_i - \Delta_i, \widetilde{p}_i + \Delta_i]$.

H.T. Nguyen et al.: Computing Stat. under Interval & Fuzzy Uncertain., SCI 393, pp. 177–180.
springerlink.com

In this case, we arrive at the following problem:

- we know the values $a_1, \ldots, a_n$;
- we know the intervals $[\underline{p}_i, \overline{p}_i]$;

- we want to find the range $[\underline{a}, \overline{a}]$ of possible values of the sum $\sum\limits_{i=1}^{n} a_i \cdot p_i$

  under the constraints $p_i \in [\underline{p}_i, \overline{p}_i]$ and $\sum\limits_{i=1}^{n} p_i = 1$.

## Algorithms

For this problem, linear-time algorithms are known; see, e.g., [52] and [130].

Let us give an example of such a linear-time algorithm. For simplicity of description, our algorithm will be a minor modification of a linear-time algorithm for computing the upper endpoint $\overline{V}$ for the variance (possible when narrowed intervals satisfy the no-subset property).

Before we describe the algorithm, let us mention that the smallest possible value $\underline{a}$ of the linear form $\sum\limits_{i=1}^{n} a_i \cdot p_i$ under given constraints is equal to $-\overline{b}$,

where $\overline{b}$ is the largest possible value of the form $\sum\limits_{i=1}^{n} b_i \cdot p_i$, with $b_i = -a_i$.

Thus, it is sufficient to describe how to compute the upper endpoint $\overline{a}$.

For simplicity, let us first consider the case when all the intervals are non-degenerate, i.e., when $\Delta_i > 0$ for all $i$.

The proposed algorithm is iterative. At each iteration of this algorithm, we have three sets:

- the set $I^-$ of all the indices $i$ from 1 to $n$ for which we already know that for the optimal vector $p$, we have $p_i = \underline{p}_i$;
- the set $I^+$ of all the indices $j$ for which we already know that for the optimal vector $p$, we have $p_j = \overline{p}_j$;
- the set $I = \{1, \ldots, n\} \setminus (I^- \cup I^+)$ of the indices $i$ for which we are still undecided.

In the beginning, $I^- = I^+ = \emptyset$ and $I = \{1, \ldots, n\}$. At each iteration we also update the values of two auxiliary quantities $E^- \stackrel{\text{def}}{=} \sum\limits_{i \in I^-} \underline{p}_i$ and $E^+ \stackrel{\text{def}}{=}$

$\sum\limits_{j \in I^+} \overline{p}_j$. In principle, we could compute these values by computing these sums. However, to speed up computations on each iteration, we update these two auxiliary values in a way that is faster than re-computing the corresponding two sums. Initially, since $I^- = I^+ = \emptyset$, we take $E^- = E^+ = 0$.

At each iteration we do the following:

- first, we compute the median $m$ of the set $I$ (median in terms of sorting by the values $a_i$);
- then, by analyzing the elements of the undecided set $I$ one by one, we divide them into two subsets $P^- = \{i : \widetilde{p}_i \leq \widetilde{p}_m\}$ and $P^+ = \{j : \widetilde{p}_j > \widetilde{p}_m\}$;

- we compute $e^- = E^- + \sum\limits_{i \in P^-} \underline{p}_i$ and $e^+ = E^+ + \sum\limits_{j \in P^+} \overline{p}_j$;
- If $e^- + e^+ > 1$, then we replace $I^-$ with $I^- \cup P^-$, $E^-$ with $e^-$, and $I$ with $P^+$.
- If $e^- + e^+ + 2\Delta_m < 1$, then we replace $I^+$ with $I^+ \cup P^+$, $E^+$ with $e^+$, and $I$ with $P^-$.
- Finally, if $e^- + e^+ \le 1 \le e^- + e^+ + 2\Delta_m$, then we replace $I^-$ with $I^- \cup (P^- - \{m\})$, $I^+$ with $I^+ \cup P^+$, $I$ with $\{m\}$, $E^-$ with $e^- - \underline{p}_m$, and $E^+$ with $e^+$.

At each iteration the set of undecided indices is divided in half. Iterations continue until we have only one undecided index $I = \{k\}$.

After this we return, as $\overline{a}$, the value of the linear function $\sum\limits_{i=1}^{n} a_i \cdot p_i$ for the vector $p$ for which $p_i = \underline{p}_i$ for $i \in I^-$, $p_j = \overline{p}_j$ for $j \in I^+$, and $p_k = 1 - e^- - e^+$ for the remaining value $k$.


## Proof of the Algorithm's Correctness

Let us assume that the indices have already been sorted by the values $a_i$, i.e., that $a_1 \le a_2 \le \ldots \le a_n$.

Let us now show that we can always select an optimal tuple in which:

- at most one $p_k$ is strictly inside the corresponding interval,
- all the values $p_i$ with $i < k$ are equal to the corresponding lower endpoints $p_i = \underline{p}_i$, and
- all the values $p_j$ with $j > k$ are equal to the corresponding upper endpoints $p_j = \overline{p}_j$.

Indeed, there is always an optimal tuple. If for some optimal tuple, we have two different values $p_i$ and $p_j$, $i < j$, for which $p_i > \underline{p}_i$ and $p_j < \overline{p}_j$, then, for an arbitrarily small $\Delta > 0$, replacing $p_i$ with $p_i - \Delta$ and $p_j$ with $p_j + \Delta$ does not change the condition $\sum\limits_{i=1}^{n} p_i = 1$, while the value of the desired linear function increases by $(a_j - a_i) \cdot \Delta$. Since $i < j$, we have $a_i \le a_j$ and thus, $a_j - a_i \ge 0$. If $a_j - a_i > 0$, then we would be able to increase the value of $a$ – which contradicts our assumption that the tuple was optimal. Thus, $a_i = a_j$. In this case, replacing $p_i$ with $p_i - \Delta$ and $p_j$ with $p_j + \Delta$ does not change the value of $a$ at all. As $\Delta$ increases, the value $p_i$ goes towards $\underline{p}_i$ and the value $p_j$ goes towards $\overline{p}_j$. We can therefore select $\Delta$ for which one of these values reaches the corresponding endpoint.

This procedure can be repeated until for every $i < j$, we have $p_i = \underline{p}_i$ or $p_j = \overline{p}_j$. This implies that we can have at most one $k$ for which $p_k$ is strictly within the corresponding interval: otherwise, if we had two such values $p_k$ and $p_{k'}$, with $k < k'$, we would have $p_k > \underline{p}_k$ and $p_{k'} < \overline{p}_{k'}$. Similarly, we can show that for all $i < k$, we have $p_i = \underline{p}_i$, and that for all $j > k$, we have $p_j = \underline{p}_j$.

In the resulting optimal tuple $p_1, \ldots, p_n$, the first elements are equal to $\underline{p}_i$, then we may have one element which is strictly inside its interval, and then we have values $p_j = \overline{p}_j$.

For the resulting vector $p = (\underline{p}_1, \ldots, \underline{p}_{k-1}, p_k, \overline{p}_{k+1}, \ldots, \overline{p}_n)$, with $\underline{p}_k \leq p_k \leq \overline{p}_k$, the condition $\sum_{i=1}^{n} p_i = 1$ implies that $\Sigma_k \leq 1 \leq \Sigma_{k-1}$, where $\Sigma_k \stackrel{\text{def}}{=} \sum_{i=1}^{k} \underline{p}_i + \sum_{j=k+1}^{n} \overline{p}_j$. When we go from $\Sigma_k$ to $\Sigma_{k+1}$, we replace a larger value $\overline{p}_{k+1}$ with a smaller value $\underline{p}_{k+1}$. Hence $\Sigma_k > \Sigma_{k+1}$. Thus there has to be exactly one $k_{\max}$ for which $\Sigma_k \leq 1 \leq \Sigma_{k-1}$.

So if we have $\Sigma_m > 1$, this means that the value $k_{\max}$ corresponding to the maximum of the linear form is $> m$. Hence for all the indices $i \leq m$ we already know that in the optimal vector $p$ we have $p_i = \underline{p}_i$. Thus these indices can be added to the set $I^-$.

If $\Sigma_{m-1} \, (= \Sigma_m + 2\Delta_m) < 1$, this means that the value $k_{\min}$ corresponding to the maximum of the linear form is $< m$. Hence for all the indices $j \geq m$ we already know that in the optimal vector $p$ we have $p_j = \overline{p}_j$. Thus these indices can be added to the set $I^+$.

Finally, if $\Sigma_m \leq 1 \leq \Sigma_{m-1}$ then this $m$ is where the maximum of the linear form is attained.

The algorithm has been justified.

*Comment.* This same algorithm can be easily applied if one of the intervals consists of a single point only. This value is plugged in and the variable is eliminated.

# Computing Entropy under Interval Uncertainty. I

Measurement results (and, more generally, estimates) are never absolutely accurate: there is always an uncertainty, the actual value $x$ is, in general, different from the estimate $\tilde{x}$. Sometimes, we know the probability of different values of the estimation error $\Delta x \overset{\text{def}}{=} \tilde{x} - x$, sometimes, we only know the interval of possible values of $\Delta x$, sometimes, we have interval bounds on the cdf of $\Delta x$. To compare different measuring instruments, it is desirable to know which of them brings more information – i.e., it is desirable to gauge the amount of information. For probabilistic uncertainty, this amount of information is described by Shannon's entropy; similar measures can be developed for interval and other types of uncertainty. In this chapter, we start analyzing the problem of estimating information amount under different types of uncertainty.

## Formulation and Analysis of the Problem

*Uncertainty is inevitable.* For each type of information that we are soliciting, there are several ways to acquire this information.

For example, if we are interested in measuring the value of a physical quantity $x$, we may use different types of sensors. No matter how accurate the sensor, the measured value $\tilde{x}$ is, in general, different from the actual value $x$ of the measured quantity.

*Types of uncertainty: in brief.* For different sensors, we have different type of information about this difference $\Delta x \overset{\text{def}}{=} \tilde{x} - x$:

In some cases, we know which values of $\Delta x$ are possible and what is the frequency of each of the different possible values. In other words, we know a probability distribution on $\Delta x$. This type of uncertainty is usually called a *probabilistic uncertainty.* It is reasonable to describe the corresponding probability distribution by a cumulative distribution function (cdf, for short) $F(t) \overset{\text{def}}{=} \text{Prob}(x \leq t)$.

H.T. Nguyen et al.: Computing Stat. under Interval & Fuzzy Uncertain., SCI 393, pp. 181–191.
springerlink.com

In other cases, the only information we have is an upper bound $\Delta$ on the measurement error. In this case, after we got the measured value $\widetilde{x}$, the only information that we have about the actual (unknown) value $x$ of the measured quantity is that $x$ belongs to the interval $[\widetilde{x} - \Delta, \widetilde{x} + \Delta]$. This is the case of *interval uncertainty*.

So far, we have described two extreme cases:

- Probabilistic uncertainty describes the case when we have a *complete* information about the probability distribution.
- Interval uncertainty corresponds to the case when we have *no* information about the probabilities.

In most practical situations, we have *some* information about the probabilities.

As we have mentioned, to get a complete description of a probability distribution, we need to know the values of cdf $F(t)$ for all possible real numbers $t$. When we have a partial information about the probabilities, this means that we only have a partial information about the values $F(t)$. In other words, for every $t$, instead of the actual; (unknown) value $F(t)$, we only know the interval $[\underline{F}(t), \overline{F}(t)]$ that contains the (unknown) actual value $F(t)$. In other words, we have a *probability box* (p-box, for short) that contains the actual (unknown) cdf $F(t)$.

In measurements, the p-box is probably the most general description of possible uncertainty. In many practical situations, however, we cannot get all the information from measurements, we must also use human expertise. The accuracy of human expertise is rarely described solely in terms of guaranteed bounds. For expert estimates, in addition to guaranteed bounds on $\Delta x$ and on $F(t)$, we also have expert estimates that provide better bounds but with limited confidence.

For example, by looking at a medical image such as an X-ray image, an expert medical doctor can guarantee that the size of the tumor is, say, between 1 and 2 cm. However, with 80% certainty, she can say that the size is between 1.2 and 1.7 cm.

To take such uncertainty into consideration, we can use fuzzy techniques. For example, a nested family of intervals corresponding to different levels of certainty forms a fuzzy number (the intervals are the $\alpha$-cuts of this fuzzy number). For p-boxes, we have, similarly, a nested family of p-boxes corresponding to different levels of certainty – i.e., a fuzzy-valued cdf.

*Need to compare different types of uncertainty.* Often, there is a need to compare different types of uncertainty. For example, we may have two sensors: one with a smaller bound on a systematic (interval) component of the measurement error, the other with the smaller bound on the standard deviation of the random component of the measurement error. If we can only afford one of these sensors, which one should we buy? Which of the two sensors brings us more information about the measured signal?

To be able to make such decisions, we must be able to compare which of the uncertainties corresponding to the two sensors carries more information – and for that, we must be able to gauge this amount of information.

*Traditional amount of information: brief reminder.* The traditional Shannon's notion of the amount of information is based on defining information as the (average) number of "yes"-"no" (binary) questions that we need to ask so that, starting with the initial uncertainty, we will be able to completely determine the object.

After each binary question, we can have 2 possible answers. So, if we ask $q$ binary questions, then, in principle, we can have $2^q$ possible results. Thus, if we know that our object is one of $n$ objects, and we want to uniquely pinpoint the object after all these questions, then we must have $2^q \geq n$. In this case, the smallest number of questions is the smallest integer $q$ that is $\geq \log_2(n)$. This smallest number is called a *ceiling* and denoted by $\lceil \log_2(n) \rceil$.

For discrete probability distributions, we get the standard formula for the average number of questions $-\sum p_i \cdot \log_2(p_i)$. For the continuous case, we can estimate the average number of questions that are needed to find an object with a given accuracy $\varepsilon$ – i.e., divide the whole original domain into sub-domains of radius $\varepsilon$ and diameter $2\varepsilon$.

For example, if we start with an interval $[a, b]$ of width $b - a$, then we need to subdivide it into $n \sim (b - a)/(2\varepsilon)$ sub-domains, so we must ask

$$\log_2(n) \sim \log_2(b - a) - \log_2(\varepsilon) - 1$$

questions. In the limit, the term that does not depend on $\varepsilon$ leads to

$$\log_2(b - a).$$

For continuous probability distributions, we get the standard Shannon's expression $\log_2(n) \sim S - \log_2(2\varepsilon)$, where $S = -\int \rho(x) \cdot \log_2 \rho(x) \, dx$.

*How to extend these formulas to p-boxes etc.? Axiomatic approach.* To extend the formulas for information to more general uncertainty, i.e., to come up with generalized information theory, several researchers use an axiomatic approach: they find properties of information, and look for generalizations that satisfy as many of these properties as possible; see, e.g. [157, 158] and [162].

This approach has led to many interesting results, but sometimes, there are several possible generalizations, so which of them should we choose?

*Our idea.* A natural idea is to choose the definition that kind of coincides with the average number of binary questions that we need to ask.

Since we want to extend the information to the case when probabilities are not known exactly, the average number of questions may also depend on which exactly distribution is actually there. So, it is reasonable to consider the worst-case average number of questions – this is in line with the definition for intervals.

*What we do in this chapter.* In this and following chapters, we describe how this idea can be transformed into a formal definition of the amount of information corresponding to different types of uncertainty, and how to compute the corresponding amounts of information.

*Comment.* Several of our results first appeared in [54, 58, 195, 355, 353].

*Traditional amount of information: detailed reminder.* Our objective is to extend estimates of the average number of binary questions from the probability distributions to a more general case. To do that, let us recall, in detail, how this number is estimated for probability distributions. The need for such a reminder comes from the fact that while most researchers are familiar with Shannon's formula for the entropy, most researchers are not aware how this formula was (or can be) derived.

*Discrete case: no information about probabilities.* Let us start with the simplest situation when we know that we have $n$ possible alternatives $A_1, \ldots, A_n$, and we have no information about the probability (frequency) of different alternatives. Let us show that in this case, the smallest number of binary questions that we need to determine the alternative is indeed $q \stackrel{\text{def}}{=} \lceil \log_2(n) \rceil$.

We have already shown that the number of questions cannot be smaller than $\lceil \log_2(n) \rceil$; so, to complete the derivation, let us show that it is sufficient to ask $q$ questions.

Indeed, let's enumerate all $n$ possible alternatives (in arbitrary order) by numbers from 0 to $n - 1$, and write these numbers in the binary form. Using $q$ binary digits, one can describe numbers from 0 to $2^q - 1$. Since $2^q \geq n$, we can thus describe each of the $n$ numbers by using only $q$ binary digits. So, to uniquely determine the alternative $A_i$ out of $n$ given ones, we can ask the following $q$ questions: "is the first binary digit 0?", "is the second binary digit 0?", etc, up to "is the $q$-th digit 0?".

*Case of a discrete probability distribution.* Let us now assume that we also know the probabilities $p_1, \ldots, p_n$ of different alternatives $A_1, \ldots, A_n$. If we are interested in an individual selection, then the above arguments show that we cannot determine the actual alternative by using fewer than $\log_2(n)$ questions. However, if we have many $(N)$ similar situations in which we need to find an alternative, then we can determine all $N$ alternatives by asking $\ll N \cdot \log_2(n)$ binary questions.

To show this, let us fix $i$ from 1 to $n$, and estimate the number of events $N_i$ in which the output is $i$.

This number $N_i$ is obtained by counting all the events in which the output was $i$, so $N_i = n_1 + n_2 + \ldots + n_N$, where $n_k$ equals to 1 if in $k$-th event the output is $i$ and 0 otherwise. The average $\mathrm{e}(n_k)$ of $n_k$ equals to $p_i \cdot 1 + (1 - p_i) \cdot 0 = p_i$. The mean square deviation $\sigma[n_k]$ is determined by the formula

$$\sigma^2[n_k] = p_i \cdot (1 - \mathrm{e}(n_k))^2 + (1 - p_i) \cdot (0 - \mathrm{e}(n_k))^2.$$

If we substitute here $\mathrm{e}(n_k) = p_i$, we get $\sigma^2[n_k] = p_i \cdot (1 - p_i)$. The outcomes of all these events are considered independent, therefore $n_k$ are independent random variables. Hence the average value of $N_i$ equals to the sum of the averages of $n_k$: $e[N_i] = e[n_1] + e[n_2] + \ldots + e[n_N] = Np_i$. The mean square deviation $\sigma[N_i]$ satisfies a likewise equation $\sigma^2[N_i] = \sigma^2[n_1] + \sigma^2[n_2] + \ldots = N \cdot p_i \cdot (1 - p_i)$, so $\sigma[N_i] = \sqrt{p_i \cdot (1 - p_i) \cdot N}$.

For big $N$ the sum of equally distributed independent random variables tends to a Gaussian distribution (the well-known *Central Limit Theorem*), therefore for big $N$, we can assume that $N_i$ is a random variable with a Gaussian distribution. Theoretically a random Gaussian variable with the average $a$ and a standard deviation $\sigma$ can take any value. However, in practice, if, e.g., one buys a voltmeter with guaranteed 0.1V standard deviation, and it gives an error 1V, it means that something is wrong with this instrument. Therefore it is assumed that only some values are practically possible. Usually a "$k$-sigma" rule is accepted that the real value can only take values from $a - k \cdot \sigma$ to $a + k \cdot \sigma$, where $k$ is 2, 3, or 4. So in our case we can conclude that $N_i$ lies between $N \cdot p_i - k \cdot \sqrt{p_i \cdot (1 - p_i) \cdot N}$ and $N \cdot p_i + k \cdot \sqrt{p_i \cdot (1 - p_i) \cdot N}$. Now we are ready for the formulation of Shannon's result.

*Comment.* In this quality control example the choice of $k$ matters, but, as we'll see, in our case the results do not depend on $k$ at all.

## Definition 24.1

- *Let a real number $k > 0$ and a positive integer $n$ be given. The number $n$ is called* the number of outcomes.
- *By a* probability distribution, *we mean a sequence $\{p_i\}$ of $n$ real numbers, $p_i \geq 0$, $\sum p_i = 1$. The value $p_i$ is called a* probability *of $i$-th event.*
- *Let an integer $N$ is given; it is called* the number of events.
- *By a* result of $N$ events *we mean a sequence $r_k$, $1 \leq k \leq N$ of integers from 1 to $n$. The value $r_k$ is called the* result of $k$-th event.
- *The total number of events that resulted in the $i$-th outcome will be denoted by $N_i$.*
- *We say that the result of $N$ events is* consistent *with the probability distribution $\{p_i\}$ if for every $i$, we have $N \cdot p_i - k \cdot \sigma_i \leq N_i \leq N + k \cdot \sigma_i$, where $\sigma_i \stackrel{\text{def}}{=} \sqrt{p_i \cdot (1 - p_i) \cdot N}$.*
- *Let's denote the number of all consistent results by $N_{cons}(N)$.*
- *The number $\lceil \log_2(N_{cons}(N)) \rceil$ will be called* the number of questions, *necessary to determine the results of $N$ events and denoted by $Q(N)$.*
- *The fraction $Q(N)/N$ will be called* the average number of questions.
- *The limit of the average number of questions when $N \to \infty$ will be called the* information.

**Theorem.** (Shannon) *When the number of events $N$ tends to infinity, the average number of questions tends to $S(p) \stackrel{\text{def}}{=} -\sum p_i \cdot \log_2(p_i)$.*

*Comments*

- Shannon's theorem says that if we know the probabilities of all the outputs, then the average number of questions that we have to ask in order to get a complete knowledge equals to the entropy of this probabilistic distribution.
- As we promised, this average number of questions does not depend on the threshold $k$.
- Since we somewhat modified Shannon's definitions, we cannot use the original proof. Our proof (and proof of other results) is given at the end of this chapter.

*Case of a continuous probability distribution.* After a finite number of "yes"-"no" questions, we can only distinguish between finitely many alternatives. If the actual situation is described by a real number, then, since there are infinitely many different possible real numbers, after finitely many questions, we can only get an approximate value of this number.

Once we fix the accuracy $\varepsilon > 0$, we can talk about the number of questions that are necessary to determine a number $x$ with this accuracy $\varepsilon$, i.e., to determine an approximate value $r$ for which $|x - r| \leq \varepsilon$.

Once an *approximate* value $r$ is determined, possible *actual* values of $x$ form an interval $[r - \varepsilon, r + \varepsilon]$ of width $2\varepsilon$. Vice versa, if we have located $x$ on an interval $[\underline{x}, \overline{x}]$ of width $2\varepsilon$, this means that we have found $x$ with the desired accuracy $\varepsilon$: indeed, as an $\varepsilon$-approximation to $x$, we can then take the midpoint $(\underline{x} + \overline{x})/2$ of the interval $[\underline{x}, \overline{x}]$.

Thus, the problem of determining $x$ with the accuracy $\varepsilon$ can be reformulated as follows: we divide the real line into intervals $[x_i, x_{i+1}]$ of width $2\varepsilon$ $(x_{i+1} = x_i + 2\varepsilon)$, and by asking binary questions, find the interval that contains $x$. As we have shown, for this problem, the average number of binary question needed to locate $x$ with accuracy $\varepsilon$ is equal to $S = -\sum p_i \cdot \log_2(p_i)$, where $p_i$ is the probability that $x$ belongs to $i$-th interval $[x_i, x_{i+1}]$.

In general, this probability $p_i$ is equal to $\int_{x_i}^{x_{i+1}} \rho(x) \, dx$, where $\rho(x)$ is the probability distribution of the unknown values $x$. For small $\varepsilon$, we have $p_i \approx 2\varepsilon \cdot \rho(x_i)$, hence $\log_2(p_i) = \log_2(\rho(x_i)) + \log_2(2\varepsilon)$. Therefore, for small $\varepsilon$, we have

$$S = -\sum \rho(x_i) \cdot \log_2(\rho(x_i)) \cdot 2\varepsilon - \sum \rho(x_i) \cdot 2\varepsilon \cdot \log_2(2\varepsilon).$$

The first sum in this expression is the integral sum for the integral

$$S(\rho) \stackrel{\text{def}}{=} -\int \rho(x) \cdot \log_2(x) \, dx$$

(this integral is called the *entropy* of the probability distribution $\rho(x)$); so, for small $\varepsilon$, this sum is approximately equal to this integral (and tends to this integral when $\varepsilon \to 0$). The second sum is a constant $\log_2(2\varepsilon)$ multiplied by an integral sum for the interval $\int \rho(x) \, dx = 1$. Thus, for small $\varepsilon$, we have

$$S \approx -\int \rho(x) \cdot \log_2(x) \, dx - \log_2(2\varepsilon).$$

So, the average number of binary questions that are needed to determine $x$ with a given accuracy $\varepsilon$, can be determined if we know the entropy of the probability distribution $\rho(x)$.

*Our results: in brief.* Of course, the abstract definition is a good idea, but the big challenge is translating this abstract definition into explicit easy-to-use analytical formulas and/or algorithms. This is what we do in this paper.

*Comment.* In our previous work [69, 286, 287] we provided such formulas for fuzzy numbers and for Dempster-Shafer knowledge bases. In this chapter, we provide similar analytical (or at least computable) formulas for the more general case of p-boxes and fuzzy-valued probability distributions.

*Partial information about probability distribution: discrete case.* In many real-life situations, instead of having *complete* information about the probabilities $p = (p_1, \ldots, p_n)$ of different alternatives, we only have *partial* information about these probabilities – i.e., we only know a *set $P$* of possible values of $p$.

If it is possible to have $p \in P$ and $p' \in P$, then it is also possible that we have $p$ with some probability $\alpha$ and $p'$ with the probability $1 - \alpha$. In this case, the resulting probability distribution $\alpha \cdot p + (1 - \alpha) \cdot p'$ is a convex combination of $p$ and $p'$. Thus, it it reasonable to require that the set $P$ contains, with every two probability distributions, their convex combinations – in other words, that $P$ is a convex set; see, e.g., [338].

**Definition 24.2**

- *By a* probabilistic knowledge, *we mean a convex set $P$ of probability distributions.*
- *We say that the result of $N$ events is* consistent *with the probabilistic knowledge $P$ if this result is consistent with one of the probability distributions $p \in P$.*
- *Let's denote the number of all consistent results by $N_{cons}(N)$.*
- *The number* $\lceil \log_2(N_{cons}(N)) \rceil$ *will be called* the number of questions, necessary to determine the results of $N$ events *and denoted by $Q(N)$.*
- *The fraction $Q(N)/N$ will be called the* average number of questions.
- *The limit of the average number of questions when $N \to \infty$ will be called the* information.

**Definition 24.2.** *By the* entropy $S(P)$ *of a probabilistic knowledge $P$, we mean the largest possible entropy among all distributions $p \in P$; $S(P) \overset{\text{def}}{=} \max\limits_{p \in P} S(p)$.*

**Proposition 24.1.** *When the number of events $N$ tends to infinity, the average number of questions tends to the entropy $S(P)$.*

*Partial information about probability distribution: continuous case.* In the continuous case, we also often encounter situations in which we only have partial information about the probability distribution; one such case is the case of $p$-boxes. In such situations, instead of knowing the *exact* probability distribution $\rho(x)$, we only know a (convex) class $P$ that contains the (unknown) distribution.

In such situations, we can similarly ask about the average number of questions that are needed to determine $x$ with a given accuracy $\varepsilon$.

Once we fix an accuracy $\varepsilon$ and a subdivision of the real line into intervals $[x_i, x_{i+1}]$ of width $2\varepsilon$, we have a discrete problem of determining the interval containing $x$. Due to Proposition 24.1, for this discrete problem, the average number of "yes"-"no" questions is equal to the largest entropy $S(p)$ among all the corresponding discrete distributions $p_i = \int_{x_i}^{x_{i+1}} \rho(x)\, dx$. As we have mentioned, for small $\varepsilon$, $S(p) \sim S(\rho) - \log_2(2\varepsilon)$, where $S(\rho) = -\int \rho(x) \cdot \log_2(\rho(x))\, dx$ is the entropy of the corresponding continuous distribution. Thus, the largest discrete entropy $S(p)$ comes from the distribution $\rho(x) \in P$ for which the corresponding (continuous) entropy $S(\rho)$ attains the largest possible value.

*Computing the amount of information.* According to the above results, the amount of information in p-box – or more generally, in a class of distributions $P$ – is equal to the largest entropy among all the distributions from the given class $P$.

Good news is that a lot of research has gone into algorithms for finding distributions with the largest entropy among different classes $P$ – largely as a part of the Maximum Entropy approach in which when we only know a class of distributions $P$, then we assume that the actual distribution is the one with the largest entropy from $P$; see, e.g., [144].

Because of this, for many classes $P$, we already know the corresponding maximum entropy distribution, so we can explicitly compute the corresponding amount of information. For classes $P$ for which the corresponding maximum entropy distribution is not known, finding such a distribution means maximizing a convex function (entropy) over a convex set $P$; it is known that maximizing a convex function over a convex set is a computationally feasible problem; see, e.g., [334].

*Problem with our definition: we need a multi-dimensional notion of information.* In our approach, we measure the information as the average number of "yes"-"no" questions that are needed to locate an object with a given accuracy.

According to our results, for a p-box, thus defined amount of information is equal to the amount of information corresponding to the distribution with the largest entropy among all the distributions from a given p-box.

So, by the above definition of the amount of information, we are not be able to distinguish between this distribution and entire p-box. This is counter-intuitive. For example, it is well known that the Gaussian distribution has the largest entropy among all the distribution with the same standard deviation

$\sigma$, but clearly, we have more information if we know that the distribution is Gaussian than if we simply know its standard deviation but not its shape.

To account for this difference, we must supplement the average number of questions by additional characteristics describing the desired amount of information. Thus, to describe the amount of information for general uncertainty, instead of a single number, we need several different numbers, which form a multi-dimensional measure of uncertainty.

In the following chapter, we explore two natural ways to implement this idea.

## Proofs

*Proof of Shannon's Theorem.* Let's first fix some values $N_i$, that are consistent with the given probabilistic distribution. Due to the inequalities that express the consistency demand, the ratio $f_i = N_i/N$ tends to $p_i$ as $N \to \infty$. Let's count the total number $C$ of results, for which for every $i$ the number of events with outcome $i$ is equal to this $N_i$. If we know $C$, we will be able to compute $N_{cons}$ by adding these $C$'s.

Actually we are interested not in $N_{cons}$ itself, but in $Q(N) \approx \log_2(N_{cons})$, and moreover, in $\lim(Q(N)/N)$. So we'll try to estimate not only $C$, but also $\log_2(C)$ and $\lim \log_2(C)/N$.

To estimate $C$ means to count the total number of sequences of length $N$, in which there are $N_1$ elements, equal to 1, $N_2$ elements, equal to 2, etc. The total number $C_1$ of ways to choose $N_1$ elements out of $N$ is well-known in combinatorics, and is equal to $\binom{N_1}{N} = \dfrac{N!}{(N_1)! \cdot (N - N_1)!}$. When we choose these $N_1$ elements, we have a problem in choosing $N_2$ out of the remaining $N - N_1$ elements, where the outcome is 2; so for every choice of 1's we have $C_2 = \binom{N_2}{N - N_1}$ possibilities to choose 2's. Therefore in order to get the total number of possibilities to choose 1's and 2's, we must multiply $C_2$ by $C_1$. Adding 3's, 4's, $\ldots$, $n$'s, we get finally the following formula for $C$:

$$C = C_1 \cdot C_2 \cdot \ldots \cdot C_{n-1} =$$

$$\frac{N!}{N_1!(N - N_1)!} \cdot \frac{(N - N_1)!}{N_2!(N - N_1 - N_2)!} \cdot \ldots = \frac{N!}{N_1! N_2! \ldots N_n!}$$

To simplify computations let's use the well-known Stirling formula $k! \sim (k/e)^k \cdot \sqrt{2\pi \cdot k}$. Then, we get

$$C \approx \frac{\left(\dfrac{N}{e}\right)^N \sqrt{2\pi \cdot N}}{\left(\dfrac{N_1}{e}\right)^{N_1} \cdot \sqrt{2\pi \cdot N_1} \cdot \ldots \cdot \left(\dfrac{N_n}{e}\right)^{N_n} \cdot \sqrt{2\pi \cdot N_n}}$$

Since $\sum N_i = N$, terms $e^N$ and $e^{N_i}$ cancel each other.

To get further simplification, we substitute $N_i = N \cdot f_i$, and correspondingly $N_i^{N_i}$ as $(N \cdot f_i)^{N \cdot f_i} = N^{N \cdot f_i} \cdot f_i^{N \cdot f_i}$. Terms $N^N$ is the numerator and $N^{N \cdot f_1} \cdot N^{N \cdot f_2} \ldots \cdot N^{N \cdot f_n} = N^{N \cdot f_1 + N \cdot f_2 + \ldots + N \cdot f_n} = N^N$ in the denominator cancel each other. Terms with $\sqrt{N}$ lead to a term that depends on $N$ as $c \cdot N^{-(n-1)/2}$. So, we conclude that

$$\log_2(C) \approx -N \cdot f_1 \cdot \log_2(f_1) - \ldots - N \cdot f_n \log_2(f_n) -$$

$$\frac{n-1}{2} \cdot \log_2(N) - \text{const.}$$

When $N \to \infty$, we have $1/N \to 0$, $\log_2(N)/N \to 0$, and $f_i \to p_i$, therefore

$$\frac{\log_2(C)}{N} \to -p_1 \cdot \log_2(p_1) - \ldots - p_n \cdot \log_2(p_n),$$

i.e., $\log_2(C)/N$ tends to the entropy of the probabilistic distribution. The proposition is proven.

*Comment.* Strictly speaking, we need to prove that the ratio $Q(N)/N = S$ also tends to this entropy. This can be done similarly to what we did in [69].

*Proof of Proposition 24.1.* By definition, a result is consistent with the probabilistic knowledge $P$ if and only if it is consistent with one of the distributions $p \in P$. Thus, the set of all the results which are consistent with $P$ can be represented as a union of the sets of all the results consistent with different probability distributions $p \in P$. In the proof of Shannon's theorem, we have shown that for each $p \in P$, the corresponding number is asymptotically equal to $\exp(N \cdot S(p))$.

To be more precise, for every $N$, the number $C$ of results with given frequencies $\{f_j\}$ ($f_j \approx p_j$) has already been computed in the proof of Shannon's theorem: $\lim (\log_2(C))/N = -\sum f_j \log_2(f_j)$.

The total number of the results $N_{cons}$ which are consistent with a given probabilistic knowledge $P$ is equal to the sum of $N_{co}$ different values of $C$ that correspond to different $f_j$. For a given $N$, there are at most $N + 1$ different values of $N_1 = N \cdot f_1$ $(0,1,\ldots,N)$, at most $N + 1$ different values of $N_2$, etc., totally at most $(N + 1)^n$ different sets of $\{f_j\}$. So, we get an inequality $C_{\max} \leq N_{cons} \leq (N + 1)^n \cdot C_{\max}$, from which we conclude that $\lim Q(N)/N = \lim \log_2(C_{\max})/N$.

Now, that we have found an asymptotics for $C$, let's compute $N_{cons}$ and $Q(N)/N$. For a given probabilistic distribution $\{p_i\}$ and every $i$, possible values of $N_i$ form an interval of length $L_i \stackrel{\text{def}}{=} 2k \cdot \sqrt{p_i \cdot (1 - p_i)} \cdot \sqrt{N}$. So there are no more than $L_i$ possible values of $N_i$. The maximum value for $p_i \cdot (1 - p_i)$ is attained when $p_i = 1/2$, therefore $p_i \cdot (1 - p_i) \leq 1/4$, and hence $L_i \leq 2k \cdot \sqrt{N/4} = (k/2) \cdot \sqrt{N}$. For every $i$ from 1 to $n$ there are at most $(k/2) \cdot \sqrt{N}$ possible values of $N_i$, so the total number of possible combinations of $N_1, \ldots, N_n$ is smaller than $((k/2) \cdot \sqrt{N})^n$. Let us denote this number of combinations by $N(p)$.

The total number $N_{cons}$ of consistent results is the sum of $N(p)$ different values of $C$ (values that correspond to $N(p)$ different combinations of $N_1$, $N_2$, ..., $N_n$). Let's denote the biggest of these values $C$ by $C_{\max}$. Since $N_{cons}$ is the sum of $N(p)$ terms, and each of these terms is not larger than the largest of them $C_{\max}$, we conclude that $N_{cons} \leq N(p) \cdot C_{\max}$. On the other hand, the sum $N_{cons}$ of non-negative integers is not smaller than the largest of them, i.e., $C_{\max} \leq N_{cons}$. Combining these two inequalities, we conclude that $C_{\max} \leq N_{cons} \leq N(p) \cdot C_{\max}$. Since $N(p) \leq ((k/2) \cdot \sqrt{N})^n$, we conclude that $C_{\max} \leq N_{cons} \leq ((k/2) \cdot \sqrt{N})^n \cdot C_{\max}$. Turning to logarithms, we find that $\log_2(C_{\max}) \leq \log_2(N_{cons}) \leq \log_2(C_{\max}) + (n/2) \cdot \log_2(N) + \text{const}$. Dividing by $N$, tending to the limit $N \to \infty$ and using the fact that $\log_2(N)/N \to 0$ and the (already proved) fact that $\log_2(C_{\max})/N$ tends to the entropy $S$, we conclude that $\lim Q(N)/N = S$.

# Computing Entropy under Interval Uncertainty. II

## Formulation and Analysis of the Problem, and the Corresponding Results and Algorithms

*Formulation of the problem.* In most practical situations, our knowledge is incomplete: there are several ($n$) different states which are consistent with our knowledge. How can we gauge this uncertainty? A natural measure of uncertainty is the average number of binary ("yes"-"no") questions that we need to ask to find the exact state. This idea is behind Shannon's *information theory*: according to this theory, when we know the probabilities $p_1, \ldots, p_n$ of different states (for which $\sum p_i = 1$), then this average number of questions is equal to $S = -\sum_{i=1}^{n} p_i \cdot \log_2(p_i)$. In information theory, this average number of question is called the *amount of information.*

In practice, we rarely know the exact values of the probabilities $p_i$; these probabilities come from experiments and are, therefore, only known with uncertainty. Usually, from the experiments, we can find *confidence intervals* $\boldsymbol{p}_i = [\underline{p}_i, \overline{p}_i]$, i.e., intervals which contain the (unknown) values $p_i$. Since $p_i \geq 0$ and $\sum p_i = 1$, we must have $\underline{p}_i \geq 0$ and $\sum \underline{p}_i \leq 1 \leq \sum \overline{p}_i$. How can we estimate the amount of information under such interval uncertainty?

For different values $p_i \in \boldsymbol{p}_i$, we get, in general, different values of the amount of information $S$. Since $S$ is a continuous function, the set of possible values of $S$ is an interval. So, to gauge the corresponding uncertainty, we must find the range $\boldsymbol{S} = [\underline{S}, \overline{S}]$ of possible values of $S$.

Thus, we arrive at the following computational problem: given $n$ intervals $\boldsymbol{p}_i = [\underline{p}_i, \overline{p}_i]$, find the range

$$\boldsymbol{S} = [\underline{S}, \overline{S}] = \left\{ -\sum_{i=1}^{n} p_i \cdot \log_2(p_i) \,\middle|\, p_i \in \boldsymbol{p}_i \,\&\, \sum_{i=1}^{n} p_i = 1 \right\}.$$

*Comment.* Some of the results presented in this chapter first appeared in [353] and [355].

H.T. Nguyen et al.: Computing Stat. under Interval & Fuzzy Uncertain., SCI 393, pp. 193–209.
springerlink.com © Springer-Verlag Berlin Heidelberg 2012

*Computation of $\overline{S}$ is feasible (takes polynomial time)*. Since the function $S$ is concave, computation of $\overline{S}$ is feasible; see [182] and [334].

The following algorithm computes $\overline{S}$ in time $O(n \cdot \log(n))$:

- First, we sort $2n$ endpoints $\underline{p}_i$ and $\overline{p}_i$ into a sequence

$$0 = p_{(0)} < p_{(1)} < p_{(2)} < \ldots < p_{(m)} < p_{(m+1)} = 1.$$

In the process of this sorting, for each $k$ from 1 to $m$, we form the sets $A_k^- = \{i : \underline{p}_i = p_{(k)}\}$ and $A_k^+ = \{i : \overline{p}_i = p_{(k)}\}$.
- Then, for each $k$ from 0 to $m$, we compute the values $M_k$, $P_k$, and $n_k$ as follows.
  - We start with $M_0 = - \sum_{i=1}^{n} \underline{p}_i \cdot \log_2(\underline{p}_i)$, $P_0 = \sum_{i=1}^{n} \underline{p}_i$, and $n_0 = n$.
  - Once we know $M_k$, $P_k$, and $n_k$, we compute the next values of these quantities as follows:

$$M_{k+1} = M_k + \sum_{j \in A_{k+1}^-} \underline{p}_j \cdot \log_2(\underline{p}_j) - \sum_{j \in A_{k+1}^+} \overline{p}_j \cdot \log_2(\overline{p}_j);$$

$$P_{k+1} = P_k - \sum_{j \in A_{k+1}^-} \underline{p}_j + \sum_{j \in A_{k+1}^+} \overline{p}_j; \quad n_{k+1} = n_k - \#(A_{k+1}^-) + \#(A_{k+1}^+).$$

- If $n_k = n$, we take $S_k = M_k$.
- If $n_k < n$, then we compute $p = \dfrac{1 - P_k}{n - n_k}$.
  - If $p \in [p_{(k)}, p_{(k+1)}]$, then we compute

$$S_k = M_k - (n - n_k) \cdot p \cdot \log_2(p).$$

  - Otherwise, we ignore this $k$.
- Finally, we find the largest of these values $S_k$ as the desired bound $\overline{S}$.

*Linear-time algorithm for computing $\overline{S}$*. It is possible to compute $\overline{S}$ in linear time. The corresponding algorithm is iterative. At each iteration of this algorithm we have three sets:

- the set $J^-$ of all the endpoints $\underline{p}_i$ and $\overline{p}_j$ for which we already know that for the optimal vector $p$ we have, correspondingly, $p_i \neq \underline{p}_i$ (for $\underline{p}_i$) or $p_j = \overline{p}_j$ (for $\overline{p}_j$);
- the set $J^+$ of all the endpoints $\underline{p}_i$ and $\overline{p}_j$ for which we already know that for the optimal vector $p$ we have, correspondingly, $p_i = \underline{p}_i$ (for $\underline{p}_i$) or $p_j \neq \overline{p}_j$ (for $\overline{p}_j$);
- the set $J$ of the endpoints $\underline{p}_i$ and $\overline{p}_j$ for which we have not yet decided whether these endpoints appear in the optimal vector $p$.

In the beginning, $J^- = J^+ = \emptyset$ and $J$ is the set of all $2n$ endpoints. At each iteration we also update the values $N^- = \#(J^-)$, $N^+ = \#(J^+)$, $E^- = \sum_{\overline{p}_j \in J^-} \overline{p}_j$, and $E^+ = \sum_{\underline{p}_i \in J^+} \underline{p}_i$. Initially, $N^- = N^+ = E^- = E^+ = 0$.

At each iteration we do the following.

- First we compute the median $m$ of the set $J$.
- Then, by analyzing the elements of the undecided set $J$ one by one, we divide them into two subsets

$$Q^- = \{p \in J : p \le m\}, \quad Q^+ = \{p \in J : p > m\}.$$

We also compute $m^+ = \min\{p : p \in Q^+\}$.

- We compute $e^- = E^- + \sum_{\overline{p}_j \in Q^-} \overline{p}_j$, $e^+ = E^+ + \sum_{\underline{p}_i \in Q^+} \underline{p}_i$,

$$n^- = N^- + \#\{\overline{p}_j \in Q^-\}, \quad n^+ = N^+ + \#\{\underline{p}_i \in Q^+\},$$

and $r = \dfrac{1 - e^- - e^+}{N - n^- - n^+}$.

- If $r < m$, then we replace $J^-$ with $J^- \cup Q^-$, $E^-$ with $e^-$, $J$ with $Q^+$, and $N^-$ with $n^-$.
- If $r > m^+$, then we replace $J^+$ with $J^+ \cup Q^+$, $E^+$ with $e^+$, $J$ with $P^-$, and $N^+$ with $n^+$.
- If $m \le r \le m^+$, then we replace $J^-$ with $J^- \cup Q^-$, $J^+$ with $J^+ \cup Q^+$, $J$ with $\emptyset$, $E^-$ with $e^-$, $E^+$ with $e^+$, $N^-$ with $n^-$, and $N^+$ with $n^+$.

At each iteration the set of undecided indices is divided in half. Iterations continue until all indices are decided. After this we return, as $\overline{S}$, the value of the entropy for the vector $x$ for which:

- $p_j = \overline{p}_j$ for indices $j$ for which $\overline{p}_j \in J^-$,
- $p_i = \underline{p}_i$ for indices $i$ for which $\underline{p}_i \in J^+$, and
- $p_i = r$ for all other indices $i$.

*Computing $\underline{S}$ is, in general, NP-hard.* Several algorithms for computing $\underline{S}$ are known; see, e.g., [1, 2, 3, 4, 5]. In the worst case, these algorithms take time that grows exponentially with $n$.

The following result shows that this exponential time is caused by the complexity of the problem.

**Proposition 25.1** *The problem of computing $\underline{S}$ is NP-hard.*

*Effective algorithms for computing $\underline{S}$ when intervals are not contained in each other.* Usually, when we know $p_i$ with some uncertainty, we know the approximate values $\widetilde{p}_i$ and the accuracy $\Delta$ of this approximation. In this case, we know that the actual (unknown) value of $p_i$ belongs to the interval $[\widetilde{p}_i - \Delta, \widetilde{p}_i + \Delta]$. Since these intervals all have the same width $2\Delta$, none of them can be a proper subset of the other. It turns out that if we restrict ourselves to intervals that satisfy this condition, then it is possible to compute $\underline{S}$ efficiently.

**Definition 25.1.** *We say that intervals $[\underline{p}_i, \overline{p}_i]$ satisfy the* no-subset property *if $[\underline{p}_i, \overline{p}_i] \not\subset (\underline{p}_j, \overline{p}_j)$ for all $i$ and $j$ (for which the intervals $\mathbf{p}_i$ and $\mathbf{p}_j$ are non-degenerate).*

*An $O(n \cdot \log_2(n))$ algorithm that computes $\underline{S}$ for all cases when the no-subset property holds*

- First, we sort $n$ intervals $\boldsymbol{p}_i$ in lexicographic order:

$$\boldsymbol{p}_1 \leq_{\text{lex}} \boldsymbol{p}_2 \leq_{\text{lex}} \cdots \leq_{\text{lex}} \boldsymbol{p}_n,$$

  where $[\underline{a}, \overline{a}] \leq_{\text{lex}} [\underline{b}, \overline{b}]$ if and only if either $\underline{a} < \overline{b}$, or $\underline{a} = \underline{b}$ and $\overline{a} \leq \overline{b}$.
- Second, for each $i$ from 1 to $n$, we compute

$$M_i = \sum_{j:j<i} f(\underline{p}_j) + \sum_{m:m>i} f(\overline{p}_m); \quad P_i = \sum_{j:j<i} \underline{p}_j + \sum_{m:m>i} \overline{p}_m.$$

  First, we compute $M_1 = \sum_{j=2}^{n} f(\overline{p}_j)$ and $P_1 = \sum_{j=2}^{n} \overline{p}_j$; then, we sequentially compute other values as

$$M_i = M_{i-1} + f(\underline{p}_{i-1}) - f(\overline{p}_i); \quad P_i = P_{i-1} + \underline{p}_{i-1} - \overline{p}_i.$$

- For every $i$, we compute $p_i = \dfrac{1 - P_i}{n - 1}$. If $p_i \in [\underline{p}_i, \overline{p}_i]$, we compute

$$S_i = M_i + f(p_i).$$

- Finally, we return the smallest of these values $S_i$ as $\underline{S}$.

*Linear-time algorithm for computing $\underline{S}$ for the case when narrowed intervals satisfy the no-subset property* For simplicity, let us consider the case when all the intervals are non-degenerate, i.e., when $\Delta_i > 0$ for all $i$.

The proposed algorithm is iterative. At each iteration of this algorithm we have three sets:

- the set $I^-$ of all the indices $i$ from 1 to $n$ for which we already know that for the optimal vector $p$, we have $p_i = \underline{p}_i$;
- the set $I^+$ of all the indices $j$ for which we already know that for the optimal vector $p$, we have $p_j = \overline{p}_j$;
- the set $I = \{1, \ldots, n\} \setminus (I^- \cup I^+)$ of the indices $i$ for which we are still undecided.

In the beginning, $I^- = I^+ = \emptyset$ and $I = \{1, \ldots, n\}$. At each iteration we also update the values of two auxiliary quantities $E^- \stackrel{\text{def}}{=} \sum_{i \in I^-} \underline{p}_i$ and $E^+ \stackrel{\text{def}}{=} \sum_{j \in I^+} \overline{p}_j$. In principle, we could compute these values by computing these sums. However, to speed up computations on each iteration, we update these two auxiliary values in a way that is faster than re-computing the corresponding two sums. Initially, since $I^- = I^+ = \emptyset$, we take $E^- = E^+ = 0$.

At each iteration we do the following:

- first, we compute the median $m$ of the set $I$ (median in terms of sorting by $\widetilde{p}_i$);

- then, by analyzing the elements of the undecided set $I$ one by one, we divide them into two subsets $P^- = \{i : \widetilde{p}_i \leq \widetilde{p}_m\}$ and $P^+ = \{j : \widetilde{p}_j > \widetilde{p}_m\}$;
- we compute $e^- = E^- + \sum_{i \in P^-} \underline{p}_i$ and $e^+ = E^+ + \sum_{j \in P^+} \overline{p}_j$;
- If $e^- + e^+ > 1$, then we replace $I^-$ with $I^- \cup P^-$, $E^-$ with $e^-$, and $I$ with $P^+$.
- If $e^- + e^+ + 2\Delta_m < 1$, then we replace $I^+$ with $I^+ \cup P^+$, $E^+$ with $e^+$, and $I$ with $P^-$.
- Finally, if $e^- + e^+ \leq 1 \leq e^- + e^+ + 2\Delta_m$, then we replace $I^-$ with

$$I^- \cup (P^- - \{m\}),$$

$I^+$ with $I^+ \cup P^+$, $I$ with $\{m\}$, $E^-$ with $e^- - \underline{p}_m$, and $E^+$ with $e^+$.

At each iteration the set of undecided indices is divided in half. Iterations continue until we have only one undecided index $I = \{k\}$. After this we return, as $\underline{S}$, the value of the entropy for the vector $p$ for which $p_i = \underline{p}_i$ for $i \in I^-$, $p_j = \overline{p}_j$ for $j \in I^+$, and $p_k = 1 - e^- - e^+$ for the remaining value $k$.

*Continuous (p-box) case: formulation of the problem and a seemingly natural solution.* As we have mentioned, in the traditional statistical approach, the uncertainty in a probability distribution is usually described by Shannon's entropy

$$S = - \int \rho(x) \cdot \log_2(\rho(x)) \, dx,$$

where $\rho(x) = F'(x)$ is the probability density function of this distribution.

In the situations when we have partial information about the probability distribution $F(x)$ – e.g., when we only know that $F(x)$ belongs to a non-degenerate p-box $\mathbf{F}(x) = [\underline{F}(x), \overline{F}(x)]$, a reasonable estimate for an arbitrary statistical characteristic $S$ is the range of possible values of $S$ over all possible distributions $F(x) \in \mathbf{F}(x)$.

It therefore seems natural to apply this approach to entropy as well – and return the range of entropy as a gauge of uncertainty of a p-box; see, e.g., [158] and [355].

*Limitations of the above (seemingly natural) solution.* The problem with the above approach is that every non-degenerate p-box includes discrete distributions, i.e., distributions which take discrete values $x_1, \ldots, x_n$ with finite probabilities. For such distributions, Shannon's entropy is $-\infty$.

Thus, for every non-degenerate p-box, the resulting interval $[\underline{S}, \overline{S}]$ has the form $[-\infty, \overline{S}]$. Thus, once the distribution with the largest entropy $\overline{S}$ is fixed, we cannot distinguish between a very narrow p-box or a very thick p-box – in both case, we end up with the same interval $[-\infty, \overline{S}]$.

It is therefore desirable to develop a new approach that would enable us to distinguish between these two cases.

*Case of p-boxes: description of the situation.* The traditional approach of interval-valued entropy does not allow us to distinguish between narrow and wide p-boxes. For a wide p-box, it is OK to make a wide interval like $[-\infty, \overline{S}]$, but for narrow p-boxes, we would like to have narrower estimates. Let us therefore consider narrow p-boxes.

Since entropy is defined for smooth (differentiable) cdfs $F(x)$, it is reasonable to start with the case when the central function of a p-box is also smooth. In other words, we consider p-boxes of the type

$$\mathbf{F}(x) = [F_0(x) - \Delta F(x), F_0(x) + \Delta F(x)],$$

where $F_0(x)$ is differentiable, with derivative $\rho_0(x) \stackrel{\text{def}}{=} F_0'(x)$, and $\Delta F(x)$ is small.

*Formulation of the problem.* For each $\varepsilon > 0$ and for each distribution $F(x) \in \mathbf{F}(x)$, we can use the above formulas to estimate the average number $S_\varepsilon(F)$ of "yes"-"no" question that we need to ask to determine the actual value with accuracy $\varepsilon$. Our objective is to compute the range $[\underline{S}, \overline{S}] = \{S_\varepsilon(F) : F \in \mathbf{F}\}$.

*Estimates.* We have mentioned earlier that asymptotically,

$$\overline{S} \sim - \int \rho_0(x) \cdot \log_2(\rho_0(x))\, dx - \log_2(2\varepsilon).$$

It turns out that for the lower bound, we have the following asymptotics:

$$\underline{S} \sim - \int \rho_0(x) \cdot \log_2(\max(2\Delta F(x), 2\varepsilon \cdot \rho_0(x)))\, dx.$$

*Comment.* This result holds when $\varepsilon$ and the width of $\Delta F$ both tends to 0. If instead we fix the width $\Delta F$ and let $\varepsilon \to 0$, then $\overline{S} \to \infty$ but $\underline{S}$ remains finite.

*Alternative approach: an entropy of determining the probability distribution.* We started with the situation when we do not know the object, we only know the probabilities of different objects, and we wanted to find out how many "yes"-"no" questions we need to find the object $x$.

In the new situation, in addition to not knowing the object $x$, we also do not know the exact probability distribution $\rho(x)$. It is therefore reasonable, in addition to finding out how many binary questions we need to find $x$, to also find out how many "yes"-"no" questions we need to find the exact probability distribution $\rho(x)$.

Of course, just like we cannot determine the real number $x$ after finitely many "yes"-"no" questions, we are not able to determine $\rho(x)$ exactly after finitely many questions, we can only obtain an approximate value of a probability distribution.

A natural way to describe a probability distribution is via its cdf $F(x)$. There are two reasons why the approximate cdf may be different from the actual one: we may get the probabilities only approximately, and we may get the values at which these probabilities are attained only approximately. It is therefore reasonable to fix two accuracy values $\varepsilon$ (accuracy with which we approximate probabilities) and $\delta$ (accuracy with which we approximate $x$) and try to find an approximation $\widetilde{F}(x)$ to $F(x)$ in which, for every $x$, we have $|\widetilde{F}(\widetilde{x}) - F(x)| \leq \varepsilon$ for some $\widetilde{x}$ for which $|\widetilde{x} - x| \leq \delta$.

When $P$ is a p-box, then, for every number $x_0$, we have the interval $[\underline{F}(x_0), \overline{F}(x_0)]$ of possible values of the probability $F(x_0) = \mathrm{Prob}(X \leq x_0)$. We want to find the actual value of $\varepsilon$ with the accuracy $\varepsilon$. We have already mentioned that this is equivalent to localizing $F(x_0)$ within an interval of width $2\varepsilon$. Within the original interval of width $w(x_0) \stackrel{\text{def}}{=} \overline{F}(x_0) - \underline{F}(x_0)$, there are $n(x_0) \stackrel{\text{def}}{=} w(x_0)/(2\varepsilon)$ such subintervals, so, to localize $F(x_0)$, we need $\sim \log_2(n(x_0)) = \log_2(w(x_0)) - \log_2(2\varepsilon)$ questions.

To get the spatial accuracy $\delta$, we need to repeat this procedure for the values $x_1$, $x_2 = x_1 + 2\delta$, etc. Overall, we thus need $\sum \log_2(w(x_i)) - \sum \log_2(2\varepsilon)$ questions. If we multiply the first sum by $2\delta$, then we get the integral sum for $\int \log_2(w(x))\,dx$; so, the first sum is $\sim \int \log_2(w(x))\,dx/(2\delta)$. The second sum is a constant that does not depend on the p-box at all.

Thus, for a p-box $[\underline{F}(x), \overline{F}(x)]$, the overall number of questions that we need to ask to determine the probability distribution $F(x)$ with a given accuracy is determined by the integral $\int \log_2(\overline{F}(x) - \underline{F}(x))\,dx$. This easy-to-compute integral can thus serve as an additional information measure for p-boxes.

*Adding fuzzy uncertainty.* The main idea behind fuzzy uncertainty is that, instead of just describing which objects are possible, we also describe, for each object, the degree to which this object is possible. For each degree of possibility $\alpha$, we can determine the set of objects that are possible with at least this degree of possibility – the *$\alpha$-cut* of the original fuzzy set. Vice versa, if we know $\alpha$-cuts for every $\alpha$, then, for each object $x$, we can determine the degree of possibility that $x$ belongs to the original fuzzy set.

A fuzzy set can be thus viewed as a nested family of its $\alpha$-cuts.

Thus, if instead of a (crisp) set $P$ of possible probability distributions (e.g., a p-box), we have a fuzzy set $\mathcal{P}$ of possible probability distributions, then we can view this information as a family of nested crisp sets $\mathcal{P}(\alpha)$ – $\alpha$-cuts of the given fuzzy set.

In this case, once we fix a measure of information $I(P)$ for crisp sets of distributions – e.g., the maximum entropy, we can then extend this measure to fuzzy sets $\mathcal{P}$ – by defining $I(\mathcal{P})$ as a fuzzy number whose $\alpha$-cut coincides with $I(\mathcal{P}(\alpha))$.

*Comment.* Instead of describing the information in a fuzzy set by a fuzzy number, we can, alternatively, interpret degree of possibility in probabilistic terms and compute the corresponding information by using probability formulas; see, e.g., [286] and [287].

## Proofs of Theoretical Results and Justifications of Algorithms

*Proof that an $O(n \cdot \log_2(n))$ algorithm for computing $\overline{S}$ is correct.* Let

$$(p_1, \ldots, p_n)$$

be the values of probabilities at which the entropy $S$ attains its maximum. The fact that $S$ attains its maximum means that if we change the values $p_i$, then the corresponding change $\Delta S$ in $S$ is non-positive: $\Delta S \leq 0$. We will use this condition for different changes in $p_i$.

For each value of $p_i$, we have three possibilities:

- this value can be strictly inside the corresponding interval $[\underline{p}_i, \overline{p}_i]$;
- this value can be at the left end of this interval, i.e., $p_i = \underline{p}_i$; and
- this value can be at the right end of this interval, i.e., $p_i = \overline{p}_i$.

Let us consider these possibilities one by one.

Let us first consider the values $p_j$ which are strictly inside the corresponding intervals. If for some $j$ and $k$, the corresponding probabilities are strictly inside the corresponding intervals, i.e., if we have $p_j \in (\underline{p}_j, \overline{p}_j)$ and $p_k \in (\underline{p}_k, \overline{p}_k)$, then for a sufficiently small real number $\Delta$, we can replace $p_j$ with $p_j + \Delta$ and $p_k$ with $p_k - \Delta$ and still get a sequence of probabilities for which $p_i \in [\underline{p}_i, \overline{p}_i]$ for all $i$ and $\sum p_i = 1$. For small $\Delta$, the corresponding change $\Delta S$ in entropy is equal to

$$\left( \frac{\partial S}{\partial x_j} - \frac{\partial S}{\partial x_j} \right) \cdot \Delta + o(\Delta) = (-\log_2(p_j) + \log_2(p_k)) \cdot \Delta + o(\Delta).$$

Since $\Delta$ can be positive or negative, the only way to have $\Delta S \leq 0$ for all small $\Delta$ is to make sure that the coefficient at $\Delta$ is equal to 0, i.e., that $-\log_2(p_j) + \log_2(p_k) = 0$. This implies that $p_j = p_k$ – i.e., that all the values $p_j$ which are inside the corresponding intervals coincide. Let us denote this common value of $p_j$ by $p$.

Let us now consider the situation when $p_j$ is at the left end of the corresponding interval, i.e., when $p_j = \underline{p}_j$. If for some other $k$, the corresponding value $p_k$ is at the right end or strictly inside the corresponding interval, then $p_k > \underline{p}_k$. In this case, we can only make a similar change $p_j \to p_j + \Delta$ and $p_k \to p_k - \Delta$ when $\Delta > 0$. Then, the requirement that $\Delta S \leq 0$ means that the coefficient at $\Delta$ should be non-positive, i.e., that $-\log_2(p_j) + \log_2(p_k) \leq 0$. Thus, we conclude that $p_k \leq p_j$. In particular, for the case when $p_k$ is inside the corresponding interval – and is, thus, equal to $p$ – we conclude that $p \leq p_j$.

Similarly, if $p_j$ is at the right end of the corresponding interval, i.e., if $p_j = \overline{p}_j$, then, for every $k$ for which $p_k > \underline{p}_k$, we conclude that $p_k \geq p_j$. In particular, we can conclude that $p_j \leq p$.

Let us now consider the case when there are some values $p_i$ strictly inside the corresponding interval, so there is a value $p$. Let us show that is we know where $p$ is located in comparison with all the endpoints $[\underline{p}_i, \overline{p}_i]$, then we can uniquely determine all the values $p_i$.

Indeed, if the entire interval $[\underline{p}_i, \overline{p}_i]$ is located to the left of $p$, i.e., if $\overline{p}_i < p$, then:

- the minimum cannot be attained strictly inside the interval – because it would have been attained at the point $p_i = p$, and we are considering the case when the entire interval $[\underline{p}_i, \overline{p}_i]$ is located to the left of $p$;
- similarly, the minimum cannot be attained for $p_i = \overline{p}_i$, because then, as we have proven, we should have $p \leq p_i$, and the entire interval $[\underline{p}_i, \overline{p}_i]$ is located to the left of $p$.

Thus, in this case, the only remaining possibility is $p_i = \overline{p}_i$.

Similarly, if the entire interval $[\underline{p}_i, \overline{p}_i]$ is located to the right of $p$, i.e., if $p < \underline{p}_i$, then $p_i = \underline{p}_i$.

If $\underline{p}_i < p < \overline{p}_i$, then, similarly, we cannot have $p_i = \underline{p}_i$ and $p_i = \overline{p}_i$, so we must have $p_i$ inside and hence, $p_i = p$.

To exploit this conclusion, let us formalize how we can describe the location of $p$ in relation to $2n$ endpoints. If we sort these endpoints $\underline{p}_i$ and $\overline{p}_i$ into a sequence $p_{(1)} \leq p_{(2)} \leq \ldots \leq p_{(2n)}$, then we divide the entire real line into $2n + 1$ "zones" $[p_{(k)}, p_{(k+1)}]$, where we denoted $p_{(0)} \overset{\text{def}}{=} 0$ and $p_{(2n+1)} \overset{\text{def}}{=} 1$.

Let us pick a zone $[p_{(k)}, p_{(k+1)}]$, and show how we can find the possibly optimal values $p_i$ (and the corresponding value of the entropy) under the assumption that the (unknown) value $p$ belongs to the this zone.

If $\overline{p}_i < p$, then we must have $\overline{p}_i \leq p_{(k)}$ – otherwise, if $\overline{p}_i > p_{(k)}$, then, since $p_{(k)}$ describe all the endpoints, we would have $\overline{p}_i \geq p_{(k+1)}$ and hence $\overline{p}_i > p$. Thus, in the optimal arrangement of probabilities, we have $p_i = \overline{p}_i$.

Similarly, if $\underline{p}_i > p$, then we have $p_i = \underline{p}_i$. For all other $i$, we have $p_i = p$. This value $p$ can be computed based on the fact that $\sum p_i = 1$.

For each of $2n + 1$ zones, we need to analyze $n$ values $p_i$; thus, for each of the zones, we need $O(n)$ computation steps. Overall, we get a quadratic algorithm for computing $\overline{S}$.

Before we describe this algorithm, we should mention that the above description only works when we actually have an index $i$ for which $p_i$ is strictly inside the corresponding interval. If no such index exists, then we can still conclude that every value $p_j = \overline{p}_j$ is smaller than or equal than every value $p_k = \underline{p}_k$. Thus, there exists a value $p$ that is greater than or equal than all $j$ for which $p_j = \overline{p}_j$ and less than or equal than all $k$ for which $p_k = \underline{p}_k$. By using this $p$, we arrive at the same conclusion about the values $p_i$.

Thus, in general, we arrive at the following quadratic-time algorithm for computing $\overline{S}$ (first described in [168]):

- First, we sort $2n$ endpoints of $n$ intervals $\boldsymbol{p}_i$ into an increasing sequence $p_{(0)} = 0 < p_{(1)} < p_{(2)} < \ldots < p_{(m)} < p_{(m+1)} = 1$. (If all the endpoints are different, then $m = 2n$, but since some endpoints may coincide, we may have $m < 2n$; in general, $m \leq 2n$.)

- Second, for every $k$ from 0 to $m-1$, we compute the following three values:

$$M_k = - \sum_{i:\overline{p}_i \le p_{(k)}} \overline{p}_i \cdot \log_2(\overline{p}_i) - \sum_{j:\underline{p}_j \ge p_{(k+1)}} \underline{p}_j \cdot \log_2(\underline{p}_j);$$

$$P_k = \sum_{i:\overline{p}_i \le p_{(k)}} \overline{p}_i + \sum_{j:\underline{p}_j \ge p_{(k+1)}} \underline{p}_j; \quad n_k = \#\{i : \overline{p}_i \le p_{(k)} \vee \underline{p}_i \ge p_{(k+1)}\}.$$

- If $n_k = n$, we take $S_k = M_k$.
- If $n_k < n$, then we compute $p = \dfrac{1 - P_k}{n - n_k}$.
  - If $p \in [p_{(k)}, p_{(k+1)}]$, then we compute $S_k = M_k - (n - n_k) \cdot p \cdot \log_2(p)$.
  - Otherwise, we ignore this $k$.
- Finally, we find the largest of these values $S_k$ as the desired bound $\overline{S}$.

Let us show that the computation time for this algorithm can be reduced to $O(n \cdot \log_2(n))$. Indeed, sorting takes $O(n \cdot \log_2(n))$ steps; see, e.g., [73]. Once we have a sorted list, we can find, for each of the $2n$ endpoints $\underline{p}_i$ and $\overline{p}_i$, where they are in this sorting. We can thus, for each of the values $p_{(j)}$, mark which endpoints coincide with this value.

The initial computation of the values $M_0$, $P_0$, and $n_0$ takes $O(n)$ steps. Once we go from $M_k$ to $M_{k+1}$ (or from $P_k$ to $P_{k+1}$), we only need to update the values corresponding to the endpoints of this zone. Overall, for all the updates, we thus need as much time as there are updated values $p_i$ overall.

Each endpoint in this arrangement changes only once, so overall, we need a linear number of steps ($2n$) to update all the values $M_k$, all the values $P_k$, and all the values $n_k$. Thus, overall, this algorithm takes time $O(n \cdot \log_2(n)) + O(n) + O(n) = O(n \cdot \log_2(n))$.

*Proof that the above fast algorithm always computes $\overline{S}$ in linear time.* Let us first prove that the fast algorithm described in the main text always computes the desired bound $\overline{S}$. Indeed, in the previous proof, we have shown that if we sort all $2n$ endpoints into a sequence $p_{(1)} \le p_{(2)} \le \ldots \le p_{(2n)}$, then for some $k = k_{\max}$ the maximum $\overline{S}$ is attained for the vector $p$ for which the following holds:

- For all indices $j$ for which $\overline{p}_j \le p_{(k)}$, we have $p_j = \overline{p}_j$.
- For all indices $i$ for which $\underline{p}_i \ge x_{(k+1)}$, we have $p_i = \underline{p}_i$.
- For all other indices, we have $p_i = \text{const}$. Since $\sum\limits_{i=1}^{n} p_i = 1$, we conclude that this constant is equal to $r_k \overset{\text{def}}{=} \dfrac{1 - E_k}{n - N_k}$, where

$$E_k = \sum_{j:\overline{p}_j \le p_{(k)}} \overline{p}_j + \sum_{i:\underline{p}_i \ge p_{(k+1)}} \underline{p}_i;$$

$$N_k = \#\{j : \overline{p}_j \le p_{(k)}\} + \#\{i : \underline{p}_i \ge p_{(k+1)}\}.$$

It can also be proven that for the optimal $k$ we have $r_k \in [p_{(k)}, p_{(k+1)}]$. These facts can proven by the same analysis (adding $\Delta p$ to one value $p_j$ and subtracting $\Delta p$ from another value $p_k$) as in our above analysis.

Let us first prove that if $r_k = \dfrac{1 - E_k}{n - N_k} \leq p_{(k+1)}$ then the similar inequality $r_{k+1} = \dfrac{1 - E_{k+1}}{n - N_{k+1}} \leq p_{(k+2)}$ holds for the next value $k$. Indeed, the given inequality $\dfrac{1 - E_k}{n - N_k} \leq p_{(k+1)}$ is equivalent to $1 - E_k \leq (n - N_k) \cdot p_{(k+1)}$.

The only difference between the sums $E_k = \displaystyle\sum_{j:\overline{p}_j \leq p_{(k)}} \overline{p}_j + \sum_{i:\underline{p}_i \geq p_{(k+1)}} \underline{p}_i$ and $E_{k+1} = \displaystyle\sum_{j:\overline{p}_j \leq p_{(k+1)}} \overline{p}_j + \sum_{i:\underline{p}_i \geq p_{(k+2)}} \underline{p}_i$ is that:

- some terms equal to $p^{(k+1)}$ may be added (if there are $j$ for which $\overline{p}_j = p_{(k+1)}$), and
- some other terms equal to to $p^{(k+1)}$ may be subtracted (if there are $i$ for which $\underline{p}_i = p_{(k+1)}$).

In general, $E_{k+1} = E_k + c_k \cdot p_{(k+1)}$ for some integer $c_k$ (positive, negative, or zero), and $N_{k+1} = N_k + c_k$. Subtracting $c_k \cdot p_{(k+1)}$ from both sides of the given inequality $1 - E_k \leq (n - N_k) \cdot p_{(k+1)}$, we conclude that $1 - E_{k+1} \leq (n - N_{k+1}) \cdot p_{(k+1)}$, i.e. that $r_{k+1} = \dfrac{1 - E_{k+1}}{n - N_{k+1}} \leq p_{(k+1)}$. Since the sequence $p_{(k)}$ is sorted, we thus conclude that $p_{(k+1)} \leq p_{(k+2)}$ and hence $r_{k+1} \leq p_{(k+2)}$.

So if the inequality $r_k \leq p_{(k+1)}$ holds for some $k$, it holds for all larger values of $k$ as well. Thus this inequality holds for all $k$ after a certain value $l_0$.

Similarly, we can prove that if the inequality $r_k \geq p_{(k)}$ holds for some $k$, then it holds for $k-1$ as well – since the only difference between $E_k$ and $E_{k-1}$ consists of adding and/or subtracting some values $p_{(k)}$. So if the inequality $r_k \geq p_{(k)}$ holds for some $k$, it holds for all smaller values of $k$ as well. Thus, this inequality holds for all $k$ until a certain value $k_0$.

Similarly to the proof about $\underline{V}$, we can prove that if there are several values $k = l_0, l_0 + 1, \ldots, k_0$ for which both inequalities hold $p_{(k)} \leq r_k \leq p_{(k+1)}$, then for these $k$, the entropy has exactly the same value.

So:

- for $k < k_{\max}$, we have $r_k > p_{(k+1)}$,
- for $k > k_{\max}$, we have $r_k < p_{(k)}$, and
- for $k = k_{\max}$ (or, to be more precise, for $l_0 \leq k \leq k_0$), we have

$$p_{(k)} \leq r_k \leq p_{(k+1)}.$$

Hence:

- if $r_k < p_{(k)}$, then we cannot have $k < k_{\max}$ and $k = k_{\max}$, hence $k > k_{\max}$;
- if $r_k > p_{(k+1)}$, then we cannot have $k > k_{\max}$ and $k = k_{\max}$, hence $k < k_{\max}$;

- if $p_{(k)} \le r_k \le p_{(k+1)}$, then we cannot have $k < k_{\min}$ and $k > k_{\min}$, hence $k = k_{\max}$.

Thus, the above algorithm finds the correct value of $k_{\max}$ and thence, the correct value of $\overline{S}$.

To complete our proof, we must show that the proposed algorithm for computing $\overline{S}$ takes linear time. Indeed, at each iteration, computing median takes linear time, and all other operations with $J$ take time $t$ linear in the number of elements $|J|$ of $J$: $t \le C \cdot |J|$ for some $C$. We start with the set $J$ of size $2n$. On the next iteration, we have a set of size $2n/2 = n$, then $n/2$, etc. Thus, the overall computation time is $\le C \cdot (2n + n + n/2 + \ldots) \le C \cdot 4n$, i.e. linear in $n$.

*Proof of Proposition 25.1, that the problem of computing $\underline{S}$ is NP-hard.* By definition, a problem is called NP-hard if every problem from the class NP can be reduced to it. To prove that a problem $\mathcal{P}$ is NP-hard, it is sufficient to reduce one of the known NP-hard problems $\mathcal{P}_0$ to $\mathcal{P}$. The reason for this is as follows: since $\mathcal{P}_0$ is known to be NP-hard, it means that every problem from the class NP can be reduced to $\mathcal{P}_0$, and since $\mathcal{P}_0$ can be reduced to $\mathcal{P}$, thus, we can deduce that every problem from the class NP can be reduced to $\mathcal{P}$.

$1^\circ$. For our proof, we will select the following *subset* problem as the known NP-hard problem $\mathcal{P}_0$: given $n$ positive integers $s_1, \ldots, s_n$, check whether there exist signs $\eta_i \in \{-1, +1\}$ for which the signed sum $\sum_{i=1}^{n} \eta_i \cdot s_i$ equals to 0.

We will eventually prove that this problem can be reduced to the problem of computing $\underline{S}$; this computational problem will be denoted by $\mathcal{P}$. However, directly proving that $\mathcal{P}_0$ can be reduced to $\mathcal{P}$ seems to be difficult. Therefore, we introduce the following auxiliary problem, denoted as $\mathcal{P}_1$: given a real number $a > 0$ and $n$ intervals $\boldsymbol{q}_1 = [\underline{q}_1, \overline{q}_1], \boldsymbol{q}_2 = [\underline{q}_2, \overline{q}_2], \ldots, \boldsymbol{q}_n = [\underline{q}_n, \overline{q}_n]$, where $\sum_{i=1}^{n} \underline{q}_i \le a \le \sum_{i=1}^{n} \overline{q}_i$ and $0 \le \underline{q}_i$ for all $i$, find the lower endpoint $\underline{L}$ of the range

$$\boldsymbol{L} = [\underline{L}, \overline{L}] = \left\{ -\sum_{i=1}^{n} q_i \cdot \log_2(q_i) \,\middle|\, q_i \in \boldsymbol{q}_i \ \& \ \sum_{i=1}^{n} q_i = a \right\}$$

*Comment.* Similarly to our problem $\mathcal{P}$, the new problem $\mathcal{P}_1$ is also about minimizing entropy $S$: the only difference is that instead of the restriction $\sum_{i=1}^{n} p_i = 1$, we have a new restriction $\sum_{i=1}^{n} q_i = a$.

$2^\circ$. To reduce $\mathcal{P}_0$ to $\mathcal{P}_1$ means that for every instance $(s_1, \ldots, s_n)$ of the problem $\mathcal{P}_0$, we can find a corresponding instance of the problem $\mathcal{P}_1$ from whose solution, we can easily check whether the desired signs $\eta_i$ in $\mathcal{P}_0$ exist.

In order to select an appropriate instance, let us first analyze the function $-q \cdot \log_2(q)$. This function is equal to 0 for $q = 0$ and for $q = 1$. It attains its maximum when

$$\frac{\partial}{\partial q}(-q \cdot \log_2(q)) = -\log_2(e) \cdot (1 + \ln(q)) = 0,$$

i.e., when $q = \dfrac{1}{e}$. The corresponding maximum is equal to $-\dfrac{1}{e} \cdot \log_2\left(\dfrac{1}{e}\right) = \dfrac{\log_2(e)}{e}$. One can easily check that the function $-q \cdot \log_2(q)$ is convex; therefore, for every real number $r$ between 0 and the maximum – i.e., for which $0 < r < \dfrac{\log_2(e)}{e}$, there exist exactly two different values $q$ for which $-q \cdot \log_2(q) = r$. Let us denote the smaller of these two values by $q^-(r)$, and the larger one by $q^+(r)$. We can check that that $0 < q^-(r) < q^+(r) < 1$ and $0 < q^+(r) - q^-(r) < 1$. As $r$ grows from 0 to its largest value, the difference $q^+(r) - q^-(r)$ decreases from 1 to 0.

Now, for each instance $(s_1, \ldots, s_n)$ of the problem $\mathcal{P}_0$, we select the corresponding instance of the problem $\mathcal{P}_1$, i.e., the intervals $[\underline{q}_i, \overline{q}_i]$ and the real number $a$, as follows:

- First, we select a positive real number $z$ for which $z \cdot \max(s_i) < 1$.
- Next, for each $i$ from 1 to $n$, we find $r_i$ for which $q^+(r_i) - q^-(r_i) = z \cdot s_i$, and take $\underline{q}_i = q^-(r_i)$ and $\overline{q}_i = q^+(r_i)$.
- Finally, we select $a = \displaystyle\sum_{i=1}^{n} \frac{\underline{q}_i + \overline{q}_i}{2}$.

It is easy to check that for thus selected values, $\underline{q}_i \geq 0$ and $\displaystyle\sum_{i=1}^{n} \underline{q}_i \leq a \leq \sum_{i=1}^{n} \overline{q}_i$.

Let $L_0 \stackrel{\text{def}}{=} -\displaystyle\sum_{i=1}^{n} \underline{q}_i \cdot \log_2(\underline{q}_i)$. We will show that $\underline{L} = L_0$ if and only if there exist signs $\eta_i$ for which $\displaystyle\sum_{i=1}^{n} \eta_i \cdot s_i = 0$.

$3°$. Let us first prove that $\underline{L} \geq L_o$.

Indeed, due to our choice of $\underline{q}_i$ and $\overline{q}_i$, the function $-q \cdot \log_2(q)$ attains the same value at the two endpoints of the interval $[\underline{q}_i, \overline{q}_i]$ and is larger everywhere inside this interval. Thus, for every $i$ and for every $q_i \in [\underline{q}_i, \overline{q}_i]$, we have $-q_i \cdot \log_2(q_i) \geq -\underline{q}_i \cdot \log_2(\underline{q}_i)$. By adding these inequalities, we conclude that

$$L = -\sum_{i=1}^{n} q_i \cdot \log_2(q_i) \geq -\sum_{i=1}^{n} -\underline{q}_i \cdot \log_2(\underline{q}_i) = L_0.$$

Since all the values of $L$ are larger than or equal to $L_0$, the smallest possible value $\underline{L}$ of the function $L$ also satisfies the inequality $\underline{L} = L_0$.

$4°$. Let us first prove that if the desired signs $\eta_i$ exist, then $\underline{L} = L_0$.

Indeed, in this case, we can select $q_i = \underline{q}_i$ when $\eta_i = -1$ and $q_i = \overline{q}_i$ when $\eta_i = 1$. Both cases can be described by a single formula

$$q_i = \frac{\underline{q}_i + \overline{q}_i}{2} + \frac{\eta_i \cdot (\overline{q}_i - \underline{q}_i)}{2} = \frac{\underline{q}_i + \overline{q}_i}{2} + \frac{\eta_i \cdot z \cdot s_i}{2}.$$

Since $-\underline{q}_i \cdot \log_2(\underline{q}_i) = -\overline{q}_i \cdot \log_2(\overline{q}_i)$, for this choice of $q_i$, we have

$$L = -\sum_{i=1}^{n} q_i \cdot \log_2(q_i) = -\sum_{i=1}^{n} \underline{q}_i \cdot \log_2(\underline{q}_i) = L_0.$$

In this case,

$$\sum_{i=1}^{n} q_i = \sum_{i=1}^{n} \left( \frac{\underline{q}_i + \overline{q}_i}{2} + \frac{\eta_i \cdot z \cdot s_i}{2} \right) =$$

$$\sum_{i=1}^{n} \frac{\underline{q}_i + \overline{q}_i}{2} + \frac{z}{2} \cdot \sum_{i=1}^{n} \eta_i \cdot s_i = \sum_{i=1}^{n} \frac{\underline{q}_i + \overline{q}_i}{2} = a.$$

Since for this choice of $q_i$, we have $L = L_0$, we can thus onclude that the smallest possible value $\underline{L}$ of $L$ cannot exceed $L_0$: $\underline{L} \leq L_0$.

We have already proven that $\underline{L} \geq L_0$, so we can conclude that $\underline{L} = L_0$.

$5^\circ$. Now let us prove that if $\underline{L} = L_0$, then the desired signs $\eta_i$ exists.

Let $q_1, \ldots, q_n$ be the values that minimize $L$, i.e., for which $L = \underline{L}$. From the equality $\underline{L} = L_0$, we will conclude that for every $i$, we have either $q_i = \underline{q}_i$ or $q_i = \overline{q}_i$. This can be proven by reduction to a contradiction: if for some $j$, we have $q_j \neq \underline{q}_j$ and $q_j \neq \overline{q}_j$, then we will get $-q_j \cdot \log_2(q_j) > -\underline{q}_j \cdot \log_2(\underline{q}_j)$. For every other $i$, we have $-q_i \cdot \log_2(q_i) \geq -\underline{q}_i \cdot \log_2(\underline{q}_i) = -\overline{q}_i \cdot \log_2(\overline{q}_i)$. By adding all these inequalities, we can conclude that

$$\underline{L} = L = -\sum_{i=1}^{n} q_i \cdot \log_2(q_i) > -\sum_{i=1}^{n} \underline{q}_i \cdot \log_2(\underline{q}_i) = L_0,$$

which contradicts to our assumption that $\underline{L} = L_0$. This contradiction shows that indeed, for every $i$, we have either $q_i = \underline{q}_i$ or $q_i = \overline{q}_i$.

Let us set $\eta_i = -1$ when $q_i = \underline{q}_i$ and $\eta_i = 1$ when $q_i = \overline{q}_i$. Then,

$$q_i = \frac{\underline{q}_i + \overline{q}_i}{2} + \frac{\eta_i \cdot z \cdot s_i}{2}.$$

From the condition $\sum q_i = a$, we now conclude that

$$a = \sum_{i=1}^{n} q_i = \sum_{i=1}^{n} \frac{\underline{q}_i + \overline{q}_i}{2} + \frac{\eta_i \cdot z \cdot s_i}{2} = a + z \cdot \sum_{i=1}^{n} \eta_i \cdot s_i,$$

hence $\sum_{i=1}^{n} \eta_i \cdot s_i = 0$.

Therefore, we have proven that the subset problem $\mathcal{P}_0$ can be reduced to the auxiliary problem $\mathcal{P}_1$. Thus, the auxiliary problem $\mathcal{P}_1$ is also NP-hard.

$6°$. To complete the proof, we need to show that the auxiliary problem $\mathcal{P}_1$ can be reduced to our $\mathcal{P}$. In other words, for every instance of the auxiliary problem $\mathcal{P}_1$, we can find the corresponding instance of the original problem $\mathcal{P}$, from whose solution we can easily find the solution to the instance of $\mathcal{P}_1$.

Indeed, let us consider an instance of the auxiliary $\mathcal{P}_1$, i.e., the intervals $[\underline{q}_i, \overline{q}_i]$ and the real number $a$ for which $\underline{q}_i \geq 0$ and $\sum_{i=1}^{n} \underline{q}_i \leq a \leq \sum_{i=1}^{n} \overline{q}_i$. As the corresponding instance of the original problem, we will take $\underline{p}_i = \dfrac{\underline{q}_i}{a}$ and $\overline{p}_i = \dfrac{\overline{q}_i}{a}$.

Possible values $p_i \in [\underline{p}_i, \overline{p}_i]$ and $q_i \in [\underline{q}_i, \overline{q}_i]$ can be obtained from each other by, correspondingly, multiplying or dividing by $a$. For each set $q_i = p_i \cdot a$, we have

$$L = -\sum_{i=1}^{n} q_i \cdot \log_2(q_i) = -\sum_{i=1}^{n} a \cdot p_i \cdot \log_2(a \cdot p_i) = -a \cdot \sum_{i=1}^{n} p_i \cdot \log_2(a \cdot p_i) =$$

$$-a \cdot \sum_{i=1}^{n} p_i \cdot \log_2(p_i) - a \cdot \log_2(a) \cdot \sum_{i=1}^{n} p_i =$$

$$-a \cdot \sum_{i=1}^{n} p_i \cdot \log_2(p_i) - a \cdot \log_2(a) = a \cdot S - a \cdot \log_2(a).$$

Thus, $L$ is an increasing function of $S$, hence the minimum $\underline{L}$ is equal to

$$\underline{L} = a \cdot \underline{S} - a \cdot \log_2(a).$$

Therefore, if we get the solution $\underline{S}$ to the above instance of our original problem $\mathcal{P}$, we will thus be able to easily compute the solution $\underline{L}$ to the corresponding instance of the auxiliary problem $\mathcal{P}_1$.

Therefore, the auxiliary problem $\mathcal{P}_1$ – whose NP-hardness we have already proven – can be reduced to the original problem $\mathcal{P}$. So, we have prove that the original problem $\mathcal{P}$ of computing $\underline{S}$ is indeed NP-hard.

*Justification of the $O(n \cdot \log_2(n))$ algorithm for computing $\underline{S}$ when intervals are not contained in each other.* It is easy to show that when we sort the intervals in lexicographic order, then both their lower endpoints $\underline{p}_i$ and upper endpoints $\overline{p}_i$ are also sorted: $\underline{p}_i \leq \underline{p}_{i+1}$ and $\overline{p}_i \leq \overline{p}_{i+1}$. (Indeed, otherwise, we would get a violation of the subset property.) Let us thus assume that the intervals are thus sorted.

Let us now show that it is sufficient to consider monotonic optimal tuples $p_1, \ldots, p_n$, for which $p_i \leq p_{i+1}$ for all $i$. Indeed, if $p_i > p_{i+1}$, then, since $\underline{p}_i \leq \overline{p}_i \leq \overline{p}_{i+1}$ and $p_i > p_{i+1} \geq \underline{p}_{i+1}$, we have $p_i \in [\underline{p}_{i+1}, \overline{p}_{i+1}]$ and similarly $p_{i+1} \in [\underline{p}_i, \overline{p}_i]$. Thus, we can swap the values $p_i$ and $p_{i+1}$ without changing the value of $S$. We can repeat this swap as many times as necessary until we get a monotonic tuple that has the exact same value $S = \underline{S}$.

Let us now show that in the optimal tuple, at most one $p_i$ can be inside the corresponding interval. Indeed, if we have two values $p_j$ and $p_k$ strictly inside their intervals, then, similarly to the case of $\overline{S}$, we can conclude that $p_j = p_k$. Now, for $p_j - \Delta = p - \Delta$ and $p_k + \Delta = p + \Delta$, the function $S$ should have a minimum at $\Delta = 0$ and thus, its second derivative relative to $\Delta$ should be non-negative. However, an explicit computation shows that this derivative is negative. Thus, our assumption is false, and at most one $p_j$ can be inside the corresponding interval.

Similar to the case of $\overline{S}$, we can now conclude that:

- if $p_j = \underline{p}_j$ and $p_m > \underline{p}_m$, then $p_j \leq p_m$; and
- if $p_m = \overline{p}_m$ and $p_j < \overline{p}_j$, then $p_m \geq p_j$.

Thus, each value $p_j = \underline{p}_j$ precede all the values $p_m = \overline{p}_m$, and the only value $p_i$ which is strictly inside the corresponding interval lies in between these values. Thus, in a monotonic optimal tuple $p_1, \ldots, p_n$, the first elements are equal to $\underline{p}_j$, then we may have one element which is strictly inside its interval, and then we have values $p_m = \overline{p}_m$.

The above algorithm tests all such (possibly optimal) sequences and finds the one for which the entropy is the largest.

*Proof that under the no-subset property, the fast algorithm always computes $\underline{S}$ in linear time.* In the previous proof, we have already shown that, if we sort the intervals $\boldsymbol{p}_i$ by their midpoints, then the minimum $\underline{S}$ is always attained at a monotonic tuple $p_1, \ldots, p_n$ in which the first elements are equal to $\underline{p}_j$, then we may have one element which is strictly inside its interval, and then we have values $p_m = \overline{p}_m$.

For the resulting vector $p = (\underline{p}_1, \ldots, \underline{p}_{k-1}, p_k, \overline{p}_{k+1}, \ldots, \overline{p}_n)$, with $\underline{p}_k \leq p_k \leq \overline{p}_k$, the condition $\sum_{i=1}^{n} p_i = 1$ implies that $\Sigma_k \leq 1 \leq \Sigma_{k-1}$, where

$$\Sigma_k \overset{\text{def}}{=} \sum_{i=1}^{k} \underline{p}_i + \sum_{j=k+1}^{n} \overline{p}_j.$$

When we go from $\Sigma_k$ to $\Sigma_{k+1}$, we replace a larger value $\overline{p}_{k+1}$ with a smaller value $\underline{p}_{k+1}$. Hence $\Sigma_k > \Sigma_{k+1}$. Thus there has to be exactly one $k_{\max}$ for which $\Sigma_k \leq 1 \leq \Sigma_{k-1}$.

So if we have $\Sigma_m > 1$, this means that the value $k_{\max}$ corresponding to the minimum of $S$ is $> m$. Hence for all the indices $i \leq m$ we already know that in the optimal vector $p$ we have $p_i = \underline{p}_i$. Thus these indices can be added to the set $I^-$.

If $\Sigma_{m-1} \, (= \Sigma_m + 2\Delta_m) < 1$, this means that the value $k_{\min}$ corresponding to the minimum of $S$ is $< m$. Hence for all the indices $j \geq m$ we already know that in the optimal vector $p$ we have $p_j = \overline{p}_j$. Thus these indices can be added to the set $I^+$.

Finally, if $\Sigma_m \le 1 \le \Sigma_{m-1}$ then this $m$ is where the minimum of $S$ is attained.

The algorithm has been justified.

The proof that the new algorithm for computing $\underline{S}$ takes linear time is similar to the proof about the linear-time algorithm for computing $\overline{S}$.

*Proof of the asymptotic formula for $\underline{S}$ for the case of p-boxes.* When we discretize the distribution, we get $p_i \approx \rho_0(x_i) \cdot \Delta x_i$, hence

$$-\sum p_i \cdot \log_2(p_i) \approx -\int \rho_0(x) \cdot \log_2(\rho_0(x) \cdot \Delta x)\, dx.$$

To minimize the entropy, we can take the discrete distribution with values $x_1, \ldots, x_n$ as far away from each other as possible. A distribution which is located at $x_i$ and $x_{i+1}$ and has 0 probability to be in between is described by a cdf $F(x)$ which is horizontal on $[x_i, x_{i+1}]$. Thus, we must select a cdf $F(x) \in \mathbf{F}(x)$ for which these horizontal segments are as long as possible. The length of a horizontal segment is bounded by the geometry of the p-box:

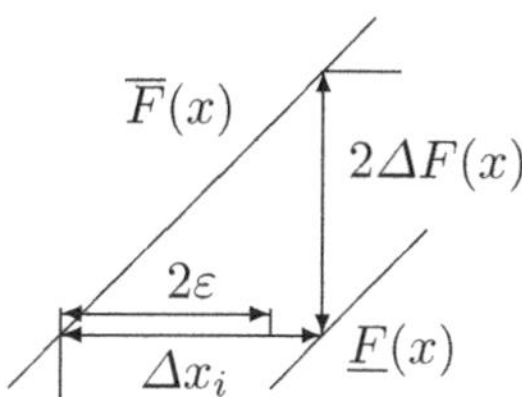


Thus, this length cannot exceed $\dfrac{2\Delta F(x)}{\rho_0(x)}$. If this length is $> 2\varepsilon$, then we can take this interval between the sequential values $x_i$. If this length is $< 2\varepsilon$, then we can still take $\Delta x_i = 2\varepsilon$. Thus, in general, we take $\Delta x_i = \max\left(\dfrac{2\Delta F(x)}{\rho_0(x)}, 2\varepsilon\right)$. Substituting this expression into the above asymptotic formula, we get the desired asymptotic for $\underline{S}$.

# Computing the Range of Convex Symmetric Functions under Interval Uncertainty

In general, a statistical characteristic $f$ can be more complex so that even computing $f$ can take much longer than linear time. For such $f$, the question is how to compute the range $[\underline{y}, \overline{y}]$ in as few calls to $f$ as possible. We show that for convex symmetric functions $f$, we can compute $\overline{y}$ in $n$ calls to $f$.

The results of this chapter first appeared in [353].

## Formulation and Analysis of the Problem and the Resulting Algorithms

*Computing the range of convex symmetric functions under interval uncertainty: formulation of the problem.* In general, a statistical characteristic $f$ can be more complex so that even computing $f$ can take much longer than linear time. For such $f$, the question is how to compute the range $[\underline{y}, \overline{y}]$ in as few calls to $f$ as possible. In this context, we can classify range-computing algorithms by this number of calls: it is reasonable to call an algorithm quadratic-time if it uses $O(n^2)$ calls, linear time if it uses $O(n)$ calls, etc.

In this section, we show that for a practically useful class of convex symmetric functions $f$, we can compute $\overline{y}$ in $n$ calls to $f$ – i.e., in the context of number of calls, in linear time.

Specifically, we consider continuous convex symmetric functions on convex symmetric sets $S \subseteq R^n$ containing a non-degenerate box $[\underline{x}_1, \overline{x}_1] \times \ldots \times [\underline{x}_n, \overline{x}_n]$, with $\underline{x}_i < \overline{x}_i$ for all $i$.

A set $S \in R^n$ is called *symmetric* if with every point

$$x = (x_1, \ldots, x_{i-1}, x_i, x_{i+1}, \ldots, x_{j-1}, x_j, x_{j+1}, \ldots, x_n) \in S$$

it also contain its arbitrary permutation; it is sufficient to require that for every $i$ and $j$, the set $S$ contain the corresponding transposition $\pi_{i,j}(x) \stackrel{\text{def}}{=} (x_1, \ldots, x_{i-1}, x_j, x_{i+1}, \ldots, x_{j-1}, x_i, x_{j+1}, \ldots, x_n)$. A set $S$ is called *convex* if for two points $x, x' \in S$ and for every real number $\alpha \in (0, 1)$, the set $S$ also contains the convex combination $\alpha \cdot x + (1 - \alpha) \cdot x'$.

H.T. Nguyen et al.: Computing Stat. under Interval & Fuzzy Uncertain., SCI 393, pp. 211–219.
springerlink.com                                                  © Springer-Verlag Berlin Heidelberg 2012

A function $f : S \to R$ is called *symmetric* if $f(x) = f(\pi_{i,j}(x))$ for every transposition $\pi_{i,j}$, and *convex* if $f(\alpha \cdot x + (a - \alpha) \cdot x') \leq \alpha \cdot f(x) + (1 - \alpha) \cdot f(x')$ for all $x, x' \in S$ and for all $\alpha \in (0, 1)$.

*Comment.* It is known that each convex function defined on an open convex set is continuous; see, e.g., [49]. So, if, e.g., the set $S$ coincides with entire space $R^n$, then we do not need to require continuity: any convex function $f : R^n \to R$ is automatically continuous.

*Examples.* Variance $\sum\limits_{i=1}^{n} (x_i - E)^2$ and entropy $-\sum\limits_{i=1}^{n} p_i \cdot \log(p_i)$ are examples of convex symmetric statistical characteristics. More general examples are higher-order even central moments $\sum\limits_{i=1}^{n} (x_i - E)^{2d}$ $(d = 1, 2, \ldots)$ and *generalized entropy functions*, i.e., functions $\sum\limits_{i=1}^{n} g(p_i)$ with convex $g(p)$.

Many important physical quantities outside statistics are also convex (or concave); see, e.g., [31, 49, 292, 343]. Some of these convex or concave characteristics are also symmetric.

*Computational complexity: what is known.* It is known that for convex functions, there exists a feasible (polynomial-time) algorithm for computing its minimum $y$ (see, e.g., [49] and [334]), but computing its maximum $\overline{y}$ is, in general, NP-hard [334]; as we have mentioned earlier, it is even NP-hard for population variance. It is therefore desirable to find feasible algorithms that solve the maximum in practically reasonable situations. For variance and entropy, such algorithms are known for the case when the inputs satisfy the following *no-subset property:* $[\underline{x}_i, \overline{x}_i] \not\subset (\underline{x}_j, \overline{x}_j)$ for all $i \neq j$.

*Algorithm for computing $\overline{y}$ with linear number of calls to $f$.* The following algorithm computes the maximum $\overline{y}$ of a given continuous symmetric convex function $f(x_1, \ldots, x_n)$ over a given box $\boldsymbol{x}_1 \times \ldots \times \boldsymbol{x}_n$ for all the cases in which the intervals $\boldsymbol{x}_i$ satisfy the no-subset property:

- First, we sort $n$ intervals $\mathbf{x}_i$ in lexicographic order:

$$\mathbf{x}_1 \leq_{lex} \mathbf{x}_2 \leq_{lex} \cdots \leq_{lex} \mathbf{x}_n.$$

- Second, for each $k$ from 0 to $n$, we compute $f(s^{(k)})$, where $s^{(k)} \stackrel{\text{def}}{=} (\underline{x}_1, \ldots, \underline{x}_k, \overline{x}_{k+1}, \ldots, \overline{x}_n)$.
- Finally, we return the largest of $n + 1$ values $f(s^{(k)})$ as $\overline{y}$.

This algorithm takes $O(n \cdot \log(n))$ steps for sorting (see, e.g., [73]), $n + 1$ calls to $f$ (to compute $n + 1$ values $f(s^{(k)})$), and $O(n)$ steps to find the largest of these $n + 1$ values. Thus, in addition to $n + 1$ calls to $f$, this algorithm takes $O(n \cdot \log(n)) + O(n) = O(n \cdot \log(n))$ computational steps.

*Comment.* If the algorithm for computing the function $f$ is feasible, i.e., takes a polynomial time $t \leq P(n)$ for some polynomial $P(n)$, then computing $\overline{y}$

can be done in time $\leq P(n) \cdot (n+1) + O(n \cdot \log(n))$ – i.e., also in polynomial time.

*Comment.* A function $f$ is convex if and only if $-f$ is concave. The minimum of $-f$ is equal to minus the maximum of $f$. Thus, the above algorithm can also be used to compute the minima of continuous symmetric concave functions over a box $[\underline{x}_1, \overline{x}_1] \times \ldots \times [\underline{x}_n, \overline{x}_n]$ whose intervals satisfy the no-subset property.

*Possibility of a faster algorithm.* The above algorithm for computing $\overline{y}$ is not always optimal. For example, computing the variance $V$ takes linear time $C \cdot n$, so $n+1$ computations of variance means $(n+1) \cdot C \cdot n \approx C \cdot n^2$ time – while the linear-time algorithm for computing $\overline{V}$ (presented in Section 2) is much faster (for large $n$).

It turns out that a speed-up is possible not only for the variance $V$, but also for several other symmetric convex functions $f$.

*Main idea behind the speed-up: some functions $f$ are easy to revise.* One of the reasons why we can speed up the computation of $\overline{V}$ is that this function is *easy to revise* in the following sense.

When we go from $s^{(k)}$ to $s^{(k+1)}$, we only change a single component $s_{k+1}$ of the point $s$, from $\overline{x}_{k+1}$ to $\underline{x}_{k+1}$. Thus, if we keep the values $M \stackrel{\text{def}}{=} \frac{1}{n} \cdot \sum_{i=1}^{n} x_i^2$

and $E = \frac{1}{n} \cdot \sum_{i=1}^{n} x_i$, then updating each of these two values means computing

the new values $E' = E - \dfrac{\overline{x}_{k+1} - \underline{x}_{k+1}}{n}$ and $M' = M - \dfrac{(\overline{x}_{k+1})^2 - (\underline{x}_{k+1})^2}{n}$, and

then computing $V' = M' - (E')^2$. All these updates take a constant number (10) of arithmetic operations (independent on $n$). Thus, overall, we need $C \cdot n$ time to compute $V(s^{(0)})$ and time $10 \cdot n$ to compute $n$ values $V(s^{(1)})$, ..., $V(s^{(n)})$. So, overall, we need time $C \cdot n + 10 \cdot n + O(n \cdot \log(n)) = O(n \cdot \log(n))$ which is, for large $n$, smaller than $C \cdot n^2$.

This idea can be applied to other "easy-to-revise" functions. To describe this result, let us first introduce two auxiliary notions: of a revised tuple and of a revisable computation scheme.

*Revised tuples.* For every tuple $x = (x_1, \ldots, x_{i-1}, x_i, x_{i+1}, \ldots, x_n)$, for every integer $i \leq n$, and for every real number $x_i'$, by a *revised tuple*, we mean a tuple $r_{i,x_i'}(x) \stackrel{\text{def}}{=} (x_1, \ldots, x_{i-1}, x_i', x_{i+1}, \ldots, x_n)$, in which the $i$-th component of $x$ is replaced by $x_i'$.

*Revisable computation schemes.* Let $f(x_1, \ldots, x_n)$ be a computable function. By a *revisable computation scheme* for computing $f(x_1, \ldots, x_n)$, we mean a tuple $\langle f_1(x_1, \ldots, x_n), \ldots, f_m(x_1, \ldots, x_n), A_f, \Delta_f \rangle$, where:

- $f_1(x)$, ..., $f_m(x)$ are computable functions;
- $A_f$ is an algorithm which, given $n$ values $x = (x_1, \ldots, x_n)$, computes $f(x)$, $f_1(x)$, ..., $f_m(x)$;

- $\Delta_f$ is an algorithm which, given the values $f(x)$, $f_1(x)$, ..., $f_m(x)$, an integer $i$, and a real number $x'_i$, computes the values $f(x')$, $f_1(x')$, ..., $f_m(x')$ for $x' = r_{i,x'_i}(x)$.

*Comment.* Of course, for every computable function $f$, there is a trivial revisable computation scheme for computing $f$: e.g., we can take $m = n$ and $f_i(x_1,\ldots,x_n) = x_i$ for all $i = 1,\ldots,m$. In this case, the algorithm $A_f$ simply means computing $f(x)$ and also returning the original values $x_1,\ldots,x_m$, while the algorithm $\Delta_f$ ignores the previous values $f(x)$, $f_1(x)$, ..., $f_m(x)$, and simply computes the new value $f(x')$ (and also returns the new values $x'_1,\ldots,x'_n$).

*Easy-to-revise functions: a description.* We are interested only in the computable functions $f$ which are "easy-to-revise", i.e., for which there exists a revisable computation scheme in which

- the algorithm $A_f$ has approximately the same computational complexity as the best known algorithm for simply computing $f(x)$, and
- the algorithm $\Delta_f$ is much faster than $A_f$.

*Comment.* In contrast to the notions of a revised tuple and a revisable computation scheme, the notion of an easy-to-revise function is somewhat informal: it depends on which algorithms for computing $f$ we know, and on how we define "approximately the same" and "much faster".

*Examples.* Let us first explain why the variance $f = V$ is indeed easy-to-revise in the sense of the above (somewhat informal) definition. For the variance, $f_1 = M$ and $f_2 = E$. Computing $f_1 = M$, $f_2 = E$, and $f = M - E^2$ takes linear time $C \cdot n$ – approximately the same time as for all known algorithms for computing variance. However, if we change one of the components $x_i$ to a different value $x'_i$, then, as we have mentioned, updating $E$ and $M$ to new values $E'$ and $M'$ and computing the new value $V' = M' - (E')^2$ takes 10 computational steps. So, here, the revising algorithm $\Delta_f$ takes 10 time steps, which, for large $n$, is much faster than $C \cdot n$.

Similarly, the mean $E$ is easy-to-revise: we need time $C \cdot n$ to compute $E$, but only $3 \ll C \cdot n$ steps to update $E$.

Higher central even moments, entropy, and generalized entropy are other examples of easy-to-revise symmetric convex functions. For example, the 4-th central moment is a linear combination of the moments $M_k \overset{\text{def}}{=} \sum_{i=1}^{n} x_i^k$ of orders $k = 1, 2, 3, 4$, and each of these four functions $m_k$ is easy to revise.

*Algorithm for computing $\overline{y}$ for easy-to-revise symmetric convex functions $f$.* For easy-to-revise functions, we can compute $\overline{y}$ as follows:

- first, we apply the algorithm $A_f$ to compute the values of $f$ and of the auxiliary functions $f_1,\ldots,f_m$ at $s^{(0)}$;

- then, for each $k$ from 1 to $n$, we apply the algorithm $\Delta_f$ to revise the values of $f, f_1, \ldots, f_m$ from $x = s^{(k-1)}$ to $x = s^{(k)}$;
- finally, we compute $\overline{y}$ as the largest of $n + 1$ values $f(s^{(k)})$.

This algorithm calls the algorithm $A_f$ once (to compute $f(s^{(0)})$, calls the revision algorithm $\Delta_f$ $n$ times, and uses $O(n \cdot \log(n))$ computational steps in addition to these calls.

*Comment.* Since $\Delta_f$ is much faster than $A_f$, and $A_f$ takes approximately the same time as computing $f$, the new algorithm for computing $\overline{y}$ takes much less time than computing $f$ for $n + 1$ tuples $s^{(0)}, \ldots, s^{(n)}$.

It is worth mentioning that this new algorithm is not always optimal: e.g., for the variance, this algorithm takes time $O(n \cdot \log(n)) + O(n) = O(n \cdot \log(n))$, but we know that we can compute $\overline{V}$ even faster: in linear time.

*Computing $\underline{y}$ and $\overline{y}$ under the constraint $\sum_{i=1}^{n} x_i = c$: a problem.* As we have mentioned earlier, for entropy, we have an additional constraint $\sum_{i=1}^{n} p_i = 1$ on the possible values of the probabilities $p_i$. The same constraint holds for computing other characteristics of probabilities such as a generalized entropy.

So, we arrive at the following problem: we know a continuous symmetric convex function $f(x_1, \ldots, x_n)$ (given as an algorithm or, equivalently, as a computer program), we know the intervals $\boldsymbol{x}_i = [\underline{x}_i, \overline{x}_i]$ that satisfy the above no-subset property, and we know the number $c$ for which we should have $\sum_{i=1}^{n} x_i = c$. Our objective is to compute the range

$$[\underline{y}, \overline{y}] = \left\{ f(x_1, \ldots, x_n) : x_1 \in \boldsymbol{x}_1, \ldots, x_n \in \boldsymbol{x}_n, \sum_{i=1}^{n} x_i = c \right\}.$$

*Computing $\underline{y}$ under the constraint $\sum_{i=1}^{n} x_i = c$.* The constraints $x_i \in \boldsymbol{x}_i$ and $\sum_{i=1}^{n} x_i = c$ describe a convex set, so we can compute the minimum $\underline{y}$ of a convex function $f$ over this set in polynomial time.

*Algorithm for computing $\overline{y}$ under the constraint $\sum_{i=1}^{n} x_i = c$ when the intervals satisfy the no-subset property.* Under the above no-subset property, the following algorithm computes $\overline{y}$ by calling $f$ once and by using $O(n)$ computational steps in addition to this call.

This algorithm is iterative. At each iteration of this algorithm, we have three sets:

- the set $I^-$ of all the indices $i$ from 1 to $n$ for we already know that for the optimal vector $x$, we have $x_i = \underline{x}_i$;
- the set $I^+$ of all the indices $j$ for which we already know that for the optimal vector $x$, we have $x_j = \overline{x}_j$;

- the set $I = \{1,\ldots,n\} - I^- - I^+$ of the indices $i$ for which we are still undecided.

In the beginning, $I^- = I^+ = \emptyset$ and $I = \{1,\ldots,n\}$. At each iteration, we also update the values of two auxiliary quantities $E^- \stackrel{\text{def}}{=} \sum_{i \in I^-} \underline{x}_i$ and $E^+ \stackrel{\text{def}}{=} \sum_{j \in I^+} \overline{x}_j$. In principle, we could compute these values by computing these sums, but to speed up computations, on each iteration, we update these two auxiliary values in a way that is faster than re-computing the corresponding two sums. Initially, since $I^- = I^+ = \emptyset$, we take $E^- = E^+ = 0$.

At each iteration, we do the following:

- first, we compute the median $m$ of the set $I$ $\big($median in terms of sorting by $\widetilde{x}_i = \dfrac{\underline{x}_i + \overline{x}_i}{2}\big)$;
- then, by analyzing the elements of the undecided set $I$ one by one, we divide them into two subsets

$$X^- = \{i : \widetilde{x}_i \le \widetilde{x}_m\} \text{ and } X^+ = \{j : \widetilde{x}_j > \widetilde{x}_m\};$$

- we compute $e^- = E^- + \sum_{i \in X^-} \underline{x}_i$ and $e^+ = E^+ + \sum_{i \in X^+} \overline{x}_i$;
- if $e^- + e^+ > c$, then we replace $I^-$ with $I^- \cup X^-$, $E^-$ with $e^-$, and $I$ with $X^+$;
- if $e^- + e^+ + 2\Delta_m < c$, then we replace $I^+$ with $I^+ \cup X^+$, $E^+$ with $e^+$, and $I$ with $X^-$;
- finally, if $e^- + e^+ \le c \le e^- + e^+ + 2\Delta_m$, then we replace $I^-$ with $I^- \cup (X^- - \{m\})$, $I^+$ with $I^+ \cup X^+$, $I$ with $\{m\}$, $E^-$ with $e^- - \underline{p}_m$, and $E^+$ with $e^+$.

At each iteration, the set of undecided indices is divided in half. Iterations continue until we have only one undecided index $I = \{k\}$, after which we return, as $\overline{y}$, the value of the function $f(x_1,\ldots,x_n)$ for the vector $x$ for which $x_i = \underline{x}_i$ for $i \in I^-$, $x_j = \overline{x}_j$ for $j \in I^+$, and $x_k = c - E^- - E^+$ for the remaining value $k$.

## Proofs

*Proof that the new algorithm for computing the maximum $\overline{y}$ of a symmetric convex function is correct.* To prove that the algorithm is correct we must show that the maximum $\overline{y}$ of the function $f$ is attained at one of the points $s^{(k)}$.

One can easily check that since the intervals $\mathbf{x}_i$ are already sorted in lexicographic order and satisfy the no-subset property, the lower endpoints and the upper endpoints are also sorted, i.e., $\underline{x}_i \le \underline{x}_{i+1}$ and $\overline{x}_i \le \overline{x}_{i+1}$ for all $i$.

The maximum of a continuous function on a bounded closed polyhedron $\boldsymbol{x}_1 \times \ldots \times \boldsymbol{x}_n$ is always attained at some point, and since the function $f$ is convex, it is attained at one of the vertices of this set [49, 292, 343], i.e., when for each $i$, we have $s_i = \underline{x}_i$ or $s_i = \overline{x}_i$.

There may be several vertices at which the maximum is attained. Out of all maximizing vertices, we choose a one $s$ with the largest length of the starting sequence of lower bounds. We will denote this length by $k$; this means that $s$ has the form $s = (\underline{x}_1, \ldots, \underline{x}_k, \overline{x}_{k+1}, \ldots)$, i.e., it starts with $k$ lower bounds and then has an upper bound at the $(k+1)$-st place. We will prove that for this point $s$, all the components $s_l$ for $l > k+1$ are upper bounds, i.e., that $s = s^{(k)}$.

Indeed, let us assume that for some $l > k+1$, the component $s_l$ of the chosen point is a lower bound: $s_l = \underline{x}_l$. We will then construct another point $s'$ at which $f$ also attains its maximum and which has a longer starting sequence of lower bounds – which contradicts to our choice of $s$.

In this construction, we will only change the $(k+1)$-st and $l$-th coordinates, so this construction can be naturally illustrated on the corresponding plane. First, we consider the bisecting line $x_{k+1} = x_l$ of the first and third quadrant and find an orthogonal line to it ($x_{k+1} + x_l = \text{const}$) which passes through $s$. The line has to intersect the interior of the rectangle and to leave it again at some point $s''$ – which is either the left or the upper face; see the following pictures, in which $s''_l \overset{\text{def}}{=} \underline{x}_l + (\overline{x}_{k+1} - \underline{x}_{k+1})$ and $s''_{k+1} \overset{\text{def}}{=} \overline{x}_{k+1} - (\overline{x}_l - \underline{x}_l)$. Let $z''$ be a point which is symmetric relative to $s''$. Since the endpoints are sorted, one can prove that $s$ is in between $s''$ and $z''$, i.e., that $s$ is a convex combination of $s''$ and $z''$.

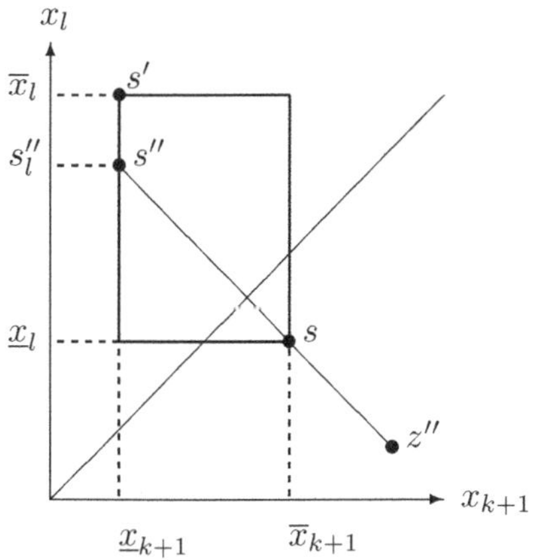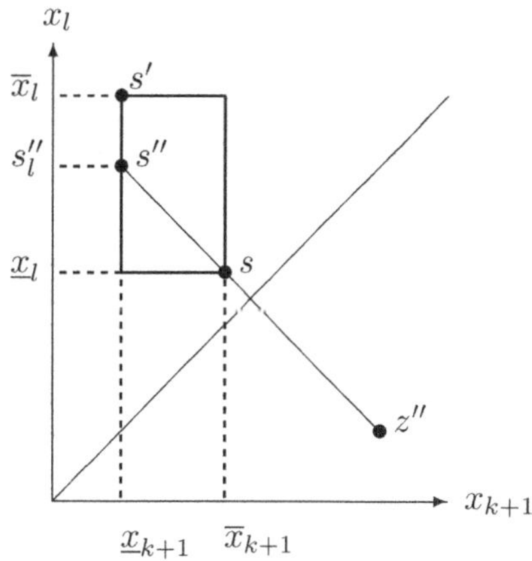

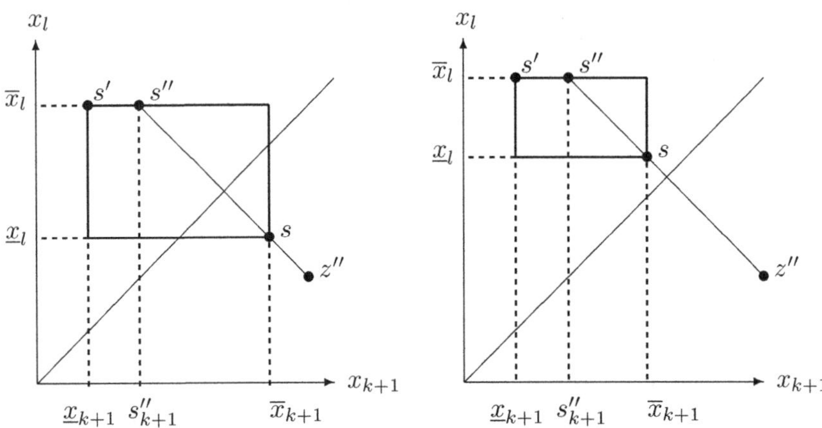

The point $s$ at which the maximum is attained is a convex combination of the points $s''$ and $z''$, i.e., $s = \alpha \cdot s'' + (1 - \alpha) \cdot z''$. Since $f$ is a convex function, we have $f(s) \leq \alpha \cdot f(s'') + (a - \alpha) \cdot f(z'')$. Due to symmetry, we have $f(s'') = f(z'')$ hence $f(s) \leq f(s'')$. Since the function $f$ attains its maximum at the point $s$, it thus attains the maximum at the point $s''$ as well.

This point $s''$ is on a straight line segment – namely, on one of the faces of the rectangle. Since the function $f$ is convex, the only way for it to attain the maximum inside the straight line segment is to attain the same maximum on both endpoints of this face, in particular, at a points $s'$ at which $s'_{k+1} = \underline{x}_{k+1}$. For this new point, we have $s'_1 = \underline{x}_1, \ldots, s'_k = \underline{x}_k$, and $s'_{k+1} = \underline{x}_{k+1}$ – which contradicts to our assumption that $k$ is the largest length of the starting lower-endpoint sequence in a maximizing point. Correctness is proven.

*Proof that the new algorithm for computing $\overline{y}$ under the constraint $\sum\limits_{i=1}^{n} x_i = c$ is correct.* Similarly to the previous proof, we can conclude that the maximum $\overline{y}$ is always attained at one of the vertices of the convex polyhedron

$$(\boldsymbol{x}_1 \times \ldots \times \boldsymbol{x}_n) \cap \left\{ x : \sum_{i=1}^{n} x_i = c \right\},$$

i.e., at a point $s$ at which for at least $n - 1$ values $s_i$, we have $s_i = \underline{x}_i$ or $s_i = \overline{x}_i$.

Similarly to the previous proof, we can also conclude that the maximum is attained at one of the points $s^{(k)} = (\underline{x}_1, \ldots, \underline{x}_{k-1}, s_k, \overline{x}_{k+1}, \ldots, \overline{x}_n)$. The value $s_k$ can determined by the condition $\sum\limits_{i=1}^{n} s_i = c$. For this value $s_k$ to be between $\underline{x}_k$ and $\overline{x}_k$, we must make sure that $\Sigma_k \leq 1 \leq \Sigma_{k-1}$, where $\Sigma_k \overset{\text{def}}{=} \sum\limits_{i=1}^{k} \underline{x}_i + \sum\limits_{j=k+1}^{n} \overline{x}_j$. Similar to the case of entropy, we can conclude that since the sums $\Sigma_k$ decrease with $k$, there is only one value $k$ for which the above inequality holds, so we can use the linear-time algorithm from the entropy case to find this value $k$. Once this value is found, we have thus found the maximizing point $s^{(k)}$ and thus, a single call to $f$ finds the desired maximum $f(s^{(k)})$.

Correctness is proven.

# 27

# Computing Statistics under Interval Uncertainty: Possibility of Parallelization

In this chapter, we show how the algorithms for estimating variance under interval and fuzzy uncertainty can be parallelized. The results of this chapter first appeared in [336].

*Need for parallelization.* Traditional algorithms for computing the population variance $V$ based on the exact values $x_1, \ldots, x_n$ take linear time $O(n)$. Algorithms for estimating variance under interval uncertainty take a larger amount of computation time – e.g., time $O(n \cdot \log(n))$. How can we speed up these computations?

If we have several processors, then it is desirable to perform these algorithms in parallel on several processors, and thus, speed up computations. In this chapter, we show how the algorithms for estimating variance under interval and fuzzy uncertainty can be parallelized.

In order to describe how to parallelize these algorithms, let us describe the existing sequential (non-parallel) algorithms for estimating the variance under interval uncertainty.

*Algorithm for computing $\overline{V}$ in the no-(proper)-subset case.* The corresponding algorithm is as follows:

- First, we sort the values $\widetilde{x}_i$ into an increasing sequence. Without losing generality, we can assume that $\widetilde{x}_1 \le \widetilde{x}_2 \le \ldots \le \widetilde{x}_n$.
- Then, for every $k$ from 0 to $n$, we compute the value $V^{(k)} = M^{(k)} - (E^{(k)})^2$ of the population variance $V$ for the vector $x^{(k)} = (\underline{x}_1, \ldots, \underline{x}_k, \overline{x}_{k+1}, \ldots, \overline{x}_n)$. (For $k = 0$, $x^{(0)} = (\overline{x}_1, \ldots, \overline{x}_n)$.)
- Finally, we compute $\overline{V}$ as the largest of $n + 1$ values $V^{(0)}, \ldots, V^{(n)}$.

To compute the values $V^{(k)}$, first, we explicitly compute $M^{(0)} = \dfrac{1}{n} \cdot \sum_{i=1}^{n} (\overline{x}_i)^2$,

$E^{(0)} = \dfrac{1}{n} \cdot \sum_{i=1}^{n} \overline{x}_i$, and $V^{(0)} = M^{(0)} - (E^{(0)})^2$. Once we know the values $M^{(k)}$ and $E^{(k)}$, we can compute

H.T. Nguyen et al.: Computing Stat. under Interval & Fuzzy Uncertain., SCI 393, pp. 221–224.
springerlink.com                                        © Springer-Verlag Berlin Heidelberg 2012

$$M^{(k+1)} = M^{(k)} + \frac{1}{n} \cdot (\underline{x}_{k+1})^2 - \frac{1}{n} \cdot (\overline{x}_{k+1})^2$$

and

$$E^{(k+1)} = E^{(k)} + \frac{1}{n} \cdot \underline{x}_{k+1} - \frac{1}{n} \cdot \overline{x}_{k+1}.$$

*Possibility of parallelization.* For large $n$, we may want to further speed up computations if we have several processors working in parallel.

In the general case, all the stages of the above algorithm can be parallelized by known techniques. In particular, Stage 3 is a particular case of a general *prefix-sum* problem, in which we must compute the values

$$a_n, \quad a_n * a_{n-1}, \quad a_n * a_{n-1} * a_{n-2}, \dots,$$

for some associative operation $*$ (in our case, $* = \max$).

*Case of potentially unlimited number of processors.* If we have a potentially unlimited number of processors, then we can do the following (see, e.g., [140], for the information on how to parallelize the corresponding stages):

- on Stage 1, we can sort the values $\widetilde{x}_i$ in time $O(\log(n))$;
- on Stage 2, we can compute the values $V^{(i)}$ (i.e., solve the prefix-sum problem) in time $O(\log(n))$;
- on Stage 3, we can compute the maximum of $V^{(i)}$ in time $O(\log(n))$.

As a result, we can check monotonicity in time

$$O(\log(n)) + O(\log(n)) + O(\log(n)) = O(\log(n)).$$

*Example.* To give the readers a better understanding on how these stages can be parallelized, let us describe, in detail, parallelization of Stage 3. In other words, let us describe how to compute the maximum of $n + 1$ given values $V^{(0)}, \dots, V^{(n)}$ in parallel.

As we have mentioned, the parallelized algorithm consists of $O(\log(n))$ steps. At the first step, we divide $n + 1$ values into pairs $(V^{(0)}, V^{(1)})$, $(V^{(2)}, V^{(3)})$, ... Since we have assumed that we have a potentially unlimited number of processors, we can allocate an individual processor to each pair – to the total of $\lceil (n+1)/2 \rceil$ processors. At the first step, each processor compares the corresponding two numbers and thus computes the maximum of this pair:

- the first processor computes the value

$$m(0, 1) \stackrel{\text{def}}{=} \max(V^{(0)}, V^{(1)});$$

- at the same time, the second processor computes the value

$$m(2, 3) \stackrel{\text{def}}{=} \max(V^{(2)}, V^{(3)});$$

- etc.

At the end of the first step, we thus have $\lceil (n+1)/2 \rceil \approx n/2$ values $m(0,1)$, $m(2,3)$, $m(4,5)$, $m(6,7)$, etc.

At the second step, we divide these $\lceil (n+1)/2 \rceil \approx n/2$ values into pairs, and compute the maximum of each pair:

- the first processor computes the value

$$m(0,3) \stackrel{\text{def}}{=} \max(m(0,1), m(2,3));$$

by definition of $m(0,1)$ and $m(2,3)$, this value is equal to

$$\max(V^{(0)}, V^{(1)}, V^{(2)}, V^{(3)});$$

- at the same time, the second processor computes the value

$$m(4,7) \stackrel{\text{def}}{=} \max(m(4,5), m(6,7));$$

by definition of $m(4,5)$ and $m(6,7)$, this value is equal to

$$\max(V^{(4)}, V^{(5)}, V^{(6)}, V^{(7)});$$

- etc.

At the end of the second step, we thus have $\approx n/4$ values $m(0,3)$, $m(4,7)$, etc., describing the maxima of four elements.

At the third step, we repeat this procedure again, and get the values $m(0,7)$, $m(8,15)$, etc., describing the maxima of $8 = 2^3$ elements.

At the $k$-th step, we get the values

$$m(0, 2^k - 1), m(2^k, 2^k + (2^k - 1)), \ldots,$$

describing the maxima of $2^k$ elements.

As soon as we get $2^k = n$, i.e., as soon as $k \approx \log_2(n)$, we get the desired maximum of all $n$ elements. Thus, we can indeed compute the desired maximum in $O(\log(n))$ steps.

*Case of a fixed number of processors.* If we have $p < n$ processors, then we can:

- on Stage 1, sort $n$ values in time $O\left( \dfrac{n \cdot \log(n)}{p} + \log(n) \right)$ ; see, e.g., [140];

- on Stage 2, compute the values $V^{(i)}$ in time $O\left( \dfrac{n}{p} + \log(p) \right)$ ; see, e.g., [46];

- on Stage 3, compute the maximum of $V^{(i)}$ in time $O\left( \dfrac{n}{p} + \log(p) \right)$.

Overall, we thus need time

$$O\left( \frac{n \cdot \log(n)}{p} + \log(n) \right) + O\left( \frac{n}{p} + \log(p) \right) + O\left( \frac{n}{p} + \log(p) \right) =$$

$$O\left( \frac{n \cdot \log(n)}{p} + \log(n) + \log(p) \right).$$

*Example.* To illustrate how this parallelization works, let us again use Stage 3. Specifically, let us show how we can use $p$ processors to compute the maximum of given $n + 1$ values $V^{(0)}, \ldots, V^{(n)}$ in parallel in time $O\left(\dfrac{n}{p} + \log(p)\right)$.

Indeed, let us divide $n + 1$ values into $p$ subgroups with $\dfrac{n + 1}{p}$ elements in each subgroup. To each of these subgroups, we assign one of the $p$ processors. Each processor computes the maximum of all its $\dfrac{n + 1}{p}$ values in time $\dfrac{n + 1}{p} = O\left(\dfrac{n}{p}\right)$, and these processors work in parallel. After that, we have $p$ values – the maximum of the first subgroup, the maximum of the second subgroup, etc.

To find the maximum of all $n + 1$ elements, it is now sufficient to find the largest of these $p$ subgroup maxima. We already know that if we have $p$ processors, then we can compute the maximum of $p$ values in parallel in time $O(\log(p))$.

Thus, we have a two-step process for computing the maximum. The first step takes time $O\left(\dfrac{n}{p}\right)$, the second step takes time $O(\log(p))$. Thus, the total computation time of this two-step process is indeed equal to $O\left(\dfrac{n}{p} + \log(p)\right)$.

# Computing Statistics under Interval Uncertainty: Case of Relative Accuracy

*Formulation of the problem.* In the previous chapters, we have shown that for many statistical characteristics $C$, computing them with a given *absolute* accuracy $\varepsilon$ – i.e., computing a value $\widetilde{C}$ for which $|\widetilde{C} - C| \leq \varepsilon$ – is NP-hard.

It turns out that if we are interested in computing these characteristics with *relative* accuracy – relative with respect to, e.g., the largest of the inputs – then it often possible to estimate these characteristics in polynomial time.

These results first appeared in [57, 176].

*Towards a new technique: back to straightforward interval computations, centered form, etc.* We would like to compute a good estimate for the range in reasonable time. As we have mentioned, there are algorithms that always compute an enclosure for the range in feasible time – straightforward interval computations, centered form, etc. However, as we have mentioned on the example of the variance (see Chapter 14), we cannot directly apply these algorithm because their application leads to excess width.

As we have shown (Chapter 7), the main reason for excess width is that intermediate results are dependent on each other, and straightforward interval computations ignore this dependence. For example, the actual range of $f(x_1) = x_1 - x_1^2$ over $\mathbf{x}_1 = [0, 1]$ is $\mathbf{y} = [0, 0.25]$. Computing this $f$ means that we first compute $x_2 := x_1^2$ and then subtract $x_2$ from $x_1$. According to straightforward interval computations, we compute $\mathbf{r} = [0, 1]^2 = [0, 1]$ and then $\boldsymbol{x}_1 - \boldsymbol{x}_2 = [0, 1] - [0, 1] = [-1, 1]$. This excess width comes from the fact that the formula for interval subtraction implicitly assumes that both $a$ and $b$ can take arbitrary values within the corresponding intervals $\boldsymbol{a}$ and $\boldsymbol{b}$, while in this case, the values of $x_1$ and $x_2$ are clearly not independent: $x_2$ is uniquely determined by $x_1$, as $x_2 = x_1^2$.

*New techniques: main idea.* A natural idea (see, e.g., [54, 55]) is to remedy the above reason why interval computations lead to excess width. Specifically, at every stage of the computations, in addition to keeping the *intervals* $\boldsymbol{x}_i$ of possible values of all intermediate quantities $x_i$, we also keep several *sets*:

- sets $\boldsymbol{x}_{ij}$ of possible values of pairs $(x_i, x_j)$;

H.T. Nguyen et al.: Computing Stat. under Interval & Fuzzy Uncertain., SCI 393, pp. 225–234.
springerlink.com © Springer-Verlag Berlin Heidelberg 2012

- if needed, sets $\boldsymbol{x}_{ijk}$ of possible values of triples $(x_i, x_j, x_k)$; etc.

In the above example, instead of just keeping two intervals $\boldsymbol{x}_1 = \boldsymbol{x}_2 = [0, 1]$, we would then also generate and keep the set $\boldsymbol{x}_{12} = \{(x_1, x_1^2) \mid x_1 \in [0, 1]\}$. Then, the desired range is computed as the range of $x_1 - x_2$ over this set – which is exactly $[0, 0.25]$.

In other words, from *interval* computations, we move to *set* computations.

In the interval computations context, the idea of representing dependence in terms of sets of possible values of tuples was first described by Shary; see, e.g., [303, 304] and references therein.

How can we propagate this set uncertainty via arithmetic operations? Let us describe this on the example of addition, when, in the computation of $f$, we use two previously computed values $x_i$ and $x_j$ to compute a new value $x_k := x_i + x_j$. In this case, we set $\boldsymbol{x}_{ik} = \{(x_i, x_i + x_j) \mid (x_i, x_j) \in \boldsymbol{x}_{ij}\}$, $\boldsymbol{x}_{jk} = \{(x_j, x_i + x_j) \mid (x_i, x_j) \in \boldsymbol{x}_{ij}\}$, and for every $l \neq i, j$, we take

$$\boldsymbol{x}_{kl} = \{(x_i + x_j, x_l) \mid (x_i, x_j) \in \boldsymbol{x}_{ij}, (x_i, x_l) \in \boldsymbol{x}_{il}, (x_j, x_l) \in \boldsymbol{x}_{jl}\}.$$

*Comment.* From the mathematical viewpoint, a subset $\boldsymbol{x}_{ij}$ of the set of all possible pairs $\boldsymbol{x}_i \times \boldsymbol{x}_j$ is a *relation*. It is therefore not surprising that processing this uncertainty is similar to processing relations in other application areas such as relational database systems; see, e.g., [333]. For example, a natural intermediate step in computing $\boldsymbol{x}_{ik}$ is when, given the relations $\boldsymbol{x}_{ia}$ and $\boldsymbol{x}_{ib}$, we form a new relation $\{(x_a, x_i, x_b) \mid (x_a, x_i) \in \boldsymbol{x}_{ai}, (x_i, x_b) \in \boldsymbol{x}_{ib}\}$. In relational algebra, this intermediate relation is called a *join* and denoted by $\boldsymbol{x}_{ai} \bowtie_i \boldsymbol{x}_{ib}$.

*From main idea to actual computer implementation.* In interval computations, we cannot represent an arbitrary interval inside the computer, we need an enclosure. Similarly, we cannot represent an arbitrary set inside a computer, we need an enclosure.

To describe such enclosures, we fix the number $C$ of granules (e.g., $C = 10$). We divide each interval $\boldsymbol{x}_i$ into $C$ equal parts $\boldsymbol{X}_i$; thus each box $\boldsymbol{x}_i \times \boldsymbol{x}_j$ is divided into $C^2$ subboxes $\boldsymbol{X}_i \times \boldsymbol{X}_j$. We then describe each set $\boldsymbol{x}_{ij}$ by listing all subboxes $\boldsymbol{X}_i \times \boldsymbol{X}_j$ which have common elements with $\boldsymbol{x}_{ij}$; the union of such subboxes is an enclosure for the desired set $\boldsymbol{x}_{ij}$.

Of course, in reality, there is no need to actually list these subboxes: to describe an arbitrary set, it is sufficient to store $10 \times 10 = 100$ bits of information describing whether each of the $10 \times 10$ subboxes belongs to the list. In other words, a set can be represented as $10 \times 10$ array of Boolean values. Similarly, for triples, we can represent the corresponding set as a 3-D array of size $10 \times 10 \times 10$, etc.

*Historical comment.* This representation of a set by the union of grid cells which intersect with this set is well known in data mining as an upper approximation in the sense of *rough set* theory; see, e.g., [279, 280].

*Possibility of improvement.* The above approach is a good way to describe generic sets, but in practice, the resulting description may be redundant.

- For example, even if we know that all the values $(x_1, x_2)$ are possible, we still need 100 Boolean values to describe this set.
- Similarly, if the set consists of all the values for which $x_1 = x_2$, then out of 100 subboxes, only 10 diagonal boxes are affected, but we still need all 100 Boolean values.

A more efficient idea is to represent sets is by using a *paving* – in the style of [142]. In this approach, we start with a $2 \times 2$ subdivision. For each of the $2 \times 2 = 4$ subboxes, we:

- mark this subbox as "in" if it is completely inside the desired set;
- mark this subbox as "out" if it is completely outside the desired set;
- otherwise, if this subbox contains both points from the desired set and point outside the desired set, we subdivide this box into $2 \times 2 = 4$ subboxes, and repeat the procedure.

As a result, we get a list consisting of boxes of different sizes – starting with larger ones and only decreasing the size when necessary.

*How to propagate set uncertainty.* The above implementation enables us to implement all arithmetic operations on data given with set uncertainty. For example, to implement $\boldsymbol{x}_{ik} = \{(x_i, x_i + x_j) \,|\, (x_i, x_j) \in \boldsymbol{x}_{ij}\}$, we take all the subboxes $\boldsymbol{X}_i \times \boldsymbol{X}_j$ that form the set $\boldsymbol{x}_{ij}$; for each of these subboxes, we enclosure the corresponding set of pairs $\{(x_i, x_i + x_j) \,|\, (x_i, x_j) \in \boldsymbol{X}_i \times \boldsymbol{X}_j\}$ into a set $\boldsymbol{X}_i \times (\boldsymbol{X}_i + \boldsymbol{X}_j)$. This set may have non-empty intersection with several subboxes $\boldsymbol{X}_i \times \boldsymbol{X}_k$; all these subboxes are added to the computed enclosure for $\boldsymbol{x}_{ik}$. Once can easily see if we start with the exact range $\boldsymbol{x}_{ij}$, then the resulting enclosure for $\boldsymbol{x}_{ik}$ is an $(1/C)$-approximation to the actual set – and so when $C$ increases, we get more and more accurate representations of the desired set.

Similarly, to find an enclosure for

$$\boldsymbol{x}_{kl} = \{(x_i + x_j, x_l) \,|\, (x_i, x_j) \in \boldsymbol{x}_{ij}, (x_i, x_l) \in \boldsymbol{x}_{il}, (x_j, x_l) \in \boldsymbol{x}_{jl}\},$$

we consider all the triples of subintervals $(\boldsymbol{X}_i, \boldsymbol{X}_j, \boldsymbol{X}_l)$ for which $\boldsymbol{X}_i \times \boldsymbol{X}_j \subseteq \boldsymbol{x}_{ij}$, $\boldsymbol{X}_i \times \boldsymbol{X}_l \subseteq \boldsymbol{x}_{il}$, and $\boldsymbol{X}_j \times \boldsymbol{X}_l \subseteq \boldsymbol{x}_{jl}$; for each such triple, we compute the box $(\boldsymbol{X}_i + \boldsymbol{X}_j) \times \boldsymbol{X}_l$; then, we add subboxes $\boldsymbol{X}_k \times \boldsymbol{X}_l$ which intersect with this box to the enclosure for $\boldsymbol{x}_{kl}$.

*First example: computing the range of $x - x$.* For $f(x) = x - x$ on $[0, 1]$, the actual range is $[0, 0]$, but straightforward interval computations lead to an enclosure $[0, 1] - [0, 1] = [-1, 1]$. In straightforward interval computations, we have $r_1 = x$ with the exact interval range $\boldsymbol{r}_1 = [0, 1]$, and we have $r_2 = x$ with the exact interval range $\boldsymbol{x}_2 = [0, 1]$. The variables $r_1$ and $r_2$ are dependent, but we ignore this dependence.

In the new approach: we have $\boldsymbol{r}_1 = \boldsymbol{r}_2 = [0, 1]$, and we also have $\boldsymbol{r}_{12}$:

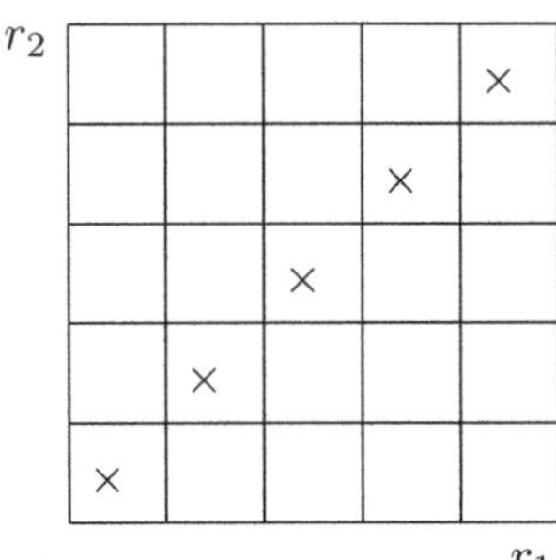

For each small box, we have $[-0.2, 0.2]$, so the union is $[-0.2, 0.2]$.

If we divide into more pieces, we get an interval closer to 0.

*Second example: computing the range of $x - x^2$.* In straightforward interval computations, we have $r_1 = x$ with the exact interval range interval $\boldsymbol{r}_1 = [0, 1]$, and we have $r_2 = x^2$ with the exact interval range $\boldsymbol{x}_2 = [0, 1]$. The variables $r_1$ and $r_2$ are dependent, but we ignore this dependence and estimate $\boldsymbol{r}_3$ as $[0, 1] - [0, 1] = [-1, 1]$.

In the new approach: we have $\boldsymbol{r}_1 = \boldsymbol{r}_2 = [0, 1]$, and we also have $\boldsymbol{r}_{12}$. First, we divide the range $[0, 1]$ into 5 equal subintervals $\boldsymbol{R}_1$. The union of the ranges $\boldsymbol{R}_1^2$ corresponding to these 5 subintervals $\boldsymbol{R}_1$ is $[0, 1]$, so $\boldsymbol{r}_2 = [0, 1]$. We divide this interval $\boldsymbol{r}_2$ into 5 equal sub-intervals $[0, 0.2]$, $[0.2, 0.4]$, etc. We now compute the set $\boldsymbol{r}_{12}$ as follows:

- for $\boldsymbol{R}_1 = [0, 0.2]$, we have $\boldsymbol{R}_1^2 = [0, 0.04]$, so only sub-interval $[0, 0.2]$ of the interval $\boldsymbol{r}_2$ is affected;
- for $\boldsymbol{R}_1 = [0.2, 0.4]$, we have $\boldsymbol{R}_1^2 = [0.04, 0.16]$, so also only sub-interval $[0, 0.2]$ is affected;
- for $\boldsymbol{R}_1 = [0.4, 0.6]$, we have $\boldsymbol{R}_1^2 = [0.16, 0.36]$, so two sub-intervals $[0, 0.2]$ and $[0.2, 0.4]$ are affected, etc.

For each possible pair of small boxes $\boldsymbol{R}_1 \times \boldsymbol{R}_2$, we have $\boldsymbol{R}_1 - \boldsymbol{R}_2 = [-0.2, 0.2]$, $[0, 0.4]$, or $[0.2, 0.6]$, so the union of $\boldsymbol{R}_1 - \boldsymbol{R}_2$ is $\boldsymbol{r}_3 = [-0.2, 0.6]$.

If we divide into more and more pieces, we get the enclosure which is closer and closer to the exact range $[0, 0.25]$.

*How to compute* $r_{ik}$. The above example is a good case to illustrate how we compute the range $r_{13}$ for $r_3 = r_1 - r_2$. Indeed, since $r_3 = [-0.2, 0.6]$, we divide this range into 5 subintervals $[-0.2, -0.04]$, $[-0.04, 0.12]$, $[0.12, 0.28]$, $[0.28, 0.44]$, $[0.44, 0.6]$.

- For $R_1 = [0, 0.2]$, the only possible $R_2$ is $[0, 0.2]$, so $R_1 - R_2 = [-0.2, 0.2]$. This covers $[-0.2, -0.04]$, $[-0.04, 0.12]$, and $[0.12, 0.28]$.
- For $R_1 = [0.2, 0.4]$, the only possible $R_2$ is $[0, 0.2]$, so $R_1 - R_2 = [0, 0.4]$. This interval covers $[-0.04, 0.12]$, $[0.12, 0.28]$, and $[0.28, 0.44]$.
- For $R_1 = [0.4, 0.6]$, we have two possible $R_2$:
  - for $R_2 = [0, 0.2]$, we have $R_1 - R_2 = [0.2, 0.6]$; this covers $[0.12, 0.28]$, $[0.28, 0.44]$, and $[0.44, 0.6]$;
  - for $R_2 = [0.2, 0.4]$, we have $R_1 - R_2 = [0, 0.4]$; this covers $[-0.04, 0.12]$, $[0.12, 0.28]$, and $[0.28, 0.44]$.
- For $R_1 = [0.6, 0.8]$, we have $R_1^2 = [0.36, 0.64]$, so three possible $R_2$: $[0.2, 0.4]$, $[0.4, 0.6]$, and $[0.6, 0.8]$, to the total of $[0.2, 0.8]$. Here, $[0.6, 0.8] - [0.2, 0.8] = [-0.2, 0.6]$, so all 5 subintervals are affected.
- Finally, for $R_1 = [0.8, 1.0]$, we have $R_1^2 = [0.64, 1.0]$, so two possible $R_2$: $[0.6, 0.8]$ and $[0.8, 1.0]$, to the total of $[0.6, 1.0]$. Here, $[0.8, 1.0] - [0.6, 1.0] = [-0.2, 0.4]$, so the first 4 subintervals are affected.

*Limitations of this approach.* The main limitation of this approach is that when we need an accuracy $\varepsilon$, we must use $\sim 1/\varepsilon$ granules; so, if we want to compute the result with $k$ digits of accuracy, i.e., with accuracy $\varepsilon = 10^{-k}$, we must consider exponentially many boxes ($\sim 10^k$). In plain words, this method is only applicable when we want to know the desired quantity with a given accuracy (e.g., 10%).

*Cases when this approach is applicable.* In practice, there are many problems when it is sufficient to compute a quantity with a given accuracy: e.g., when we detect an outlier, we usually do not need to know the variance with a high accuracy, an accuracy of 10% is more than enough.

Let us describe the case when interval computations do not lead to the exact range, but set computations do – of course, the range is "exact" modulo accuracy of the actual computer implementations of these sets.

*Example: estimating variance under interval uncertainty.* Suppose that we know the intervals $\mathbf{x}_1, \ldots, \mathbf{x}_n$ of possible values of $x_1, \ldots, x_n$, and we need to compute the range of the variance $V = \dfrac{1}{n} \cdot M - \dfrac{1}{n^2} \cdot E^2$, where $M \stackrel{\text{def}}{=} \sum\limits_{i=1}^{n} x_i^2$ and $E \stackrel{\text{def}}{=} \sum\limits_{i=1}^{n} x_i$.

A natural way to to compute $V$ is to compute the intermediate sums $M_k \stackrel{\text{def}}{=} \sum\limits_{i=1}^{k} x_i^2$ and $E_k \stackrel{\text{def}}{=} \sum\limits_{i=1}^{k} x_i$. We start with $M_0 = E_0 = 0$; once we know the pair $(M_k, E_k)$, we compute $(M_{k+1}, E_{k+1}) = (M_k + x_{k+1}^2, E_k + x_{k+1})$. Since the values of $M_k$ and $E_k$ only depend on $x_1, \ldots, x_k$ and do not depend on $x_{k+1}$, we can conclude that if $(M_k, E_k)$ is a possible value of the pair and $x_{k+1}$ is a possible value of this variable, then $(M_k + x_{k+1}^2, E_k + x_{k+1})$ is a possible value of $(M_{k+1}, E_{k+1})$. So, the set $\mathbf{p}_0$ of possible values of $(M_0, E_0)$ is the single point $(0, 0)$; once we know the set $\mathbf{p}_k$ of possible values of $(M_k, E_k)$, we can compute $\mathbf{p}_{k+1}$ as $\{(M_k + x^2, E_k + x) \mid (M_k, E_k) \in \mathbf{p}_k, x \in \mathbf{x}_{k+1}\}$. For $k = n$, we will get the set $\mathbf{p}_n$ of possible values of $(M, E)$; based on this set, we can then find the exact range of the variance $V = \dfrac{1}{n} \cdot M - \dfrac{1}{n^2} \cdot E^2$.

What $C$ should we choose to get the results with an accuracy $\varepsilon \cdot V$? On each step, we add the uncertainty of $1/C$; to, after $n$ steps, we add the inaccuracy of $n/C$. Thus, to get the accuracy $n/C \approx \varepsilon$, we must choose $C = n/\varepsilon$.

What is the running time of the resulting algorithm? We have $n$ steps; on each step, we need to analyze $C^3$ combinations of subintervals for $E_k$, $M_k$, and $x_{k+1}$. Thus, overall, we need $n \cdot C^3$ steps, i.e., $n^4/\varepsilon^3$ steps. For fixed accuracy $C \sim n$, so we need $O(n^4)$ steps – a polynomial time, and for $\varepsilon = 1/10$, the coefficient at $n^4$ is still $10^3$ – quite feasible.

For example, for $n = 10$ values and for the desired accuracy $\varepsilon = 0.1$, we need $10^3 \cdot n^4 \approx 10^7$ computational steps – "nothing" for a Gigaherz ($10^9$ operations per second) processor on a usual PC. For $n = 100$ values and the same desired accuracy, we need $10^4 \cdot n^4 \approx 10^{12}$ computational steps, i.e., $10^3$ seconds (15 minutes) on a Gigaherz processor. For $n = 1000$, we need $10^{15}$ steps, i.e., $10^6$ computational steps – 12 days on a single processor or a few hours on a multi-processor machine.

In comparison, the exponential time $2^n$ needed in the worst case for the exact computation of the variance under interval uncertainty, is doable ($2^{10} \approx 10^3$ step) for $n = 10$, but becomes unrealistically astronomical ($2^{100} \approx 10^{30}$ steps) already for $n = 100$.

*Comment.* When the accuracy increases $\varepsilon = 10^{-k}$, we get an exponential increase in running time – but this is OK since, as we have mentioned, the problem of computing variance under interval uncertainty is, in general, NP-hard.

*Other statistical characteristics.* Similar algorithms can be presented for computing many other statistical characteristics. For example, for every integer

$d > 2$, the corresponding higher-order central moment $C_d = \dfrac{1}{n} \cdot \sum\limits_{i=1}^{n} (x_i - \overline{x})^d$ is a linear combination of $d$ moments $M^{(j)} \overset{\text{def}}{=} \sum\limits_{i=1}^{n} x_i^j$ for $j = 1, \ldots, d$; thus, to find the exact range for $C_d$, we can keep, for each $k$, the set of possible values of $d$-dimensional tuples $(M_k^{(1)}, \ldots, M_k^{(d)})$, where $M_k^{(j)} \overset{\text{def}}{=} \sum\limits_{i=1}^{k} x_i^j$. For these computations, we need $n \cdot C^{d+1} \sim n^{d+2}$ steps – still a polynomial time.

Another example is covariance $\text{Cov} = \dfrac{1}{n} \cdot \sum\limits_{i=1}^{n} x_i \cdot y_i - \dfrac{1}{n^2} \cdot \sum\limits_{i=1}^{n} x_i \cdot \sum\limits_{i=1}^{n} y_i$. To compute covariance, we need to keep the values of the triples $(\text{Cov}_k, X_k, Y_k)$, where $\text{Cov}_k \overset{\text{def}}{=} \sum\limits_{i=1}^{k} x_i \cdot y_i$, $X_k \overset{\text{def}}{=} \sum\limits_{i=1}^{k} x_i$, and $Y_k \overset{\text{def}}{=} \sum\limits_{i=1}^{k} y_i$. At each step, to compute the range of

$$(\text{Cov}_{k+1}, X_{k+1}, Y_{k+1}) = (\text{Cov}_k + x_{k+1} \cdot y_{k+1}, X_k + x_{k+1}, Y_k + y_{k+1}),$$

we must consider all possible combinations of subintervals for $\text{Cov}_k$, $X_k$, $Y_k$, $x_{k+1}$, and $y_{k+1}$ – to the total of $C^5$. Thus, we can compute covariance in time $n \cdot C^5 \sim n^6$.

Similarly, to compute correlation $\rho = \text{Cov}/\sqrt{V_x \cdot V_y}$, we can update, for each $k$, the values of $(C_k, X_k, Y_k, X_k^{(2)}, Y_k^{(2)})$, where $X_k^{(2)} = \sum\limits_{i=1}^{k} x_i^2$ and $Y_k^{(2)} = \sum\limits_{i=1}^{k} y_i^2$ are needed to compute the variances $V_x$ and $V_y$. These computations take time $n \cdot C^7 \sim n^8$.

*Systems of ordinary differential equations (ODEs) under interval uncertainty.* A general system of ODEs has the form $\dot{x}_i = f_i(x_1, \ldots, x_m, t)$, $1 \leq i \leq m$. Interval uncertainty usually means that the exact functions $f_i$ are unknown, we only know the expressions of $f_i$ in terms of parameters, and we have interval bounds on these parameters.

There are two types of interval uncertainty: we may have global parameters whose values are the same for all moments $t$, and we may have noise-like parameters whose values may different at different moments of time — but always within given intervals. In general, we have a system of the type

$$\dot{x}_i = f_i(x_1, \ldots, x_m, t, a_1, \ldots, a_k, b_1(t), \ldots, b_l(t)),$$

where $f_i$ is a known function, and we know the intervals $\mathbf{a}_j$ and $\mathbf{b}_j(t)$ of possible values of $a_i$ and $b_j(t)$.

*Example.* For example, the case of a differential inequality when we only know the bounds $\underline{f}_i(x_1, \ldots, x_n, t)$ and $\overline{f}_i(x_1, \ldots, x_n, t)$ on $f_i(x_1, \ldots, x_n, t)$ can be described as

$$\widetilde{f}_i(x_1, \ldots, x_n, t) + b_1(t) \cdot \Delta(x_1, \ldots, x_n, t),$$

where

$$\widetilde{f}_i(x_1, \ldots, x_n, t) \stackrel{\text{def}}{=} \frac{\underline{f}_i(x_1, \ldots, x_n, t) + \overline{f}_i(x_1, \ldots, x_n, t)}{2},$$

$$\Delta(x_1, \ldots, x_n, t) \stackrel{\text{def}}{=} \frac{\overline{f}_i(x_1, \ldots, x_n, t) - \underline{f}_i(x_1, \ldots, x_n, t)}{2},$$

and $\boldsymbol{b}_1(t) = [-1, 1]$.

*Solving systems of ordinary differential equations (ODEs) under interval uncertainty.* For the general system of ODEs, Euler's equations take the form

$$x_i(t + \Delta t) = x_i(t) + \Delta t \cdot f_i(x_1(t), \ldots, x_m(t), t, a_1, \ldots, a_k, b_1(t), \ldots, b_l(t)).$$

Thus, if for every $t$, we keep the set of all possible values of a tuple $(x_1(t), \ldots, x_m(t), a_1, \ldots, a_k)$, then we can use the Euler's equations to get the exact set of possible values of this tuple at the next moment of time.

The reason for exactness is that the values $x_i(t)$ depend only on the previous values $b_j(t - \Delta t)$, $b_j(t - 2\Delta t)$, etc., and not on the current values $b_j(t)$.

To predict the values $x_i(T)$ at a moment $T$, we need $n = T/\Delta t$ iterations.

To update the values, we need to consider all possible combinations of $m + k + l$ variables $x_1(t), \ldots, x_m(t), a_1, \ldots, a_k, b_1(t), \ldots, b_l(t)$; so, to predict the values at moment $T = n \cdot \Delta t$ in the future for a given accuracy $\varepsilon > 0$, we need the running time $n \cdot C^{m+k+l} \sim n^{k+l+m+1}$. This is is still polynomial in $n$.

*Other possible cases when our approach is efficient.* Similar computations can be performed in other cases when we have an iterative process where a fixed finite number of variables is constantly updated.

In such problems, there is an additional factor which speeds up computations. Indeed, in the modern computers, fetching a value from the memory, in general, takes much longer than performing an arithmetic operation. To decrease this time, computers have a hierarchy of memories – from registers from which the access is the fastest, to cash memory (second fastest), etc. Thus, to take full use of the speed of modern processors, we must try our best to keep all the intermediate results in the registers. In the problems in which, at each moment of time, we can only keep (and update) a small current values of the values, we can store all these values in the registers – and thus, get very fast computations (only the input values $x_1, \ldots, x_n$ need to be fetched from slower-to-access memory locations).

*Comment.* The discrete version of the class of problems when we have an iterative process where a fixed finite number of variables is constantly updated is described in [320], where efficient algorithms are proposed for solving these discrete problems – such as propositional satisfiability. The use of this idea for interval computations was first described in Chapter 12 of [182].

*Additional advantage of our technique: possibility to take constraints into account.* Traditional formulations of the interval computation problems assume that we can have arbitrary tuples $(x_1, \ldots, x_n)$ as long as $x_i \in \boldsymbol{x}_i$ for all $i$. In practice, we may have additional constraints on $x_i$. For example, we may know that $x_i$ are observations of a smoothly changing signal at consequent moments of time (or that the values $x_i$ and $x_{i+1}$ of a geophysical density at nearby two nearby points cannot differ much). In such cases, we know that $|x_i - x_{i+1}| \le \varepsilon$ for some small known value $\varepsilon > 0$. Such constraints are easy to take into account in our approach.

For example, if know that $\boldsymbol{x}_i = [-1, 1]$ for all $i$ and we want to estimate the value of a high-frequency Fourier coefficient $f = x_1 - x_2 + x_3 - x_4 + \ldots - x_{2n}$, then usual interval computations lead to an enclosure $[-2n, 2n]$, while, for small $\varepsilon$, the actual range for the sum $(x_1 - x_2) + (x_3 - x_4) + \ldots$ where each of $n$ differences is bounded by $\varepsilon$, is much narrower: $[-n \cdot \varepsilon, n \cdot \varepsilon]$ (and for $x_i = i \cdot \varepsilon$, these bounds are actually attained).

Computation of $f$ means computing the values $f_k = x_1 - x_2 + \ldots + (-1)^{k+1} \cdot x_k$ for $k = 1, \ldots$ At each stage, we keep the set $\boldsymbol{s}_k$ of possible values of $(f_k, x_k)$, and use this set to find

$$\boldsymbol{s}_{k+1} = \{((f_k + (-1)^k \cdot x_{k+1}, x_{k+1}) \mid (f_k, x_k) \in \boldsymbol{s}_k \ \& \ |x_k - x_{k+1}| \le \varepsilon\}.$$

In this approach, when computing $f_{2k}$, we take into account that the value $x_{2k}$ must be $\varepsilon$-close to the value $x_k$ and thus, that we only add $\le \varepsilon$. Thus, our approach leads to almost exact bounds – modulo implementation accuracy $1/C$.

In this simplified example, the problem is linear, so we could use linear programming to get the exact range, but set computations work for similar non-linear problems as well.

*Toy example with a constraint.* The problem is to find the range of $r_1 - r_2$ when $\boldsymbol{r}_1 = [0, 1]$, $\boldsymbol{r}_2 = [0, 1]$, and $|r_1 - r_2| \le 0.1$. Here, the actual range is $[-0.1, 0.1]$, but straightforward interval computations return $[0, 1] - [0, 1] = [-1, 1]$.

In the new approach, first, we describe the constraint in terms of subboxes:

$r_2$

|   |   |   |   | × | × |
|---|---|---|---|---|---|
|   |   |   | × | × | × |
|   |   | × | × | × |   |
|   | × | × | × |   |   |
| × | × |   |   |   |   |

$r_1$

Next, we compute $\boldsymbol{R}_1 - \boldsymbol{R}_2$ for all possible pairs and take the union. The result is $[-0.6, 0.6]$.

If we divide into more pieces, we get the enclosure closer to $[-0.1, 0.1]$.

*Possible extension to p-boxes and classes of probability distributions.* A practically important situation is when, for each $z$, instead of the exact value of the cumulative distribution function $F(z)$, we know an interval $\boldsymbol{F}(z) = [\underline{F}(z), \overline{F}(z)]$ of possible values of $F(z)$; such an "interval-valued" cdf is, as we have mentioned earlier, called a *probability box*, or a *p-box*, for short.

In general, once we know the classes $\mathcal{F}_i$ of possible distributions for $x_i$, and a data processing algorithms $f(x_1, \ldots, x_n)$, we would like to know the class $\mathcal{F}$ of possible resulting distributions for $y = f(x_1, \ldots, x_n)$.

For problems like systems of ODEs, it is sufficient to keep, and update, for all $t$, the set of possible joint distributions for the tuple $(x_1(t), \ldots, a_1, \ldots)$.

We would like to estimate the values with some accuracy $\varepsilon \sim 1/C$ and the probabilities with the similar accuracy $1/C$. To describe a distribution with this uncertainty, we divide both the $x$-range and the probability ($p$-) range into $C$ granules, and then describe, for each $x$-granule, which $p$-granules are covered. Thus, we enclose this set into a finite union of p-boxes which assign, to each of $x$-granules, a finite union of $p$-granule intervals.

A general class of distributions can be enclosed in the union of such p-boxes. There are finitely many such assignments, so, for a fixed $C$, we get a finite number of possible elements in the enclosure.

We know how to propagate uncertainty via simple operations with a finite amount of p-boxes (see, e.g., [97]), so for ODEs we get a polynomial-time algorithm for computing the resulting p-box for $y$.

*For p-boxes, we need further improvements to make this method practical.* Formally, the above method is polynomial-time. However, it is not yet practical beyond very small values of $C$. Indeed, in the case of interval uncertainty, we needed $C^2$ or $C^3$ subboxes. This amount is quite feasible even for $C = 10$.

To describe a p-subbox, we need to attach one of $C$ probability granules to each of $C$ $x$-granules; these are $\sim C^C$ such attachments, so we need $\sim C^C$ subboxes. For $C = 10$, we already get an unrealistic $10^{10}$ increase in computation time.

# Part III

# Towards Computing Statistics under Interval and Fuzzy Uncertainty: Gauging the Quality of the Input Data

# How Reliable Is the Input Data?

In traditional interval computations, we assume that the interval data corresponds to guaranteed interval bounds, and that fuzzy estimates provided by experts are correct. In practice, measuring instruments are not 100% reliable, and experts are not 100% reliable, we may have estimates which are "way off", intervals which do not contain the actual values at all. Usually, we know the percentage of such outlier un-reliable measurements. However, it is desirable to check that the reliability of the actual data is indeed within the given percentage. The problem of checking (gauging) this reliability is, in general, NP-hard; in reasonable cases, there exist feasible algorithms for solving this problem.

The results of this chapter first appeared in [208].

## Formulation and Analysis of the Problem, and the Corresponding Results and Algorithms

*Reliability of interval data.* In interval computations, i.e., in processing interval data, we usually assume that all the measuring instruments functioned correctly, and that all the resulting intervals

$$[\widetilde{x} - \Delta, \widetilde{x} + \Delta]$$

indeed contain the actual value $x$.

In practice, nothing is 100% reliable. There is a certain probability that a measurement instrument malfunctions. As a result, when we repeatedly measure the same quantity several times, we may have a certain number of measurement results (and hence intervals) which are "way off", i.e., which do not contain the actual value at all.

For example, when we measure the temperature, we will usually get values which are close to the actual temperature, but once in a while the thermometer will not catch the temperature at all, and return a meaningless value like 0. It may be the fault of a sensor, and/or it may be a fault of the processor

H.T. Nguyen et al.: Computing Stat. under Interval & Fuzzy Uncertain., SCI 393, pp. 237–242.
springerlink.com   © Springer-Verlag Berlin Heidelberg 2012

which processes data from the sensor. Such situations are rare, but when we process a large amount of data, it is typical to encounter some outliers.

Such outliers can ruin the results of data processing. For example, if we compute the average temperature in a given geographic area, then averaging the correct measurement results would lead a good estimate, but if we add an outlier, we can get a nonsense result. For example, based on the measurements of temperature in El Paso in Summer resulting in 95, 100, and 105, we can get a meaningful value

$$\frac{95 + 100 + 105}{3} = 100.$$

However, if we add an outlier 0 to this set of data points, we get a misleading estimate

$$\frac{95 + 100 + 105 + 0}{4} = 75$$

creating the false impression of El Paso climate.

A natural way to characterize the reliability of the data is to set up the bound on the probability $p$ of such outliers. Once we know the value $p$, then, out of $n$ results of measuring the same quantity, we can dismiss $k \stackrel{\text{def}}{=} p \cdot n$ largest values and $k$ smallest values, and thus make sure that the outliers do not ruin the results of data processing.

*Need to gauge the reliability of interval data.* Where does the estimate $p$ for data reliability come from? The main idea of gauging this value comes from the fact that if we measure the same quantity several times, and all measurements are correct (no outliers), then all resulting intervals $\boldsymbol{x}^{(1)}, \ldots, \boldsymbol{x}^{(n)}$ contain the same (unknown) value $x$ – and thus, their intersection is non-empty.

If we have an outlier, then it is highly probably that this outlier will be far away from the actual value $x$ – and thus, the intersection of the resulting $n$ intervals (including intervals coming from outliers) will be empty.

In general, if the percentage of outliers does not exceed $p$, then we expect that out of $n$ given intervals, at least $n - k$ of these intervals (where $k \stackrel{\text{def}}{=} p \cdot n$) correspond to correct measurements and thus, have a non-empty intersection.

So, to check whether our estimate $p$ for reliability is correct, we must be able to check whether out of $n$ given intervals, $n - k$ have a non-empty intersection.

*Need to gauge reliability of interval data: multi-D case.* In the previous text, we considered a simplified situation in which each measuring instrument measures exactly one quantity. In practice, a measuring instrument often measure several different quantities $x_1, \ldots, x_d$. Due to uncertainty, after the measurement, for each quantity $x_i$, we have an interval $\boldsymbol{x}_i$ of possible values. Thus, the set of all possible values of the tuple $x = (x_1, \ldots, x_d)$ is a *box*

$$X = \boldsymbol{x}_1 \times \ldots \times \boldsymbol{x}_d = \{(x_1, \ldots, x_d) : x_1 \in \boldsymbol{x}_1, \ldots, x_d \in \boldsymbol{x}_d\}.$$

In this multi-D case, if all the measurements are correct (no outliers), all the corresponding boxes $X^{(1)}, \ldots, X^{(n)}$ contain the actual (unknown) tuple and thus, the intersection of all these boxes is non-empty.

Thus, to check whether our estimate $p$ for reliability is correct, we must be able to check whether out of $n$ given boxes, $n - k$ have a non-empty intersection.

*How to gauge reliability of fuzzy data.* In the fuzzy case, several experts estimate the value of the desired (1-D or multi-D) quantity $x$. Each of such estimates means that in addition to the (wider) "guaranteed" interval or box $X(0)$ (about which the expert is 100% confident that it contains the actual value of $x$) we also have narrower intervals (boxes) $X(\alpha)$ which contain $x$ with certainty $1 - \alpha$.

If all experts are right, then at least all the guaranteed boxes $X(0)$ should contain the actual value $x$. Thus, in this situation, the boxes $X(0)$ corresponding to different experts must have a non-empty intersection. In practice, some experts may be wrong; as a result, the corresponding boxes may be way off, and the intersection of all the experts' boxes may turn out to be empty.

It is reasonable to gauge the reliability of the experts (and, correspondingly, the reliability of the resulting fuzzy data) by the probability $p$ that an expert is wrong. For example, if $p = 0.1$, this means that we expect 90% of the experts to provide us with correct bounds $X(0)$. In this case, we expect that out of all the boxes provided by the experts, we can select 90% of them in such a way that the intersection of these selected boxes will be non-empty.

For boxes $X(\alpha)$ which are known with smaller certainty, the experts themselves agree that these boxes may not cover the actual value $x$ – and thus, the intersection of all such boxes can also turn out to be false. To describe the related reliability, we must know, for every $\alpha$, the probability $p$ that the corresponding box $X(\alpha)$ does not contain the actual value $x$. For example, if for $\alpha = 0.5$, we have $p = 0.3$, this means that we expect 70% of the experts' boxes $X(0.5)$ to contain the (unknown) actual value $x$. In this case, we expect that out of all the boxes $X(0.5)$ based on expert estimates, we can select 70% of them in such a way that the intersection of these selected boxes will be non-empty.

To check whether the data fits these reliability estimates, we must therefore be able to check whether out of $n$ given boxes, $n - k$ have a non-empty intersection.

*Resulting computational problem: box intersection problem.* Thus, both in the interval and in the fuzzy cases, we need to solve the following computational problem:

- Given:
  - integers $d$, $n$, and $k$; and
  - $n$ $d$-dimensional boxes
$$X^{(j)} = [\underline{x}_1^{(j)}, \overline{x}_1^{(j)}] \times \ldots \times [\underline{x}_n^{(j)}, \overline{x}_n^{(j)}],$$
    $j = 1, \ldots, n$, with rational bounds $\underline{x}_i^{(j)}$ and $\overline{x}_i^{(j)}$.

- *Check:* whether we can select $n - k$ of these $n$ boxes in such a way that the selected boxes have a non-empty intersection.

*First result: the box intersection problem is NP-complete.* The first result related to this problem is that in general, the above computational problem is NP-hard.

**Proposition 29.1.** *The box intersection problem is NP-complete.*

*Case of fixed dimension: efficient algorithm for gauging reliability.* In general, when we allow unlimited dimension $d$, the box intersection problem (computational problem related to gauging reliability) is computationally difficult (NP-hard).

In practice, however, the number $d$ of quantities measured by a sensor is small: e.g.,

- a GPS sensor measures 3 spatial coordinates;
- a weather sensor measures (at most) 5: temperature, atmospheric pressure, and the 3 dimensions of the wind vector, etc.

It turns out that if we limit ourselves to the case of a fixed dimension $d$, then we can solve the above computational problem in polynomial time $O(n^d)$; see, e.g., [120].

Indeed, for each of $d$ dimensions $x_i$ ($1 \leq i \leq d$), the corresponding $n$ intervals have $2n$ endpoints $\underline{x}_i^{(j)}$ and $\overline{x}_i^{(j)}$. Let us show if there exists a vector $x$ which belongs to $\geq n - k$ boxes $X^{(j)}$, then there also exists another point $y$ with this property in which every coordinate $y_i$ coincides with one of the endpoints. Indeed, if for some $i$, the value $x_i$ is not an endpoint, then we can take the closest endpoint as $y_i$. One can easily check that this change will keep the vector is all the boxes $X^{(j)}$.

Thus, to check whether there exists a vector $x$ that belongs to at least $n - k$ boxes $X^{(j)}$, it is sufficient to check whether there exist a vector formed by endpoints which satisfies this property. For each vector $y = (y_1, \ldots, y_d)$ and for each box $X^{(j)}$, it takes $d = O(1)$ steps to check whether $y \in X^{(j)}$. After repeating this check for all $n$ boxes, we thus check whether this vector $y$ satisfies the desired property in time $n \cdot O(1) = O(n)$.

For each of $d$ dimensions, there are $2n$ possible endpoints; thus, there are $(2n)^d$ possible vectors $y$ formed by such endpoints. For each of these vectors, we need time $O(n)$, so the overall computation time for this procedure takes time $O(n) \cdot (2n)^d = O(n^{d+1})$ – i.e., indeed time which grows polynomially with $n$.

*Remaining problem.* In the previous section, we have shown that for a bounded dimension $d$, we can solve the box intersection problem in polynomial time. However, as we have mentioned, polynomial time does not always mean that the algorithm is practically feasible.

For example, for a meteorological sensor, the dimension $d$ is equal to 5, so we need $n^6$ computational steps. For $n = 10$, we get $10^6$ steps, which is easy to perform. For $n = 100$, we need $100^6 = 10^{12}$ steps which is also doable – especially on a fast computer. However, for a very reasonable amount of $n = 10^3 = 1000$ data points, the above algorithm takes $1000^6 = 10^{18}$ computational steps – which already requires a long time, and for $n = 10^4$ data points, the algorithm takes a currently practically impossible amount of $10^{24}$ computational steps.

It is therefore desirable to speed up the computations.

## Proof

As we have mentioned in the main text, in gauging reliability, it is important to be able to solve the following box intersection problem:

- Given: a set of $n$ $d$-dimensional boxes, and a number $k < n$.
- Check: is there a vector $x$ which belongs to at least $n - k$ of these $n$ boxes?

This box intersection problem obviously in NP: it is easy to check that a given vector $x$ belongs to each of the boxes, and thus, to check whether it belongs to at least $n - k$ of the boxes. So we only need a proof of NP-hardness.

The proof is by reduction from the following auxiliary "limited clauses" problem which has been proved to be NP-complete:

- Given: a 2-CNF formula $F$ and a number $k$,
- check: is there a Boolean vector which satisfies at most $k$ clauses of $F$.

This problem was proved to be NP-complete in [159] (see also [16], p. 456).

As we have mentioned in Chapter 8, to prove the NP-hardness of our box intersection problem, it is therefore sufficient to be able to reduce this "limited clauses" problem to the box intersection problem.

Indeed, suppose that we are given a 2-CNF formula $F$. Let us denote the number of Boolean variables in this formula by $d$, and the overall number of clauses in this formula $F$ by $n$. Based on the formula $F$, let us build a set of $n$ $d$-dimensional boxes, one for each clause. If clause $C_i$ contains Boolean variables $z_{i1}$ and $z_{i2}$ variables, then the $i$-th box $X^{(i)}$ has sides $[0, 1]$ in all dimensions except in the dimensions associated with variables $z_{i1}$ and $z_{i2}$. For those two dimensions, the side is:

- $[0, 0]$ if the variable occurs positively in the clause (i.e., if the clause contains the positive literal $z_{ij}$), and
- $[1, 1]$ is the variable occurs negatively in the clause (i.e., if the clause contains the negative literal $\neg z_{ij}$).

According to the construction:

- for a clause $z_{i1} \lor z_{i2}$, a vector $x$ belongs to the box

$$X^{(i)} = \ldots \times [0, 1] \times [0, 0] \times [0, 1] \times \ldots \times [0, 1] \times [0, 0] \times [0, 1] \times \ldots$$

if and only of $x_{i1} = 0$ and $x_{i2} = 0$;

- for a clause $z_{i1} \vee \neg z_{i2}$, a vector $x$ belongs to the box $X^{(i)}$ if and only if $x_{i1} = 0$ and $x_{i2} = 1$;
- for a clause $\neg z_{i1} \vee z_{i2}$, a vector $x$ belongs to the box $X^{(i)}$ if and only if $x_{i1} = 1$ and $x_{i2} = 0$;
- for a clause $\neg z_{i1} \vee \neg z_{i2}$, a vector $x$ belongs to the box $X^{(i)}$ if and only if $x_{i1} = 1$ and $x_{i2} = 1$.

The claim is that there exists a vector $x$ which belongs to at least $n - k$ of these $n$ boxes if and only if there is a Boolean vector $z$ which satisfies at most $k$ clauses of the formula $F$.

Suppose that there exists a vector $x$ which belongs to at least $n - k$ of these $n$ boxes. According to our construction, each box $X^{(i)}$ comes from a clause $C_i$ that contains variables $z_{i1}$ and $z_{i2}$. For each box $X^{(i)}$ to which the vector $x$ belongs, make $z_{i1} = $ "false" if the box has $[0, 0]$ on the side associated with variable $z_{i1}$. Similarly, we make $z_{i2} = $ "false" if the box has $[0, 0]$ on the side associated with variable $z_{i2}$. Because of the way the boxes were build, the Boolean vector we build will make the clause associated with the box corresponding box $X^{(i)}$ false.

For example, if the clause is $z_{i1} \vee z_{i2}$, then the box will have $[0, 0]$ for the sides associated with both variable, so they will be both assigned the "false" Boolean value, making the clause false. This means that the Boolean formula built will make at least $n - k$ clauses become false. This formula will satisfy at most $k = n - (n - k)$ clauses.

In the opposite direction, if there is a Boolean vector $z$ which satisfies at most $k$ clauses of the formula $F$, build a vector $x = (x_1, \ldots, x_n)$ which has value:

- $x_i = 0$ in dimension $i$ if the Boolean variable $z_i$ associated with this dimension is false, and
- $x_i = 1$ otherwise.

One can check that for this arrangement, $x \in X^{(i)}$ if and only if the original Boolean vector $z$ made the corresponding clause $C_i$ false.

Since the Boolean vector $z$ satisfies at most $k$ clauses of the formula $F$, it makes at least $n - k$ clauses false. This means that the vector $x$ that we have built will belong to all the boxes associated with at least $n - k$ clauses that are false.

The reduction is proven, and so is NP-hardness.

# How Accurate Is the Input Data?

Different models can be used to describe real-life phenomena: deterministic, probabilistic, fuzzy, models in which we have interval-valued or fuzzy-valued probabilities, etc. Models are usually not absolutely accurate. It is therefore important to know how accurate is a given model. In other words, it is important to be able to measure a mismatch between the model and the empirical data. In this chapter, we describe an approach of measuring this mismatch which is based on the notion of utility, the central notion of utility theory.

The main results of this chapter first appeared in [206]. In one of the following application chapters (Chapter 35), we show that a similar approach can be used to measure the loss of privacy.

## Formulation and Analysis of the Problem, and the Corresponding Results

*Models are usually approximate.* In most areas of science and engineering, we only have *approximate* models for the real-world phenomena, i.e., models which are not 100% accurate. Since the models are approximate, their predictions are also only approximate.

*It is desirable to gauge the accuracy of a model.* In order to understand how accurate are the models' predictions, we need to know how accurate are the models themselves.

An ideal way to gauge the quality of a model is to compare it with the empirical data, i.e., to *validate* this model.

*Simplest case: deterministic phenomena.* Let us start with the simplest situation, when we have a deterministic phenomenon and we have a deterministic model which describes this phenomenon. In this situation, we can simply compare the measured value of the desired quantity with the values predicted by the model.

H.T. Nguyen et al.: Computing Stat. under Interval & Fuzzy Uncertain., SCI 393, pp. 243–248.
springerlink.com                                    © Springer-Verlag Berlin Heidelberg 2012

In such a situation, the difference between the actual and predicted values is a reasonable measure of a mismatch between the real-life phenomenon and the model.

*Real-life situation: non-deterministic phenomena.* In real life, many phenomena are non-deterministic. For such phenomena, we cannot predict the exact values of the corresponding quantities; at best, we can predict the *probabilities* of different values of these quantities.

To validate such models, we must therefore compare the predicted probability distribution with the empirical probability distribution. In such situations, it is not completely clear how we can measure the mismatch between the corresponding probability distributions, i.e., how we can gauge the validity of the probabilistic models.

*Additional complexity and relation to fuzzy techniques.* In practice, the situation is even more complex. Based on a finite sample of real-life events, we cannot uniquely determine the corresponding empirical distribution: we can only provide, with different degrees of confidence, bounds on the corresponding probabilities.

In other words, for each event, instead of a single value of its probability, we get a nested family of confidence intervals corresponding to different levels of uncertainty. Nested families are, in effect, equivalent to fuzzy numbers; see, e.g., [90, 156, 246, 252], so hopefully, techniques for processing fuzzy numbers will be helpful in this comparison.

*What we do in this chapter.* In this chapter, we mainly consider the case of probability distributions. The last section discusses the possibility of extending these results to a more general case of interval-valued probability distributions (p-boxes) and nested (= fuzzy) families of such interval-valued objects.

*Utility approach: a reminder.* In decision making theory, it is proven that under certain reasonable assumptions, a person's preferences are defined by his or her *utility function* $U(x)$ which assigns to each possible outcome $x$ a real number $U(x)$ called *utility*; see, e.g., [150] and [285].

In many real-life situations, a person's choice $s$ does not determine the outcome uniquely, we may have different outcomes $x_1, \ldots, x_n$ with probabilities, correspondingly, $p_1, \ldots, p_n$. For such a choice, we can describe the utility $U(s)$ associated with this choice as the expected value of the utility of outcomes:

$$U(s) = E[U(x)] = p_1 \cdot U(x_1) + \ldots + p_n \cdot U(x_n).$$

Among several possible choices, a user selects the one for which the utility is the largest: a possible choice $s$ is preferred to a possible choice $s'$ (denoted $s > s'$) if and only if

$$U(s) > U(s').$$

In the general case, when we have a (1-dimensional) probability distribution with the cumulative distribution function (cdf) $F(x)$, the utility is described by a similar formula

$$U(s) = E[U(x)] = \int U(x)\, dF(x).$$

In particular, in the continuous case, when we have a probability distribution with the probability density (pdf) $\rho(x)$, the utility is described by a formula

$$U(s) = E[U(x)] = \int U(x) \cdot \rho(x)\, dx.$$

*Application of utility approach to the problem of measuring a mismatch between probability distributions.* Since our preferences are characterized by the utility values, it is reasonable to measure mismatch by the possible decrease in utility. Specifically, let $F_1(x)$ denote the cdf of the model, and $F_2(x)$ denote the (usually unknown) cdf of the actual distribution. Similarly, in the continuous case, we will denote the pdf corresponding to the model by $\rho_1(x)$ and the actual pdf by $\rho_2(x)$.

In these terms, if we make a decision based on the model distribution $F_1(x)$, then the expect value of utility is

$$U_1 = \int U(x)\, dF_1(x).$$

Since the actual distribution is different, the actual value of the expected utility is equal to

$$U_2 = \int U(x)\, dF_2(x).$$

If the actual expected utility is smaller than the what we planned, i.e., if $U_2 < U_1$, then we have a loss caused by the mismatch. It is therefore reasonable to characterize the mismatch by this loss $U_1 - U_2$.

This loss describes the effect of the mismatch on a specific problem characterized by a specific utility function $U(x)$. Usually, a model is used for *many* different applications, with many different utility functions. In some applications, the difference between the two probability distribution may be irrelevant for our objectives; in this case, there is no loss. In other situations, this different may lead to a significant loss.

It is reasonable to gauge the mismatch by the worst possible loss caused by this mismatch.

*Which functions $U(x)$ should we consider.* In different situations, we may have different utility functions $U(x)$ that describe the dependence of a (predicted) gain on the (unknown) actual value of the corresponding parameter $x$.

This prediction only makes sense only if we can predict $U(x)$ for each situation with a reasonable accuracy, e.g., with an accuracy $\varepsilon > 0$. Measurements are never 100% accurate, and measurement of $x$ are not exception. Let us denote by $\delta$ the accuracy with which we measure $x$, i.e., the upper bound on the (absolute value of) the difference $\Delta x \stackrel{\text{def}}{=} \widetilde{x} - x$ between the measured value $\widetilde{x}$ and the (unknown) actual value $x$. Due to this difference, the estimated value $U(\widetilde{x})$ is different from the ideal prediction $U(x)$. Usually, measurement errors $\Delta x$ are small, so we can expand the prediction inaccuracy

$$\Delta U \stackrel{\text{def}}{=} U(\widetilde{x}) - U(x) = U(x + \Delta x) - U(x)$$

in Taylor series in $\Delta x$ and ignore quadratic and higher order terms in this expansion, leading to $\Delta U \approx U'(x) \cdot \Delta x$, where $U'(x)$ denotes the derivative of the utility function $U(x)$.

Since the largest possible value of $\Delta x$ is $\delta$, the largest possible value for $\Delta U$ is thus $|U'(x)| \cdot \delta$. Since this value should not exceed $\varepsilon$, we thus conclude that $|U'(x)| \cdot \delta \leq \varepsilon$, i.e., that $|U'(x)| \leq M \stackrel{\text{def}}{=} \varepsilon/\delta$.

Thus, we arrive at the following definition.

**Definition 30.1.** *Let $F_1(x)$ and $F_2(x)$ be two probability distributions on a real line, and let $M > 0$ be a real number. By the degree of mismatch $d_M(F_1, F_2)$ between the distributions, we mean the largest possible value of the difference*

$$\int U(x)\, dF_1(x) - \int U(x)\, dF_2(x)$$

*over all possible functions $U(x)$ for which $|U'(x)| \leq M$ for all $x$.*

**Proposition 30.1.** *For every two distributions,*

$$d_M(F_1, F_2) = M \cdot \int |F_1(x) - F_2(x)|\, dx.$$

*Comment.* In view of this result, it is reasonable to measure the mismatch between two probability distributions by the following $L_1$-*metric*:

$$d(F_1, F_2) \stackrel{\text{def}}{=} \int |F_1(x) - F_2(x)|\, dx.$$

This metric was indeed proposed and successfully used in model validation [108, 109]. The above result shows that this metric is not only reasonable, it follows from the general decision theory-motivated utility-based approach.

*Extension to p-boxes.* In practice, based on the empirical data, we cannot uniquely determine the corresponding probabilities $F(x)$. Instead, we can have confidence intervals $[\underline{F}(x), \overline{F}(x)]$ that contain the (unknown) values of these probabilities; see, e.g., [305, 337]. Such an interval-valued function that assigns, to every real number $x$, the corresponding interval $[\underline{F}(x), \overline{F}(x)]$ is called a *p-box*; see, e.g., [97]. Once we fix the confidence level, we thus have a *p-box* that contains all probability distributions which are consistent with the given empirical data.

In this situation, when the empirical data is describe by a p-box, how can we describe to what extent a given probability model is consistent with the empirical data? If the model $F(x)$ fits within the p-box $\boldsymbol{F}$ ($F \in \boldsymbol{F}$), i.e., if $F(x) \in [\underline{F}(x), \overline{F}(x)]$ for all $x$, this means that we have a perfect match.

In general, it is reasonable to define the degree of mismatch as the smallest possible mismatch between a model $F_1$ and distributions from a given p-box:

$$d(F_1, \boldsymbol{F}) = \min_{F_2 \in \boldsymbol{F}} d(F_1, F_2).$$

A model itself is not necessarily formulated in precise probabilistic terms. For example, we can say that according to our model, we have a normal distribution with the mean between $-0.1$ and $0.1$. In this case, a model is also naturally described by a p-box. In such situations, it is reasonable to define the mismatch between the p-box $\boldsymbol{F}_1$ describing the model and the p-box $\boldsymbol{F}_2$ describing the empirical distribution as the smallest possible mismatch between the probability distributions $F_1 \in \boldsymbol{F}_1$ and $F_2 \in \boldsymbol{F}_2$:

$$d(\boldsymbol{F}_1, \boldsymbol{F}_2) = \min_{F_1 \in \boldsymbol{F}_1, F_2 \in \boldsymbol{F}_2} d(F_1, F_2).$$

*Case of fuzzy uncertainty.* Instead of fixing a single confidence level, it is reasonable to consider confidence intervals $\boldsymbol{F}^{(\alpha)}(x)$ corresponding to different confidence levels $\alpha$. The resulting nested family of intervals can be naturally viewed as a fuzzy number for which these intervals are $\alpha$-cuts; see, e.g., [90, 156, 246, 252]. Alternatively, the probabilities $F(x)$ may be given by experts and thus, can be naturally represented as fuzzy numbers. In both case, it is reasonable to characterize the mismatch between the corresponding "fuzzy-valued" probability distributions $\boldsymbol{F}_1$ and $\boldsymbol{F}_2$ as a function that assigns, to every level $\alpha$, the degree of mismatch between the corresponding $\alpha$-cuts:

$$d^{(\alpha)}(\boldsymbol{F}_1, \boldsymbol{F}_2) = \min_{F_1 \in \boldsymbol{F}_1^{(\alpha)}, F_2 \in \boldsymbol{F}_2^{(\alpha)}} d(F_1, F_2).$$

## Proofs

*Proof of Proposition 30.1.* The desired difference $\Delta U = U_1 - U_2$ can be reformulated as the integral

$$\Delta U = \int U(x)\,(dF_1(x) - dF_2(x)).$$

Integrating this expression by parts, we conclude that

$$\Delta U = \int (F_1(x) - F_2(x)) \cdot U'(x)\,dx.$$

Since $|U'(x)| \leq M$, we conclude that

$$(F_1(x) - F_2(x)) \cdot U'(x) \leq |(F_1(x) - F_2(x)) \cdot U'(x)| =$$

$$|F_1(x) - F_2(x)| \cdot |U'(x)| \leq M \cdot |F_1(x) - F_2(x)|$$

and thus,

$$\Delta U = \int (F_1(x) - F_2(x)) \cdot U'(x)\, dx \le$$

$$\int M \cdot |F_1(x) - F_2(x)|\, dx = M \cdot \int |F_1(x) - F_2(x)|\, dx.$$

So, we have $\Delta U \le d_M(F_1, F_2)$ for all possible utility functions $U(x)$. Thus, the largest possible value of $\Delta U$ cannot exceed $d_M(F_1, F_2)$.

Let us now show that the largest possible value of $\Delta U$ is actually equal to $d_M(F_1, F_2)$, i.e., the value $\Delta U = d_M(F_1, F_2)$ is attained for some utility function $U(x)$. Indeed, as such a utility function, we can take

$$U(x) \stackrel{\text{def}}{=} \int_{-\infty}^{x} M \cdot \text{sign}(F_1(t) - F_2(t))\, dt,$$

where $\text{sign}(u)$ is defined as usual:

- when $u > 0$, we define $\text{sign}(u) = +1$;
- when $u < 0$, we define $\text{sign}(u) = -1$;
- when $u = 0$, we define $\text{sign}(u) = 0$.

For this utility function, $U'(x) = M \cdot \text{sign}(F_1(x) - F_2(x))$ and thus, $|U'(x)| \le M$ for all $x$. On the other hand, for this function,

$$\Delta U = \int (F_1(x) - F_2(x)) \cdot U'(x)\, dx =$$

$$\int (F_1(x) - F_2(x)) \cdot M \cdot \text{sign}(F_1(x) - F_2(x))\, dx.$$

For each value $u$, we have $u \cdot \text{sign}(u) = |u|$. Thus,

$$\Delta U = \int M \cdot |F_1(x) - F_2(x)|\, dx,$$

i.e., indeed, $\Delta U = d_M(F_1, F_2)$. The proposition is proven.

# Part IV

---

# Applications

# 31

# From Computing Statistics under Interval and Fuzzy Uncertainty to Practical Applications: Need to Propagate the Statistics through Data Processing

*Need for data processing.* In many areas of science and engineering, we are interested in a quantity $y$ which is difficult (or even impossible) to measure directly. For example, it is difficult to directly measure the distance to a faraway star or the amount of oil in an oil well. To estimate this quantity, we can:

- measure auxiliary easier-to-measure (or to estimate) quantities $x_1, \ldots, x_n$ which are related to $y$ by a known dependence $y = f(x_1, \ldots, x_n)$, and then
- use the results $\widetilde{x}_1, \ldots, \widetilde{x}_n$ of measuring (or estimating) $x_i$ to compute the estimate $\widetilde{y} = f(\widetilde{x}_1, \ldots, \widetilde{x}_n)$ for $y$.

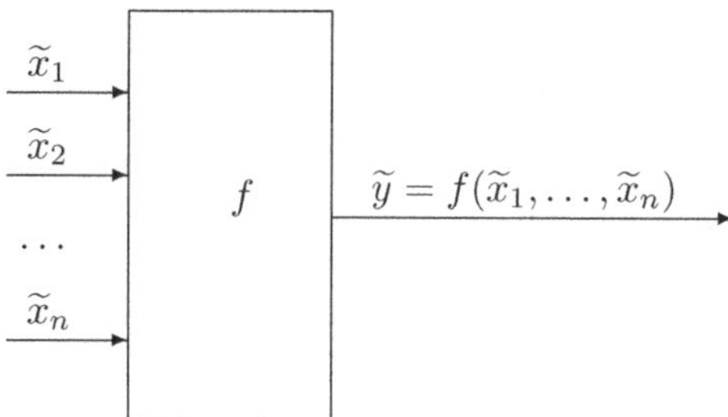


In some cases, the dependence $f$ is described by an explicit formula. For example, we can measure the distance $y$ to the Moon by sending a strong laser signal and measuring the time by which the reflected signal comes back to Earth. In this example, the distance is equal to $y = \dfrac{1}{2} \cdot x_1 \cdot x_2$, where:

- $x_1$ is the travel time, and
- $x_2$ is the speed of light.

Here, $n = 2$ and $f(x_1, x_2) = \dfrac{1}{2} \cdot x_1 \cdot x_2$. Once we have the measured value $\widetilde{x}_1$ of the travel time and the measured value $\widetilde{x}_2$ of the speed of light, we can estimate the distance as $\widetilde{y} = \dfrac{1}{2} \cdot \widetilde{x}_1 \cdot \widetilde{x}_2$.


H.T. Nguyen et al.: Computing Stat. under Interval & Fuzzy Uncertain., SCI 393, pp. 251–259.
springerlink.com        © Springer-Verlag Berlin Heidelberg 2012

In many practical cases, we do not have such an explicit expression. Instead, the dependence between $y$ and $x_i$ is described by a complex system of equations (e.g., differential equations). For example, for a geophysical problem (like the problem of estimating the amount of oil in a well), the dependence means that the values $y$ and $x_i$ correspond to the same solution of the system of differential equations describing seismic wave propagation:

- $y$ is the corresponding amount of oil, and
- $x_i$ is the time that it takes for a seismic to travel between its origin and the sensor that measures the $i$-th seismic signal.

In such cases, we usually have an algorithm for solving the corresponding system, i.e., for computing $y$ in terms of $x_i$. This algorithm will be denoted by the same letter $f$, so that $f(\widetilde{x}_1, \ldots, \widetilde{x}_n)$ means the result of applying the algorithm $f$ to the measured values $\widetilde{x}_1, \ldots, \widetilde{x}_n$.

Similarly, if we want to predict the future value of a physical quantity $y$, we:

- estimate the current values of this and/or other physical quantities $x_1, \ldots, x_n$, and
- use the estimates $\widetilde{x}_1, \ldots, \widetilde{x}_n$ and the known relation $y = f(x_1, \ldots, x_n)$ between the current and the future values to estimate the predicted value of $y$ as $\widetilde{y} = f(\widetilde{x}_1, \ldots, \widetilde{x}_n)$.

The computations related to both problems constitute *data processing* – the main use of computers in science and engineering.

*Need to take uncertainty into account.* In the case of data processing, we start with measurement or estimation results $\widetilde{x}_1, \ldots, \widetilde{x}_n$. Measurements and estimates are never exact. There is a non-zero difference $\Delta x_i \stackrel{\text{def}}{=} \widetilde{x}_i - x_i$ between the (approximate) measurement result (estimates) $\widetilde{x}_i$ and the (unknown) actual value $x_i$ of the $i$-th quantity $x_i$. This difference is called the *measurement (estimation) error*. The result $\widetilde{y} = f(\widetilde{x}_1, \ldots, \widetilde{x}_n)$ of applying the algorithm $f$ to the measurement results (estimates) $\widetilde{x}_i$ is, in general, different from the result $y = f(x_1, \ldots, x_n)$ of applying this algorithm to the actual values $x_i$. Thus, our estimate $\widetilde{y}$ is, in general, different from the actual value $y$ of the desired quantity: $\Delta y \stackrel{\text{def}}{=} \widetilde{y} - y \neq 0$.

In many practical applications, it is important to know not only the desired estimate for the quantity $y$, but also how accurate this estimate is. For example, in geophysical applications, it is not enough to know that the amount of oil in a given oil field is about 100 million tons: it is also important to know how accurate this estimate is.

If the amount is $100 \pm 10$, this means that the estimates are good enough, and we should start exploring this oil field.

On the other hand, if it is $100 \pm 200$, this means that it is quite possible that the actual value of the desired quantity $y$ is 0, i.e., that there is no oil at all. In this case, it may be prudent to perform additional measurements before we invest a lot of money into drilling oil wells.

The situation becomes even more critical in medical emergencies: it is not enough to have an estimate of blood pressure or body temperature to make a decision (e.g., whether to perform a surgery), it is important that even with the measurement uncertainty, we are sure about the diagnosis – and if we are not, maybe it is desirable to perform more accurate measurements.

It is therefore desirable to find out the uncertainty $\Delta y$ caused by the uncertainties $\Delta x_i$ in the inputs:

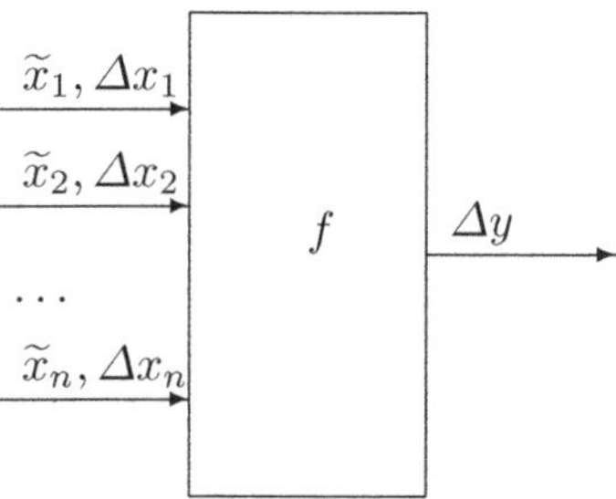


*Comment.* We assumed that the relation $f$ provides the *exact* relation between the variables $x_1, \ldots, x_n$, and the desired value $y$. If so, then, in the ideal case in which we plug in the actual (unknown) values of $x_i$ into the algorithm $f$, we get the exact value $y = f(x_1, \ldots, x_n)$ of $y$.

In many real-life situations, the relation $f$ between $x_i$ and $y$ is only *approximately* known. In this case, even if we know the exact values of $x_i$, substituting these values into the approximate function $f$ will not provide us with the exact value of $y$. In such situations, there is even more uncertainty in $y$:

- first, there is an uncertainty in $y$ caused by the the uncertainty in the inputs;
- second, there is a *model uncertainty* caused by the fact that the known algorithm $f$ only provides an approximate description of the dependence between the inputs and the output.

A model uncertainty has to be estimated separately and added to the uncertainty caused by the measurement errors.

*Uncertainty of the results of direct measurements.* To estimate the uncertainty $\Delta y$ caused by the measurement uncertainties $\Delta x_i$, we need to have some information about these original uncertainties $\Delta x_i$. We have already mentioned, in Part I, that often, this uncertainty is described by probabilities or intervals. These situations corresponds to the two extreme cases:

- probabilistic uncertainty means that we have the full information about the corresponding probabilities, while

- interval uncertainty means that we have no information about the probabilities at all – other than the upper bounds on the measurement errors.

In practice, we often have *partial* information about the probabilities. In the
case of this partial information, we only have a partial knowledge of the
corresponding statistical characteristics. In other words, instead of the exact
values of the moments, cdf $F(x)$, etc., we only have *intervals* of possible values
of these characteristics. How can we represent this partial information?

*Case of cdf.* If we use cdf $F(x)$ to represent a distribution, then full information corresponds to the case when we know the exact value of $F(x)$ for every
$x$. Partial information means:

- either that we only know approximate values of $F(x)$ for all $x$, i.e., that
  for every $x$, we only know the interval that contains $F(x)$; in this case, we
  get a *p-box* [97];
- or that we only know the values of $F(x)$ for some $x$, i.e, that we only
  know the values $F(x_1)$, ..., $F(x_i)$, ..., $F(x_n)$ for finitely many values
  $x = x_1, \ldots, x_i, \ldots, x_n$; in this case, we have a *histogram*.

It is also possible that we know only approximate values of $F(x)$ for some $x$;
in this case, we have an *interval-valued histogram.*

*Case of moments.* If we use moments to represent a distribution, then partial
information means that we either know the exact values of finitely many
moments, or that we know intervals of possible values of several moments.

*How to process partial information about probabilities: formulation of the
problem.* In this Part, we consider the general data processing problem:

- we know the estimates $\widetilde{x}_1, \ldots \widetilde{x}_n$ for several quantities $x_1, \ldots, x_n$;
- we know an algorithm $y = f(x_1, \ldots, x_n)$ that relate the values $x_1, \ldots, x_n$
  of the input quantities with the value $y$ of the desired quantity;
- by applying this algorithm to the known estimates, we get the estimate
  $\widetilde{y} = f(\widetilde{x}_1, \ldots, \widetilde{x}_n)$ for the desired quantity $y$.

Since the estimates $\widetilde{x}_i$ are, in general, different from the (unknown) actual
values $x_i$, the estimate $\widetilde{y}$ is also, in general, different from the desired value
$y$. To find out how accurate is our estimate $\widetilde{y}$, we must use the available
information about the uncertainties $\Delta x_i = \widetilde{x}_i - x_i$ into the information about
$\Delta y = \widetilde{y} - y$.

Let us describe how this can be done for all different types of partial
information about probabilities described above.

*First case: complete information about the probabilities.* The actual values
$x_i = \widetilde{x}_i - \Delta x_i$ are unknown. However, when we know the probability distributions for the uncertainties $\Delta x_i$, we thus know the probability distribution
for the values $x_i$ as well. Thus, we can determine the probability distribution
for the value $y = f(x_1, \ldots, x_n) = f(\widetilde{x}_1 - \Delta x_1, \ldots, \widetilde{x}_n - \Delta x_n)$ and hence, for
$\Delta y = \widetilde{y} - y$.

A natural way to describe this distribution is by using *Monte-Carlo simulation* method. In this method, for an arbitrary number of simulations $N$, we simulate the values $\Delta y^{(k)}$, $1 \le k \le N$, as follows:

- first, we simulate the random value $\xi_1^{(k)}, \ldots, \xi_n^{(k)}$ distributed according to the same distributions as the actual values $\Delta x_i$;
- then, we simulate the actual values $x_i^{(k)}$ as $x_i^{(k)} = \widetilde{x}_i - \xi_i^{(k)}$;
- finally, we simulate the actual value $\Delta y^{(k)}$ as $\Delta y^{(k)} = \widetilde{y} - f(x_1^{(k)}, \ldots, x_n^{(k)})$.

The values $\Delta y^{(k)}$ form a statistical sample from the desired distribution. Thus, based on this sample, we can estimate all the statistical characteristics of this distribution; see, e.g., Lodwick et al. [204].

*Case of partial information about probabilities: general idea.* Partial information means that instead of knowing all the probabilities, we only have some partial information about several statistical characteristics, such as the moments, the values of the cdf, etc. To be more precise, instead of knowing the exact value of the corresponding characteristic $c$, we only know the intervals $[\underline{c}, \overline{c}]$ of its possible values: $\underline{c} \le c \le \overline{c}$.

Most statistical characteristics linearly depend on the corresponding probabilities. For example, for a discrete distribution, in which we have values $v_1, \ldots, v_m$ with probabilities $p_1, \ldots, p_m$, the mean is equal to $\mathrm{e}[v] = \sum\limits_{i=1}^{m} p_i \cdot v_i$, the second moment is equal to $\mathrm{e}[v] = \sum\limits_{i=1}^{m} p_i \cdot v_i^2$, the cdf is equal to $F(x) = \sum\limits_{i:v_i \le x} p_i$, etc. Thus, the partial information about these statistical characteristics can be described as a system of linear inequalities in terms of the unknown probabilities $p_i$. For example, the bounds $[\underline{E}, \overline{E}]$ on the mean can be described as

$$\underline{E} \le \sum_{i=1}^{m} p_i \cdot v_i \le \overline{E}.$$

Under these inequalities – that represent the available information about the inputs – we want to find the range of possible values of different statistical characteristics of $y = f(x_1, \ldots, x_n)$. In other words, we want to find the smallest and the largest possible values of the desired characteristic under the linear inequality.

The desired characteristic is also usually linear in $p_i$, so we arrive at the following problem: optimize the value of a linear function under a system of linear inequalities. This problem is known as *Linear Programming.* There exist efficient methods for solving this problem, and these methods have been effectively applied to solve the corresponding problems of processing partial information about probabilities; see, e.g., Berleant et al. [32, 33, 34, 35, 36, 37, 39, 40, 41, 296, 367].

*Comment.* Of course, this method cannot be directly applied to statistical characteristic like variance or correlation which are *not* linear functions of the corresponding probabilities.

*Simplified example: discrete case.* To illustrate the above idea, let us consider a simplified example in which $f(x_1, x_2) = x_1 + x_2$, and both variables $x_i$ are *discrete*:

- the variable $x_1$ takes possible values $v_{11}, \ldots, v_{1m}$, and
- the variable $x_2$ takes possible values $v_{21}, \ldots, v_{2m}$.

Let us assume that we known the bounds $\underline{E}_1$ and $\overline{E}_1$ on the first moment of $x_1$ and the bounds $\underline{E}_2$ and $\overline{E}_2$ on the first moment of $x_2$. Based on this information, for several values $y_0$, we want to find the range of possible values of the cdf $F(y_0) = \mathrm{Prob}(y \leq y_0)$. Let us show how this problem can be reformulated as a linear programming problem.

The unknowns here are the values $p_{ij}$, the probabilities that $x_1 = v_{1i}$ and $x_2 = v_{2j}$. These unknown values must be non-negative and satisfy the condition

$$\sum_{i=1}^{m} \sum_{j=1}^{m} p_{ij} = 1.$$

The information about the mean value of $x_1$ can be described as

$$\underline{E}_1 \leq \sum_{i=1}^{m} \sum_{j=1}^{m} p_{ij} \cdot v_{1i} \leq \overline{E}_1.$$

Similarly, the information about the mean value of $x_2$ can be described as

$$\underline{E}_2 \leq \sum_{i=1}^{m} \sum_{j=1}^{m} p_{ij} \cdot v_{2j} \leq \overline{E}_2.$$

Under these constraints, we must find the set of possible values of the quantity

$$F(y_0) = \sum_{(i,j):v_{1i}+v_{2j} \leq y_0} p_{ij}.$$

*Realistic case: continuous input variables.* In the above formulation, we considered the simplified *discrete* case, in which each input variable can only take finitely many values. In reality, inputs variables are *continuous*, they can take all possible real values from a certain interval.

In some cases, we know the *histogram*, i.e., we divide the range of possible values of each variable $x_i$ into finite many sub-intervals (zones), and we know the lower and upper bounds on the probability of $x_i$ to be in each of these zones. In this case, linear programming-type approach can also be used; see, e.g., [32, 33, 34, 35, 36, 37, 39, 40, 41, 296, 367].

In other cases, we know the bounds on the cdfs, i.e., the p-boxes. Corresponding algorithms are given in Ferson [97] and Ferson et al. [105]. In situations when we know bounds on the moments, we can use methods from Granvilliers et al. [122] and Kreinovich [169].

*Case study: first moments.* Let us describe, in detail, the case when we know the first moments of the corresponding distributions.

This situation has a perfect practical sense because, in some practical situations, in addition to the lower and upper bounds on each random variable $x_i$, we know the bounds $\boldsymbol{E}_i = [\underline{E}_i, \overline{E}_i]$ on its mean $E_i$. Indeed, in measurement practice (see, e.g., [283]), the overall measurement error $\Delta x$ is usually represented as a sum of two components:

- a *systematic* error component $\Delta_s x$ which is defined as the expected value $\mathrm{e}[\Delta x]$, and
- a *random* error component $\Delta_r x$ which is defined as the difference between the overall measurement error and the systematic error component: $\Delta_r x \overset{\mathrm{def}}{=} \Delta x - \Delta_s x$.

In addition to the bound $\Delta$ on the overall measurement error, the manufacturers of the measuring instrument often provide an upper bound $\Delta_s$ on the systematic error component: $|\Delta_s x| \leq \Delta_s$.

This additional information is provided because, with this additional information, we not only get a bound on the accuracy of a single measurement, but we also get an idea of what accuracy we can attain if we use repeated measurements to increase the measurement accuracy. Indeed, the very idea that repeated measurements can improve the measurement accuracy is natural: we measure the same quantity by using the same measurement instrument several $(N)$ times, and then take, e.g., an arithmetic average
$$\bar{x} = \frac{\widetilde{x}^{(1)} + \ldots + \widetilde{x}^{(k)} + \ldots + \widetilde{x}^{(N)}}{N}$$ of the corresponding measurement results $\widetilde{x}^{(1)} = x + \Delta x^{(1)}, \ldots, \widetilde{x}^{(k)} = x + \Delta x^{(k)}, \ldots, \widetilde{x}^{(N)} = x + \Delta x^{(N)}$.

- If systematic error is the only error component, then all the measurements lead to exactly the same value $\widetilde{x}^{(1)} = \ldots = \widetilde{x}^{(k)} = \ldots = \widetilde{x}^{(N)}$, and averaging does not change the value – hence does not improve the accuracy.
- On the other hand, if we know that the systematic error component is 0, i.e., $\mathrm{e}[\Delta x] = 0$ and $\mathrm{e}[\widetilde{x}] = x$, then, as $N \to \infty$, the arithmetic average tends to the actual value $x$. In this case, by repeating the measurements sufficiently many times, we can determine the actual value of $x$ with an arbitrary given accuracy

In general, by repeating measurements sufficiently many times, we can arbitrarily decrease the random error component and thus attain accuracy as close to $\Delta_s$ as we want.

When this additional information is given, then, after we performed a measurement and got a measurement result $\widetilde{x}$, then not only we get the information that the actual value $x$ of the measured quantity belongs to the interval $\boldsymbol{x} = [\widetilde{x} - \Delta, \widetilde{x} + \Delta]$, but we can also conclude that the expected value of $x = \widetilde{x} - \Delta x$ (which is equal to $\mathrm{e}[x] = \widetilde{x} - \mathrm{e}[\Delta x] = \widetilde{x} - \Delta_s x$) belongs to the interval $\boldsymbol{E} = [\widetilde{x} - \Delta_s, \widetilde{x} + \Delta_s]$.

If we have this information for every $x_i$, then, in addition to the interval $\boldsymbol{y}$ of possible values of $y$, we would also like to know the interval of possible values of $\mathrm{e}[y]$. This additional interval will hopefully provide us with the information on how repeated measurements can improve the accuracy of this indirect measurement. Thus, we arrive at the following problem:

*Precise formulation of the problem.* Given an algorithm computing a function $f(x_1, \ldots, x_i, \ldots, x_n)$ from $R^n$ to $R$, and values $\underline{x}_1, \overline{x}_1, \ldots, \underline{x}_i, \overline{x}_i, \ldots, \underline{x}_n, \overline{x}_n, \underline{E}_1, \overline{E}_1, \ldots, \underline{E}_i, \overline{E}_i, \ldots, \underline{E}_n, \overline{E}_n$, we want to find

$$\underline{E} \stackrel{\text{def}}{=} \min\{\mathrm{e}[f(x_1, \ldots, x_i, \ldots, x_n)] \mid \text{all distributions of } (x_1, \ldots, x_i, \ldots, x_n)$$

$$\text{for which } x_1 \in [\underline{x}_1, \overline{x}_1], \ldots, x_i \in [\underline{x}_i, \overline{x}_i], \ldots, x_n \in [\underline{x}_n, \overline{x}_n], \tag{31.1}$$

$$\mathrm{e}[x_1] \in [\underline{E}_1, \overline{E}_1], \ldots, \mathrm{e}[x_i] \in [\underline{E}_i, \overline{E}_i], \ldots, \mathrm{e}[x_n] \in [\underline{E}_n, \overline{E}_n]\};$$

and $\overline{E}$ which is the maximum of $\mathrm{e}[f(x_1, \ldots, x_n)]$ for all such distributions.

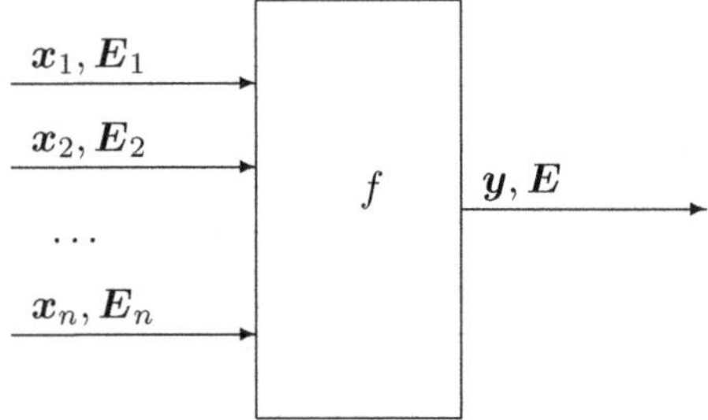



In addition to considering all possible distributions, we can also consider the case when all the variables $x_i$ are independent.

*Algorithms for solving the problem: case of exactly known moments.* The main idea behind straightforward interval computations can be applied here as well. Namely, first, we find out how to solve this problem for the case when $n = 2$ and $f(x_1, x_2)$ is one of the standard arithmetic operations. Then, once we have an arbitrary algorithm $f(x_1, \ldots, x_n)$, we parse it and replace each elementary operation on real numbers with the corresponding operation on quadruples $(\underline{x}, \underline{E}, \overline{E}, \overline{x})$.

To implement this idea, we must therefore know how to solve the above problem for elementary operations.

For *addition*, the answer is simple. Since $\mathrm{e}[x_1 + x_2] = \mathrm{e}[x_1] + \mathrm{e}[x_2]$, if $y = x_1 + x_2$, there is only one possible value for $E = \mathrm{e}[y]$: the value $E = E_1 + E_2$. This value does not depend on whether we have correlation or nor, and whether we have any information about the correlation. Thus, $\boldsymbol{E = E_1 + E_2}$.

Similarly, the answer is simple for *subtraction:* if $y = x_1 - x_2$, there is only one possible value for $E = \mathrm{e}[y]$: the value $E = E_1 - E_2$. Thus, $\boldsymbol{E = E_1 - E_2}$.

For *multiplication*, if the variables $x_1$ and $x_2$ are independent, then

$$\mathrm{e}[x_1 \cdot x_2] = \mathrm{e}[x_1] \cdot \mathrm{e}[x_2].$$

Hence, if $y = x_1 \cdot x_2$ and $x_1$ and $x_2$ are independent, there is only one possible value for $E = e[y]$: the value $E = E_1 \cdot E_2$; hence $\boldsymbol{E} = \boldsymbol{E}_1 \cdot \boldsymbol{E}_2$.

The first non-trivial case is the case of multiplication in the presence of possible correlation. When we know the exact values of $E_1$ and $E_2$, the solution to the above problem is as follows (see, e.g., [122, 169]): For multiplication $y = x_1 \cdot x_2$, when we have no information about the correlation,

$$\underline{E} = \max(p_1 + p_2 - 1, 0) \cdot \overline{x}_1 \cdot \overline{x}_2 + \min(p_1, 1 - p_2) \cdot \overline{x}_1 \cdot \underline{x}_2 +$$

$$\min(1 - p_1, p_2) \cdot \underline{x}_1 \cdot \overline{x}_2 + \max(1 - p_1 - p_2, 0) \cdot \underline{x}_1 \cdot \underline{x}_2; \tag{31.2}$$

$$\overline{E} = \min(p_1, p_2) \cdot \overline{x}_1 \cdot \overline{x}_2 + \max(p_1 - p_2, 0) \cdot \overline{x}_1 \cdot \underline{x}_2 +$$

$$\max(p_2 - p_1, 0) \cdot \underline{x}_1 \cdot \overline{x}_2 + \min(1 - p_1, 1 - p_2) \cdot \underline{x}_1 \cdot \underline{x}_2, \tag{31.3}$$

where $p_i \stackrel{\text{def}}{=} (E_i - \underline{x}_i)/(\overline{x}_i - \underline{x}_i)$.

For the *inverse* $y = 1/x_1$, the finite range is possible only when $0 \notin \boldsymbol{x}_1$. Without losing generality, we can consider the case when $0 < \underline{x}_1$. In this case, the range of possible values of $E$ is $\boldsymbol{E} = [1/E_1, p_1/\overline{x}_1 + (1 - p_1)/\underline{x}_1]$.

Similar formulas can be produced for max and min, and also for the cases when there is a strong correlation between $x_i$: namely, when $x_1$ is (non-strictly) increasing or decreasing in $x_2$.

*Algorithms for solving the problem: general case.* For multiplication (under no assumption about correlation), if we only know the intervals of possible values of $E_i$, then to find $\underline{E}$, it is sufficient to consider the following combinations of $p_1$ and $p_2$:

- $p_1 = \underline{p}_1$ and $p_2 = \underline{p}_2$; $p_1 = \underline{p}_1$ and $p_2 = \overline{p}_2$; $p_1 = \overline{p}_1$ and $p_2 = \underline{p}_2$; $p_1 = \overline{p}_1$ and $p_2 = \overline{p}_2$;
- $p_1 = \max(\underline{p}_1, 1 - \overline{p}_2)$ and $p_2 = 1 - p_1$ (if $1 \in \boldsymbol{p}_1 + \boldsymbol{p}_2$); and
- $p_1 = \min(\overline{p}_1, 1 - \underline{p}_2)$ and $p_2 = 1 - p_1$ (if $1 \in \boldsymbol{p}_1 + \boldsymbol{p}_2$).

The smallest value of $\underline{E}$ for all these cases is the desired lower bound $\underline{E}$.

To find $\overline{E}$, it is sufficient to consider the following combinations of $p_1$ and $p_2$:

- $p_1 = \underline{p}_1$ and $p_2 = \underline{p}_2$; $p_1 = \underline{p}_1$ and $p_2 = \overline{p}_2$; $p_1 = \overline{p}_1$ and $p_2 = \underline{p}_2$; $p_1 = \overline{p}_1$ and $p_2 = \overline{p}_2$;-
- $p_1 = p_2 = \max(\underline{p}_1, \underline{p}_2)$ (if $\boldsymbol{p}_1 \cap \boldsymbol{p}_2 \neq \emptyset$); and
- $p_1 = p_2 = \min(\overline{p}_1, \overline{p}_2)$ (if $\boldsymbol{p}_1 \cap \boldsymbol{p}_2 \neq \emptyset$).

The largest value of $\overline{E}$ for all these cases is the desired upper bound $\overline{E}$.

*Important open problems.* What if, in addition to intervals and first moments, we also know second moments? This problem is important for design of computer chips.

What if, in addition to moments, we also know p-boxes?

# Applications to Bioinformatics

*Formulation of the practical problem.* In cancer research, it is important to find out the genetic difference between the cancer cells and the healthy cells.

In the ideal world, we should be able to have a sample of cancer cells, and a sample of healthy cells, and thus directly measure the concentrations $c$ and $h$ of a given gene in cancer and in healthy cells. In reality, it is very difficult to separate the cells, so we have to deal with samples that contain both cancer and normal cells.

Let $y_i$ denote the result of measuring the concentration of the gene in $i$-th sample, and let $x_i$ denote the percentage of cancer cells in $i$-th sample. Then, we should have $x_i \cdot c + (1 - x_i) \cdot h \approx y_i$ (approximately equal because there are measurement errors in measuring $y_i$).

It is worth mentioning that this system can be somewhat simplified if instead of $c$, we consider a new variable $a \stackrel{\text{def}}{=} c - h$. In terms of the new unknowns $a$ and $h$, the system takes the following form: $a \cdot x_i + h \approx y_i$.

*Ideal case.* Let us first consider an idealized case in which we know the exact percentages $x_i$. In this case, we can find the desired values $c$ and $h$ by solving a system of linear equations $x_i \cdot c + (1 - x_i) \cdot h \approx y_i$ with two unknowns $c$ and $h$.

The errors of measuring $y_i$ are normal independent identically distributed random variables. So, to estimate $a$ and $h$, we can use the Least Squares Method (LSM)

$$\text{Minimize} \quad \sum_{i=1}^{n} (a \cdot x_i + h - y_i)^2,$$

according to which $a = \dfrac{C_{x,y}}{V_x}$ and $h = E_y - a \cdot E_x$, where $E_x = \dfrac{1}{n} \cdot \sum_{i=1}^{n} x_i$ is the population mean, $V_x = \dfrac{1}{n-1} \cdot \sum_{i=1}^{n} (x_i - E_x)^2$ is the population variance, and

H.T. Nguyen et al.: Computing Stat. under Interval & Fuzzy Uncertain., SCI 393, pp. 261–264.
springerlink.com 

$$C_{xy} = \frac{1}{n-1} \cdot \sum_{i=1}^{n} (x_i - E_x) \cdot (y_i - E_y)$$

is the population covariance. Once we know $a = c - h$ and $h$, we can then estimate $c$ as $a + h$.

*Need to take interval uncertainty into account* (see, e.g., [172, 188, 173]). The problem is that the concentrations $x_i$ come from experts who manually count different cells, and experts can only provide interval bounds on the values $x_i$ such as $x_i \in [0.7, 0.8]$. Different values of $x_i$ in the corresponding intervals lead to different values of $a$ and $h$. It is therefore desirable to find the range of $a$ and $h$ corresponding to all possible values $x_i \in [\underline{x}_i, \overline{x}_i]$.

This problem is a particular case of the general problem that we analyze in this book: how to efficiently deduce the statistical information from interval data.

*In general, this problem is NP-hard.* We have already proven that, in general, the problem of computing the ranges of statistical characteristic under interval uncertainty is NP-hard – it is NP-hard even for the variance.

It is possible to prove that computing the range of the ratio $C_{x,y}/V_x$ is also an NP-hard problem; the proof is presented at the end of this chapter.

*Linear approximation.* One of the known approximate techniques is linearization, when we approximate the statistic $S$ with the linear terms in its Taylor expansion:

$$S \approx S_{\mathrm{lin}} = S_0 - \sum_{i=1}^{n} S_i \cdot \Delta x_i,$$

where $S_0 \stackrel{\text{def}}{=} S(\widetilde{x}_1, \ldots, \widetilde{x}_n)$, $S_i \stackrel{\text{def}}{=} \dfrac{\partial S}{\partial x_i}(\widetilde{x}_1, \ldots, \widetilde{x}_n)$, and $\Delta x_i \stackrel{\text{def}}{=} \widetilde{x}_i - x_i$. For the linear function, we get the exact formula for the range: $\boldsymbol{S} = [S_0 - \Delta_S, S_0 + \Delta_S]$, where $\Delta_S \stackrel{\text{def}}{=} \sum_{i=1}^{n} |S_i| \cdot \Delta_i$.

Let us apply this general formula to our case. Let $\widetilde{x}_i = (\underline{x}_i + \overline{x}_i)/2$ be the midpoint of the $i$-th interval, and let $\Delta_i = (\overline{x}_i - \underline{x}_i)/2$ be the half-width of this interval. For $a$, we have

$$\frac{\partial a}{\partial x_i} = \frac{1}{(n-1) \cdot V_x} \cdot (y_i - E_y - 2a \cdot x_i + 2a \cdot E_x).$$

We can use the formula $E_y = a \cdot E_x + h$ to simplify this expression, resulting in

$$\Delta_a = \frac{1}{(n-1) \cdot V_x} \cdot \sum_{i=1}^{n} |\delta y_i - a \cdot \delta x_i| \cdot \Delta_i,$$

where we denoted $\delta y_i \stackrel{\text{def}}{=} y_i - a \cdot x_i - h$ and $\delta x_i \stackrel{\text{def}}{=} x_i - E_x$.

Since $h = E_y - a \cdot E_x$, we have $\dfrac{\partial h}{\partial x_i} = -\dfrac{\partial a}{\partial x_i} \cdot E_x - \dfrac{1}{n} \cdot a$, so

$$\Delta_h = \sum_{i=1}^{n} \left| \frac{\partial h}{\partial x_i} \right| \cdot \Delta_i.$$

*Prior estimation of the resulting accuracy.* The above formulas provide us with the accuracy *after* the data has been processed. It is often desirable to have an estimate *prior* to measurements, to make sure that we will get $c$ and $h$ with desired accuracy.

The difference $\delta y_i$ is a measurement error, so it is normally distributed with 0 mean and standard deviation $\sigma_y$ corresponding to the accuracy of measuring $y_i$. The difference $\delta x_i$ is distributed with 0 mean and standard deviation $\sqrt{V_x}$. For estimation purposes, it is reasonable to assume that the values $\delta x_i$ are also normally distributed. It is also reasonable to assume that the errors in $x_i$ and $y_i$ are uncorrelated, so the linear combination $\delta y_i - a \cdot \delta x_i$ is also normally distributed, with 0 mean and variance $\sigma_y^2 + a^2 \cdot V_x$. It is also reasonable to assume that all the values $\Delta_i$ are approximately the same: $\Delta_i \approx \Delta$.

For normal distribution $\xi$ with 0 mean and standard deviation $\sigma$, the mean value of $|\xi|$ is equal to $\sqrt{2/\pi} \cdot \sigma$. Thus, the absolute value $|\Delta y_i - a \cdot \Delta x_i|$ of the above combination has a mean value $\sqrt{2/\pi} \cdot \sqrt{\sigma_y^2 + a^2 \cdot V_x}$. Hence, the expected value of $\Delta_a$ is equal to

$$\frac{2}{\pi} \cdot \frac{\sqrt{\sigma_y^2 + a^2 \cdot V_x} \cdot \Delta}{V_x}.$$

Since measurements are usually more accurate than expert estimates, we have $\sigma_y^2 \ll V_x$, hence

$$\Delta_a \approx \frac{2}{\pi} \cdot a \cdot \Delta.$$

Similar estimates can be given for $\Delta_h$.

*What to do in the general case.* Linearization is not always acceptable. Sometimes, the intervals are wide, so that quadratic terms cannot be ignored.

When the linear approximation is not accurate enough, we can use the fact that, as we saw in the previous part of this book, efficient algorithms are known for computing the ranges in reasonable situations. So, we can compute the interval ranges for $C_{xy}$ and for $V_x$ and divide the resulting ranges for the desired bioinformatics-related parameters $a = c - h$ and $h$.

*Proof that finding the exact range the ratio $C_{x,y}/V_x$ is NP-hard.* Our proof is similar to the proof that computing the range for the variance is NP-hard: namely, we reduce a partition problem (known to be NP-hard) to our problem. In the partition problem, we are given $m$ positive integers $s_1, \ldots, s_m$, and we must check whether there exist values $\varepsilon_i \in \{-1, 1\}$ for which $\displaystyle\sum_{i=1}^{m} \varepsilon_i \cdot$

$s_i = 0$. We will reduce this problem to the following problem: $n = m + 2$, $y_1 = \ldots = y_m = 0$, $y_{m+1} = 1$, $y_{m+2} = -1$, $x_i = [-s_i, s_i]$ for $i \leq m$, $x_{m+1} = 1$, and $x_{m+2} = -1$. In this case, $E_y = 0$, so

$$C_{x,y} = \frac{1}{n-1} \cdot \sum_{i=1}^{n} x_i \cdot y_i - \frac{n}{n-1} \cdot E_x \cdot E_y = \frac{2}{m+2}.$$

Therefore, $C_{x,y}/V_x \to \min$ if and only if $V_x \to \max$.

Here, $V_x = \dfrac{1}{m+1} \cdot \left( \sum_{i=1}^{m} x_i^2 + 2 \right) - \dfrac{m+2}{m+1} \cdot \left( \dfrac{1}{m+2} \cdot \sum_{i=1}^{m} x_i \right)^2$. Since $|x_i| \leq$

$s_i$, we always have $V_x \leq V_0 \overset{\text{def}}{=} \dfrac{1}{m+1} \cdot \left( \sum_{i=1}^{m} s_i^2 + 2 \right)$, and the only possibility

to have $V_x = V_0$ is when $x_i = \pm s_i$ for all $i$ and $\sum x_i = 0$. Thus, $V_x = V_0$ if and only if the original partition problem has a solution. Hence, $C_{x,y}/V_x = \dfrac{2}{\sum s_i^2 + 2}$ if and only if the original instance of the partition problem has a solution.

The reduction is proven, so our problem is indeed NP-hard.

*Comment.* In this proof, we consider the case when the values $x_i$ can be negative and larger than 1, while in bioinformatics, $x_i$ is always between 0 and 1. However, we can easily modify this proof: First, we can shift all the values $x_i$ by the same constant to make them positive; shift does not change neither $C_{x,y}$ nor $V_x$. Second, to make the positive values $\leq 1$, we can then re-scale the values $x_i$ ($x_i \to \lambda \cdot x_i$), thus multiplying $C_{x,y}/V_x$ by a known constant.

As a result, we get new values $x_i' = \dfrac{1}{2} \cdot (1 + x_i/K)$, where $K \overset{\text{def}}{=} \max s_i$, for which $x_i' \in [0, 1]$ and the problem of computing $C_{x,y}/V_x$ is still NP-hard.

# 33
## Applications to Computer Science: Optimal Scheduling for Global Computing

In many practical situations, in particular in many bioinformatics problems, the amount of computations is so huge that the only way to perform these computations in reasonable time is to distribute them between multiple processors. The more processors we engage, the faster the resulting computations; thus, in addition to processor exclusively dedicated to this job, systems often use idle time on other processors. The use of these otherwise engaged processors adds additional uncertainty to computations.

How should we schedule the computational tasks so as to achieve the best utilization of the computational resources? Because of the presence of uncertainty, this scheduling problem is very difficult not only to solve but even to formalize (i.e., to describe in precise terms). In this chapter, we provide the first steps towards formalizing and solving this scheduling problem.

The main result of this chapter first appeared in [12].

## Formulation and Analysis of the Problem, and the Corresponding Results and Algorithms

*Supercomputing.* At present, most useful computations are performed on individual computers. However, there are practical problems which require orders of magnitude more computations than a regular computer can perform. To perform such computations, we need what is often called a "supercomputer".

Such problems include processing DNA data and other relevant bioinformatics data, weather prediction and climate analysis, etc. For example, in bioinformatics, one of the most time-consuming tasks is to look for known patterns in a long DNA or RNA sequence.

*Supercomputing in the past.* In this chapter, we will analyze scheduling in *global computing*. To explain the idea (and the necessity) of global computing, it is important to explain how the concept of supercomputing has evolved in the last decades.

H.T. Nguyen et al.: Computing Stat. under Interval & Fuzzy Uncertain., SCI 393, pp. 265–275.
springerlink.com

In the past, the ability to use supercomputers to simulate such things as nuclear weapons design was an important part of military confrontation. As a result, special classified technology was used to design supercomputers, technology that was not allowed on regular individual computers.

*Supercomputing at present.* Since the end of the Cold War, military restrictions no longer serve as a serious limitation to the mass-produced computer technology. As a result, the current PCs are almost as fast as specially designed computer processors.

Hence, many existing supercomputers are designed by connecting regular off-the-shelf computer processors together.

*Global computing.* Since regular computers are almost as powerful as any processor within a supercomputer, a natural idea is to use idle cycles of the regular computers to perform high-throughput computations. This idea enables us, in effect, to build a powerful supercomputer out of the existing computers – and we do not even need to own them, it is enough to use their idle cycles. Of course, due to communication time, this idea may not always work for real-time computations where we need, e.g., to predict the path of a upcoming dangerous storm. However, in many scientific computations, a communications-related delay of a day or two may be quite reasonable – as long as we do eventually perform all the necessary computations.

This idea started in the 1990s with SETI@Home, where global computing was used to process signals from radio telescopes in search for messages from extra-terrestrial intelligence. At present, this idea is actively used in mainstream research. For example, our group has developed easy-to-install web browser extension tools [43, 44, 327, 328, 365] which, in effect, enable computers to work together. The resulting networks are already being used for bioinformatics applications [326] and [329].

*One of the main problems of global computing: scheduling under uncertainty.* A serious problem in global computing is scheduling; see, e.g., [6, 129, 160, 317, 318, 323]. A similar problem occurs when we combine several processors into a single supercomputer, but there, usually, all the processors are similar, and we are in complete control of them. The corresponding scheduling problem is computationally difficult, but it is well formulated, without any serious uncertainty.

In contrast, in global computing, we are connecting computers of different types, some of which we own, some of which we don't own (so we can only use their idle cycles). We do not have an exact understanding of how the resulting collaboration affects computation time, how much time is available as idle cycles, etc. In other words, to make an efficient use of resources in global computing, we must perform scheduling under uncertainty. The existing scheduling tools for such scheduling are still imperfect [6, 129, 160, 317, 318, 323].

*We need all types of uncertainty.*

- In some cases, we have *interval uncertainty*: e.g., we may know that a certain processing step takes between 5 and 10 minutes on a given computer.

- In other cases, we have *probabilistic* uncertainty: e.g., based on the past experience, we may know the mean processing value – or we may even know the probability distribution for computation time.
- We can also have *expert* estimates for computation time, such as "the time is usually much faster than 10 minutes" which are natural to describe in *fuzzy* terms.

In scheduling, we need to take into account all these three types of uncertainty.

Let us show how the problem of selecting the optimal schedule under these types of uncertainty can be described in precise terms.

*Formalizing the main objective: first try.* The need for parallelization comes from the fact that computing the original task on a sequential machine takes too long a time.

From this viewpoint, a reasonable objective of parallelization is to minimize the overall computation time $t$.

*Formalizing the main objective: complications caused by uncertainty.* In global computing, we use idle time of otherwise engaged computers. This idle time depends on whether (and to what extent) these computers are engaged in other computations. Thus, for the same schedule, the actual computation time $t$ may differ from situation to situation.

So, we may get different computation times with different probabilities.

*Formalizing the main objective: second try.* Due to uncertainty, we cannot guarantee the *exact* value of the computation time. Moreover, with some (hopefully small) probability, the actual computation time may turn out be be very large.

If this probability is small enough, then the situation is quite tolerable: indeed, for every computer (even a dedicated one), there is always a probably of hardware failure which would make the computations impossible.

It is therefore reasonable to select a tolerable probability of failure $\varepsilon$, and to gauge each schedule by the time $t_0$ during which this schedule completes computation with probability $1 - \varepsilon$.

Then, we select the schedule for which this time $t_0$ is the smallest.

*Formalizing the main objective: additional complications caused by uncertainty.* In the idealized case when we know the probabilities of all possible engagements of different computers, we can simulate the involved network of computers and find the probability that the task will be performed in any time period $t$. In such idealized situation, we can then find the value $t_0$ for which the probability of success is $1 - \varepsilon$, and select the schedule for which the value is the smallest.

In reality, we do not have a full knowledge of the corresponding probabilities. Because of this incomplete knowledge, for a given schedule, we cannot uniquely predict the probability that under schedule, the original task will be performed in time $t$. The actual probability of success may depend on the parameters which are unknown to us.

*Formalizing the main objective: final idea.* We have mentioned that we cannot exactly predict the actual probability with which a given plan will succeed in time $t$. For different possible probability distributions, we may have different probabilities.

Our objective is to guarantee that the computations are done. Thus, a reasonable measure of the schedule's quality is the time $t_0$ by which we can *guarantee* that the computations finish with the probability $1 - \varepsilon$

*Formalizing the main objective: resulting formalization.* We want to select the schedule for which the time $t_0$ during which computations are guaranteed to finish with probability $\geq 1 - \varepsilon$.

*Need to compute the time of guaranteed completion.* In view of this objective, to select the optimal schedule, we must be able, for each schedule, to compute the time $t_0$ during which computations are guaranteed to finish with probability $\geq 1 - \varepsilon$.

Let us now describe what information we can use to compute this time, and how we can use this information.

*Need for predicting the guaranteed computation time: reminder.* We have argued that a reasonable way to select a computation schedule is to select a schedule for which the guaranteed (with probability $\geq 1 - \varepsilon$) computation time $t_0$ is the smallest. Thus, to find the optimal schedule, we must be able to compute this guaranteed computation time.

*What is a schedule.* The main idea of parallelization is that the original time-consuming task into jobs (subtasks) which can be performed independently. A schedule describes how exactly this parallelization is performed, i.e., how exactly the original task is divided into subtasks, and which processor is assigned which subtask.

*Example from bioinformatics.* As we have mentioned, in bioinformatics, one of the most time-consuming tasks is to look for known patterns in a long DNA or RNA sequence. This task can be parallelized if:

- we divide the original sequence into pieces,
- assign each piece to a different computer, and
- ask the corresponding computer to search for the desired pattern within its piece.

Of course, the pieces must overlap – otherwise this procedure may miss the pattern if it happens that this pattern is split between two neighboring pieces.

As soon as all the jobs are done, the original task is performed.

*An expression for the overall time in terms of times of subtasks.* The overall computation time of the parallelized procedure can be defined as $t = t_e - t_s$, where:

- $t_s$ is the moment when the original task was submitted by the user to the submit point, and

- $t_E$ is the moment when all the jobs (subtasks) have been performed and their results have been returned to the submit point.

For every job $i$, let $t_i$ denote the time from $t_s$ to the moment when the results of this job are returned to the submit point. The original task is done when the last of these jobs is performed, the job which takes the largest amount of time. Thus $t = \max_i t_i$, where the maximum is taken over all the jobs $i$.

For each task $i$, we need some time to send it off to a currently idle processor, process it there, and then send the results back. The overall time $t_i$ is therefore equal to the sum $\sum_j t_{ij}$ of the times of these steps. The overall computation time is thus equal to the longest of these times, i.e., to

$$t = \max_i \sum_j t_{ij}.$$

*Comment.* For bioinformatics problems, each subtask is performed on a single processor, hence each job time $t_i$ is the sum of the times $t_{ij}$ corresponding to three above-described steps. The possibility of using a single processor for each subtask is due to the fact that each subtask is reasonably short, so the probability that the auxiliary processor remains idle during these computations remains high.

In other application areas, it may not be possible to subdivide the original task into parallelizable short subtasks; the subtasks are much longer. In this case, there is a high probability that a processor would stop being idle before the subtask is completed, and the subtask will not be finished. To avoid this situation, it is reasonable to subdivide this subtask into several sequential steps, and assign each step to a different processor:

- the first processor performs the first step, then return the results to the submit point;
- these results will then be sent to the second processor, to perform the second step of the subtask, etc.

In this case, the overall time $t_i$ for computing the $i$-th job can be described by a similar formula $t_i = \sum_j t_{ij}$, but now we can have more than three steps $j$.

In view of this possibility, in the following text, we will consider the general case of possibly $> 3$ steps $j$.

*We only have partial information about the times $t_{ij}$.* We have described a formula that relates the desired computation time $t$ with the times $t_{ij}$ of performing different steps.

If we knew the exact values of $t_{ij}$, then we could use the above formula to compute the exact value of the overall computation time. If we knew the probability distribution for each of the times $t_{ij}$, then we could find the probabilities of different values of $t$; in particular, we would be able to find the probability that $t$ is below the given value $t_0$.

In reality, in most cases, we do not know the exact value $t_{ij}$, and we only partial knowledge about the corresponding probabilities.

*What types of partial information about the times $t_{ij}$ do we have?* We can safely assume that different values $t_{ij}$ are statistically independent.

- For some times $t_{ij}$, we know the probability distribution.
- For other times $t_{ij}$, we know the bounds $\underline{t}_{ij} \leq t_{ij} \leq \overline{t}_{ij}$ and the mean $e[t_{ij}]$.
- In other cases, we have fuzzy information about the bounds and means.

We would like to use this information to estimate the guaranteed computation time.

*From the computational viewpoint, it is sufficient to consider interval uncertainty.* In the fuzzy case, to describe the corresponding uncertainty about $t_{ij}$, for each value $t$ of the time $t_{ij}$, we describe the degree $\mu_{ij}(t)$ to which this value is possible.

For each degree of certainty $\alpha$, we can determine the set of values of $t_{ij}$ that are possible with at least this degree of certainty – the $\alpha$-*cut* $t_{ij}(\alpha) \overset{\text{def}}{=} \{t \,|\, \mu_{ij}(t) \geq \alpha\}$ of the original fuzzy set. In many practical cases, this $\alpha$-cut is an interval.

Vice versa, if we know $\alpha$-cuts for every $\alpha$, then, for each value $t$, we can determine the degree of possibility that $t$ belongs to the original fuzzy set for $t_{ij}$; see, e.g., [90] and [252]. A fuzzy set can be thus viewed as a nested family of its $\alpha$-cuts.

A *fuzzy number* can be defined as a fuzzy set for which all $\alpha$-cuts are intervals.

So, if instead of an interval $[\underline{t}_{ij}, \overline{t}_{ij}]$ of possible values of the time $t_{ij}$, we have a fuzzy number $\mu_{ij}(t)$ of possible values, then we can view this information as a family of nested intervals $t_{ij}(\alpha)$ ($\alpha$-cuts of the given fuzzy sets).

Our objective is then to compute the fuzzy number $t_0$ corresponding to the desired time. In this case, for each level $\alpha$, the corresponding $\alpha$-cut of the desired fuzzy number can be computed based on the $\alpha$-cuts $t_{ij}(\alpha)$ of the corresponding input fuzzy sets. The resulting nested intervals form the fuzzy number for the desired time $t_0$.

So, e.g., if we want to describe 10 different levels of uncertainty, then we must solve 10 interval computation problems. Thus, from the computational viewpoint, it is sufficient to produce an efficient algorithm for the interval case.

*Towards a mathematical formulation of the problem.* Let us observe that the function $t = \max_i \sum_j t_{ij}$ which describes the dependence of the overall computation time $t$ on the times $t_{ij}$ is non-negative and convex.

Let us recall that a function $f : R^m \to R$ is called *convex* if

$$f(\alpha \cdot x + (1 - \alpha) \cdot y) \leq \alpha \cdot f(x) + (1 - \alpha) \cdot f(y)$$

for every $x, y \in R^m$ and for every $\alpha \in (0, 1)$. It is known that the maximum of several linear functions is convex, so our function is indeed convex.

*Our objective.* We want to find the smallest possible value $t_0$ such that for all possible distributions consistent with the known information, we have $t \leq t_0$ with the probability $\geq 1 - \varepsilon$ (where $\varepsilon > 0$ is a given small probability).

*What information we can use.* We assume that different values $t_{ij}$ are statistically independent:

- About some of the variables $t_{ij}$, we know their exact statistical characteristics.
- About some other variables $t_{ij}$, we only know their interval ranges $[\underline{t}_{ij}, \overline{t}_{ij}]$ and their means $E_{ij}$.

*Additional property: the dependency is non-degenerate.* We only have partial information about the probability distribution of the variables $t_{ij}$. For each possible probability distribution $p$, we can find the largest value $t_p$ for which, for this distribution, $t \leq t_p$ with probability $\geq 1 - \varepsilon$. The desired value $t_0$ is the largest of the values $t_p$ corresponding to different probability distributions $p$: $t_0 = \sup_{p \in \mathcal{P}} t_p$, where $\mathcal{P}$ denotes the class of probability distributions $p$ which are consistent with the known information.

If we learn some additional information about the distribution of $t_{ij}$ – e.g., if we learn that $t_{ij}$ actually belongs to a proper subinterval of the original interval $[\underline{t}_{ij}, \overline{t}_{ij}]$ – we thus decrease the class $\mathcal{P}$ of distributions $p$ which are consistent with this information, to a new class $\mathcal{P}' \subset \mathcal{P}$. Since the class has decreased, the new value $t_0' = \sup_{p \in \mathcal{P}'} t_p$ is the maximum over a smaller set and thus, cannot be larger than the original value $t_0$: $t_0' \leq t_0$.

From the purely mathematical viewpoint, it is, in principle, possible that the desired value $t_0$ does not actually depend on some of the variables $t_{ij}$. In this case, if we narrow down the interval of possible values of the corresponding variable $t_{ij}$, this will not change the resulting value $t_0$.

In our problem, however, it is reasonable to assume that the dependence of $t_0$ on $t_{ij}$ is *non-degenerate* in the sense that every time we narrow down one of the intervals $[\underline{t}_{ij}, \overline{t}_{ij}]$, the resulting value $t_0$ actually decreases: $t_0' < t_0$.

As a result, we arrive at the following problem.

*Formulation of the problem and the main result.*

GIVEN:

- a finite set of $M$ pairs of integers $(i, j)$, and its subset $F$;
- a real number $\varepsilon > 0$;
- a convex non-negative function

$$t = F(t_{11}, t_{12}, \ldots);$$

- probability distributions for variables $t_{ij}$ with $(i,j) \in F$ – e.g., given in the form of cumulative distribution function (cdf) $F_{ij}(t)$;
- intervals $\mathbf{t}_{ij} = [\underline{t}_{ij}, \overline{t}_{ij}]$ and values $E_{ij}$ corresponding to $(i,j) \notin F$.

TAKE: all possible joint probability distributions on $R^M$ for which:

- all $N$ random variables $t_{ij}$ are independent;
- for all $(i,j) \in F$, the variable $t_{ij}$ has a given distribution $F_{ij}(t)$;
- for each $(i,j) \notin F$, $t_{ij} \in \mathbf{t}_{ij}$ with probability 1 and the mean value of $t_{ij}$ is equal to $E_{ij}$.

FIND: the smallest possible value $t_0$ such that for all possible distributions consistent with the known information, we have $t \stackrel{\text{def}}{=} F(t_{11}, t_{12}, \ldots) \le t_0$ with probability $\ge 1 - \varepsilon$.

PROVIDED: that the problem is *non-degenerate* in the sense that if we narrow down one of the intervals $\mathbf{t}_{ij}$, the value $t_0$ decreases.

The following result explains how we can compute this value $t_0$.

**Proposition 33.1.** *The desired value $t_0$ is attained when for each $(i,j) \notin F$, we use a 2-point distribution for $t_{ij}$, in which:*

- $t_{ij} = \underline{t}_{ij}$ *with probability* $\underline{p}_{ij} \stackrel{\text{def}}{=} \dfrac{\overline{t}_{ij} - E_{ij}}{\overline{t}_{ij} - \underline{t}_{ij}}$.
- $t_{ij} = \overline{t}_{ij}$ *with probability* $\overline{p}_{ij} \stackrel{\text{def}}{=} \dfrac{E_{ij} - \underline{t}_{ij}}{\overline{t}_{ij} - \underline{t}_{ij}}$.

*Comment.* This proposition was first proven in [269] and [270] for a different computer-related application – to chip design; see Chapter 37.

*Resulting algorithm for computing $t_0$.* Because of the above Proposition, we can compute the desired value $t_0$ by using the following Monte-Carlo simulation:

- We set each value $t_{ij}$, $(i,j) \notin F$, to be equal:
  - to $\overline{t}_{ij}$ with probability $\overline{p}_{ij}$ and
  - to the value $\underline{t}_{ij}$ with the probability $\underline{t}_{ij}$.
- We simulate the values $t_{ij}$, $(i,j) \in F$, as random variables distributed according to the distributions $F_{ij}(x)$.
- For each simulation $s$, $1 \le s \le N_i$, we get the simulated values $t_{ij}^{(s)}$, and then, a value $t^{(s)} = F(t_{11}^{(s)}, t_{12}^{(s)}, \ldots)$. We then sort the resulting $N_i$ values $t^{(s)}$ into an increasing sequence

$$t_{(1)} \le t_{(2)} \le \ldots \le t_{(N_i)},$$

and take, as $t_0$, the $N_i \cdot (1 - \varepsilon)$-th term $t_{(N_i \cdot (1-\varepsilon))}$ in this sorted sequence.

# Proofs

*Proof of Proposition 33.1*

$1°$. By definition, $t_0$ is the largest value of $t_p$ over all possible distributions $p \in \mathcal{P}$. This means that for the given $t_0$, for all possible distributions $p \in \mathcal{P}$, we have $\mathrm{Prob}(t \le t_0) \ge 1 - \varepsilon$. Let $p \in \mathcal{P}$ be the "worst-case" distribution, i.e., the distribution for which the probability $\mathrm{Prob}(t \le t_0)$ is the smallest. Let us show that this "worst case" occurs when all variables $t_{ij}$ with $(i, j) \notin F$ have the 2-point distributions described in the Proposition.

$2°$. Let us fix a pair $(i_0, j_0) \notin F$ and show that in the "worst case", $t_{i_0 j_0}$ indeed has the desired 2-point distribution.

Let us fix the distributions for all other $t_{ij}$, $(i, j) \notin F$ as in the worst case. Then, the fact that the probability $\mathrm{Prob}(t \le t_0)$ is the smallest means that if we replace the worst-case distribution for $t_{i_0 j_0}$ with some other distribution, we can only increase this probability. In other words, when we correspondingly fix the distributions for $t_{ij}$, $(i, j) \ne (i_0, j_0)$, the probability $\mathrm{Prob}(t \le t_0)$ attains the smallest possible value at the desired distribution for $t_{i_0 j_0}$.

$3°$. The distribution for $t_{i_0 j_0}$ is located on an interval $\boldsymbol{t}_{i_0 j_0} = [\underline{t}_{i_0 j_0}, \overline{t}_{i_0 j_0}]$, i.e., on a set with infinitely many points. However, with an arbitrary large value $N$ (and thus, for an arbitrarily small discretization error $\delta = (\overline{t}_{i_0 j_0} - \underline{t}_{i_0 j_0})/N$), we can assume that all the distributions are located on a finite grid of values

$$v_0 \stackrel{\mathrm{def}}{=} \underline{t}_{i_0 j_0}, \quad v_1 \stackrel{\mathrm{def}}{=} \underline{t}_{i_0 j_0} + \delta, \quad v_2 \stackrel{\mathrm{def}}{=} \underline{t}_{i_0 j_0} + 2\delta, \ldots, v_N = \overline{t}_{i_0 j_0}.$$

The smaller $\delta$, the better this approximation. Thus, without losing generality, we can assume that the distribution of $t_{i_0 j_0}$ is located on finitely many points $v_k$.

$4°$. In this approximation, the probability distribution for $t_{i_0 j_0}$ can be described by the probabilities $q_k \stackrel{\mathrm{def}}{=} p_{i_0 j_0}(v_k)$ of different values $v_k$.

$5°$. The minimized probability $\mathrm{Prob}(t \le t_0)$ can be described as the sum of the probabilities of different combinations of $t_{ij}$ over all the combinations for which $t = F(t_{11}, t_{12}, \ldots) \le t_0$. We assumed that all the variables $t_{ij}$ are independent. Thus, the probability of each combination of $t_{ij}$ is equal to the product of the corresponding probabilities $p_{11}(t_{11}) \cdot p_{12}(t_{12}) \cdot \ldots$ Since the probability distributions for $t_{ij}$, $(i, j) \ne (i_0, j_0)$, are fixed, the minimized probability is thus a linear combination of probabilities $p_{i_0 j_0}(v_k)$, i.e., of the probabilities $q_k$. In other words, the minimized probability has the form $\sum_{k=0}^{N} c_k \cdot q_k$ for some coefficients $c_k$.

$6°$. By describing the probability distribution on $t_{i_0 j_0}$ via the probabilities $q_k = p_{i_0 j_0}(v_k)$ of different values $v_k \in [\underline{t}_{ij}, \overline{t}_{ij}]$, we automatically restrict ourselves to distributions which are located on this interval. The only restrictions that we have on the probability distribution of $t_{i_0 j_0}$ is that:

- it is a probability distribution, i.e., $q_k \geq 0$ for all $k$ and $\sum_{k=0}^{N} q_k = 1$, and

- the mean value of this distribution is equal to $E_{i_0 j_0}$, i.e., $\sum_{k=0}^{N} q_k \cdot v_k = E_{i_0 j_0}$.

Thus, the worst-case distribution for $t_{i_0 j_0}$ is a solution to the following linear programming problem:

Minimize

$$\sum_{k=0}^{N} c_k \cdot q_k$$

under the constraints

$$\sum_{k=0}^{N} q_k = 1,$$

$$\sum_{k=0}^{N} q_k \cdot v_k = E_{i_0 j_0},$$

$$q_k \geq 0, \quad k = 0, 1, 2, \ldots, N.$$

$7^\circ$. It is known that the solution to a linear programming problem is always attained at a vertex of the corresponding constraint set.

In other words, in the solution to the linear programming problem with $N+1$ unknowns $q_0, q_1, \ldots, q_N$, at least $N+1$ constraints are equalities. Since we already have 2 equality constraints, this means that out of the remaining constraints $q_k \geq 0$, at least $N-1$ are equalities. In other words, this means that in the optimal distribution, all but two values of $q_k = p_{i_0 j_0}(v_k)$ are equal to 0.

Thus, the "worst-case" distribution for $t_{i_0 j_0}$ is located on 2 points $v$ and $v'$ within the interval $[\underline{t}_{i_0 j_0}, \overline{t}_{i_0 j_0}]$.

$8^\circ$. Let us prove, by reduction to a contradiction, that these two points cannot be different from the endpoints of this interval.

Indeed, let us assume that they are different. Without losing generality, we can assume that $v \leq v'$. Then, this "worst-case" distribution is actually located on the proper subinterval $[v, v'] \subset [\underline{t}_{i_0 j_0}, \overline{t}_{i_0 j_0}]$ of the original interval $t_{i_0 j_0}$.

Since the maximum $t_0$ of $t_p$ is attained on this distribution, replacing the original interval $t_{i_0 j_0}$ with its proper subinterval $[v, v']$ would not change the value $t_0$ – while our assumption of non-degeneracy states that such a replacement would always lead to a smaller value $t_0$. This contradiction shows that the values $v$ and $v'$ – on which the worst-case distribution is located – have to be endpoints of the interval $[\underline{t}_{i_0 j_0}, \overline{t}_{i_0 j_0}]$.

$9°$. In other words, we conclude that the worst-case distribution is located at 2 points: $\underline{t}_{i_0 j_0}$ and $\overline{t}_{i_0 j_0}$.

Such a distribution is uniquely determined by the probabilities $\underline{p}_{i_0 j_0}$ and $\overline{p}_{i_0 j_0}$ of these two points. Since the sum of these probabilities is equal to 1, it is sufficient to describe one of these probabilities, e.g., $\overline{p}_{i_0 j_0}$; then, $\underline{p}_{i_0 j_0} = 1 - \overline{p}_{i_0 j_0}$. The condition that the mean of $t_{i_0 j_0}$ is $E_{i_0 j_0}$, i.e., that

$$\underline{p}_{i_0 j_0} \cdot \underline{t}_{i_0 j_0} + \overline{p}_{i_0 j_0} \cdot \overline{t}_{i_0 j_0} = (1 - \overline{p}_{i_0 j_0}) \cdot \underline{t}_{i_0 j_0} + \overline{p}_{i_0 j_0} \cdot \overline{t}_{i_0 j_0} = E_{i_0 j_0},$$

uniquely determines $\overline{p}_{i_0 j_0}$ (and hence $\underline{p}_{i_0 j_0}$) – exactly by the expression from the Proposition.

The Proposition is proven.

# Applications to Information Management: How to Estimate Degree of Trust

In this chapter, we use the probabilistic and interval uncertainty to estimate the degree of trust in an agent. Some of these results first appeared in [56, 141].

## Results and Algorithms

In the traditional approach to trust, we either trust an agent or not. If we trust an agent, we allow this agent full access to a particular task. For example, I trust my bank to handle my account; the bank (my agent) outsources money operations to another company (sub-agent), etc. The problem with this approach is that I may have only 99.9% trust in bank, bank in its contractor, etc. As a result, for long chains, the probability of a security leak may increase beyond any given threshold. To resolve this problem, we must keep track of trust probabilities.

Let us describe this idea in precise terms. We have a finite set $A$; its elements are called *agents*. For some pairs $(a, b)$ of agents, we know that an agent $a$ has some degree of direct trust in an agent $b$. We will denote the set of all such pairs by $E$. For each pair $(a, b) \in E$, we know the probability $p_0(a, b)$ with which $a$ directly trusts $b$. We can estimate this probability, e.g., as a frequency in which a similar trust was justified.

Our objective is to describe, for given two agents $f$ and $s$, the resulting probability $p_t(f, s)$ with which the agent $f$ should trust the agent $s$.

Let us show how this problem can be described in precise terms. The pair $(A, E)$, where $E \subseteq A \times A$, forms a *graph* in which agents are nodes and possible trust pairs are edges. To each edge $(a, b)$, we associate a value $p_0(a, b) \in [0, 1]$.

Some of the trusts may be misplaced: an agent $a$ may trust an agent $b$ with a certain probability, but $b$ may be misusing $a$'s trust. Let $E' \subseteq E$ denote the (unknown) set of pairs in which the trust is justified; we will call this set the *actual trust set*.

We do not know for sure who is trustworthy and who is not, so at best, we can find some information about the probabilities $p(E')$ of different trust

H.T. Nguyen et al.: Computing Stat. under Interval & Fuzzy Uncertain., SCI 393, pp. 277–281.
springerlink.com

sets $E'$. First, these probabilities must add up to 1: $\sum_{E'} p(E') = 1$. Second, for every pair $(a, b) \in E$, the probability that $a$ directly trusts $b$, i.e., the probability that the edge $(a, b)$ belongs to the actual trust set $E'$, should be equal to $p_0(a, b)$: $\sum_{E':(a,b) \in E'} p(E') = p_0(a, b)$.

Once the probability distribution $p(E')$ is fixed, we can determine the probability $p_t(f, s)$ with which $f$ should trust $s$ as the probability that in the actual trust set $E'$, there is a path leading from $f$ to $s$. If we denote the existence of such a path by $f \xrightarrow{E'} s$, then the desired probability $p_t(f, s)$ can be described as $\sum_{E':f \xrightarrow{E'} s} p(E')$.

We may have different probability distributions $p(E')$ which are consistent with the data $p_0(a, b)$; for different distributions, we may have different values of the trust $p_t(f, s)$. In security situations, it is desirable to find the *guaranteed* level of trust, i.e., the smallest possible value of $p_t(f, s)$ over all possible probability distributions which are consistent with the data $p_0(a, b)$. We will denote this smallest possible value by $\underline{p}_t(f, s)$; in these terms, our objective is to compute $\underline{p}_t(f, s)$.

This problem can be viewed as a particular case of the general problem of dealing with probabilities in expert systems. Indeed, here, for every agent $a$, we have a statement "$a$" meaning that $f$ trusts $a$. We have a fact $f \rightarrow$ meaning that $f$ trust himself. For each edge $(a, b) \in E$ (meaning that $a$ trusts $b$), we have a rule $b \leftarrow a$ (meaning that if $f$ trusts $a$, he should also trust $b$), which holds with probability $p_0(a, b)$. The query is "$s$?" – i.e., with what probability should we trust $s$.

Let us show that in this particular problem, we can efficiently compute the desired probability. Namely, let us define the *length* ("distrust") of an edge as $d_0(a, b) \overset{\text{def}}{=} 1 - p_0(a, b)$. We can naturally extend this definition to *paths*, i.e., sequences $(a_0, \ldots, a_n)$ in which $(a_i, a_{i+1}) \in E$ for all $i$. Namely, the *length* $\ell(\gamma)$ of a path $\gamma = (a_0, \ldots, a_n)$ is defined as usual: $\ell(\gamma) \overset{\text{def}}{=} \sum_{i=0}^{n-1} d_0(a_i, a_{i+1})$. The length of the shortest path from $f$ to $s$ is defined as follows:

$$\underline{d}_t(f, s) \overset{\text{def}}{=} \min\{\ell(\gamma) \mid \gamma \text{ is a path from } f \text{ to } s\}.$$

**Proposition.** $\underline{p}_t(f, s) = \max(1 - \underline{d}_t(f, s), 0)$.

So, we can use Dijkstra's algorithm (see, e.g., [73]) to find the shortest path in a graph, and then compute $\underline{p}_t(f, s)$.


## Proof


$1°$. Let us first show that for every distribution $p(E')$ that is consistent with the given information, we have $d_t(f, s) \leq \underline{d}_t(f, s)$, where $d_t(f, s) \overset{\text{def}}{=} 1 - p_t(f, s)$.

Let $\gamma_0 = (a_0, a_1, \ldots, a_n)$ be the shortest path from $a_0 = f$ to $a_n = s$. Then, by definition of the function $\underline{d}_t(f, s)$, we have

$$\underline{d}_t(f, s) = d_0(a_0, a_1) + \ldots + d_0(a_{n-1}, a_n), \tag{9}$$

where $(a_i, a_{i+1}) \in E$ for all $i$. If all the connections $(a_i, a_{i+1})$ are in the random graph $E'$, then in $E'$, there exists a path from $f = a_0$ to $a_n = s$ – namely, the path $\gamma_0$. So, if there is no path from $s$ to $f$, then at least one of the connections $(a_i, a_{i+1})$ is not present in the random graph.

Let us denote:

- the event that there is no path from $f$ to $s$ by $N_t(f, s)$, and
- the event that there is no direct connection from $a_i$ to $a_{i+1}$ by $N_0(a_i, a_{i+1})$.

Then:

- the probability of $N_t(f, s)$ is equal to $d_t(f, s)$, while
- the probability of each event $N_0(a_i, a_{i+1})$ is equal to $d_0(a_i, a_{i+1})$.

In these terms, the above logical conclusion takes the following form: if the event $N_t(f, s)$ occurs, then at least one of the events $N_0(a_i, a_{i+1})$ must have occurred:

$$N_t(f, s) \supset (N_0(a_0, a_1) \vee \ldots \vee N_0(a_{n-1}, a_n)).$$

Therefore, the probability $d_t(f, s)$ of the event $N_t(f, s)$ cannot exceed the probability of the disjunction:

$$d_t(f, s) \leq P(N_0(a_0, a_1) \vee \ldots \vee N_0(a_{n-1}, a_n)).$$

It is well known that the probability of a disjunction

$$S_1 \vee \ldots \vee S_n$$

of arbitrary $n$ events $S_1, \ldots, S_n$ cannot exceed the sum of the corresponding probabilities $p(S_1) + \ldots + p(S_n)$. In our case, this means that $d_t(f, s)$ cannot exceed the sum of the probabilities $p(N_0(a_0, a_1)) + \ldots + p(N_0(a_{n-1}, a_n))$, i.e., that $d_t(f, s) \leq d_0(a_0, a_1) + \ldots + d_0(a_{n-1}, a_n)$.

Due to (9), this means that $d_t(f, s) \leq \underline{d}_t(f, s)$.

$2°$. From $d_t(f, s) = 1 - p_t(f, s) \leq \underline{d}_t(f, s)$, it follow that $p_t(f, s) \geq 1 - \underline{d}_t(f, s)$. Since $p_t(f, s)$ is a probability, it is a non-negative number, so

$$p_t(f, s) \geq \max(1 - \underline{d}_t(f, s), 0).$$

In other words, for every distribution $p(E')$ that is consistent with the given information, the corresponding probability $p_t(f, s)$ is larger than or equal to $\max(1 - \underline{d}_t(f, s), 0)$. Thus, the infimum $\underline{p}_t(f, s)$ of all such values $p_t(f, s)$ is also smaller than or equal to this number, i.e.,

$$\underline{p}_t(f, s) \geq \max(1 - \underline{d}_t(f, s), 0).$$

$3°$. To complete the proof, we will produce an example of a probability distribution $p(E')$ for which $p_t(f,s) \leq \max(1 - \underline{d}_t(f,s), 0)$. Then, for $\underline{p}_t(f,s) \leq p_t(f,s)$, we will get $\underline{p}_t(f,s) \leq \max(1 - \underline{d}_t(f,s), 0)$, hence

$$\underline{p}_t(f,s) = \max(1 - \underline{d}_t(f,s), 0).$$

To describe the corresponding distribution, we will use the standard projection $\pi$ of the real line $\mathbb{R}$ onto the interval $[0,1)$ that assigns to each real number $x$ its fractional part $\pi(x) \stackrel{\text{def}}{=} x - \lfloor x \rfloor$. This projection has a simple geometric interpretation: we interpret the interval $[0,1)$ as a circle of circumference 1, and we "wrap up" the real line around this circle.

It is easy to see that in this wrapping, the length of an interval is preserved as long as it does not exceed 1. Thus, for any interval $I = [x,y] \subseteq \mathbb{R}$ of length $y - x \leq 1$, its projection $\pi(I)$ is either an interval, or a pair of intervals, and the total length of the set $\pi(I)$ is equal to the length $y - x$ of the original interval.

The corresponding distribution is located on the graphs $E(\omega)$ corresponding to different real numbers $\omega \in [0,1)$; these graphs will be describe below. The probability of different graphs $E(\omega)$ is described by the uniform distribution on the interval $[0,1)$.

The graphs $E(\omega)$ are described as follows. For every two nodes $a$ and $b$ for which $(a,b) \in E$, we consider the interval $I(a,b) \stackrel{\text{def}}{=} [\underline{d}_t(f,a), \underline{d}_t(f,a) + d_0(a,b)]$ of length $d_0(a,b)$. For every $\omega \in [0,1)$, the edge $(a,b)$ belongs to the graph $E(\omega)$ if and only if $\omega \notin \pi(I(a,b))$.

$3.1°$. Let us prove that thus defined distribution is indeed consistent with the original information, i.e., with all the given values $p_0(a,b)$.

Indeed, since the distribution on $\omega$ is uniform, the probability that $\omega \in \pi(I(a,b))$ is equal to the total length of the set $\pi(I(a,b))$. Due to the above property of the projection, this total length is equal to the length of the original interval $I(a,b)$, i.e., to $d_0(a,b)$. Thus, the probability that $\omega \in \pi(I(a,b))$ is equal to $d_0(a,b)$. Therefore, the probability of the opposite event $\omega \notin \pi(I(a,b))$ is equal to $1 - d_0(a,b)$. By definition of $d_0(a,b)$, the value $1 - d_0(a,b)$ is exactly $p_0(a,b)$.

So, the probability that the edge $(a,b)$ belongs to the graph $E(\omega)$ is exactly $p_0(a,b)$.

$3.2°$. Let us prove that for every path $\gamma = (a_0, \ldots, a_n)$ that starts at $a_0 = f$, if all the edges $(a_0, a_1), \ldots, (a_{n-1}, a_n)$ from this path belong to the graph $E(\omega)$, then $\omega \geq \underline{d}_t(a_0, a_n)$.

We will prove this statement by induction over the length $n$ of the path.

$3.2.1°$. *Induction base.* For $n = 0$, we have $\underline{d}_t(a_0, a_0) = 0$, so the desired inequality is clearly true.

$3.2.2°$. *Induction step.* Let us assume that the desired property is true for $n$, and we have added one more edge $(a_n, a_{n+1})$ to the path $\gamma$. Let us prove that the desired property holds for the extended path.

If all $n + 1$ edges $(a_0, a_1), \ldots, (a_{n-1}, a_n), (a_n, a_{n+1})$ belong to the graph $E(\omega)$, then the first $n$ edges also belong to the graph. By the induction assumption, this means that $\omega \geq \underline{d}_t(f, a_n)$.

If $\underline{d}_t(f, a_n) \geq 1$, then – since $\omega$ is from the interval $[0, 1)$ – this inequality means that no such $\omega$ exists; so, the graph $E(\omega)$ simply cannot contain all the edges from the path $\gamma$. In this case, the statement that we are trying to prove is trivially true for an enlarged path – because false implies everything.

It is therefore sufficient to only consider the case when $\underline{d}_t(a_0, a_n) < 1$. In this case, by definition of $E(\omega)$, the fact that $(a_n, a_{n+1}) \in E$ means that

$$\omega \notin \pi([\underline{d}_t(f, a_n), \underline{d}_t(f, a_n) + d_0(a_n, a_{n+1})]).$$

Since $\underline{d}_t(f, a_n) < 1$, the projection $\pi(\ldots)$ starts with the value $\underline{d}_t(f, a_n)$. Whether this projection is a single interval or two intervals depends on whether $\underline{d}_t(f, a_n) + d_0(a_n, a_{n+1}) < 1$ or not. Let us consider both possibilities:

- If $\underline{d}_t(f, a_n) + d_0(a_n, a_{n+1}) \geq 1$, then the projection contains all the values from $\underline{d}_t(f, a_n)$ to 1. Since $\omega \geq \underline{d}_t(f, a_n)$ and $\omega$ cannot belong to this projection, we conclude that no such $\omega$ is possible, so the desired property is trivially true.
- If $\underline{d}_t(f, a_n) + d_0(a_n, a_{n+1}) < 1$, then the projection coincides with the original interval

$$[\underline{d}_t(f, a_n), \underline{d}_t(f, a_n) + d_0(a_n, a_{n+1})].$$

Since $\omega \geq \underline{d}_t(f, a_n)$ and $\omega$ cannot belong to this projection, we conclude that

$$\omega \geq \underline{d}_t(f, a_n) + d_0(a_n, a_{n+1}).$$

Let us continue the analysis of the second case. By definition, $\underline{d}_t(f, a_{n+1})$ is the length of the shortest path from $f$ to $a_{n+1}$. In particular, when we add the edge $(a_n, a_{n+1})$ to the shortest path from $f$ to $a_n$, we conclude that

$$\underline{d}_t(f, a_{n+1}) \leq \underline{d}_t(f, a_n) + d_0(a_n, a_{n+1}).$$

Hence, the inequality $\omega \geq \underline{d}_t(f, a_n) + d_0(a_n, a_{n+1})$ implies that

$$\omega \geq \underline{d}_t(f, a_{n+1}).$$

In both cases, the induction step is proven, and so is the result.

$3.3°$. Due to Part 3.2 of this proof, if there is a path from $f$ to $s$ in a graph $E(\omega)$, then $\omega \geq \underline{p}_t(f, s)$. Thus, the probability $p_t(f, s)$ that there exists such a path does not exceed the probability that a uniformly distributed number $\omega \in [0, 1)$ is $\geq \underline{p}_t(f, s)$. In other words,

$$p_t(f, s) \leq \max(1 - \underline{p}_t(f, s), 0).$$

The statement is proven, so the above algorithm has been justified.

# Applications to Information Management: How to Measure Loss of Privacy

In this chapter, we use the experience of measuring a degree of mismatch between probability models, p-boxes, etc., described in Chapter 30, to measure loss of privacy. Some of our privacy-related results first appeared in [58].

## Formulation and Analysis of the Problem, and the Corresponding Results

*Measuring loss of privacy is important.* Privacy means, in particular, that we do not disclose all information about ourselves. If some of the originally undisclosed information is disclosed, some privacy is lost. To compare different privacy protection schemes, we must be able to gauge the resulting loss of privacy.

*Seemingly natural idea: measuring loss of privacy by the acquired amount of information.* Since privacy means that we do not have complete information about a person, a seemingly natural idea is to gauge the loss of privacy by the amount of new information that we gained about this person.

*Why information is not always a perfect measure of loss of privacy.* In our opinion, the amount of new information is not always a good measure of the loss of privacy because it does not distinguish between:

- crucial information that may seriously affect a person, and
- irrelevant information – that may not affect a person at all.

To make a distinction between these two types of information, let us estimate potential financial losses caused by the loss of privacy.

*Example when loss of privacy can lead to a financial loss.* As an example, let us consider how a person's blood pressure $x$ affects the premium that this person pays for his or her health insurance.

From the previous experience, insurance companies can deduce, for each value of blood pressure $x$, the expected (average) value of the medical expenses $f(x)$ of all individuals with this particular value of blood pressure.

H.T. Nguyen et al.: Computing Stat. under Interval & Fuzzy Uncertain., SCI 393, pp. 283–287.
springerlink.com 

So, when the insurance company knows the exact value $x$ of a person's blood pressure, it can offer this person an insurance rate $F(x) \stackrel{\text{def}}{=} f(x) \cdot (1 + \alpha)$, where $\alpha$ is the general investment profit. Indeed:

- If an insurance company offers higher rates, then its competitor will be able to offer lower rates and still make a profit.
- On the other hand, if the insurance company is selling insurance at a lower rate, then it will not earn enough profit, and investors will pull their money out and invest somewhere else.

To preserve privacy, we only keep the information that the blood pressure of all individuals from a certain group is between two bounds $L$ and $U$, and we do not know have any additional information about the blood pressure of different individuals. Under this information, how much will the insurance company charge to insure people from this group?

Based on the past experience, the insurance company is able to deduce the relative frequency of different values $x \in [L, U]$, e.g., in the form of the corresponding probability density $\rho(x)$. In this case, the expected medical expenses of an average person from this group are equal to

$$E[f(x)] \stackrel{\text{def}}{=} \int \rho(x) \cdot f(x) \, dx.$$

Thus, the insurance company will insure the person for a cost of

$$E[F(x)] = \int \rho(x) \cdot F(x) \, dx.$$

Let us now assume that for some individual, the privacy is lost, and for this individual, we know the exact value $x_0$ of his or her blood pressure. For this individual, the company can now better predict its medical expenses as $f(x_0)$ and thus, offer a new rate $F(x_0) = f(x_0) \cdot (1 + \alpha)$. When

$$F(x_0) > E[F(x)],$$

the person whose privacy is lost also experiences a financial loss

$$F(x_0) - E[F(x)].$$

We will use this financial loss to gauge the loss of privacy.

*Need for a worst-case comparison.* In the above example, there is a financial loss only if the person's blood pressure $x_0$ is worse than average. A person whose blood pressure is lower than average will only benefit from reduced insurance rates.

However, in a somewhat different situation, if the person's blood pressure is smaller (better) than average, this person's loss or privacy can also lead to a financial loss. For example, an insurance company may, in general, pay for a preventive medication that lowers the risk of heart attacks – and of the resulting huge medical expenses. The higher the blood pressure, the larger the risk of a heart attack. So, if the insurance company learns that a certain individual has a lower-than-average blood pressure and thus, a lower-than-average risk of a heart attack, this risk may not justify the expenses on the preventive medication. Thus, due to a privacy loss, the individual will have to pay for this potentially beneficial medication from his/her own pocket – and thus, also experience a financial loss.

So, to gauge a privacy loss, we must consider not just a single situation, but several different situations, and gauge the loss of privacy by the worst-case financial loss caused by this loss of privacy.

*Which functions $F(x)$ should we consider.* Similarly to Chapter 30, we should consider functions $F(x)$ for which $|F'(x)| \leq M$ for some given number $M > 0$.

*Resulting definitions.* Thus, we arrive at the following definition:

**Definition 35.1.** *Let $P$ be a class of probability distributions on a real line, and let $M > 0$ be a real number. By the* amount of privacy $A(P)$ *related to $P$, we mean the largest possible value of the difference $F(x_0) - \int \rho(x) \cdot F(x)\, dx$ over:*

- *all possible values $x_0$,*
- *all possible probability distributions $\rho \in P$, and*
- *all possible functions $F(x)$ for which $|F'(x)| \leq M$ for all $x$.*

The above definition involves taking a maximum over all distributions $\rho \in P$ which are consistent with the known information about the group to which a given individual belongs. In some cases, we know the exact probability distribution, so the family $P$ consists of only one distribution. In other situations, we may not know this distribution. For example, we may only know that the value of $x$ is within the interval $[L, U]$, and we do not know the probabilities of different values within this interval. In this case, the class $P$ consists of all distributions which are located on this interval (with probability 1).

When we learn new information about this individual, we thus reduce the group and hence, change from the original class $P$ to a new class $Q$. This change, in general, decreases the amount of privacy.

In particular, when we learn the exact value $x_0$ of the parameter, then the resulting class of distribution reduces to a single distribution concentrated on this $x_0$ with probability 1 – for which $F(x_0) - \int \rho(x) \cdot F(x)\, dx = 0$ and thus, the privacy is 0. In this case, we have a 100% loss of privacy – from the original value $A(P)$ to 0. In other cases, we may have a partial loss of privacy.

In general, it is reasonable to define the *relative loss of privacy* as a ratio

$$\frac{A(P) - A(Q)}{A(P)}. \tag{35.1}$$

In other words, it is reasonable to use the following definition:

**Definition 35.2**

- *By a privacy loss, we mean a pair $\langle P, Q \rangle$ of classes of probability distributions.*
- *For each privacy loss $\langle P, Q \rangle$, by the* measure of a privacy loss, *we mean the ratio (35.1).*

*Comment.* At first glance, it may sound as if these definitions depend on an (unknown) value of the parameter $M$. However, it is easy to see that the actual measure of the privacy loss does not depend on $M$:

**Proposition 35.1.** *For each pair $\langle P, Q \rangle$, the measure of the privacy loss is the same for all $M > 0$.*

*The new definition of privacy loss is in good agreement with intuition.* Let us show that the new definition adequately describes the difference between learning that the parameter is in the lower half of the original interval and that the parameter is even.

**Proposition 35.2.** *Let $[l, u] \subseteq [L, U]$ be intervals, let $P$ be the class of all probability distributions located on the interval $[L, U]$, and let $Q$ be the class of all probability distributions located on the interval $[l, u]$. For this pair $\langle P, Q \rangle$, the measure of the privacy loss is equal to $1 - \dfrac{u - l}{U - L}$.*

*Comment.* In particular, if we start with an interval $[L, U]$, and then we learn that the actual value $x$ is in the lower half $[L, (L+U)/2]$ of this interval, then we get a 50% privacy loss.

What about the case when we assume that $x$ is even? Similarly to the proof of Proposition 35.2, one can prove that if both $L$ and $U$ are even, and $Q$ is the class of all distributions $\rho(x)$ which are located, with probability 1, on even values $x$, we get $A(Q) = A(P)$. Thus, the even-values restriction lead to a 0% privacy loss.

Thus, the new definition of the privacy loss is indeed in good agreement with our intuition.

## Proofs

*Proof of Proposition 35.1.* To prove this proposition, it is sufficient to show that for each $M > 0$, the measure of privacy loss is the same for this $M$ and for $M_0 = 1$. Indeed, for each function $F(x)$ for which $|F'(x)| \leq M$ for all $x$, for the re-scaled function $F_0(x) \stackrel{\text{def}}{=} F(x)/M$, we have $|F_0'(x)| \leq 1$ for all $x$, and

$$F(x_0) - \int \rho(x) \cdot F(x)\, dx =$$

$$M \cdot \left( F_0(x_0) - \int \rho(x) \cdot F_0(x)\, dx \right). \tag{35.2}$$

Vice versa, if $|F_0'(x)| \leq 1$ for all $x$, for the re-scaled function $F(x) \stackrel{\text{def}}{=} M \cdot F_0(x)$, we have $|F'(x)| \leq M$ for all $x$, and (35.2). Thus, the maximized values corresponding to $M$ and $M_0 = 1$ different by a factor $M$. Hence, the resulting amounts of privacy $A(P)$ and $A_0(P)$ corresponding to $M$ and $M_0$ also differ by a factor $M$: $A(P) = M \cdot A_0(P)$. Substituting this expression for $A(P)$ (and a similar expression for $A(Q)$) into the definition (35.1), we can therefore conclude that $\dfrac{A(P) - A(Q)}{A(P)} = \dfrac{A_0(P) - A_0(Q)}{A_0(P)}$, i.e., that the measure of privacy is indeed the same for $M$ and $M_0 = 1$. The proposition is proven.

*Proof of Proposition 35.2.* Due to Proposition 35.1, for computing the measure of the privacy loss, it is sufficient consider the case $M = 1$. Let us show that for this $M$, we have $A(P) = U - L$.

Let us first show that for every $x_0 \in [L, U]$, for every probability distribution $\rho(x)$ on the interval $[L, U]$, and for every function $F(x)$ for which $|F'(x)| \leq 1$, the privacy loss $F(x_0) - \int \rho(x) \cdot F(x)\, dx$ does not exceed $U - L$.

Indeed, since $\int \rho(x)\, dx = 1$, we have

$$F(x_0) = \int \rho(x) \cdot F(x_0)\, dx$$

and hence,

$$F(x_0) - \int \rho(x) \cdot F(x)\, dx = \int \rho(x)\, (F(x_0) - F(x))\, dx.$$

Since $|F'(x)| \leq 1$, we conclude that

$$|F(x_0) - F(x)| \leq |x_0 - x|.$$

Both $x_0$ and $x$ are within the interval $[L, U]$, hence $|x_0 - x| \leq U - L$, and $|F(x_0) - F(x)| \leq U - L$. Thus, the average value $\int \rho(x) \cdot (F(x_0) - F(x))\, dx$ of this difference also cannot exceed $U - L$.

Let us now show that there exists a value $x_0 \in [L, U]$, a probability distribution $\rho(x)$ on the interval $[L, U]$, and a function $F(x)$ for which $|F'(x)| \leq 1$, for which the privacy loss $F(x_0) - \int \rho(x) \cdot F(x)\, dx$ is exactly $U - L$. As such an example, we take $F(x) = x$, $x_0 = U$, and $\rho(x)$ located at a point $x = L$ with probability 1. In this case, the privacy loss is equal to $F(U) - F(L) = U - L$.

Similarly, we can prove that $A(Q) = u - l$, so we get the desired measure of the privacy loss. The proposition is proven.

# Application to Signal Processing: Using 1-D Radar Observations to Detect a Space Explosion Core among the Explosion Fragments

A radar observes the result of a space explosion. Due to radar's low horizontal resolution, we get a 1-D signal $x(t)$ representing different 2-D slices. Based on these slices, we must distinguish between the body at the core of the explosion and the slowly out-moving fragments. We propose new algorithms for processing this 1-D data. Since these algorithms are time-consuming, we also exploit the possibility of parallelizing these algorithms.

Some results from this chapter first appeared in [82].

*Formulation of the problem.* Most astronomical processes are slow; however, sometimes, space explosions happen: starts become supernovae, planetoids are torn apart by tidal and gravitational forces, etc. Even the Universe itself is currently viewed as a result of such an explosion – the Big Bang.

From the astrophysical viewpoint, these explosions are very important, because, e.g., supernovae explosions is how heavy metals spread around in the Universe.

The explosion processes are very rare and very fast, so unless they are very powerful and spectacular – like an explosion of a nearby supernovae that happened in 1054 – they are very difficult to observe. As a result, space explosion processes often go unnoticed.

What we do observe in most cases is the *result* of the space explosion, i.e., the explosion core   the remainder of the original celestial body   surrounded by the explosion fragments. The most well known example of such a result is the Crab Nebula formed after the 1054 supernovae explosion.

In order to better understand the corresponding physical process, it is extremely important to identify the explosion core.

In space, there is not much friction, so, due to inertia, most of the fragments travel with approximately the same speed as in the beginning of the explosion. Dividing the distance between the two fragments by their relative speed, we can determine – reasonably accurately – when the explosion occurred (this is how we know that the supernovae in the Crab Nebulae exploded in the year 1054). At that explosion time, all the fragments and the core were located

H.T. Nguyen et al.: Computing Stat. under Interval & Fuzzy Uncertain., SCI 393, pp. 289–298.
springerlink.com                                     © Springer-Verlag Berlin Heidelberg 2012

at the same point, so it is difficult to distinguish between the core and the fragments.

In general, we have a 2-D (and sometimes even 3-D) image of the result of the explosion. In such situations, detecting the explosion core is an image processing problem.

However, there is one important case when we only have 1-D data. In this case, we cannot use image processing techniques, we have to use techniques for processing 1-D data – i.e., DSP techniques.

This is the case of nearby space explosions, when the radar is the main source of information. A radar sends a pulse signal toward an object; this signal reflects from the object back to the station. We can measure, very accurately, the overall time that the signal traveled, which gives us the distance to the object. We can also measure the velocity, or, to be more precise, the rate with which the distance changes. It is, however, very difficult to separate the signals from different fragments located at the same distance.

As a result, what we observe is a 1-D signal $s(t)$, where each value $s(t)$ represents the intensity of the reflection from all the fragments located at distance $c \cdot t$ from the radar – i.e., from the 2-D slice corresponding to this distance. Based on these slices, we must distinguish between the body at the core of the explosion and the (slowly expanding) fragments.

In this paper, we describe a new method of identifying a core based on the slice observations.

*Repeated signal measurements at several different moments of time $T_k$.* At first glance, there may seem to be no difference between the signals reflected by the fragments and the signal reflected by the core. However, in the process of an explosion, fragments usually start rotating fast, at random rotation frequencies, with random phases. As a result, the signals reflected from the fragments oscillate, while the signal from the original core practically does not change.

As a result, the reflected signals change with time. Therefore, it makes sense to measure the signal $s(t)$ not just once, but at several consequent moments of time, i.e., to consider the signals $s_1(t), \ldots, s_N(t)$ measured at moments $T_1 < \ldots < T_N$, and use the difference between the dynamic character of the fragments and the static character of the core to identify the core.

*Relating measurements performed at different moments of time $T_k \neq T_l$: the corresponding $t$-scales are linearly related.* In order to compare signals measured at different moments of time $T_k \neq T_l$, we must identify the layers measured at different moments of time.

Let $T_0$ be the moment of explosion, and let $x_0$ be the initial distance between the radar and the core (and the fragments) at that initial moment of time $T_0$. We assume that our coordinate system has the radar as its origin, and that the $x$ axis is the axis in the direction of the analyzed "cloud". For each fragment $i$, let $v_x^{(i)}$ be the $x$-component of the velocity of $i$-th fragment (velocity relative to the radar). Hence, at moment $T_k$, the $x$-coordinate of

$i$-th fragment in our coordinate system – i.e., its distance from the radar – is equal to $x^{(i)}(T_k) = x_0 + v_x^{(i)} \cdot (T_k - T_0)$. Therefore, the radar signal reflected from this fragment corresponds to the time

$$t_k^{(i)} = \frac{x_k^{(i)}}{c} = \frac{x_0}{c} + v_x^{(i)} \cdot \frac{T_k - T_0}{c}. \tag{36.1}$$

Similarly, when we repeat the radar measurement at time $T_l \neq T_k$, the radar signal reflected from the $i$-th fragment corresponds to the time

$$t_l^{(i)} = \frac{x_0}{c} + v_x^{(i)} \cdot \frac{T_l - T_0}{c}. \tag{36.2}$$

What is the relation between the corresponding times $t_k^{(i)}$ and $t_l^{(i)}$? From the equation (36.1), we conclude that

$$v_x^{(i)} = \frac{c \cdot t_k^{(i)} - x_0}{T_k - T_0}.$$

Substituting this expression into the formula (36.2), we conclude that

$$t_l^{(i)} = \frac{x_0}{c} + \frac{c \cdot t_k^{(i)} - x_0}{T_k - T_0} \cdot \frac{T_l - T_0}{c} = a_{kl} \cdot t_k^{(i)} + b_{kl}, \tag{36.3}$$

where

$$a_{kl} = \frac{T_l - T_0}{T_k - T_0} > 0$$

and

$$b_{kl} = \frac{x_0}{c} - \frac{x_0}{T_k - T_0} \cdot \frac{T_l - T_0}{c}$$

do not depend on $i$.

In other words, the $t$-scales of the signals $s_k(t)$ and $s_l(t)$ are related by a linear dependence $t_k \to t_l = a_{kl} \cdot t_k + b_{kl}$.

*How can we experimentally find the coefficients of this linear relation?* At each moment of time $T_k$, we get the observed signal $s_k(t)$. Let $\underline{t}_k$ be the smallest time at which we get some reflection from the fragments cloud, and let $\bar{t}_k$ be the largest time at which we observe the radar reflection from this cloud. This means that there is a fragment $i$ for which $t_k^{(i)} = \underline{t}_k$, there is a fragment $j$ for which $t_k^{(j)} = \bar{t}_k$, and for every other fragment $f$, the corresponding moment of time is in between $\underline{t}_k$ and $\bar{t}_k$: $t_k^{(f)} \in [\underline{t}_k, \bar{t}_k]$.

As we have mentioned, for every other observation $T_l$, the relation between the corresponding times $t_k^{(i)}$ and $t_l^{(i)}$ is linear, with a positive coefficient $a_{kl}$. Since $a_{kl} > 0$, the corresponding linear functions $t \to a_{kl} \cdot t + b_{kl}$ is monotonically increasing. Thus, the value $t_l$ is the smallest for the same fragment $i$ for which $t_k$ was the smallest. Hence, $\underline{t}_l = t_l^{(i)} = a_{kl} \cdot t_k^{(i)} + b_{kl}$, i.e.,

$$\underline{t}_l = a_{kl} \cdot \underline{t}_k + b_{kl}. \tag{36.4}$$

Similarly,

$$\bar{t}_l = a_{kl} \cdot \bar{t}_k + b_{kl}. \tag{36.5}$$

The values $\underline{t}_k$, $\bar{t}_k$, $\underline{t}_l$, and $\bar{t}_l$ are directly observable. Thus, by solving the system of two linear equations (36.4) and (36.5) with 2 unknowns, we get explicit expressions for $a_{kl}$ and $b_{kl}$ in terms of these observable values:

$$a_{kl} = \frac{\bar{t}_l - \underline{t}_l}{\bar{t}_k - \underline{t}_k}; \quad b_{kl} = \frac{\bar{t}_k \cdot \underline{t}_l - \underline{t}_k \cdot \bar{t}_l}{\bar{t}_k - \underline{t}_k}.$$

*How can we transform signals $s_k(t)$ and $s_l(t)$ to the same scale?* Our main idea is that after we measure the fragments cloud at two different moments of time $T_k$ and $T_l$, we should compare the values $s_k(t)$ and $s_l(t)$ corresponding to the same fragments.

We know that for each moment of time $t$, the value $s_k(t)$ describes the same fragment(s) as the value $s_l(t')$, where $t' = a_{kl} \cdot t + b_{kl}$. We also know how to experimentally determine the coefficients $a_{kl}$ and $b_{kl}$. So, to make the desired comparison easier, it is reasonable to "re-scale" the signals to the same $t$-scale, so that the compared values correspond to exactly the same value $t$. In other words, we would like to generate a re-scaled signal

$$\widetilde{s}_l(t) \stackrel{\text{def}}{=} s_l(a_{kl} \cdot t + b_{kl}). \tag{36.6}$$

If the measurements were absolutely accurate, i.e., if we had the values $s_k(t)$ corresponding to each individual time $t$, then such a re-scaling would be easy: we could simply explicitly use the formula (36.6).

In real life, however, each value $s_l(t)$ corresponds not just to a single time $t$, but to the entire "bin" of values, from some value $\underline{t}$ to the value $\underline{t} + \Delta t$, where $\Delta t$ is the accuracy with which the radar can measure the time $t$ (in other words, $\Delta t = \Delta x/c$, where $\Delta x$ is the accuracy with which the radar can measure the distance). In other words, what we actually observe is a sequence of values $\ldots, s((i-1) \cdot \Delta t), s(i \cdot \Delta t), s((i+1) \cdot \Delta), \ldots$ Crudely speaking, each observed value $s(i \cdot \Delta t)$ represent the overall intensity of all the fragments for which the actual reflection time $t = x/c$ is in the interval

$$I_i \stackrel{\text{def}}{=} [(i - 0.5) \cdot \Delta t, (i + 0.5) \cdot \Delta t]. \tag{36.7}$$

Because of this discreteness, we cannot directly use the formula (36.6) to match the signals: Indeed, from the moment $T_k$ to the moment $T_l$, the cloud slightly expands. At the moment $T_k$, the value $s_k(i \cdot \Delta t)$ is the overall intensity of all the fragments for which $t_k$ belongs to the interval (36.6) of width $\Delta t$. At moment $T_l$, the times $t_l = a_{kl} \cdot t_k + b_{kl}$ corresponding to these same fragments occupy a wider interval – of width $a_{kl} \cdot \Delta t > \Delta t$. Thus, these fragments are no longer in the same bin, they may be in different bins.

How can we match the values? A natural idea is to use linear extrapolation. In other words, to estimate $\widetilde{s}(t)$ for $t = i \cdot \Delta t$, we apply the linear transformation $a_{kl} \cdot t + b_{kl}$ to the interval $I_i$. The resulting interval $\widetilde{I}_i$ consists of several parts from different intervals $I_j$. As $\widetilde{s}_l(t)$, we take a linear combination of the corresponding values $s_l(j \cdot \Delta t)$, with weights proportional to the relative length $|\widetilde{I}_i \cap I_j|/\Delta t$ of the intersection $\widetilde{I}_i \cap I_j$:

$$\widetilde{s}_l(i \cdot \Delta t) \stackrel{\text{def}}{=} \sum_j \frac{|\widetilde{I}_i \cap I_j|}{\Delta t} \cdot s_l(j \cdot \Delta t).$$

For example, if $\widetilde{I}_i$ consists of the entire interval $I_j$, 0.1 of $I_{j-1}$, and 0.05 of $I_{i-1}$, then $\widetilde{s}_l(i \cdot \Delta t)$ is equal to:

$$0.1 \cdot s_l((i-1) \cdot \Delta t) + s_l(i \cdot \Delta t) + 0.05 \cdot s_l((i+1) \cdot \Delta t).$$

In the following text, we will assume that the signals $s_i(t)$ have already been thus rescaled.

*Algorithm: main idea.* Each layer ("bin") contains several fragments. These fragments oscillate with random (uncorrelated) frequencies and phases; the overall signal $x(t)$ is the sum of the reflections from all these fragments. Due to the central limit theorem, the resulting overall signal $x(t)$ is approximately normally distributed with some mean $E(t)$ and variance $V(t)$.

If a layer only contains fragments, then, due to the independence assumption, $E(t) \approx n(t) \cdot E$ and $V(t) \approx n(t) \cdot V$, where $n(t)$ is the (unknown) number of fragment in layer $t$, and $E$ and $V$ are the mean and variance corresponding to each fragment. Therefore, for each such layer, $E(t) \approx (E/V) \cdot V(t)$.

For a layer that also contains the core, we have $E(t) \approx E_c + N(t) \cdot E$ and $V(t) \approx N(t) \cdot V$, where $E_c$ is the intensity of the core (since the core is supposed to be not rotating fast, its signal does not change with time, so the corresponding variance is negligible). Thus, for this layer,

$$E(t) \approx E_c + (E/V) \cdot V(t).$$

So, for the core, $E(t)/V(t) \gg E/V$.

Therefore, crudely speaking, our best guess for the core location is the point $t$ for which the ratio $E(t)/V(t)$ is the largest.

This is, of course, a very naive description of the idea. Let us see how this idea can be described in more adequate DSP terms.

*Motivations for the main distribution formula.* The intensity $I_i(t)$ of each fragment $i$ depends on time. Let $a_i = \lim_{T \to \infty} T^{-1} \cdot \int_0^T I_i(t)\, dt$ denote the average intensity over time, and let $b_i = \lim_{T \to \infty} T^{-1} \cdot \int_0^T (I_i(t) - a_i)^2\, dt$.

In the ensemble of fragments, let $a_0$ be the mean of $a_i$, let $A_0$ be the variance of $a_i$, let $b_0$ be the mean of $b_i$, and let $B_0$ be the mean of $a_i$. Then, according to the main idea, we can assume that $E(t)$ is normally distributed with the mean $n(t) \cdot a_0$ and the variance $n(t) \cdot A_0$, and $V(t)$ is normally distributed with the mean $n(t) \cdot b_0$ and the variance $n(t) \cdot B_0$.

We assumed the layers to be independent. As a result, we arrive at the following formula for the resulting probability distribution:

$$\rho = \prod_{t=1}^{N} \frac{1}{\sqrt{2\pi \cdot n(t) \cdot A_0}} \cdot \exp\left(-\frac{(E(t) - n(t) \cdot a_0)^2}{2n(t) \cdot A_0}\right) \times$$

$$\prod_{t=1}^{N} \frac{1}{\sqrt{2\pi \cdot n(t) \cdot B_0}} \cdot \exp\left(-\frac{(V(t) - n(t) \cdot b_0)^2}{2n(t) \cdot B_0}\right),$$

with the proviso that for the layer $t = t_0$ containing the core, we have

$$E(t) - E_c - n(t) \cdot a_0$$

instead of $E(t) - n(t) \cdot a_0$.

Based on the experimental data $E(t)$ and $V(t)$, we must find estimates for the parameters $a_0$, $b_0$, $A_0$, $B_0$, $n(t)$, $t_0$, and $E_c$ – and what we are really interested in is $t_0$. In accordance with the Maximum Likelihood Method (MLM), we must find the values of these parameters for which $\rho \to \max$. As usual in statistics, it is convenient to replace the problem of maximizing $\rho$ with a mathematically equivalent problem of minimizing a simpler function $\psi \stackrel{\text{def}}{=} -\ln(\rho)$, i.e., in our case,

$$\psi = \sum_{t=1}^{N} \frac{(E(t) - n(t) \cdot a_0)^2}{2n(t) \cdot A_0} + \sum_{t=1}^{N} \frac{(V(t) - n(t) \cdot b_0)^2}{2n(t) \cdot B_0} +$$

$$\sum_{t=1}^{N} \ln(n(t)) + \frac{N}{2} \cdot \log(A_0) + \frac{N}{2} \cdot \log(B_0). \tag{36.8}$$

*First case: when we know the parameters that characterize fragment distribution.* Let us start with the simplest case when we know the values of the parameters $a_0$, $b_0$, $A_0$, and $B_0$ that describe the distribution of fragments. In this case, differentiating by $n(t)$ and equating the derivative to 0, we conclude that

$$-\frac{1}{2n(t)^2}\left(\frac{E(t)^2}{A_0} + \frac{V(t)^2}{B_0}\right) + \frac{1}{2}\left(\frac{a_0^2}{A_0} + \frac{b_0^2}{B_0}\right) + \frac{1}{n(t)} = 0.$$

The first two terms are approximately independent on the number of fragments $n(t)$, the third term $1/n(t)$ is much smaller (since we have many fragments). So, we can safely ignore the their term and conclude that $n(t) = \|v_t\|/\|v_0\|$, where we denoted

$$v_t \stackrel{\text{def}}{=} \left(\frac{E(t)}{\sqrt{A_0}}, \frac{V(t)}{\sqrt{B_0}}\right); \quad v_0 \stackrel{\text{def}}{=} \left(\frac{a_0}{\sqrt{A_0}}, \frac{b_0}{\sqrt{B_0}}\right),$$

and $\|(v_a, v_b)\| = \sqrt{v_a^2 + v_b^2}$ denotes the length of the vector $v = (v_a, v_b)$. Substituting this expression for $n(t)$ into the corresponding part of (36.8), i.e., into

$$\psi(t) \stackrel{\text{def}}{=} \frac{(E(t) - n(t) \cdot a_0)^2}{2n(t) \cdot A_0} + \frac{(V(t) - n(t) \cdot b_0)^2}{2n(t) \cdot B_0} +$$

$$\ln(n(t)) = \frac{1}{2n(t)} \cdot \left( \frac{E(t)^2}{A_0} + \frac{V(t)^2}{B_0} \right) -$$

$$\left( \frac{E(t) \cdot a_0}{A_0} + \frac{V(t) \cdot a_0}{A_0} \right) + \frac{n(t)}{2} \cdot \left( \frac{a_0^2}{A_0} + \frac{b_0^2}{B_0} \right) + \ln(n(t)),$$

we conclude that $\psi(t) \approx \psi_0(t)$, where

$$\psi_0(t) \stackrel{\text{def}}{=} \|v_t\| \cdot \|v_0\| - v_t \cdot v_0, \tag{36.9}$$

and $v_t \cdot v_0$ denotes the dot (scalar) product. ($\approx$ because we use the approximate value for $n(t)$.)

For $t = t_0$, due to the presence of an additional variable $E_c$, we get $\psi(t_0) \approx 0$. Thus,

$$\psi = (N/2) \cdot (\log(A_0) + \log(B_0)) + \sum_{t=1}^{N} \psi_0(t) - \psi_0(t_0).$$

Thus, $\psi$ is the smallest if and only if $\psi(t_0)$ is the largest. Therefore, we arrive at the following algorithm for locating the core:

- First, we re-scale the signals $s_k(t)$ into $\widetilde{s}_k(t)$ so that the same value $t$ corresponds to the same fragments.
- For each $t$, we compute the sample average $E(t)$ and the sample variance $V(t)$ of the values $\widetilde{s}_k(t)$.
- For each $t$, we compute $v_t$ and $\psi_0(t)$, and find $t_0$ for which $\psi_0(t_0) = m \stackrel{\text{def}}{=} \max_t \psi_0(t)$.

How reliable is this estimate? We are interested in the value of a single variable $t_0$, and we know that for one variable, 95% of the values are within $2\sigma$ from the mean, and 99.9% are within $3\sigma$. In terms of $\psi = \ln(\rho)$, the mean corresponds to its minimum, the $2\sigma$ deviation means difference $(2\sigma)^2/(2\sigma^2) = 2$ from the minimum, and $3\sigma$ deviation means the difference of $(3\sigma)^2/(2\sigma^2) = 4.5$ from the minimum. Thus, with reliability 95%, we conclude that the core is among those $t$ for which $\psi_0(t) \geq m - 2$, and that with reliability 99.9%, the core is among those $t$ for which $\psi_0(t) \geq m - 4.5$.

*General case.* The value (36.8) does not change if we re-scale all the parameters: $n(t) \to K \cdot n(t)$, $a_0 \to a_0/K$, $b_0 \to b_0/K$, $A_0 \to A_0/K$, and $B_0 \to B_0/K$, for any $K > 0$. W.l.o.g., we can therefore assume that $a_0 = 1$.

Differentiating (36.8) by $a_0$, we conclude that

$$a_0 = \frac{\sum E(t)}{\sum n(t)}.$$

Similarly,

$$b_0 = \frac{\sum V(t)}{\sum n(t)}.$$

Since $a_0 = 1$, we thus get

$$b_0 = \frac{\sum V(t)}{\sum E(t)}.$$

Differentiating by $A_0$, we conclude that

$$A_0 = \frac{1}{N} \sum_t \frac{(E(t) - n(t) \cdot a_0)^2}{n(t)} =$$

$$\frac{1}{N} \left( \sum_t \frac{E(t)^2}{n(t)} - \sum_t E(t) \right) \tag{36.10}$$

and similarly,

$$B_0 = \frac{1}{N} \left( \sum_t \frac{V(t)^2}{n(t)} - b_0 \cdot \sum_t V(t) \right). \tag{36.11}$$

If we denote $\lambda \stackrel{\text{def}}{=} A_0/B_0$, then the above formula for $n(t)$ takes the form $n(t)^2 = (E(t)^2 + \lambda \cdot V^2(t))/(1 + \lambda \cdot b_0^2)$. Substituting this expression into (36.10) and (36.11) and using the fact that $A_0 = \lambda \cdot B_0$, we conclude that

$$\sum_t \frac{E(t)^2}{\sqrt{E(t)^2 + \lambda \cdot V(t)^2}} \cdot \sqrt{1 + \lambda \cdot b_0^2} - \sum_t E(t) =$$

$$\sum_t \frac{\lambda \cdot V(t)^2}{\sqrt{E(t)^2 + \lambda \cdot V(t)^2}} \cdot \sqrt{1 + \lambda \cdot b_0^2} - b_0 \cdot \left( \sum_t V(t) \right)$$

with the only unknown $\lambda$. After we find $\lambda$ from this equation, we can thus find $A_0$, $B_0$, and hence, the desired $t_0$.

To test our technique, we simulated an explosion with randomly distributed fragments. On this simulation, the above algorithm does detect the core.

*Possibility of parallelization.* In the above algorithms, processing values corresponding to bin $i$ uses only measurement only from this bin and from the neighboring bins. Therefore, if we have several processors working in parallel (see, e.g., [140]), we can speed up the computations by having each processor process a section of bins. For example, for 2 processors, the first can handle bins 1 to $N/2 + n$, and the second all the bins from $N/2 - n$ to $N$, where $n$ is the number of neighboring bins that we need to take into consideration.

*Multiple explosions: case of a very accurate radar.* Sometimes, the observed fragments cloud comes not from a single explosion, but from several consequent explosions. How can we then determine the core?

Let us show that when the radar is accurate enough, so that we can distinguish between individual fragments, the problem of determining the core becomes even easier than in the case of a single explosion.

First, we observe that if the radar is that accurate, then, by making observations at very close moments of time $T_1$, $T_2$, etc., we can *trace* individual fragments. Indeed, at the initial moment $T_1$, we identify fragments by the times $t_1^{(1)} < t_1^{(2)} < \ldots$ at which the corresponding signal $s_1(t)$ is non-zero. At the next moment $T_2$, we can find the times $t_2$, $t_2'$, $\ldots$ corresponding to the fragments as the times $t$ for which $s_2(t) \neq 0$. When the time difference $T_2 - T_1$ is so small that the relative motion of a fragment is smaller than the distance between different fragments, we can identify, for each fragment $i$, the corresponding time $t_2^{(i)}$ as the closest to $t_1^{(i)}$ among all observed values $t_2$, $t_2'$, $\ldots$

For a single explosion, a linear formula (36.3) relates $t_2^{(i)}$ and $t_1^{(i)}$; the corresponding slope $a_{kl}$ depends on the moment $T_0$ of the explosion. If two explosions occurred at moments $T_0$ and $T_0'$, we get similar linear formulas for the fragments of each explosion, with two slopes $a_{kl} \neq a_{kl}'$. Thus, by plotting the dependence of $t_2^{(i)}$ on $t_1^{(i)}$, we will get two straight lines with different slopes. The core belongs to both families of fragments. Thus, the core can be determined as the fragment $i_0$ that lies at the intersection of the two corresponding straight lines.

For two explosions, we can determine both lines and easily find the intersection. For numerous explosions, we will have many straight lines, and finding all of them may be computationally difficult; so, we need a different idea.

The dependence of $a_k$ on $T_0$ is monotonic, so in such situations, the 2-D points $t^{(i)} \stackrel{\text{def}}{=} (t_1^{(i)}, t_2^{(i)})$ occupy a zone between two straight lines with different slope $\underline{a} < \overline{a}$ corresponding to the first and the last explosions; geometrically, it is a 2-D cone with the core's value $t^{(i_0)}$ as the vertex. Since we have numerous explosions, we can conclude that the corresponding pairs fill the entire cone.

Let us show that the core can be determined as the only value $i$ for which

$$\max_{j:\, t_1^{(j)} < t_1^{(i)}} t_j^{(2)} < \min_{j:\, t_1^{(j)} > t_1^{(i)}} t_j^{(2)}. \tag{36.12}$$

Let us first consider the case $i = i_0$. For each of the corresponding straight lines, the dependence of $t_2^{(i)}$ on $t_1^{(i)}$ is monotonically increasing; since the core $i_0$ belongs to all the lines, we can therefore conclude that if $t_1^{(j)} < t_1^{(i_0)}$, then we have $t_2^{(j)} < t_2^{(i_0)}$, and if $t_1^{(j)} > t_1^{(i_0)}$, then we have $t_2^{(j)} > t_2^{(i_0)}$ – which implies (36.12).

If $t_1^{(i)} > t_1^{(i_0)}$, then the maximum in the left side of the formula (36.12) corresponds to the largest possible slope $\overline{a}_{kl}$ and is therefore equal to $t_2^{(i_0)} + \overline{a}_{kl} \cdot (t_1^{(i)} - t_1^{(i_0)})$. On the other hand, the minimum in the right side of the formula (36.12) corresponds to the smallest possible slope slope $\underline{a}_{kl}$ and is therefore equal to $t_2^{(i_0)} + \underline{a}_{kl} \cdot (t_1^{(i)} - t_1^{(i_0)})$ – which is clearly smaller than the maximum in the left side of (36.12).

Similarly, (36.12) cannot occur for $t_i^{(i)} < t_1^{(i_0)}$.

# Applications to Computer Engineering: Timing Analysis of Computer Chips

In chip design, one of the main objectives is to decrease its clock cycle. On the design stage, this time is usually estimated by using worst-case (interval) techniques, in which we only use the bounds on the parameters that lead to delays. This analysis does not take into account that the probability of the worst-case values is usually very small; thus, the resulting estimates are over-conservative, leading to unnecessary over-design and under-performance of circuits. If we knew the *exact* probability distributions of the corresponding parameters, then we could use Monte-Carlo simulations (or the corresponding analytical techniques) to get the desired estimates. In practice, however, we only have *partial* information about the corresponding distributions, and we want to produce estimates that are valid for all distributions which are consistent with this information.

In this chapter, we describe general techniques that allows us, in particular, to provide such estimates for the clock time. The results of this chapter first appeared in [269, 270, 341].

*Decreasing clock cycle: a practical problem.* In chip design, one of the main objectives is to decrease the chip's clock cycle. It is therefore important to estimate the clock cycle on the design stage.

The clock cycle of a chip is constrained by the maximum path delay over all the circuit paths $D \stackrel{\text{def}}{=} \max(D_1, \ldots, D_N)$, where $D_i$ denotes the delay along the $i$-th path. Each path delay $D_i$ is the sum of the delays corresponding to the gates and wires along this path. Each of these delays, in turn, depends on several factors such as the variation caused by the current design practices, environmental design characteristics (e.g., variations in temperature and in supply voltage), etc.

*Traditional (interval) approach to estimating the clock cycle.* Traditionally, the delay $D$ is estimated by using the worst-case analysis, in which we assume that each of the corresponding factors takes the worst possible value (i.e., the value leading to the largest possible delays). As a result, we get the time delay that corresponds to the case when all the factors are at their worst.

H.T. Nguyen et al.: Computing Stat. under Interval & Fuzzy Uncertain., SCI 393, pp. 299–304.
springerlink.com                                    © Springer-Verlag Berlin Heidelberg 2012

*It is necessary to take probabilities into account.* The worst-case analysis does not take into account that different factors come from independent random processes. As a result, the probability that all these factors are at their worst is extremely small. For example, there may be slight variations of delay time from gate to gate, and this can indeed lead to gate delays. The worst-case analysis considers the case when all these random variations lead to the worst case; since these variations are independent, this combination of worst cases is highly unprobable.

As a result, the current estimates of the chip clock time are over-conservative, over up to 30% above the observed clock time. Because of this over-estimation, the clock time is set too high – i.e., the chips are usually over-designed and under-performing; see, e.g., [48, 67, 68, 264, 265, 266, 267, 268]. To improve the performance, it is therefore desirable to take into account the probabilistic character of the factor variations.

*Robust statistical methods are needed.* If we knew the *exact* probability distributions of the corresponding parameters, then we could use Monte-Carlo simulations (or the corresponding analytical techniques) to get the desired estimates. In practice, however, we only have *partial* information about the corresponding distributions. For a few parameters, we know the exact distribution, but for most parameters, we only know the mean and some characteristic of the deviation from the mean – e.g., the interval that is guaranteed to contain possible values of this parameter.

In principle, we could pick up some distributions which are consistent with this partial information – e.g., truncated normal distributions, compute the maximum delays $D$ corresponding to all these distributions, and then take the largest $D_{\max}$ of these computed maximum delays $D$ as the clock time. This procedure will guarantee that the path delay $D$ does not exceed the clock time if the actual distribution is one of the picked ones. However, it is quite possible that some other possible distributions (different from the ones we picked), the corresponding path delay $D$ is larger than $D_{\max}$. As a result, we may be underestimating the clock time. If we set the clock time too low, we may have operations that did not have time to finish before the next cycle starts – and this is even worse than overestimating.

It is therefore desirable to provide bounds that work for all the distributions which are consistent with the given information. In statistics, estimates which are guaranteed for all distributions from some non-parametric class are called *robust* (see, e.g., [138]). In these terms, our objective is to provide robust statistical estimates for the clock time.

*How the desired delay $D$ depends on the parameters.* The variations in the each gate delay $d$ are caused by the difference between the actual and the nominal values of the corresponding parameters. It is therefore desirable to describe the resulting delay $d$ as a function of these differences $x_1, \ldots, x_n$. Since these differences are usually small, we can safely ignore quadratic (and higher order) terms in the Taylor expansion of the dependence of $d$ on $x_j$

and assume that the dependence of each delay $d$ on these differences can be described by a linear function.

As a result, each path delay $D_i$ – which, as we have mentioned, is the sum of delays at different gates and wires – can also be described as a linear function of these differences, i.e., as $D_i = a_i + \sum_{j=1}^{n} a_{ij} \cdot x_j$ for some coefficients $a_i$ and $a_{ij}$.

Thus, the desired maximum delay $D = \max_i D_i$ has the form

$$D = \max_i \left( a_i + \sum_{j=1}^{n} a_{ij} \cdot x_j \right). \tag{37.1}$$

*Comment.* As in Chapter 33, this dependence is convex.

*Our objective.* We want to find the smallest possible value $y_0$ such that for all possible distributions consistent with the known information, we have $y \le y_0$ with the probability $\ge 1 - \varepsilon$ (where $\varepsilon > 0$ is a given small probability).

*What information we can use.* What information can we use for these estimations? We can safely assume that different factors $x_j$ are statistically independent. About some of the variables $x_j$, we know their exact statistical characteristics; about some other variables $x_j$, we only know their interval ranges $[\underline{x}_j, \overline{x}_j]$ and their means $E_j$.

*Observation.* We get the exact same mathematical problem as in Chapter 33 in which we discussed optimal scheduling for global computing. Similarly to that case, it is reasonable to assume that the dependency is non-degenerate. Thus, to solve our problem, we can use the algorithm described (and justified) in Chapter 33.

*Formulation of the problem and the main result.*

GIVEN:

- natural numbers $n$, and $k \le n$;
- a real number $\varepsilon > 0$;
- a function $y = F(x_1, \ldots, x_n)$ (algorithmically defined) such that for every combination of values $x_{k+1}, \ldots, x_n$, the dependence of $y$ on $x_1, \ldots, x_k$ is convex;
- $n - k$ probability distributions $x_{k+1}, \ldots, x_n$ – e.g., given in the form of cumulative distribution function (cdf) $F_j(x)$, $k + 1 \le j \le n$;
- $k$ intervals $\boldsymbol{x}_1, \ldots, \boldsymbol{x}_k$, and
- $k$ values $E_1, \ldots, E_k$,

such that for every $x_1 \in [\underline{x}_1, \overline{x}_1], \ldots, x_k \in [\underline{x}_k, \overline{x}_k]$, we have $F(x_1, \ldots, x_n) \ge 0$ with probability 1.

TAKE: all possible joint probability distributions on $R^n$ for which:

- all $n$ random variables are independent;
- for each $j$ from 1 to $k$, $x_j \in \mathbf{x}_j$ with probability 1 and the mean value of $x_j$ is equal to $E_j$;
- for $j > k$, the variable $x_j$ has a given distribution $F_j(x)$.

FIND: the smallest possible value $y_0$ such that for all possible distributions consistent with the known information, we have $y \stackrel{\text{def}}{=} F(x_1, \ldots, x_n) \leq y_0$ with probability $\geq 1 - \varepsilon$.

PROVIDED: that the problem is *non-degenerate* in the sense that if we narrow down one of the intervals $\mathbf{x}_j$, the value $y_0$ decreases.

**Proposition 37.1.** *The desired value $y_0$ is attained when for each $j$ from 1 to $k$, we use a 2-point distribution for $x_j$, in which:*

- $x_j = \underline{x}_j$ *with probability* $\underline{p}_j \stackrel{\text{def}}{=} \dfrac{\overline{x}_j - E_j}{\overline{x}_j - \underline{x}_j}.$
- $x_j = \overline{x}_j$ *with probability* $\overline{p}_j \stackrel{\text{def}}{=} \dfrac{E_j - \underline{x}_j}{\overline{x}_j - \underline{x}_j}.$

*Resulting algorithm for computing $y_0$.* Because of Proposition 37.1, we can compute the desired value $y_0$ by using the following Monte-Carlo simulation:

- We set each value $x_j$, $1 \leq j \leq k$, to be equal to $\overline{x}_j$ with probability $\overline{p}_j$ and to the value $\underline{x}_j$ with the probability $\underline{p}_j$.
- We simulate the values $x_j$, $k < j \leq n$, as random variables distributed according to the distributions $F_j(x)$.
- For each simulation $s$, $1 \leq s \leq N_i$, we get the simulated values $x_j^{(s)}$, and then, a value $y^{(s)} = F(x_1^{(s)}, \ldots, x_n^{(s)})$. We then sort the resulting $N_i$ values $y^{(s)}$ into an increasing sequence

$$y_{(1)} \leq y_{(2)} \leq \cdots \leq y_{(N_i)},$$

and take, as $y_0$, the $N_i \cdot (1 - \varepsilon)$-th term $y_{(N_i \cdot (1 - \varepsilon))}$ in this sorted sequence.

*Comment about Monte-Carlo techniques.* Before presenting the algorithm for computing the upper bound on $y_0$, let us remark that some readers may feel uncomfortable with the use of Monte-Carlo techniques. This discomfort comes from the fact that in the *traditional* statistical approach, when we know the exact probability distributions of all the variables, Monte-Carlo methods – that simply simulate the corresponding distributions – are inferior to analytical methods. This inferiority is due to two reasons:

- First, by design, Monte-Carlo methods are approximate, while analytical methods are usually exact.

- Second, the accuracy provided by a Monte-Carlo method is, in general, proportional to $\sim 1/\sqrt{N_i}$, where $N_i$ is the total number of simulations. Thus, to achieve reasonable quality, we often need to make a lot of simulations – as a result, the computation time of a Monte-Carlo method becomes much longer than for an analytical method.

In *robust* statistics, there is often an additional reason to be uncomfortable about using Monte-Carlo methods:

- Practitioners use these methods by selecting a finite set of distributions from the infinite class of all possible distributions, and running simulations for the selected distributions.
- Since we do not test all the distributions, this practical heuristic approach sometimes misses the distributions on which the minimum or maximum of the corresponding distribution is actually attained.

In our case, we also select a finite collection of distributions from the infinite set. However, in contrast to the heuristic (un-justified) selection – which is prone to the above criticism, our selection is *justified*. Proposition 37.1 *guarantees* that the values corresponding to the selected distributions indeed provide the desired value $y_0$ – the largest over all possible distributions $p \in \mathcal{P}$.

In such situations, where a justified selection of Monte-Carlo methods is used to solve a problem of robust statistics, such Monte-Carlo methods often lead to *faster* computations than known analytical techniques. The speed-up caused by using such Monte-Carlo techniques is one of the main reasons why they were invented in the first place – to provide fast estimates of the values of multi-dimensional integrals. Many examples of efficiency of these techniques are given, e.g., in [288]; in particular, examples related to estimating how the uncertainty of inputs leads to uncertainty of the results of data processing are given in [331].

*Comment about non-linear terms.* In the formula (37.1), we ignored quadratic and higher order terms in the dependence of each path time $D_i$ on the parameters $x_j$. It is known that the maximum $D = \max_i D_i$ of convex functions $D_i$ is always convex. So, according to Proposition 37.1, the above algorithm will work if we take quadratic terms into consideration – provided that each dependence $D_i(x_1, \ldots, x_k, \ldots)$ is still convex.

*Conclusions.* In chip design, one of the main objectives is to decrease its clock cycle.

On the design stage, this time is usually estimated by using worst-case (interval) techniques, in which we only use the bounds on the parameters that lead to delays. This analysis does not take into account that the probability of the worst-case values is usually very small; thus, the resulting estimates are over-conservative, leading to unnecessary over-design and under-performance of circuits. Instead of the largest possible value of the delay, it is reasonable to determine the clock time as the time $y_0$ for which the probability that the actual delay $y$ exceeds $y_0$ does not exceed a given small value $\varepsilon$.

If we knew the *exact* probability distributions of the corresponding parameters, then we could use Monte-Carlo simulations (or the corresponding analytical techniques) to get the desired value $y_0$. In practice, however, we only have *partial* information about the corresponding distributions, and we want to produce the value $y_0$ which is valid for all distributions which are consistent with this information.

In this chapter, we describe a general technique that allows us, in particular, to compute this value $y_0$. This technique uses Monte-Carlo simulations with specially selected "worst-case" distributions, distributions for which the delay is provably largest among all distributions from the given class. Thus, to guarantee that $\text{Prob}(y \leq y_0) \geq 1 - \varepsilon$ *for all* distributions from the given class, it is sufficient to check this inequality for the selected "worst-case" distributions.

# Applications to Mechanical Engineering: Failure Analysis under Interval and Fuzzy Uncertainty

One of the main objective of mechanics of materials is to predict when the material experiences fracture (fails), and to prevent this failure. With this objective in mind, it is desirable to use it ductile materials, i.e., materials which can sustain large deformations without failure. Von Mises criterion enables us to predict the failure of such ductile materials. To apply this criterion, we need to know the exact stresses applied at different directions. In practice, we only know these stresses with interval or fuzzy uncertainty. In this chapter, we describe how we can apply this criterion under such uncertainty, and how to make this application computationally efficient.

Its main results first appeared in [358].

## Formulation and Analysis of the Problem, and the Corresponding Results and Algorithms

*Basics of mechanics of materials: the notion of stress.* When a force is applied to a material, this material deforms and at some point breaks down. We can gauge the effect of the force by the *stress*, the force per unit area. The larger the stress, the larger the deformation; at some point, larger stress leads to a breakdown.

*Case of small stress: elastic (reversible) deformations.* When the stress is small, no irreversible damage occurs, all deformations are *reversible*.

The original shape of the material (i.e., the shape in the absence of stress) is the one to which the undamaged material reverts. Thus, under small stress, the material returns to its original shape once the force is no longer applied. Such reversible deformation which return to the original shape is called *elastic*.

*Case of larger stress: irreversible (plastic) deformation.* An increased level of stress causes irreversible damage in the material. In this case, after the force is no longer applied, the material does not return to its original shape. Such irreversible deformation is called *plastic* or *yielding*.

H.T. Nguyen et al.: Computing Stat. under Interval & Fuzzy Uncertain., SCI 393, pp. 305–316.
springerlink.com        © Springer-Verlag Berlin Heidelberg 2012

*From plastic deformation to failure.* Under plastic deformation, there is an irreversible damage to the material, but this damage occurs on the microlevel. On the macrolevel, the material may be slightly misshapen and somewhat twisted, but it is still intact and it can still serve its purpose.

However, as the stress increases, it causes macrodamage too: the material experiences fractures. In many mechanical designs, the fractured material can no longer fulfil its duties, so it is usually said that this material *fails.*

*Predicting failure is extremely important.* Material failure can have catastrophic consequences. As a result, it is extremely important to predict when a material can fail.

In many practical situations, it is also important to know when the yielding starts, because while the yielding itself is usually not catastrophic, the resulting irreversible damage start weakening the mechanical construction can lead to a failure in the long run.

*Case of 1-D stress.* Let us first consider the simplest situation of a 1-D stress, when the force is only applied in one direction. In this case, as we have mentioned, both the yielding and the failure start when the stress becomes large enough. In other words, for 1-D stress, for each material, there are two thresholds:

- the threshold $\sigma_y$ after which yielding starts, and
- the threshold $\sigma_f > \sigma_y$ after which the material fails.

*Ductile materials and their practical importance.* In practical applications, it is desirable to use materials which can sustain large deformations without failure.

This is not always possible: e.g., some materials such as ceramics fail almost immediately after the yielding starts.

However, many other materials can sustain large plastic deformations without failure. Such materials are called *ductile.* Examples of ductile materials include ductile metals such as copper, silver, gold, and steel; it is possible to deform these ductile metals into wire without breaking them.

In view of many important applications of ductile materials, it is necessary to predict when they fail.

*Case of general (3-D) stress: importance.* We have mentioned that for 1-D stress, it is easy to predict when a material fails: when the stress exceeds its failure threshold.

In real life, situations in which the force comes from only one direction are rare. Usually, have a combination of stresses coming from different directions. It is therefore important to be able to predict when a material fails under such 3-D stress.

*Case of general (3-D) stress: formulation of the problem.* A general 3-D stress can be described as a combination of three stresses $\sigma_1$, $\sigma_2$, and $\sigma_3$ applied at three orthogonal directions. It is therefore desirable to be able, given the three stresses $\sigma_i$, to be able to predict when a ductile material fails under these stresses.

*First solution to this problem was provided by Maxwell.* The first solution to this important problem was provided by Maxwell (of the electromagnetic equations fame) in the 1860s. As we will see, Maxwell's formulas are still used to predict the material's failure.

Because of the continuing practical importance of Maxwell's solution, in this section, we will briefly reproduce Maxwell's derivations – to make the resulting formulas more understandable. (Of course, our rendering of Maxwell's derivation will be somewhat modernized.)

Those readers who are already familiar with failure mechanics and with the von Mises criterion (and with its motivations) are welcome skip this section.

*Need for an appropriate combination of stresses: physical motivations.* In the 1-D case, the corresponding stress $\sigma$ provides a numerical measure of how stressed the material is: when this stress exceeds a given threshold $\sigma_f$, the material fails.

In the 3-D case, we have three different stresses $\sigma_1$, $\sigma_2$, and $\sigma_3$. Informally, all these three stresses contribute to the "overall stress". When this "combined stress" exceeds a certain threshold, the material fails. Thus, to be able to predict when a material fails, we must be able to find out how this "combined stress" depends on the individual stresses $\sigma_i$.

Maxwell's main idea is thus to combine the there stresses $\sigma_1$, $\sigma_2$, and $\sigma_3$ into a single numerical criterion $f(\sigma_1, \sigma_2, \sigma_3)$ that would decide when a material fails. To be more precise, Maxwell assumed that there exists a threshold value $f_0$ such that:

- when $f(\sigma_1, \sigma_2, \sigma_3) < f_0$, the material remains intact (i.e., undamaged on the macro level);
- when $f(\sigma_1, \sigma_2, \sigma_3) \geq f_0$, the material fails.

*Need for an appropriate combination of stresses: mathematical motivations.* The existence of a combination function $f(\sigma_1, \sigma_2, \sigma_3)$ is motivated not only by physics, it can also be justified on purely mathematical grounds.

Specifically, for every material and for every triple $(\sigma_1, \sigma_2, \sigma_3)$ of the corresponding stresses, we can check whether the corresponding combination of stresses indeed leads to a failure. Thus, in the 3-D space $R^3$, we have a set $S$ of all possible combinations which lead a failure, and its complement, a set of all combinations which do not lead to a failure.

From the mathematical viewpoint, for every set $S \subseteq R^3$, there exist a function $f : R^3 \to R$ and a real number $f_0$ such that:

- $f(x) \geq f_0$ for all the points $x \in S$ and
- $f(x) < f_0$ for all $x \notin S$.

For example, as the desired function $f$, we can take a characteristic function of the set $S$, i.e., a function for which $f(x) = 1$ for $x \in S$ and $f(x) = 0$ for $x \notin S$. For this function, the above separation property occurs for $f_0 = 1$.

In the general case, for an arbitrary set $S$, this function has to be discontinuous. However, for well-behaved sets, we can select this function to be continuous (see, e.g., [153]), and if the boundary is smooth, we can have a smooth function $f(x)$.

*Main ideas behind Maxwell's solution.* Maxwell's solution is based on two ideas widely used in physics applications:

- on the mathematical idea of ignoring higher order terms in the Taylor expansion, and
- on the physical idea of symmetry.

*Ignoring higher order terms in the Taylor expansion: details.* In physics, most dependencies are smooth (differentiable). In general, a smooth function $f(\sigma_1, \sigma_2, \sigma_3)$ can be expanded into Taylor series in $\sigma_i$:

$$f(\sigma_1, \sigma_2, \sigma_3) = a_0 + \sum_{i=1}^{3} a_i \cdot \sigma_i + \sum_{i=1}^{3} \sum_{j=1}^{3} a_{ij} \cdot \sigma_i \cdot \sigma_j + \ldots$$

for appropriate coefficients $a_0$, $a_i$, $a_{ij}$, ...

We are interested in ductile materials, i.e., materials that can sustain reasonably large stresses without failing. However, even for the best of such materials, these large stresses are much smaller than the stresses that we can potentially apply. So, we can consider the values $\sigma_i$ to be reasonable small and do what physicists usually do – ignore higher order terms in the above expansion.

*First try: linear approximation.* A natural first approximation is when we ignore quadratic and higher order terms. In this case, we get the following reasonable linear approximation to the desired function $f(\sigma_1, \sigma_2, \sigma_3)$:

$$f(\sigma_1, \sigma_2, \sigma_3) = a_0 + \sum_{i=1}^{3} a_i \cdot \sigma_i.$$

*Due to symmetry, linear approximation leads to average stress.* Our main physical idea is to use symmetry. There is nothing special about each direction, hence the coefficients $a_i$ corresponding to different directions must be equal: $a_1 = a_2 = a_3$. Thus, the resulting formula reduced to

$$f(\sigma_1, \sigma_2, \sigma_3) = a_0 + a_1 \cdot \sum_{i=1}^{3} \sigma_i,$$

or, equivalently, to

$$f(\sigma_1, \sigma_2, \sigma_3) = a_0 + 3a_1 \cdot \left( \frac{1}{3} \cdot \sum_{i=1}^{3} \sigma_i \right).$$

Thus, the "combined stress" $f$ is proportional to the average stress, and the condition $f \geq f_0$ means that the average stress should exceed a certain threshold.

*Maxwell's observation: the above linearized solution contradicts to physical symmetry.* We have deduced the above solution by using the *mathematical* idea of symmetry. However, in this situation, there is also a *physical* symmetry.

Namely, suppose that we have a perfectly spherical body and we apply the exact pressure from all three directions, i.e., we have $\sigma_1 = \sigma_2 = \sigma_3$. In this case, we have a perfectly symmetric body (invariant with respect to arbitrary rotations around its center) and a perfectly symmetric stress. In a deterministic system, it is thus reasonable to expect that the system will preserve its symmetry.

One can easily see that a fracture is a violation of symmetry. Thus, we can conclude that in this perfectly symmetric case, we should not expect any fractures at all. However, according to our linearized criterion, when $\sigma_1 = \sigma_2 = \sigma_3$, we have $f = a_0 + 3a_1 \cdot \sigma_1$, so for sufficiently large stresses, we should have fracture.

Thus, the above linearized solution contradicts to physical symmetry.

Similarly, any kind of irreversible damage on the microlevel is also bound to violate symmetry, so we should not expect the absolutely symmetric stress to cause any damage at all. Thus, we should have $a_1 = 0$, i.e., $a_1 = a_2 = a_3 = 0$.

*Comment.* In practice, fractures do occur even in the symmetric case when all the stresses are equal; however, they occur at a much higher level of stress than when we have different stresses at different directions. So, in the first approximation, we can safely assume that when all three stresses $\sigma_i$ are equal, there will be no failure.

*From linear to quadratic approximation.* We have concluded that due to physical symmetry, there are no linear terms in the Taylor expansion of the function $f$: $a_i = 0$. This means that we cannot ignore quadratic terms in the Taylor expansion of the function $f$. A natural next idea is therefore to take quadratic terms into account and to ignore cubic and higher order terms in the expansion of $f$. In this case, we arrive at the following approximate expression for $f$:

$$f(\sigma_1, \sigma_2, \sigma_3) = a_0 + \sum_{i=1}^{3} \sum_{j=1}^{3} a_{ij} \cdot \sigma_i \cdot \sigma_j.$$

*Using mathematical symmetry.* Due to mathematical symmetry, this expression should not change if we swap two directions (i.e., $1 \leftrightarrow 2$ or $1 \leftrightarrow 3$). Because of this invariance requirement:

- all the values $a_{ii}$ should be equal to each other – hence equal to $a_{11}$, and
- all the values $a_{ij}$, $i \neq j$, should be equal to each other – hence equal to $a_{12}$.

Thus, we conclude that

$$f(\sigma_1, \sigma_2, \sigma_2) = a_0 + a_{11} \cdot \sum_{i=1}^{3} \sigma_i^2 + a_{12} \cdot \sum_{i \neq j} \sigma_i \cdot \sigma_j.$$

*Using physical symmetry.* In the physically symmetric case, when all the stresses coincide $\sigma_1 = \sigma_2 = \sigma_3 = \sigma$, we should not have any combined stress. In this case, the above formula leads to $f = a_0 + (3a_{11} + 6a_{12}) \cdot \sigma^2$, so we conclude that $a_{11} = -2a_{12}$. Thus, $f = a_0 - a_{12} \cdot V$, where we denoted

$$V(\sigma_1, \sigma_2, \sigma_3) \stackrel{\text{def}}{=} 2\sigma_1^2 + 2\sigma_2^2 + 2\sigma_3^2 - 2\sigma_1 \cdot \sigma_2 - 2\sigma_2 \cdot \sigma_2 - 2\sigma_3 \cdot \sigma_1 =$$

$$(\sigma_1 - \sigma_2)^2 + (\sigma_2 - \sigma_3)^2 + (\sigma_3 - \sigma_1)^2.$$

*Towards the final formula.* Since $f$ linearly depends on $V$, the failure condition $f \geq f_0$ is equivalent to $V \geq V_0$ for an appropriate the threshold $V_0$.

The threshold $V_0$ can be found out by considering the case of 1-D stress. In this case, e.g., when $\sigma_1 \neq 0$ and $\sigma_2 = \sigma_3 = 0$, we know that the failure occurs when $\sigma_1 \geq \sigma_f$. In this case, $V = 2\sigma_1^2$, so the failure occurs when $V$ reaches the level $V = \sigma_f^2$. Thus, $V_0 = 2\sigma_f^2$.

*Maxwell's approach: final criterion.* According to the Maxwell's formula, the material fails when $V \geq 2\sigma_f^2$, where

$$V \stackrel{\text{def}}{=} (\sigma_1 - \sigma_2)^2 + (\sigma_1 - \sigma_3)^3 + (\sigma_2 - \sigma_3)^2.$$

*Maxwell's mathematical solution becomes von Mises empirical failure criterion.* In 1913, von Mises experimentally confirmed that for many ductile materials, Maxwell's formula predicts failure well. Because of this confirmation, Maxwell's 1860s mathematical hypothesis is now known as a physically justified empirical fact called *von Mises criterion.*

According to this criterion, a ductile material fails under the general combination of three stresses $\sigma_1$, $\sigma_2$, and $\sigma_3$ applied at three orthogonal directions when $V \geq 2\sigma_f^2$, where $V \stackrel{\text{def}}{=} (\sigma_1 - \sigma_2)^2 + (\sigma_1 - \sigma_3)^3 + (\sigma_2 - \sigma_3)^2$.

A similar criterion $V \geq 2\sigma_y^2$ can also predict when the yielding starts. For details, see, e.g., [335].

*Need to take interval uncertainty into account.* In real life, we only know the values $\sigma_i$ with uncertainty.

*Case of interval uncertainty.* In some cases, we only know the bounds $\underline{\sigma}_i$ and $\overline{\sigma}_i$ on the actual (unknown) value of stress. In other words, we only know the interval $[\underline{\sigma}_i, \overline{\sigma}_i]$ that contains the actual (unknown) value $\sigma_i$.

*Main problem: checking whether a material can fail.* Different values $\sigma_i$ from the intervals $[\underline{\sigma}_i, \overline{\sigma}_i]$ lead, in general, to different values of the composite criterion $V$. Since the dependence of $V$ on $\sigma_i$ is continuous, in general, possible values of $V$ form an interval $[\underline{V}, \overline{V}]$.

There are three possible situations:

- if $\overline{V} < 2\sigma_f^2$, this means that all possible value of $V$ are below the failure threshold, so the material will not fail;
- if $\underline{V} \leq 2\sigma_f^2 < \overline{V}$, this means that the material may fail; on the other hand, it may survive without failure;

- if $2\sigma_f^2 \leq \underline{V}$, this means that all possible value of $V$ are above the failure threshold, so the material will fail.

In most practical situations, we are interesting in checking whether a material will not fail. To guarantee that the material will not fail, we must check that $\overline{V} < 2\sigma_f^2$. In other words, it is necessary to find the upper bound $\overline{V}$ of the set of all possible values of $V$ and check whether $\overline{V} < 2\sigma_f^2$.

In view of this need, in this chapter, we will design and analyze algorithms for computing $\overline{V}$.

*Comment.* In some situations, when we are analyzing the reason for the actual failure, we may want to check whether mechanical failure could have been a reason. In these situations, it is also desirable to compute the lower bound $\underline{V}$.

*Case of expert knowledge.* In some practical situations, we only have expert estimates describing possible stress values. These expert estimates are often described in terms of natural language. In such situations, it is reasonable to use fuzzy sets to formalize the expert knowledge.

From the computational viewpoint, as we have mentioned in Chapter 4, the case of fuzzy uncertainty can be indeed reduced to the case of interval uncertainty.

*Computing V is equivalent to computing variance.* From the mathematical viewpoint, $V$ is proportional to the sample variance of the observations $\sigma_i$.

*Computing variance under interval uncertainty: what is known.* The problem of computing sample variance under interval uncertainty has been thoroughly analyzed; see Part II of this book. In particular, it is shown that in general, the problem of computing the corresponding upper bound $\overline{V}$ is computationally difficult (NP-hard). Crudely speaking, NP-hard means that in some cases, we (most probably) have to spend exponential time $\sim c^n$ to solve this problem; for exact definitions, see, e.g., [117], [274], and Chapter 8.

It is also known that the upper bound $\overline{V}$ is always attained when each of the values $\sigma_i$ takes one of the extreme values $\underline{\sigma}_i$ or $\overline{\sigma}_i$; see, e.g., [101] and [102].

In other words, to compute $\overline{V}$, it is sufficient to consider $2^n$ possible combinations of values $\underline{\sigma}_i$ and $\overline{\sigma}_i$.

*Conclusion for von Mises criterion.* For von Mises criterion, the above result means that to compute $\overline{V}$, it is sufficient to consider $2^3 = 8$ possible combinations of values $\underline{\sigma}_i$ and $\overline{\sigma}_i$.

*Can we speed up computations?* In the above algorithm, we compute the expression $V$ eight times. Each computation of $V$ takes:

- 3 subtractions (to compute $\sigma_i - \sigma_j$),
- 3 multiplications (to compute the squares), and
- 2 additions (to compute $V$),

to the total of $3 \cdot 8 = 24$ multiplications and $(2 + 3) \cdot 8 = 40$ additions/subtractions.

Because of the practical importance of this problem, a natural question is: can we compute $\overline{V}$ faster? In this chapter, we will show that a speed up is indeed possible.

*Can we speed up: general result.* Let us first consider the general problem of estimating variance under interval uncertainty. We will prove that in general, we only need to consider $2^n - 2$ cases to find the upper bound for the variance, because the maximum is never attained when all the bounds are upper or all the bounds are lower. We also prove that, in general, we cannot pick fewer than $2^n - 2$ combinations.

In this section, we consider the general case: we have $n$ intervals $[\underline{x}_i, \overline{x}_i]$, and we want to compute the range $[\underline{V}, \overline{V}]$ of the population variance

$$V = \frac{1}{n} \sum_{i=1}^{n} (x_i - E)^2,$$

where $x_i \in [\underline{x}_i, \overline{x}_i]$, where $E \overset{\text{def}}{=} \frac{1}{n} \sum_{i=1}^{n} x_i$. It was previously known that to compute $\overline{V}$, it is sufficient to compute the value of $V$ for $2^n$ possible combinations $(x_1^{\varepsilon_1}, \ldots, x_n^{\varepsilon_n})$, where $\varepsilon_i \in \{-, +\}$, $x_i^+ \overset{\text{def}}{=} \overline{x}_i$, and $x_i^- \overset{\text{def}}{=} \underline{x}_i$. The value $\overline{V}$ is equal to the largest of the resulting $2^n$ values $V(x_1^{\varepsilon_1}, \ldots, x_n^{\varepsilon_n})$.

**Proposition 38.1.** *For every set of intervals* $[\underline{x}_1, \overline{x}_1]$, $\ldots$, $[\underline{x}_n, \overline{x}_n]$, *the value* $\overline{V}$ *is equal to the maximum of* $2^n - 2$ *values* $V(x_1^{\varepsilon_1}, \ldots, x_n^{\varepsilon_n})$ *for all* $(\varepsilon_1, \ldots, \varepsilon_n)$ *for which* $(\varepsilon_1, \ldots, \varepsilon_n) \neq (+, \ldots, +)$ *and* $(\varepsilon_1, \ldots, \varepsilon_n) \neq (-, \ldots, -)$.

**Proposition 38.2.** *For every tuple* $(\varepsilon_1, \ldots, \varepsilon_n)$ *for which* $(\varepsilon_1, \ldots, \varepsilon_n) \neq (+, \ldots, +)$ *and* $(\varepsilon_1, \ldots, \varepsilon_n) \neq (-, \ldots, -)$, *there exist* $n$ *intervals* $[\underline{x}_1, \overline{x}_1]$, $\ldots$, $[\underline{x}_n, \overline{x}_n]$ *for which the maximum* $\overline{V}$ *is only attained at the given tuple* $(\varepsilon_1, \ldots, \varepsilon_n)$ *and not attained at any other such* $\pm$ *tuple.*

*Comment.* This result is similar to the ones presented in [170].

*Conclusion for the von Mises case.* In the von Mises case, the above idea reduces the number of values $V$ to compute from 8 to 6. Thus, we only need $3 \cdot 6 = 18$ multiplications and $(2 + 3) \cdot 6 = 30$ additions/subtractions to compute the upper endpoint $\overline{V}$ corresponding to von Mises criterion.

*von Mises case: possibility of further speed up.* The possibility speed up comes from the fact that each value $V$ is the sum of 3 terms $(\sigma_i - \sigma_j)^2$.

For each of these 3 terms, there are only 4 options, corresponding to two choices of $\sigma_i = \underline{\sigma}_i$ and $\sigma_i = \overline{\sigma}_i$, and to the two similar choices for $\sigma_j$. For each choice, we need one subtraction to compute $\sigma_i - \sigma_j$ and one multiplication to compute the square. Thus, to compute the values of 4 options for each of these 3 terms, we need $3 \cdot 4 = 12$ subtractions and $3 \cdot 4 = 12$ multiplications.

To compute the values of all 6 expressions, we need to add 3 terms. Each computations takes 2 additions, so we need $6 \cdot 12 = 12$ additions. Thus, overall, we perform 12 multiplications and 24 additions/subtractions.

This is almost half of what we had originally.

*Detailed description.*

- First, we compute four squares

$$(\underline{\sigma}_1 - \underline{\sigma}_2)^2, \quad (\underline{\sigma}_1 - \overline{\sigma}_2)^2, \quad (\overline{\sigma}_1 - \underline{\sigma}_2)^2, \text{ and } (\overline{\sigma}_1 - \overline{\sigma}_2)^2.$$

- Then, we compute four squares

$$(\underline{\sigma}_2 - \underline{\sigma}_3)^2, \quad (\underline{\sigma}_2 - \overline{\sigma}_3)^2, \quad (\overline{\sigma}_2 - \underline{\sigma}_3)^2, \text{ and } (\overline{\sigma}_2 - \overline{\sigma}_3)^2.$$

- We compute four squares

$$(\underline{\sigma}_3 - \underline{\sigma}_1)^2, \quad (\underline{\sigma}_3 - \overline{\sigma}_1)^2, \quad (\overline{\sigma}_3 - \underline{\sigma}_1)^2, \text{ and } (\overline{\sigma}_3 - \overline{\sigma}_1)^2.$$

- Finally, we compute the six sums

$$(\underline{\sigma}_1 - \underline{\sigma}_2)^2 + (\underline{\sigma}_2 - \overline{\sigma}_3)^2 + (\overline{\sigma}_3 - \underline{\sigma}_1)^2;$$

$$(\underline{\sigma}_1 - \overline{\sigma}_2)^2 + (\overline{\sigma}_2 - \underline{\sigma}_3)^2 + (\underline{\sigma}_3 - \underline{\sigma}_1)^2;$$

$$(\underline{\sigma}_1 - \overline{\sigma}_2)^2 + (\overline{\sigma}_2 - \overline{\sigma}_3)^2 + (\overline{\sigma}_3 - \underline{\sigma}_1)^2;$$

$$(\overline{\sigma}_1 - \underline{\sigma}_2)^2 + (\underline{\sigma}_2 - \underline{\sigma}_3)^2 + (\underline{\sigma}_3 - \overline{\sigma}_1)^2;$$

$$(\overline{\sigma}_1 - \underline{\sigma}_2)^2 + (\underline{\sigma}_2 - \overline{\sigma}_3)^2 + (\overline{\sigma}_3 - \overline{\sigma}_1)^2;$$

$$(\overline{\sigma}_1 - \overline{\sigma}_2)^2 + (\overline{\sigma}_2 - \underline{\sigma}_3)^2 + (\underline{\sigma}_3 - \overline{\sigma}_1)^2.$$

- The largest of these six sums is the desired value $\overline{V}$.

*Important practical cases.* For a material in general shape, stresses can be of the same size. In practice, we often have a linear or a planar shape. In such cases, stresses in the direction of the shape are usually much larger than in the other directions:

- for a planar shape, we have $\sigma_3 \ll \sigma_1$ and $\sigma_3 \ll \sigma_2$; in precise terms, we have $\overline{\sigma}_3 < \underline{\sigma}_1$ and $\overline{\sigma}_3 < \underline{\sigma}_2$;
- for a linear shape when $\sigma_2 \ll \sigma_1$ and $\sigma_3 \ll \sigma_1$; in precise terms, $\overline{\sigma}_2 < \dfrac{1}{2} \cdot \underline{\sigma}_1$ and $\overline{\sigma}_3 < \dfrac{1}{2} \cdot \underline{\sigma}_1$.

*Planar case: analysis.* Here, $\sigma_1 \geq \underline{\sigma}_1 > \overline{\sigma}_3 \geq \sigma_3$ hence $\sigma_1 > \sigma_3$; similarly, $\sigma_2 > \sigma_3$, hence $\sigma_1 + \sigma_2 + \sigma_3 > 3\sigma_3$ and $E > \sigma_3$. As in the above proofs, we conclude that the maximum is attained when $\sigma_3 = \underline{\sigma}_1$. Due to Proposition 1, for $\overline{V}$, at least one of $\sigma_1$ and $\sigma_2$ is the lower bound, so we only need to compute $V$ for 3 tuples.

*Planar case: algorithm.* Here, we compute 3 values $(\sigma_1 - \sigma_2)^2$ (excluding both $\overline{x}_i$), 2 values $(\underline{\sigma}_3 - \sigma_1)^2$, and 2 values $(\underline{\sigma}_3 - \sigma_1)^2$ – to the total of 7 multiplications and 7 subtractions. After that, we need $3 \cdot 3 = 6$ additions to compute the needed 3 values of $V$. Overall, we need 7 multiplications and 13 additions/substractions.

*Linear case: analysis.* In this case, $\sigma_2 < \sigma_1$, $\sigma_3 < \sigma_1$, hence $E < \sigma_1$. Similarly, we have $\sigma_1 > 2\sigma_2$ hence $\sigma_1 + \sigma_2 + \sigma_3 > 3\sigma_2$ and $E > \sigma_2$ – and similarly $E > \sigma_2$. So, maximum is attained for $\overline{\sigma}_1$, $\underline{\sigma}_2$, and $\underline{\sigma}_3$.

*Linear case: algorithm.* So, in the linear shape, we only need to compute a single value

$$V = (\overline{\sigma}_1 - \underline{\sigma}_2)^2 + (\underline{\sigma}_2 - \underline{\sigma}_3)^3 + (\underline{\sigma}_3 - \overline{\sigma}_1)^2,$$

with 3 multiplications and 5 additions/substractions.

## Proofs

*Proof of Proposition 38.1.* Let us prove that to compute $\overline{V}$, there is no need to consider the tuple $(\overline{x}_1, \ldots, \overline{x}_n)$. If one of the intervals $[\underline{x}_i, \overline{x}_i]$ is *degenerate*, i.e., $\underline{x}_i = \overline{x}_i$, then this fact is trivially true because this same tuple can also be expressed in a different way, as

$$(\overline{x}_1, \ldots, \overline{x}_{i-1}, \underline{x}_i, \overline{x}_{i+1}, \ldots, \overline{x}_n).$$

So, to complete the proof, it is sufficient to consider the case when all the intervals are non-degenerate, i.e., when $\underline{x}_i < \overline{x}_i$ for all $i$.

Let $i_0$ be an index for which $\overline{x}_{i_0}$ is the smallest of the $n$ values $\overline{x}_i$. Let us show that in this case, replacing $\overline{x}_{i_0}$ in the tuple $(\overline{x}_1, \ldots, \overline{x}_n)$ with a slightly smaller value $x_{i_0}$ will increase $V$ – and thus, the maximum of the variance $V$ cannot be attained at the original all-maxima tuple $(\overline{x}_1, \ldots, \overline{x}_n)$.

Let us consider two cases: when all the upper endpoints $\overline{x}_i$ are the same and when some are different. If they are all the same, then for the all-maxima tuple, $V = 0$. If we replace one of them by a smaller value $x_{i_0} < \overline{x}_{i_0}$, some values $x_i$ will become different and we will get $V > 0$.

If some of the values $\overline{x}_i$ are different, then some of them are larger than the smallest bound $\overline{x}_{i_0}$ and thus, the average $E$ of the upper endpoints is also larger than $x_{i_0} = \overline{x}_{i_0}$: $x_{i_0} < E$.

It is known that

$$V = \frac{1}{n} \cdot \sum_{i=1}^{n} x_i^2 - E^2.$$

Thus,

$$\frac{\partial V}{\partial x_{i_0}} = \frac{1}{n} \cdot (2x_{i_0} - 2E) = \frac{2}{n} \cdot (x_{i_0} - E).$$

Since $x_{i_0} < E$, this derivative is negative, and thus, for slightly smaller values of $x_{i_0} < \overline{x}_{i_0}$, we will get larger values of $V$. So, in the non-degenerate case, the maximum $\overline{V}$ cannot be attained at an all-maxima tuples.

Similarly, we can prove that to compute $\overline{V}$, there is no need to consider the tuple $(\underline{x}_1, \ldots, \underline{x}_n)$. If one of the intervals $[\underline{x}_i, \overline{x}_i]$ is *degenerate*, i.e., $\underline{x}_i = \overline{x}_i$, then this fact is trivially true because this same tuple can also be expressed in a different way, as

$$(\underline{x}_1, \ldots, \underline{x}_{i-1}, \overline{x}_i, \underline{x}_{i+1}, \ldots, \underline{x}_n).$$

So, to complete the proof, it is sufficient to consider the case when all the intervals are non-degenerate, i.e., when $\underline{x}_i < \overline{x}_i$ for all $i$.

Let $i_0$ be an index for which $\underline{x}_{i_0}$ is the largest of the $n$ values $\underline{x}_i$. When all the lower endpoints $\underline{x}_i$ are the same, then for the all-minima tuple, $V = 0$. If we replace one of them by a larger value $x_{i_0} > \underline{x}_{i_0}$, some values $x_i$ will become different and we will get $V > 0$.

If some of the values $\underline{x}_i$ are different, then some of them are smaller than the largest bound $\underline{x}_{i_0}$ and thus, the average $E$ of the lower endpoints is also smaller than $x_{i_0} = \underline{x}_{i_0}$: $E < x_{i_0}$. In this case, $\dfrac{\partial V}{\partial x_{i_0}} = \dfrac{2}{n} \cdot (x_{i_0} - E) > 0$.
Since $x_{i_0} > E$, this derivative is positive, and thus, for slightly larger values of $x_{i_0} > \underline{x}_{i_0}$, we will get larger values of $V$. The proposition is proven.

*Proof of Proposition 38.2.* Let $(\varepsilon_1, \ldots, \varepsilon_n)$ be a tuple which is different from $(+, \ldots, +)$ and $(-, \ldots, -)$.

Let us fix some $\delta > 0$ (its exact value will be determined later), and let us take $[\underline{x}_i, \overline{x}_i] = [-1, -1 + \delta]$ when $\varepsilon_i = -$ and $[\underline{x}_i, \overline{x}_i] = [1 - \delta, 1]$ when $\varepsilon_i = +$. Since the tuple is different from all pluses, at least of these intervals is negative. For all intervals, $x_i \leq 1$, and for at least one negative interval, we have $x_i \leq -1 + \delta$. Thus, for the average $E = \dfrac{x_1 + \ldots + x_n}{n}$, we conclude that

$$E \leq \frac{(n-1) \cdot 1 + (-1 + \delta)}{n} = \frac{n - 2 + \delta}{n} = 1 - \frac{2 - \delta}{n}.$$

If $\dfrac{2 - \delta}{n} > \delta$, i.e., equivalently, if $2 - \delta > n \cdot \delta$, $2 > (n + 1) \cdot \delta$ and $\delta < \dfrac{2}{n+1}$, then we have $E < 1 - \delta$. Hence, for all $x_i$ from the positive intervals, we have $E < x_i$.

To guarantee this inequality, let us take $\delta = 1/(n + 1)$.

Similarly, since the tuple is different from all pluses, at least of these intervals is positive. For all intervals, $x_i \geq -1$, and for at least one positive interval, we have $x_i \leq 1 - \delta$. Since $\delta < \dfrac{2}{n+1}$, then we have $E > -1 + \delta$. Hence, for all $x_i$ from the negative intervals, we have $E > x_i$.

From the previous proof, we already know that $\dfrac{\partial V}{\partial x_i} = \dfrac{2}{n} \cdot (x_i - E)$. For $x_i$ from positive intervals, this derivative is positive, so $V$ is strictly increasing and its maximum is attained only when $x_i = \overline{x}_i$. Similarly, for $x_i$ from positive intervals, this derivative is negative, so $V$ is strictly decreasing and its maximum is attained only when $x_i = \underline{x}_i$. So, the maximum $\overline{V}$ is only attained for the values corresponding to the given tuple. The proposition is proven.

# Applications to Geophysics: Inverse Problem

In many real-life situations, we have several types of uncertainty: measurement uncertainty can lead to probabilistic and/or interval uncertainty, expert estimates come with interval and/or fuzzy uncertainty, etc. In many situations, in addition to measurement uncertainty, we have prior knowledge coming from prior data processing and/or prior knowledge coming from prior interval constraints.

In this chapter, on the example of the seismic inverse problem, we show how to combine these different types of uncertainty.

*In evaluations of natural resources and in the search for natural resources, it is very important to determine Earth structure.* Our civilization greatly depends on the things we extract from the Earth, such as fossil fuels (oil, coal, natural gas), minerals, and water. Our need for these commodities is constantly growing, and because of this growth, they are being exhausted. Even under the best conservation policies, there is (and there will be) a constant need to find new sources of minerals, fuels, and water.

The only sure-proof way to guarantee that there are resources such as minerals at a certain location is to actually drill a borehole and analyze the materials extracted. However, exploration for natural resources using indirect means began in earnest during the first half of the 20th century. The result was the discovery of many large relatively easy to locate resources such as the oil in the Middle East.

However, nowadays, most easy-to-access mineral resources have already been discovered. For example, new oil fields are mainly discovered either at large depths, or under water, or in very remote areas – in short, in the areas where drilling is very expensive. It is therefore desirable to predict the presence of resources as accurately as possible before we invest in drilling.

From previous exploration experiences, we usually have a good idea of what type of structures are symptomatic for a particular region. For example, oil and gas tend to concentrate near the top of natural underground domal structures. So, to be able to distinguish between more promising and less promising locations, it is desirable to determine the structure of the Earth

H.T. Nguyen et al.: Computing Stat. under Interval & Fuzzy Uncertain., SCI 393, pp. 317–329.
springerlink.com                                    © Springer-Verlag Berlin Heidelberg 2012

at these locations. To be more precise, we want to know the structure at different depths $z$ at different locations $(x, y)$.

*Data that we can use to determine the Earth structure.* In general, to determine the Earth structure, we can use different measurement results that can be obtained without actually drilling the boreholes: e.g., gravity and magnetic measurements, analyzing the travel-times and paths of seismic ways as they propagate through the earth, etc.

To get a better understanding of the Earth structure, we must rely on *active* seismic data – in other words, we must make artificial explosions, place sensors around them, and measure how the resulting seismic waves propagate. The most important information about the seismic wave is the *travel-time* $t_i$, i.e., the time that it takes for the wave to travel from its source to the sensor. To determine the geophysical structure of a region, we measure seismic travel times and reconstruct velocities at different depths from these data. The problem of reconstructing this structure is called the *seismic inverse problem.*

*Known algorithms for solving the seismic inverse problem: description, successes, limitations.* We want to find the values of the velocity $v(\mathbf{x})$ at different 3-D points $\mathbf{x}$. Based on the finite number of measurements, we can only reconstruct a finite number of parameters. So, we use a rectangular grid structure to divide the 3-D volume into box-shaped cells. We assume that the value of the velocity $v_j$ is the same within each cell, and we reconstruct the velocities $v_j$ within different cells.

*Algorithm for the forward problem: brief description.* Once we know the velocities $v_j$ in each cell $j$, we can then determine the paths which seismic waves take. Seismic waves travel along the shortest path – shortest in terms of time. It can be easily determined that for such paths, within each cell, the path is a straight line, and on the border between the two cells with velocities $v$ and $v'$, the direction of the path changes in accordance with Snell's law $\dfrac{\sin(\varphi)}{v} = \dfrac{\sin(\varphi')}{v'}$, where $\varphi$ and $\varphi'$ are the angles between the paths and the line orthogonal to the border between the cells. (If this formula results in $\sin(\varphi') > 1$, this means that this wave cannot penetrate into the neighboring cell at all; instead, it bounces back into the original cell with the same angle $\varphi$.)

In particular, we can thus determine the paths from the source to each sensor. The travel-time $t_i$ along $i$-th path can then be determined as the sum of travel-times in different cells $j$ through which this path passes: $t_i = \sum_j \dfrac{\ell_{ij}}{v_j}$, where $\ell_{ij}$ denotes the length of the part of $i$-th path within cell $j$.

This formula becomes closer to linear if we replace the original unknowns – velocities $v_j$ – by their inverses $s_j \stackrel{\text{def}}{=} \dfrac{1}{v_j}$, called *slownesses*. In terms of slownesses, the formula for the travel-time takes the simpler form

$$t_i = \sum_j \ell_{ij} \cdot s_j.$$

It is worth mentioning, however, that the resulting system of equations is *not* linear in the unknowns $s_j$. Indeed, the actual geometry of the shortest path between the two given points depends on the actual values of the velocities $v_j$ – i.e., equivalently, on the slownesses $s_j$. Thus, the lengths $\ell_{ij}$ of the segments of these shortest paths also depend on the slownesses $s_1, \ldots, s_m$. To be more precise, we should therefore explicitly take this dependence into account and re-write the above system as $t_i = \sum_j \ell_{ij}(s_1, \ldots, s_m) \cdot s_j$ for an appropriate non-linear dependence $\ell_{ij}(s_1, \ldots, s_m)$.

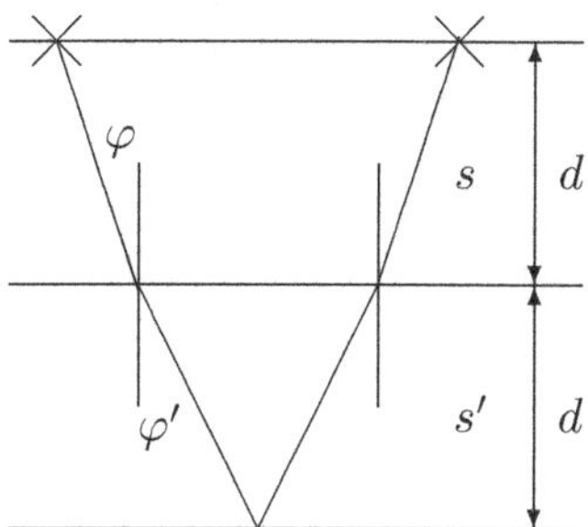


*Algorithm for the inverse problem: general description.* There are several algorithms for solving this inverse problem; see, e.g., [136, 277, 366]. The most widely used is the following iterative algorithm proposed by John Hole [136].

At each stage of this algorithm, we have some approximation to the desired slownesses. We start with some reasonable initial slownesses, and we hope that after several iterations, we will be able to get slownesses which are much closer to the actual values.

At each iteration, we first use the currently known slownesses $s_j$ to find the corresponding paths from the source to each sensor. Based on these paths, we compute the predicted values $t_i = \sum_j \ell_{ij} \cdot s_j$ of travel-times.

Since the currently known slownesses $s_j$ are only approximately correct, the travel-times $t_i$ (which are predicted based on these slownesses) are approximately equal to the measured travel-times $\widetilde{t}_i$; there is, in general, a discrepancy $\Delta t_i \stackrel{\text{def}}{=} \widetilde{t}_i - t_i \neq 0$. It is therefore necessary to use these discrepancies to update the current values of slownesses, i.e., replace the current values $s_j$ with corrected values $s_j + \Delta s_j$. The objective of this correction is to eliminate (or at least decrease) the discrepancies $\Delta t_i \neq 0$. In other words, the objective is to make sure that for the corrected values of the slowness, the predicted travel-times are closer to $\widetilde{t}_i$.

Of course, once we have changed the slownesses, the shortest paths will also change; however, if the current values of slownesses are reasonable, the differences in slowness are not large, and thus, paths will not change much. Thus, in the first approximation, we can assume that the paths are the same, i.e., that for each $i$ and $j$, the length $\ell_{ij}$ remains the same. In this approximation, the new travel-times are equal to $\sum \ell_{ij} \cdot (s_j + \Delta s_j)$. The desired condition is then $\sum \ell_{ij} \cdot (s_j + \Delta s_j) = \widetilde{t}_i$. Subtracting the formula $t_i = \sum_j \ell_{ij} \cdot s_j$ from this expression, we conclude that the corrections $\Delta s_j$ must satisfy the following system of (approximate) linear equations: $\sum \ell_{ij} \cdot \Delta s_j \approx \Delta t_i$.

Solving this system of linear equations is not an easy task, because we have many observations and many cell values and thus, many unknowns, and for a system of linear equations, computation time to solve it grows as a cube $c^3$ of the number of variables $c$. So, instead of the standard methods for solving a system of linear equations, researchers use special faster geophysics-motivated techniques (described below) for solving the corresponding systems. These methods are described, in detail, in the next subsection.

Once we solve the corresponding system of linear equations, we compute the updated values $\Delta s_j$, compute the new (corrected) slownesses $s_j + \Delta s_j$, and repeat the procedure again. We stop when the discrepancies become small; usually, we stop when the mean square error $\dfrac{1}{n} \sum\limits_{i=1}^{n} (\Delta t_i)^2$ no longer exceeds a given threshold. This threshold is normally set up to be equal to the measurement noise level, so that we stop iterations when the discrepancy between the model and the observations falls below the noise level – i.e., when, for all practical purposes, the model is adequate.

*Algorithm for the inverse problem: details.* Let us describe, in more detail, how the above auxiliary linear system of equations with unknown $\Delta s_j$ is usually solved. In other words, for a given cell $j$, how do we find the correction $\Delta s_j$ to the current value of slowness $s_j$ in this cell?

Let us first consider the simplified case when there is only path, and this path is going through the $j$-th cell. In this case, cells through which this path does not go do not need any correction. To find the corrections $\Delta s_j$ for all the cells $j$ through which this path goes, we only have one equation $\sum_j \ell_{ij} \cdot \Delta s_j = \Delta t_i$. The resulting system of linear equations is clearly under-determined: we have a single equation to find the values of several variables $\Delta s_j$. Since the system is under-determined, we have a infinite number of possible solutions. Our objective is to select the most geophysical reasonable of these solutions.

For that, we can use the following idea. Our single observation involves several cells; we cannot distinguish between the effects of slownesses in different cells, we only observe the overall effect. Therefore, there is no reason to assume that the value $\Delta s_j$ in one of these cells is different from the values in other cells. It is thus reasonable to assume that all these values are close to each other: $\Delta s_j \approx \Delta s_{j'}$. The least squares method enables us to describe this assumption as minimization of the objective function $\sum_{j,j'} (\Delta s_j - \Delta s_{j'})^2$ under the condition that $\sum \ell_{ij} \cdot \Delta s_j = \Delta t_i$. The minimum is attained when all the values $\Delta s_j$ are equal. Substituting these equal values into the equation $\sum_j \ell_{ij} \cdot \Delta s_j = \Delta t_i$, we conclude that $L_i \cdot \Delta s = \Delta t_i$, where $L_i = \sum_j \ell_{ij}$ is the overall length of $i$-th path. Thus, in the simplified case in which there is only one path, to the slowness of each cell $j$ along this path, we add the same value $\Delta s_j = \dfrac{\Delta t_i}{L_i}$.

Let us now consider the realistic case in which there are many paths, and moreover, for many cells $j$, there are many paths $i$ which go through the corresponding cell. For a given cell $j$, based on each path $i$ passing through this cell, we can estimate the correction $\Delta s_j$ by the corresponding value $\Delta s_{ij} \stackrel{\text{def}}{=} \dfrac{\Delta t_i}{L_i}$. Since there are usually several paths going through the $j$-th cell, we have, in general, several different estimates $\Delta s_j \approx \Delta s_{ij}$. Again, the least squares approach leads to $\sum_i (\Delta s_j - \Delta s_{ij})^2 \to \min$, hence to $\Delta s_j$ as the arithmetic average of the values $\Delta s_{ij}$.

*Comment.* To take into account that paths with larger $\ell_{ij}$ provide more information, researchers also used weighted average, with weight increasing with $\ell_{ij}$.

*Successes of the known algorithms.* The known algorithms have been actively used to reconstruct the slownesses, and, in many practical situations, they have led to reasonable geophysical models.

*Limitations of the known algorithms.* Often, the velocity model that is returned by the existing algorithm is not geophysically meaningful: e.g., it predicts velocities outside of the range of reasonable velocities at this depth. To avoid such situations, it is desirable to incorporate the expert knowledge into the algorithm for solving the inverse problem.

In our previous papers [18, 19, 151], we described how to do it. Specifically, we proposed a $O(c \log(c))$ time algorithm for taking interval prior knowledge into account.

In this chapter, we provide a detailed motivation for that algorithm, and we use this motivation to design a new, faster, linear-time ($O(c)$) for solving this problem.

*Interval prior knowledge.* For each cell $j$, a geophysicist often provides us with his or her estimate of possible values of the corresponding slowness $s_j$. Often, this estimates comes in the form of an interval $[\underline{s}_j, \overline{s}_j]$ that is guaranteed to contain the (unknown) actual value of slowness.

It is desirable to modify Hole's algorithm in such a way that on all iterations, slownesses $s_j$ stay within the corresponding intervals. Such a modification is described in [18, 19, 151].

*Analysis of the problem and our main idea.* Once we know the current approximations $s_j^{(k)}$ to slownesses, then, along each path $i$, we want to find the corrections $\Delta s_{ij}$ which provide the desired compensation, i.e., for which

$$\sum_{j=1}^{c} \ell_{ij} \cdot \Delta s_{ij} = \Delta t_i. \tag{39.1}$$

In Hole's algorithm, we select $\Delta s_{ij} = \dfrac{\Delta t_i}{L_i}$. With the additional knowledge, we may not be able to do this, because we want to make sure that the corrected values of slowness stay within the corresponding intervals

$$\underline{s}_j \le s_j^{(k)} + \Delta s_{ij} \le \overline{s}_j, \tag{39.2}$$

i.e., equivalently, that

$$\underline{\Delta}_j \le \Delta s_{ij} \le \overline{\Delta}_j, \tag{39.3}$$

where $\underline{\Delta}_j \overset{\text{def}}{=} \underline{s}_j - s_j^{(k)}$ and $\overline{\Delta}_j \overset{\text{def}}{=} \overline{s}_j - s_j^{(k)}$. Since $s_j^{(k)} \in [\underline{s}_j, \overline{s}_j]$, we conclude that $\underline{\Delta}_j \le 0$ and $\overline{\Delta}_j \ge 0$ – i.e., all lower endpoints are non-positive and all upper endpoints are non-negative.

How can we achieve this goal?

For each cell $j$, after an iteration of, say, Hole's algorithm, we have a corrected value of the slowness $s_j^{(k+1)} = s_j^{(k)} + \Delta s_{ij}$ which approximates the actual (unknown) slowness $s_j$: $s_j \approx s_j^{(k+1)}$. We also know that $s_j$ should be located in the interval $[\underline{s}_j, \overline{s}_j]$. Similar to our previous analysis, it is therefore reasonable to use the Least Squares Method to combine these two piece of information: i.e., we look for the value $s_j \in [\underline{s}_j, \overline{s}_j]$ for which the square $(s_j - s_j^{(k)})^2$ is the smallest possible. In geometric terms, we look for the value within the given interval $[\underline{s}_j, \overline{s}_j]$ which is the closest to $s_j^{(k+1)}$. Thus:

- If the value $s_j^{(k+1)}$ is already within the interval, we keep it intact.
- If the value $s_j^{(k+1)}$ is to the left of the interval, i.e., if $s_j^{(k+1)} < \underline{s}_j$, then the closest point from the interval is its left endpoint $\underline{s}_j$.
- Similarly, if the value $s_j^{(k+1)}$ is to the right of the interval, i.e., if $s_j^{(k+1)} > \overline{s}_j$, then the closest point from the interval is its right endpoint $\overline{s}_j$.

In other words, e.g., for $\Delta t_i > 0$, we first find the universal value $\Delta s$ and then, for those $j$ for which $\Delta s > \overline{\Delta}_j$, we replace this value with $\overline{\Delta}_j$.

As a result, we arrive at the values $\Delta s_{ij}$ which are all equal to $\Delta s$ – except for those values for which $\overline{\Delta}_j < \Delta s$; for these values, $\Delta s_{ij} = \overline{\Delta}_j$.

*Complications coming from a straightforward application of this idea.* Originally, before we took interval prior knowledge into account, we had a full compensation for $\Delta t_i$. Now that we decreased some slownesses $\Delta s_{ij}$, the resulting value of $\sum_{j=1}^{c} \ell_{ij} \cdot \Delta s_{ij}$ is, in general, smaller than $\Delta t_i$. Thus, there is a remaining discrepancy $\Delta t_i' \overset{\text{def}}{=} \Delta t_i - \sum_{j=1}^{c} \ell_{ij} \cdot \Delta s_{ij} > 0$.

To eliminate this discrepancy, we need to repeat the same procedure: divide $\Delta t_i'$ by $L_i$ and again cut down those slownesses that start going outside the corresponding intervals. Because of this cutting down, we may still get some discrepancy remaining, etc.

So, if we apply this idea in a straightforward way, we may need a large number of iterations to fully compensate for the original travel time discrepancy. The need for a large number of iterations leads to a drastic increase in computation time – which, for the seismic inverse problems, is already large.

It is therefore desirable to avoid these iterations and directly come up with a solution which provides the needed compensation of the travel time and at the same time, keeps all the corrected slownesses within the corresponding intervals.

*Formulation of the problem in precise terms.* For $\Delta t_i > 0$, we would like to find a value $\Delta s > 0$ such that if we take $\Delta s_{ij} = \Delta s$ for all $j$ for which $\Delta s \leq \overline{\Delta}_j$ and $\Delta s_{ij} = \overline{\Delta}_j$ for all other $j$, then we will satisfy the equation (39.1).

For $\Delta t_i < 0$, we would like to find a value $\Delta s < 0$ such that if we take $\Delta s_{ij} = \Delta s$ for all $j$ for which $\Delta s \geq \underline{\Delta}_j$ and $\Delta s_{ij} = \underline{\Delta}_j$ for all other $j$, then we will satisfy the equation (39.1).

*Analysis of the problem.* In the desired solution, we have $\Delta s_{ij} = \overline{\Delta}_j$ for the values $j$ for which $\overline{\Delta}_j$ is smaller than a certain threshold.

This desired solution is easier to describe if we first soft all the values $\overline{\Delta}_j$ into a non-decreasing sequence

$$\overline{\Delta}_{(1)} \leq \overline{\Delta}_{(2)} \leq \ldots \leq \overline{\Delta}_{(c)}.$$

Then, in the desired solution, there is some index $p$ for which $\Delta s_{i(j)} = \overline{\Delta}_{(j)}$ for all $j \leq p$. The common value $\Delta s$ for the indices $j > p$ can be found from the condition (39.1), i.e., from the condition that $A_p + \mathcal{L}_p \cdot \Delta s = \Delta t_i$, where we denoted $A_p \overset{\text{def}}{=} \sum_{i=1}^{p} \ell_{(i)j} \cdot \overline{\Delta}_{(j)}$ and $\mathcal{L}_p \overset{\text{def}}{=} \sum_{j=p+1}^{c} \ell_{i(j)}$. Therefore, we will get

$$\Delta s = \frac{\Delta t_i - A_p}{\mathcal{L}_p}.$$

For the correctly selected index $p$, all values $\overline{\Delta}_{(j)}$ for which we "cut off" must be smaller than this $\Delta s$, and all the other values $\overline{\Delta}_{(j)}$ must be larger than (or equal to) this $\Delta s$. Since the values $\overline{\Delta}_{(j)}$ are sorted in increasing order, it is sufficient to check that $\overline{\Delta}_{(p)} < \Delta s \leq \overline{\Delta}_{(p+1)}$.

If for some $p$, we get $\Delta s > \overline{\Delta}_{(p+1)}$, this means that need to cut some more – otherwise, for $j = p + 1$, we will still have the value outside the desired interval. On the other hand, if we get $\Delta s \leq \overline{\Delta}_{(p)}$, then there was no reason to cut off at $p$-th level – so we need to cut less.

*Designing an algorithm.* This analysis can be naturally be translated into an algorithm. First, we sort the values $\overline{\Delta}_j$; sorting takes time $O(c \cdot \log(c))$; see, e.g., [73]. Then, for every $p$ from 0 to $n$, we compute the value $\Delta s = \dfrac{\Delta t_i - A_p}{\mathcal{L}_p}$

and check whether $\overline{\Delta}_{(p)} < \Delta s \leq \overline{\Delta}_{(p+1)}$. Once we know $A_p$, computing $A_{p+1}$ takes just one step – since we need to add one term to the sum. Thus, we to compute all $c$ such values, we perorm $O(c)$ steps – to the total of $O(c \cdot \log(c)) + O(c) = O(c \cdot \log(c))$. So, we arrive at the following algorithm.

*Resulting algorithm.* It is sufficient to describe the case when $\Delta t_i > 0$ (the case when $\Delta t_i < 0$ is treated similarly). In this case, we first sort all $c$ values $\overline{\Delta}_j$ along the $i$-th path into a non-decreasing sequence

$$\overline{\Delta}_{(1)} \leq \overline{\Delta}_{(2)} \leq \dots \leq \overline{\Delta}_{(c)}.$$

Then, for every $p$ from 0 to $c$, we compute the values $A_p$ and $\mathcal{L}_p$ as follows: $A_0 = 0$, $\mathcal{L}_0 = L_i$,

$$A_p = A_{p-1} + \ell_{i(p)} \cdot \overline{\Delta}_{(p)}, \quad \mathcal{L}_p = \mathcal{L}_{p-1} - \ell_{i(p)}.$$

After that, for each $p$, we compute $\Delta s = \dfrac{\Delta t_i - A_p}{\mathcal{L}_p}$ and check whether $\overline{\Delta}_{(p)} < \Delta s \leq \overline{\Delta}_{(p+1)}$. Once this condition is satisfied, we take $\Delta s_{i(j)} = \overline{\Delta}_{(j)}$ for $j \leq p$, and $\Delta s_{i(j)} = \Delta s$ for $j > p$.

When $\Delta t_i < 0$, we similarly sort the values $\underline{\Delta}_j$ into a decreasing sequence, and find $p$ so that the first $p$ corrections are "maxed out" to $\underline{\Delta}_j$, and the rest $c - p$ corrections are determined from the condition $\Delta s = \dfrac{\Delta t_i - A_p}{\mathcal{L}_p}$.

*Comment.* Once we have computed these corrections for all the paths, then for each cell $j$, we take the average (or weighted average) of all the corrections coming from all the paths which pass through this cell.

*Example showing efficiency (and feasibility) of the new approach.* Let us consider a simple example of two vertical layers of height $d$ (see above picture), with $s > s'$. We assume that the structure below the second layer is so heavy that all the signals simply bounce back from the bottom of the second layer (in real geological situations, this is what happens, e.g., at the Moho surface). For simplicity, we consider only one signal.

Usually, the closer to the surface, the more information we have about the layer. In this example, we assume that we know $s$ exactly, but we only know an approximate value $\widetilde{s}'$ for $s'$ ($\Delta s' \overset{\text{def}}{=} \widetilde{s}' - s' \neq 0$). We start with the known values $s$ and $\widetilde{s}'$ and perform iterations following both the original Hole's algorithm and the new interval method.

When the angles $\varphi$ and $\varphi'$ are small ($\varphi \ll 1$, $\varphi' \ll 1$), then $\sin(\varphi) \approx \varphi$, $\sin(\varphi') \approx \varphi'$, and we can analytically trace the computations; for details, see [18]. For example, the horizontal distance between the source and the sensor is $2d \cdot (\tan(\varphi) + \tan(\varphi')) \approx 2d \cdot (\varphi + \varphi')$.

In the original Hole's algorithm, the discrepancy in the travel times is uniformly divided between the whole path. As a result, we replace the original approximate slowness $\widetilde{s}\,' = s' + \Delta s'$ with a more accurate estimate $s' + \dfrac{\Delta s'}{2}$. Hence, the approximation error decreases by a factor of 2. So, e.g., in 7 iterations, we can reduce this error to $< 1\%$ level.

In the new method, we take into account that the value $s$ is already known, i.e., that it is within the given interval $[s, s]$. In this case, the entire discrepancy is corrected by changing only the value $s'$. Hence, we get the correct value $s'$ in a single iteration.

*Case of interval prior knowledge: a new linear time algorithm.* As we have mentioned, the original Hole's code formulas are related to minimize the variance under a linear constraint (39.1).

In general, the problem of minimizing variance under interval uncertainty has many other practical applications beyond geophysics. (The only difference is that in most applications, there is no linear constraint similar to (39.1)). In particular, this general problem has application in geophysics; see, e.g., [255] and [256].

For this general problem, we have also proposed a linear time algorithm (see above). In this chapter, we show that a similar linear-time algorithm can be proposed for the case when we want to minimize the variance under an additional linear constraint.

*An auxiliary algorithm behind the existing linear-time algorithm.* The linear-time algorithm for estimating variance is based on the known fact that we can compute the median of a set of $n$ elements in linear time; see, e.g., [73].

*A new linear-time algorithm.* The proposed algorithm is iterative. At each iteration of this algorithm, we have three sets:

- the set $J^-$ of all the indices $j$ from 1 to $c$ for which we already know that in the desired solution, the corresponding value $\Delta s_{ij}$ will be cut off (i.e., $\Delta s_{ij} = \overline{\Delta}_j$);
- the set $J^+$ of all the indices $j$ for which we already know that in the desired solution, the corresponding value $\Delta s_{ij}$ will *not* be cut off (i.e., $\Delta s_{ij} < \overline{\Delta}_j$);
- the set $J = \{1, \ldots, c\} - J^- - J^+$ of the indices $j$ for which we are still undecided.

In the beginning, $J^- = J^+ = \emptyset$ and $J = \{1, \ldots, c\}$. At each iteration, we also update the values of two auxiliary quantities $A^- \overset{\text{def}}{=} \sum\limits_{j \in J^-} \ell_{ij} \cdot \overline{\Delta}_j$ and $\mathcal{L}^+ \overset{\text{def}}{=} \sum\limits_{j \in J^+} \ell_{ij}$. In principle, we could compute these values by computing these sums, but to speed up computations, on each iteration, we update these two auxiliary values in a way that is faster than re-computing the corresponding two sums. Initially, since $J^- = J^+ = \emptyset$, we take $A^- = \mathcal{L}^+ = 0$.

At each iteration, we do the following:

- first, we compute the median $m$ of the set $J$ (median in terms of sorting by $\overline{\Delta}_j$);

- then, by analyzing the elements of the undecided set $J$ one by one, we divide them into two subsets

$$P^- \stackrel{\text{def}}{=} \{j : \overline{\Delta}_j \le \overline{\Delta}_m\}, \quad P^+ \stackrel{\text{def}}{=} \{j : \overline{\Delta}_j > \overline{\Delta}_m\};$$

- we compute $a^- \stackrel{\text{def}}{=} A^- + \sum_{j \in P^-} \ell_{ij} \cdot \overline{\Delta}_j$ and

$$\ell^+ \stackrel{\text{def}}{=} \mathcal{L}^+ + \sum_{j \in P^+} \ell_{ij};$$

- then, we compute $\Delta s = \dfrac{\Delta_i - a^-}{\ell^+}$; also, among all the values from $P^+$, we select the smallest value, which we will denote by $\overline{\Delta}_{(p+1)}$;
- if $\Delta s > \overline{\Delta}_{(p+1)}$, then we replace $J^-$ with $J^- \cup P^-$, $A^-$ with $a^-$, and $J$ with $P^+$;
- if $\Delta s \le \overline{\Delta}_m$, then we replace $J^+$ with $J^+ \cup P^+$, $\mathcal{L}^+$ with $\ell^+$, and $J$ with $P^-$;
- finally, if $\overline{\Delta}_m < \Delta s \le \overline{\Delta}_{(p+1)}$, then we replace $J^-$ with $J^- \cup P^-$, $J^+$ with $J^+ \cup P^+$, and $J$ with $\emptyset$.

At each iteration, the set of undecided indices is divided in half. Iterations continue until all indices are decided, after which we return $\Delta s_{ij} = \overline{\Delta}s_j$ when $\overline{\Delta}_j \le \overline{\Delta}_m$ and $\Delta s_{ij} = \Delta s$ otherwise.

*Proof that the new algorithm for computing $\overline{V}$ takes linear time.* At each iteration, computing median takes linear time, and all other operations with $J$ take time $t$ linear in the number of elements $|J|$ of $J$: $t \le C \cdot |J|$ for some constant $C$. We start with the set $J$ of size $c$; on the next iteration, we have a set of size $c/2$, then $c/4$, etc. Thus, the overall computation time is $\le C \cdot (c + c/2 + c/4 + \ldots) \le C \cdot 2c$, i.e., linear in $c$.

*Case of probabilistic prior knowledge.* Often, prior information comes from processing previous observations of the region of interest. In this case, before our experiments, for each cell $j$, we know a prior (approximate) slowness value $\widetilde{s}_j$, and we know the accuracy (standard deviation) $\sigma_j$ of this approximate value $\widetilde{s}_j$. It is known that this prior information can lead to much more accurate velocity models; see, e.g., [211]. How can we modify Hole's algorithm so that it takes this prior information into account?

Due to the prior knowledge, for each cell $j$, the ratio $\dfrac{(s_j^{(k)} + \Delta s_{ij}) - \widetilde{s}_j}{\sigma_j}$ is normally distributed with 0 mean and variance 1. Since each path $i$ consists of a reasonable number of cells, we can thus conclude that the sample variance of this ratio should be close to $\sigma_j$, i.e., that

$$\frac{1}{n} \cdot \sum_{j=1}^{c} \frac{((s_j^{(k)} + \Delta s_{ij}) - \widetilde{s}_j)^2}{\sigma_j^2} = 1. \tag{39.4}$$

So, to find the corrections $\Delta s_{ij}$, we must minimize the objective function (variance)

$$V \stackrel{\text{def}}{=} \frac{1}{n} \cdot \sum_{j=1}^{c} \Delta s_{ij}^2 - \left( \frac{1}{n} \cdot \sum_{j=1}^{c} \Delta s_{ij} \right)^2 . \tag{39.5}$$

under the constraints (39.1) and (39.4).

By applying the Lagrange multiplier method to this problem, we can reduce this problem to the unconstrained minimization problem

$$\frac{1}{n} \cdot \sum_{j=1}^{c} \Delta s_{ij}^2 - \left( \frac{1}{n} \cdot \sum_{j=1}^{c} \Delta s_{ij} \right)^2 +$$

$$\lambda \cdot \left( \sum_{j=1}^{c} \ell_{ij} \cdot \Delta s_{ij} - \Delta t_i \right) +$$

$$\mu \cdot \frac{1}{n} \cdot \sum_{j=1}^{c} \frac{(s_j^{(k)} + \Delta s_{ij} - \widetilde{s}_j)^2}{\sigma_j^2} \to \min . \tag{39.6}$$

Differentiating this equation by $\Delta s_{ij}$ and equating the derivative to 0, we conclude that

$$\frac{2}{n} \cdot \Delta s_{ij} - \frac{2}{n} \cdot \overline{\Delta s} + \lambda \cdot \ell_{ij} + \frac{2\mu}{n \cdot \sigma_j^2} \cdot (s_j^{(k)} + \Delta s_{ij} - \widetilde{s}_j) = 0,$$

where

$$\overline{\Delta s} \stackrel{\text{def}}{=} \frac{1}{n} \cdot \sum_{j=1}^{c} \Delta s_{ij}. \tag{39.7}$$

Once we fix $\lambda$, $\mu$, and $\overline{\Delta s}$, we get an explicit expression for the values $\Delta s_{ij}$. Substituting these expressions into the equations (39.1), (39.4), and (39.7), we get an easy-to-solve system of 3 non-linear equations with 3 unknowns, which we can solve, e.g., by using Newton's method.

Now, instead of explicit formulas for a transition from $s_j^{(k)}$ to $s_j^{(k+1)}$, we need a separate iteration process – so the computation time is somewhat larger, but we get a more geophysically meaningful velocity map – that takes prior knowledge into account.

*Case of multiple-type prior knowledge.* In many real-life problems, it is difficult or even impossible to directly measure the desired physical quantities. In such situations, we measure other quantities, which are related to the desired ones by known formulas, and then reconstruct the values of the desired quantities from these measurement results.

The reconstructed values of the desired quantities are sometimes outside the range of what an expert would consider reasonable. In such situations, it is desirable to describe the expert's knowledge (about what is reasonable) as a precisely formulated constraint on the desired values, and to incorporate these constraints into the reconstruction process.

In the previous sections, we have shown that different types of expert knowledge can be naturally formalized in interval, fuzzy, and probabilistic terms. We also showed, on the example of the seismic inverse problem, how each of these types of expert knowledge can be used in the solution process.

*Practical need for multiple-type prior knowledge.* Previously, we (implicitly) assumed that we have only one type of expert knowledge – e.g., only interval knowledge, or only fuzzy knowledge, etc. In some practical situations, however, we may have multiple-type expert knowledge: e.g., one expert provides interval bounds, another expert provides probabilistic knowledge, etc.

This multiple-type prior knowledge is especially important for *cyberinfrastructure.* The main objective of cyberinfrastructure is to be able to seamlessly move data between different databases (where this data is stored in different formats), to feed the combined data into a remotely located program (which may require yet another data format), and to return the result to the user; see, e.g., [9, 152, 307]. It is also important to gauge the quality and accuracy of this result. We often have different models for describing uncertainty of different databases and programs; it is therefore important to be able to consider multiple-type prior knowledge; see, e.g., [119] and [207].

*How to use multiple-type prior knowledge in the seismic inverse problem.* We have mentioned that in the traditional approach, we minimize (39.5) under the constraint (39.1). Different types of prior knowledge mean adding constraints on $\Delta s_{ij}$. Probabilistic prior knowledge is naturally formalized as a constraint (39.4), and interval prior knowledge is naturally formalized as a constraint (39.2). Thus, when both probabilistic and interval prior knowledge are present, we must minimize (39.5) under the constraints (39.1), (39.2), and (39.4).

If we replace the equality in (39.4) by an inequality ($\leq 1$ instead of $=$ 1), then we get a problem of minimizing a convex function under convex constraints, a problem for which there are known efficient algorithms; see, e.g., [334].

For example, we can use a method of alternating projections, in which we first add a correction that satisfy the first constraint, then the additional correction that satisfies the second constraint, etc. In our case, we first add equal values of $\Delta s_{ij}$ to satisfy the constraint (39.5), then we restrict the values to the nearest points from the interval $[\underline{s}_j, \overline{s}_j]$ – to satisfy the constraint (39.2), and after that, find the extra corrections that satisfy the condition (39.4), after which we repeat the cycle again until the process converges.

*Conclusion.* The chapter deals with the difficult *seismic inverse problem,* in which a 3-D field (velocities of the seismic waves) has to be reconstructed.

The classical approach is to transform this problem into a huge non-linear system of equation and to use iterative techniques to solve the problem. Often, the classical approach leads to solutions that are not realistic. However, the expert has an idea of what he should not get and he can express this idea as a set of constraints. The main contribution of the chapter is to add these additional knowledge, given by the expert, to the classical approach, inside the iterative method.

# Part V

# Beyond Interval and Fuzzy Uncertainty

# Need to Go Beyond Interval and Fuzzy Uncertainty

*Types of uncertainty that we analyzed so far.* In the previous chapters, we described the uncertainty of inputs – and the resulting uncertainty in the values of the statistical characteristics and, more generally, the result of data processing. To characterize this uncertainty, we used the following three types of information.

First, we used the information about which values are possible and which values are not possible. In general, such an information can be described by a set. Since most real-world processes are continuous, the set of all possible values is usually connected. In the 1-D case, this means that we have an *interval*. In the multi-D case, if we have interval bounds on each of the variables, we have a box.

Second, in addition to the information about which values are possible, we used the information about the relative frequency (probability) of different possible values. To describe this information, we used the corresponding statistical characteristics such as moments or values of cdf $F(x) = \mathrm{Prob}(X \leq x)$ – or, alternatives, the interval bounds on the values of these characteristics: bounds on the moments and bounds on the cdf values (i.e., p-boxes).

Third, we used information provided by the experts. This information was described in terms of fuzzy *degrees* $\mu(x)$ – usually, numbers from the interval $[0, 1]$ – that describe the expert's confidence that different values $x$ arc possiblc.

*Need to go beyond these types of uncertainty.* For all these three types of information, there is a need to go beyond the above descriptions.

For *set* information, in addition to the interval bounds on each variables $x_1, \ldots, x_n$, we may have additional information: e.g., we may know that the actual values should satisfy a constraint $g(x_1, \ldots, x_n) \leq g_0$. As we have mentioned earlier, usually, we know the approximate values of $x_i$, so we can safely replace the function $g(x_1, \ldots, x_n)$ by, e.g., the first two terms in its Taylor expansion. In this case, the constraint becomes quadratic, and – in a realistic case when this constraint describes a bounded set – the set of all the tuples $x = (x_1, \ldots, x_n)$ that satisfy this constraint forms an *ellipsoid*. In this

H.T. Nguyen et al.: Computing Stat. under Interval & Fuzzy Uncertain., SCI 393, pp. 333–334.
springerlink.com                     © Springer-Verlag Berlin Heidelberg 2012

case, in addition to knowing that the actual tuple $x$ belongs to the box, we also know that it belongs to the ellipsoid – i.e., that the set of possible values of this tuple is an *intersection* of the box and the ellipsoid. This situation is analyzed in Chapter 41.

Another need to go beyond interval boxes comes from the fact that intervals are motivated by continuity, but some processes in nature – such as phase transitions – are, from the practical viewpoint, discontinuous. This situation is analyzed in Chapter 42.

For *probabilistic* information, we may also have an additional information about the corresponding probability distribution $F(x)$. This additional information can range from vague to very precise:

- We may simply know that this dependence is smooth, without having any more detailed knowledge; this situation is analyzed in Chapter 43.
- In addition to knowing that $F(x)$ is smooth – i.e., that its derivative $F'(x)$ (= a probability density function) is bounded – we sometimes also know the bounds on $F'(x)$; this situation is analyzed in Chapter 44.
- Sometimes, we even know the analytical expression for $F(x)$ – with the parameters which are only known with uncertainty. For example, in practice, when the observed signal is caused by a joint effect of many small components, it is reasonable to assume that the distribution is normal – but the parameters of this normal distribution are only known with uncertainty; this situation is analyzed in Chapter 45.

Finally, for *fuzzy* information, we assumed that we have exact numerical degrees describing expert uncertainty. This is, of course, a simplifying assumption. In practice, an expert can, at best, provide bounds (i.e., an interval) or his or her degree of certainty – or even produce a *fuzzy* degree of certainty (such as "about 0.6"). Situations with interval-valued fuzzy degrees are analyzed in Chapter 46, and the situations with more general fuzzy-valued degrees (called *type 2*) are analyzed in Chapter 47.

# Beyond Interval Uncertainty: Taking Constraints into Account

For *set* information, in addition to the interval bounds on each variables $x_1, \ldots, x_n$, we may have additional information: e.g., we may know that the actual values should satisfy a constraint $g(x_1, \ldots, x_n) \leq g_0$. As we have mentioned earlier, usually, we know the approximate values of $x_i$, so we can safely replace the function $g(x_1, \ldots, x_n)$ by, e.g., the first two terms in its Taylor expansion. In this case, the constraint becomes quadratic, and – in a realistic case when this constraint describes a bounded set – the set of all the tuples $x = (x_1, \ldots, x_n)$ that satisfy this constraint forms an *ellipsoid*. In this case, in addition to knowing that the actual tuple $x$ belongs to the box, we also know that it belongs to the ellipsoid – i.e., that the set of possible values of this tuple is an *intersection* of the box and the ellipsoid. Such a situation is analyzed in this chapter.

## Formulation and Analysis of the Problem, and Corresponding Results and Algorithms

In many real-life situations, we do not know the probability distribution of measurement errors $(\Delta x_1, \ldots, \Delta x_n)$, we only know the upper bounds $\Delta_i$ on these errors. In such situations, once we know the measurement results $\widetilde{x}_1, \ldots, \widetilde{x}_n$, we can only conclude that the actual (unknown) values of the quantity $x_i$ belongs to the interval $x_i = [\widetilde{x}_i - \Delta_i, \widetilde{x}_i + \Delta_i]$. Based on this interval uncertainty, we want to find the range of possible values of the desired quantity $y = f(x_1, \ldots, x_n)$. In general, computing this range is an NP-hard problem, but in the linear approximation when $f = \widetilde{y} + \sum_{i=1}^{n} c_i \, \Delta x_i$, we have a linear time algorithm for computing the range.

In other situations, we know the ellipsoid that contains the actual values $(\Delta x_1, \ldots, \Delta x_n)$; in the reasonable case of "independent" variables, we have

H.T. Nguyen et al.: Computing Stat. under Interval & Fuzzy Uncertain., SCI 393, pp. 335–347.
springerlink.com

an ellipsoid $E$ of the type $\sum\limits_{i=1}^{n} \dfrac{\Delta x_i^2}{\sigma_i^2} \le r^2$. In this case, we also have a linear time algorithm for computing the range of a linear function $f$.

In some cases, however, we have a combination of interval and ellipsoid uncertainty. In this case, the actual values $(\Delta x_1, \ldots, \Delta x_n)$ belong to the *intersection* of the box $\boldsymbol{x}_1 \times \ldots \times \boldsymbol{x}_n$ and the ellipsoid. In general, estimating the range over the intersection enables us to get a narrower range for $f$. In this chapter, we provide two algorithms for estimating the range of a linear function over an intersection in linear time: a simpler $O(n \log(n))$ algorithm and a (somewhat more complex) linear time algorithm. Both algorithms can be extended to the $l^p$-case, when instead of an ellipsoid we have a set $\sum\limits_{i=1}^{n} \dfrac{|\Delta x_i|^p}{\sigma_i^p} \le r^p$.

*Interval uncertainty: brief reminder.* Measurements are never 100% accurate; hence, the measurement result $\widetilde{x}_i$ is, in general, different from the actual (unknown) value $x_i$ of the corresponding quantity. Traditional engineering approach to processing measurement uncertainty assumes that we know the probability distribution of measurement errors $\Delta x_i := \widetilde{x}_i - x_i$.

In many practical situations, however, we do not know these probability distributions. In particular, in many real-life situations, we only know the upper bound $\Delta_i$ on the (absolute value of the) measurement error: $|\Delta x_i| \le \Delta_i$. In such situations, the only information that we get about the actual (unknown) value $x_i$ after the measurement is that $x_i$ belongs to the interval $\boldsymbol{x}_i = [\widetilde{x}_i - \Delta_i, \widetilde{x}_i + \Delta_i]$.

*Data processing under interval uncertainty: brief reminder.* In addition to the values of the measured quantities $x_1, \ldots, x_n$, we often need to know the values of other quantities which are related to $x_i$ by a known dependence $y = f(x_1, \ldots, x_n)$. When we know $x_i$ with interval uncertainty, i.e., when we know that $x_i \in \boldsymbol{x}_i$, then the only conclusion about $y$ is that $y$ belongs to the range $\{f(x_1, \ldots, x_n) \,|\, x_1 \in \boldsymbol{x}_1, \ldots, x_n \in \boldsymbol{x}_n\}$ of the function $f(x_1, \ldots, x_n)$ over the box $\boldsymbol{x}_1 \times \ldots \times \boldsymbol{x}_n$.

*Data processing: linear approximation.* In general, computing this range is NP-hard – even for quadratic functions $f$; see, e.g., [182]. However, in many practical situations, the measurement errors are small, thus, the intervals $\boldsymbol{x}_i$ are narrow, and so, on the box $\boldsymbol{x}_1 \times \ldots \times \boldsymbol{x}_n$, we can safely replace the original function $f(x_1, \ldots)$ by the first two terms in its Taylor expansion:

$$f(x_1, \ldots, x_n) = \widetilde{y} + \sum_{i=1}^{n} c_i \, \Delta x_i, \text{ where } y_0 := f(\widetilde{x}_1, \ldots, \widetilde{x}_n) \text{ and } c_i := \frac{\partial f}{\partial x_i}.$$

For such linear functions, the range is equal to $[\widetilde{y} - \Delta, \widetilde{y} + \Delta]$, where $\Delta = \sum\limits_{i=1}^{n} |c_i| \, \Delta_i$. The maximum value $\Delta$ of the difference $f - \widetilde{y} = \sum\limits_{i=1}^{n} c_i \, \Delta x_i$ is attained when $\Delta x_i = \Delta_i$ for $c_i \ge 0$ and $\Delta x_i = -\Delta_i$ for $c_i < 0$; correspondingly, the smallest value $-\Delta$ is attained when $\Delta x_i = -\Delta_i$ for $c_i \ge 0$ and $\Delta x_i = \Delta_i$ for $c_i < 0$.

Once we know the derivatives $c_i$ and the bounds $\Delta_i$, the value $\Delta$ describing the desired range can be computed in linear time $O(n)$.

*Comment.* To get a guaranteed enclosure for $y$, we must add to this linear range an interval $[-\delta, \delta]$ which bounds the second and higher order terms in the Taylor expansion; this is, in effect, what is known in interval computations as mean value form; see, e.g., [142, 229, 240, 241]. Asymptotically, $\delta = O(\Delta_i^2)$, so we get an asymptotically exact enclosure for the range in linear time.

*Ellipsoid uncertainty: a brief reminder.* In some cases, the information about the values $\Delta x_1, \ldots, \Delta x_n$ comes not as a bound on the values $\Delta x_i$ themselves, but rather as a bound $z \leq z_0$ on some quantity $z = g(\Delta x_1, \ldots, \Delta x_n)$ which depends on $\Delta x_i$.

When the measurement errors are small, we can expand the function $g$ into a Taylor series and keep only the lowest terms in this expansion. In particular, if we keep quadratic terms, we get a quadratic zone $g(\Delta x_1, \ldots, \Delta x_n) \leq z_0$. If this zone is a bounded set, then it describes an ellipsoid. In this case, the only information about the tuple $\Delta x = (\Delta x_1, \ldots, \Delta x_n)$ is that it belongs to this ellipsoid.

Another situation when we get such an ellipsoid uncertainty is when measurement errors are independent normally distributed random variables, with 0 mean and standard deviation $\sigma_i$. In this case, the probability density is described by the known formula $\rho(\Delta x) = \text{const} \exp\left( -\sum_{i=1}^{n} \dfrac{\Delta x_i^2}{2\sigma_i^2} \right)$. This probability density $\rho(\Delta x)$ is everywhere positive; thus, in principle, an arbitrary tuple $\Delta x$ is possible. In practical statistics, however, tuples with very low probability density $\rho(\Delta x)$ are considered impossible.

For example, in 1-dimensional case, we have a "three sigma" rule: values for which $|\Delta x_i| > 3\sigma_i$ are considered to be impossible. In multi-dimensional case, it is natural to choose some threshold $t > 0$, and consider only tuples for which $\rho(\Delta x) \geq t$ as possible ones. This formula is equivalent to $\ln(\rho(\Delta x)) \geq \ln(t)$. For Gaussian distribution, this equality takes the form $\sum_{i=1}^{n} \dfrac{\Delta x_i^2}{\sigma_i^2} \leq r^2$ for some appropriate value $r$ – i.e., the form of an ellipsoid. The sum is $\chi^2(n)$ distributed, with expectation $n$ and standard deviation $\sqrt{n}$, so here, $r^2 = n + O(\sqrt{n})$ is a natural choice. In this chapter, we will consider ellipsoids of this type.

*Comment.* If the measurement errors are small but *not independent*, then we also have an ellipsoid, but with a general definite quadratic form in the left-hand side of the inequality.

Ellipsoids are also known to be the *optimal* approximation sets for different problems with respect to several reasonable optimality criteria; see, e.g., [111, 201]. Ellipsoid error estimates are actively used in different applications; see, e.g., [27, 61, 62, 110, 113, 258, 298, 299].

*Data processing under ellipsoid uncertainty: linear approximation.* The range of a linear function $\sum\limits_{i=1}^{n} c_i \, \Delta x_i$ over an ellipsoid can be easily computed by using, e.g., the Lagrange multiplier method. First, one can easily check that the maximum of a linear function is attained at the border of the ellipsoid, i.e., when $\sum\limits_{i=1}^{n} \dfrac{\Delta x_i^2}{\sigma_i^2} = r^2$. Maximizing the linear function $\sum\limits_{i=1}^{n} c_i \, \Delta x_i$ under the above constraint is equivalent to solving the unconstrained optimization problem $\sum\limits_{i=1}^{n} c_i \, \Delta x_i + \lambda \sum\limits_{i=1}^{n} \dfrac{\Delta x_i^2}{\sigma_i^2}$, where $\lambda$ is the Lagrange multiplier. Differentiating with respect to $\Delta x_i$ and equating derivatives to 0, we conclude that the maximum value $\Delta$ of the linear function is attained when $\Delta x_i = \alpha \, c_i \sigma_i^2$ for some $\alpha$. Here, the parameter $\alpha$ is determined by the condition that $\sum\limits_{i=1}^{n} \dfrac{\Delta x_i^2}{\sigma_i^2} = r^2$ – i.e., that $\alpha^2 \sum\limits_{i=1}^{n} c_i^2 \, \sigma_i^2 = r^2$ and $\alpha = r/\sqrt{\sum c_i^2 \sigma_i^2}$. The smallest possible value $-\Delta$ of this function is attained when $\Delta x_i = -\alpha \, c_i \sigma_i^2$.

The corresponding value $\Delta$ is equal to $\Delta = r\sqrt{\sum c_i^2 \sigma_i^2}$. This value can also be computed in linear time.

*Need for combining interval and ellipsoid uncertainty.* In some practical cases, we have a combination of interval and ellipsoid uncertainty. For example, in the statistical case, we may have an ellipsoid bound and also the 3 sigma bound $|\Delta x_i| \le 3\sigma_i$ for each measurement error.

In this case, the actual values $(\Delta x_1, \ldots, \Delta x_n)$ belong to the *intersection* of the box $x_1 \times \ldots \times x_n$ and the ellipsoid.

In general, the smaller the set over which we estimate the range of a given function, the narrower the resulting range. It is therefore desirable to be able to estimate the range of a linear function $\sum\limits_{i=1}^{n} c_i \, \Delta x_i$ over such an intersection.

*What we do in this chapter: main result.* In this chapter, we provide two algorithms for estimating the range of a linear function over an intersection in linear time: a simpler $O(n \log(n))$ algorithm and a (somewhat more complex) linear time algorithm.

*From ellipsoids to generalized ellipsoids.* We have mentioned that ellipsoids correspond to normal distributions. In many practical cases, the distribution of the measurement errors is different from normal; see, e.g., [259, 261, 283]. In many such cases, we have a distribution of the type

$$\rho(\Delta x_i) = \text{const} \, \exp\left( -\sum_{i=1}^{n} \frac{|\Delta x_i|^p}{k\sigma_i^p} \right)$$

for some value $p \neq 2$ [259]. For this distribution, the condition

$$\rho(\Delta x) = \rho_1(\Delta x_1)\ldots\rho_n(\Delta x_n) \geq t$$

takes the form $\displaystyle\sum_{i=1}^{n} \frac{|\Delta x_i|^p}{\sigma_i^p} \leq r^p$ for some value $r$.

The corresponding $l^p$-methods have been successfully used in data processing; see, e.g., [89] and [324].

It is therefore reasonable to consider such *generalized ellipsoids* as well. For a generalized ellipsoid, the Lagrange approach to maximizing a linear function $\displaystyle\sum_{i=1}^{n} c_i\,\Delta x_i$ leads to

$$\sum_{i=1}^{n} c_i\,\Delta x_i + \lambda \sum_{i=1}^{n} \frac{|\Delta x_i|^p}{\sigma_i^p} \to \max,$$

$$c_i + \lambda p \cdot \mathrm{sign}(\Delta x_i)\frac{|\Delta x_i|^{p-1}}{\sigma_i^p} = 0,$$

and hence, for $p > 1$, to

$$\Delta x_i = \alpha \cdot \mathrm{sign}(c_i)|c_i|^{1/(p-1)}\sigma_i^{p/(p-1)}$$

for some constant $\alpha$. Here, the parameter $\alpha$ is determined by the condition that $\displaystyle\sum_{i=1}^{n} \frac{|\Delta x_i|^p}{\sigma_i^p} = r^p$ – i.e., that $\displaystyle\alpha^p \sum_{i=1}^{n} |c_i|^{p/(p-1)}\sigma_i^{p/(p-1)} = r^p$ and

$$\alpha = r/\sqrt[p]{\sum |c_i|^{p/(p-1)}\sigma_i^{p/(p-1)}}.$$

The smallest possible value $-\Delta$ of this function is attained when

$$\Delta x_i = -\alpha \cdot \mathrm{sign}(c_i)|c_i|^{1/(p-1)}\sigma_i^{p/(p-1)}.$$

The corresponding value $\Delta$ is equal to

$$\Delta = r \left(\sum_{i=1}^{n} |c_i|^{p/(p-1)}\sigma_i^{p/(p-1)}\right)^{(p-1)/p}.$$

This value can also be computed in linear time.

*Need for combining interval and generalized ellipsoid uncertainty.* Similarly to the case $p = 2$, it is desirable to estimate the range of a linear function $\displaystyle\sum_{i=1}^{n} c_i\,\Delta x_i$ over an intersection of a box and a generalized ellipsoid. In this chapter, we will consider this problem for $p > 1$.

*Analysis of the problem: general form of the optimal tuple.* In the general case, we want to find the maximum and the minimum of a linear function $\displaystyle\sum_{i=1}^{n} c_i\,\Delta x_i$ over an intersection of generalized ellipsoid and a box. In order to describe an algorithm for computing the maximum and minimum, let us first describe the general properties of the tuples $\Delta x$ for which these maximum and minimum are attained.

**Definition 41.1.** *By a generalized ellipsoid $E$, we mean a set of all the tuples $\Delta x = (\Delta x_1, \ldots, \Delta x_n)$ which satisfy the inequality $\sum_{i=1}^{n} \dfrac{|\Delta x_i|^p}{\sigma_i^p} \leq r^p$, where $p$, $r$, and $\sigma_i$ are positive real numbers.*

We want to find the maximum and the minimum of a linear function on the intersection $I = E \cap B$ of a generalized ellipsoid and a box

$$B = [-\Delta_1, \Delta_1] \times \ldots \times [-\Delta_n, \Delta_n].$$

Without losing generality, we can assume that all the coefficients $c_i$ of a linear function are non-negative. Indeed, if $c_i < 0$ for some $i$, then we can simply replace the original variable $\Delta x_i$ with a new variable $\Delta x_i' = -\Delta x_i$. After this replacement, the expressions for the ellipsoid $E$ and for the box $B$ remain the same, but the corresponding coefficient $c_i$ becomes positive.

Under this assumption, one can easily see that the maximum of a linear function $\sum c_i \, \Delta x_i$ with $c_i \geq 0$ is attained when $\Delta x_i \geq 0$ for all $i$. We then get the following result.

**Proposition 41.1.** *The maximum of a linear function $\sum_{i=1}^{n} c_i \, \Delta x_i$ with $c_i \geq 0$ over an intersection of a box $B = [-\Delta_1, \Delta_1] \times \ldots \times [-\Delta_n, \Delta_n]$ and a generalized ellipsoid $\sum_{i=1}^{n} \dfrac{|\Delta x_i|^p}{\sigma_i^p} \leq r^p$ is attained, for some value $\alpha$, at a tuple*

$$\Delta x_i = \min(\Delta_i, \, \alpha \, c_i^{1/(p-1)} \sigma_i^{p/(p-1)}).$$

*Observation.* This expression has an interesting relation to the corresponding expressions for the box and for the generalized ellipsoid. Indeed, let us recall that for the box, the maximum is attained for $\Delta x_i = \Delta_i$; for the generalized ellipsoid, the maximum is attained when $\Delta x_i = \alpha \, c_i^{1/(p-1)} \sigma_i^{p/(p-1)}$. According-ing to Proposition 41.1, for the intersection of the box and the generalized ellipsoid, the optimal tuple can be, crudely speaking, obtained by taking a component-wise minimum of the tuple maximizing the box and the tuple maximizing the generalized ellipsoid.

Of course, this is not exactly the component-wise minimum because the value $\alpha$ corresponding to maximizing the linear form over the intersection $E \cap B$ may be different from the value $\alpha$ corresponding to maximizing over the generalized ellipsoid $E$.

*Comment.* For general (not necessarily non-negative) coefficients $c_i$, we get

$$\Delta x_i = \text{sign}(c_i) \cdot \min(\Delta_i, \, \alpha \, |c_i|^{1/(p-1)} \sigma_i^{p/(p-1)}).$$

*Analysis of the problem: how to find $\alpha$.* According to our result, once we know the value of the parameter $\alpha$, we will be able to find all the values $\Delta x_i$ from the optimal tuple, and thus, find the largest possible value $\Delta$ of the desired linear function $\sum\limits_{i=1}^{n} c_i \Delta x_i$.

Writing $z_i := \dfrac{\Delta_i}{|c_i|^{1/(p-1)}\sigma_i^{p/(p-1)}}$, the dependence of $|\Delta x_i|$ on $\alpha$ can be described as follows:

- If $\alpha\,|c_i|^{1/(p-1)}\sigma_i^{p/(p-1)} < \Delta_i$, i.e., if $\alpha < z_i$, then we take $|\Delta x_i| = \alpha\,|c_i|^{1/(p-1)}\sigma_i^{p/(p-1)}$.
- On the other hand, if $\alpha\,|c_i|^{1/(p-1)}\sigma_i^{p/(p-1)} \geq \Delta_i$, i.e., if $\alpha \geq z_i$, then we take $|\Delta x_i| = \Delta_i$.

So, if we sort the indices by the value $z_i$, into a sequence $z_1 \leq z_2 \ldots \leq z_n$, then the maximizing tuple have the form

$$\Delta x = (\mathrm{sign}(c_1) \cdot \Delta_1, \ldots, \mathrm{sign}(c_t) \cdot \Delta_t,$$

$$\alpha\,\mathrm{sign}(c_{t+1}) \cdot |c_{t+1}|^{1/(p-1)}\sigma_{t+1}^{p/(p-1)}, \ldots, \alpha\,\mathrm{sign}(c_n) \cdot |c_n|^{1/(p-1)}\sigma_n^{p/(p-1)})$$

for some threshold value $t$ for which $z_t \leq \alpha < z_{t+1}$.

How do we find this threshold value $t$? In principle, it is possible that the optimal solution is attained when $\Delta x_i = \pm\Delta_i$ for all $i$. In this case, the generalized ellipsoid contains the whole box. In all other cases, the value $\alpha$ must be determined by the condition that the optimal tuple is on the surface of the generalized ellipsoid, i.e., that

$$\sum_{i=1}^{t} \frac{\Delta_i^p}{\sigma_i^p} + \alpha^p \sum_{j=t+1}^{n} |c_i|^{p/(p-1)}\sigma_j^{p/(p-1)} = r^p,$$

or, equivalently,

$$\sum_{i=1}^{n} \frac{(\min(\Delta_i, \alpha\,|c_i|^{1/(p-1)}\sigma_i^{p/(p-1)}))^p}{\sigma_i^p} = r^p.$$

The left-hand side of this equality is an increasing function of $\alpha$. Thus, to find the proper value of $k$, it is sufficient to check all the values $\alpha = z_1, \ldots, z_n$.

If for some $k$, we get

$$\sum_{i=1}^{k} \frac{\Delta_i^p}{\sigma_i^p} + z_k^p \sum_{j=k+1}^{n} |c_j|^{p/(p-1)}\sigma_j^{p/(p-1)} > r^p,$$

this means that we need to decrease $\alpha$, i.e., that we should have fewer values $\Delta x_i = \pm\Delta_i$ – in other words, this means that $t < k$.

On the other hand, if for some $k$, we get

$$\sum_{i=1}^{k} \frac{\Delta_i^p}{\sigma_i^p} + z_k^p \sum_{j=k+1}^{n} |c_j|^{p/(p-1)} \sigma_i^{p/(p-1)} \leq r^p,$$

this means that $t \geq k$.

So, we can find the desired threshold $t$ as the largest index $k$ for which for $\alpha = z_k$, the left-hand side of the above equality is still less than or equal to $r^p$; due to monotonicity with respect to $\alpha$, this value $t$ can be found by bisection.

Once we find this threshold value $t$, we can then find $\alpha$ from the equation

$$\sum_{i=1}^{t} \frac{\Delta_i^p}{\sigma_i^p} + \alpha^p \sum_{j=t+1}^{n} |c_j|^{p/(p-1)} \sigma_j^{p/(p-1)} = r^p,$$

i.e., $\alpha^p = \dfrac{r^p - E^-}{E^+}$, where $E^- := \sum\limits_{i=1}^{t} \dfrac{\Delta_i^p}{\sigma_i^p}$ and $E^+ := \sum\limits_{j=t+1}^{n} |c_j|^{p/(p-1)} \sigma_j^{p/(p-1)}$.

After that, we can uniquely determine the optimal tuple $\Delta x_i$ and thus the desired maximal value $\Delta = \sum\limits_{i=1}^{k} |c_i| \cdot \Delta_i + \alpha \sum\limits_{j=t+1}^{n} |c_j|^{p/(p-1)} \sigma_j^{p/(p-1)}$.

So, we arrive at the following algorithms for computing $\Delta$.

*A simpler $O(n \log(n))$ algorithm.* First, we check whether the generalized ellipsoid contains the box, i.e., whether $\sum\limits_{i=1}^{n} \dfrac{\Delta_i^p}{\sigma_i^p} \leq r^p$. If this is the case, then the desired maximum is equal to $\sum\limits_{i=1}^{n} |c_i| \Delta_i$. If this is not the case, then we apply our algorithm.

In this algorithm, we first sort the indices in the increasing order by $z_i$.

After this sorting, we apply the following iterative algorithm. At each iteration of this algorithm, we have two numbers:

- the number $i^-$ such that for all indices $i \leq i^-$, we already know that for the optimal tuple $\Delta x$, we have $|\Delta x_i| = \Delta_i$;
- the number $i^+$ of all the indices $j \geq i^+$ for which we already know that for the optimal tuple $\Delta x$, we have $|\Delta x_j| < \Delta_j$.

In the beginning, $i^- = 0$ and $i^+ = n + 1$. At each iteration, we also update the value of two auxiliary quantities $E^- := \sum\limits_{i=1}^{i^-} \dfrac{\Delta_i^p}{\sigma_i^p}$ and $E^+ :=$ $\sum\limits_{j=i^+}^{n} |c_j|^{p/(p-1)} \sigma_j^{p/(p-1)}$.

In principle, on each iteration, we could compute these sums "from scratch"; however, to speed up computations, on each iteration, we update these auxiliary values in a way that is faster than re-computing the corresponding sums.

Initially, since $i^- 0$ and $i^+ = n + 1$, we take $E^- = E^+ = 0$.

At each iteration, we do the following:

- first, we compute the midpoint $m = (i^- + i^+)/2$;

- we compute $e^- := \sum\limits_{i=i^-+1}^{m} \dfrac{\Delta_i^p}{\sigma_i^p}$ and $e^+ := \sum\limits_{j=m+1}^{i^+-1} |c_j|^{p/(p-1)} \sigma_j^{p/(p-1)}$;

- if $E^- + e^- + z_m^p \left(E^+ + e^+\right) > r^p$, then we replace $i^+$ with $m+1$ and $E^+$ with $E^+ + e^+$;

- if $E^- + e^- + z_m^p \left(E^+ + e^+\right) \leq r^p$, then we replace $i^-$ with $m$ and $E^-$ with $E^- + e^-$.

At each iteration, the set of undecided indices is divided in half. Iterations continue until all indices are decided, after which we compute $\alpha$ from the condition that $E^- + \alpha^p E^+ = r^p$, i.e., as $\alpha^p := \dfrac{r^p - E^-}{E^+}$. Once we know $\alpha$, we compute the maximizing tuple $|\Delta x_i| = \min(\Delta_i, \alpha\,|c_i|^{1/(p-1)} \sigma_i^{p/(p-1)})$ and then, the desired maximum $\sum\limits_{i=1}^{n} |c_i|\,|\Delta x_i|$.

*Computational complexity of the above algorithm.* Sorting takes time

$$O(n\,\log(n));$$

see, e.g., [73].

After this, at each iteration, all the operations with indices from $i^-$ to $i^+$ take time $t$ linear in the number of such indices: $t \leq C \cdot (i^+ - i^-)$ for some $C$. We start with the set of indices of full size $n$; on the next iteration, we have a set of size $n/2$, then $n/4$, etc. Thus, after sorting, the overall computation time is $\leq C \cdot (n + n/2 + n/4 + \ldots) \leq C \cdot 2n$, i.e., linear in $n$. So, the overall computation time is indeed $O(n\,\log(n)) + O(n) = O(n\,\log(n))$.

*Comment.* This algorithm works for an even more general case.

In some cases, we have distributions $\rho_i(\Delta x_i) = \rho_0\left(\dfrac{|\Delta x_i|}{\sigma_i}\right)$ for a different function $\rho_0(x)$. In this case, similar arguments lead to a generalized ellipsoid of the type $\sum\limits_{i=1}^{n} \psi\left(\dfrac{|\Delta x_i|}{\sigma_i}\right) \leq r_0$, where $\psi(x) := -\ln(\rho_0(x))$. The above algorithm can be extended to the case of strictly convex smooth functions $\psi(x)$ for which both this function, its derivative, and the corresponding inverse functions can be computed in polynomial time. This class includes the $l^p$-functions $\psi(x) = |x|^p$ with $p > 1$ as particular cases.

*Main idea behind the linear time algorithm.* Our second algorithm is similar to the above $O(n\,\log(n))$ algorithm. In that algorithm, the only non-linear-time part was sorting. To avoid sorting, in the second algorithm, we use the known fact that we can compute the median of a set of $n$ elements in linear time (see, e.g., [73]). (Our use of median is similar to the one from [52] and [132]).

*Our linear time algorithm is only efficient to large $n$.* It is worth mentioning that while asymptotically, the linear time algorithm for computing the median is faster than sorting, this median computing algorithm is still rather complex – so, for small $n$, sorting is faster than computing the median.

This is the reason why in this chapter, we present two different algorithms – both algorithms are practically useful:

- for large $n$, the linear time algorithm is faster;
- however, for small $n$, the $O(n \log(n))$ algorithm is faster.

Let us now describe the linear time algorithm.

*Algorithm.* First, we check whether the generalized ellipsoid contains the box, i.e., whether $\sum\limits_{i=1}^{n} \dfrac{\Delta_i^p}{\sigma_i^p} \leq r^p$. If this is the case, then the desired maximum is equal to $\sum\limits_{i=1}^{n} c_i \, \Delta_i$. If this is not the case, then we perform the following iterations.

At each iteration, we have three sets:

- the set $I^-$ of all the indices $i$ from 1 to $n$ for which we already know that for the optimal tuple $\Delta x$, we have $|\Delta x_i| = \Delta_i$;
- the set $I^+$ of all the indices $j$ for which we already know that for the optimal tuple $\Delta x$, we have $|\Delta x_j| < \Delta_j$;
- the set $I = \{1, \ldots, n\} - I^- - I^+$ of the indices $i$ for which we are still undecided.

In the beginning, $I^- = I^+ = \emptyset$ and $I = \{1, \ldots, n\}$. At each iteration, we also update the value of two auxiliary quantities $E^- := \sum\limits_{i \in I^-} \dfrac{\Delta_i^p}{\sigma_i^p}$ and $E^+ :=$ $\sum\limits_{j \in I^+} |c_j|^{p/(p-1)} \sigma_j^{p/(p-1)}$.

In principle, we could compute this value by computing this sum of squares, but to speed up computations, on each iteration, we update this auxiliary value in a way that is faster than re-computing the corresponding sum.

Initially, since $I^- = I^+ = \emptyset$, we take $E^- = E^+ 0$.

At each iteration, we do the following:

- first, we compute the median $m$ of the set $I$ (median in terms of sorting by $z_i$);
- then, by analyzing the elements of the undecided set $I$ one by one, we divide them into two subsets $P^- = \{i : z_i \leq z_m\}$ and $P^+ = \{j : z_j > z_m\}$;
- we compute $e^- = \sum\limits_{i \in P^-} \dfrac{\Delta_i^p}{\sigma_i^p}$ and $e^+ := \sum\limits_{j \in P^+} |c_j|^{p/(p-1)} \sigma_j^{p/(p-1)}$;
- if $E^- + e^- + z_m^p \, (E^+ + e^+) > r^p$, then we replace $I^+$ with $I^+ \cup P^+$, $I$ with $P^-$, and $E^+$ with $E^+ + e^+$;
- if $E^- + e^- + z_m^p \, (E^+ + e^+) \leq r^p$, then we replace $I^-$ with $I^- \cup P^-$, $I$ with $P^+$, and $E^-$ with $E^- + e^-$.

At each iteration, the set of undecided indices is divided in half. Iterations continue until all indices are decided, after which we compute $\alpha$ from the condition that $E^- + \alpha^p E^+ = r^p$, i.e., as $\alpha^p := \dfrac{r^p - E^-}{E^+}$. Once we know $\alpha$, we compute the maximizing tuple $|\Delta x_i| = \min(\Delta_i, \alpha\, |c_i|^{1/(p-1)} \sigma_i^{p/(p-1)})$ and then, the desired maximum $\sum\limits_{i=1}^{n} |c_i|\, |\Delta x_i|$.

*Computational complexity of the above algorithm.* Let us show that this algorithm indeed takes linear time. Indeed, at each iteration, computing median takes linear time, and all other operations with $I$ take time $t$ linear in the number of elements $|I|$ of $I$: $t \le C \cdot |I|$ for some $C$. We start with the set $I$ of size $n$; on the next iteration, we have a set of size $n/2$, then $n/4$, etc. Thus, the overall computation time is $\le C \cdot (n + n/2 + n/4 + \ldots) \le C \cdot 2n$, i.e., linear in $n$.

## Proofs

*Proof of Proposition 41.1.* Let $\Delta x_i$ be an optimal (maximizing) tuple.

If there are indices $i$ and $j$ for which $\Delta x_i < \Delta_i$ and $\Delta x_j < \Delta_j$, then, for sufficiently small real numbers $\varepsilon_i$ and $\varepsilon_j$, we can replace $\Delta x_i$ with $\Delta x_i + \varepsilon_i$, $\Delta x_j$ with $\Delta x_j + \varepsilon_j$, and still stay within the intervals $[0, \Delta_i]$ and $[0, \Delta_j]$ – i.e., within the box $B$. Let us select the changes $\varepsilon_i$ and $\varepsilon_j$ in such a way that the sum $s := \dfrac{|\Delta x_i|^p}{\sigma_i^p} + \dfrac{|\Delta x_j|^p}{\sigma_j^p}$ remain unchanged – then we will stay within the generalized ellipsoid as well.

For small $\varepsilon_i$ and $\varepsilon_j$, we have

$$\frac{(\Delta x_i + \varepsilon_i)^p}{\sigma_i^p} + \frac{(\Delta x_j + \varepsilon_j)^p}{\sigma_i^p} =$$

$$\frac{(\Delta x_i)^p}{\sigma_i^p} + \frac{(\Delta x_j)^p}{\sigma_i^p} + \frac{p\,\varepsilon_i\, \Delta x_i^{p-1}}{\sigma_i^p} + \frac{p\,\varepsilon_j\, \Delta x_j^{p-1}}{\sigma_j^p} + o(\varepsilon_i).$$

Thus, to make sure that $s$ does not change, we must select $\varepsilon_j$ for which

$$\frac{\varepsilon_i\, \Delta x_i^{p-1}}{\sigma_i^p} + \frac{\varepsilon_j\, \Delta x_j^{p-1}}{\sigma_j^p} = o(\varepsilon_i),$$

i.e.,

$$\varepsilon_j = -\varepsilon_i\, \frac{\Delta x_i^{p-1}}{\Delta x_j^{p-1}}\, \frac{\sigma_j^p}{\sigma_i^p} + o(\varepsilon_i).$$

The resulting change in the maximized linear function is equal to $c_i \varepsilon_i + c_j \varepsilon_j$. Substituting the expression for $\varepsilon_j$ in terms of $\varepsilon_i$, we conclude that this change is equal to

$$\varepsilon_i \left( c_i - c_j \, \frac{\Delta x_i^{p-1}}{\Delta x_j^{p-1}} \, \frac{\sigma_j^p}{\sigma_i^p} \right) + o(\varepsilon_i).$$

If the coefficient at $\varepsilon_i$ was positive, then we could take a small positive $\varepsilon_i$ and further increase the value of the linear function – which contradicts our selection of the tuple $\Delta x_i$ for which the maximum is attained. Similar, if the coefficient at $\varepsilon_i$ was negative, then we could take a small negative $\varepsilon_i$ and further increase the value of the linear function. Thus, this coefficient cannot be positive and cannot be negative – hence it must be equal to 0. So,

$$c_i - c_j \, \frac{\Delta x_i^{p-1}}{\Delta x_j^{p-1}} \, \frac{\sigma_j^p}{\sigma_i^p} = 0,$$

or, equivalently,

$$\frac{\Delta x_i^{p-1}}{c_i \sigma_i^p} = \frac{\Delta x_j^{p-1}}{c_j \sigma_j^p}.$$

This equality holds for every two indices for which $\Delta x_i < \Delta_i$ and $\Delta x_j < \Delta_j$; thus, for all such indices, the above ratio has the same value. Let us denote this common ratio by $r_0$; then, we conclude that $\dfrac{\Delta x_i^{p-1}}{c_i \sigma_i^p} = r_0$ and hence, that

$$\Delta x_i = \alpha \, c_i^{1/(p-1)} \sigma_i^{p/(p-1)},$$

where we denoted $\alpha := r_0^{1/(p-1)}$.

If $\Delta x_i < \Delta_i$ and $\Delta x_j = \Delta_j$, then we can similarly change $\Delta x_i$ and $\Delta x_j$, but only the changes for which $\varepsilon_j < 0$ will keep us inside the box. Since the sign of $\varepsilon_j$ is opposite to the sign of $\varepsilon_i$, we thus conclude that we can only take $\varepsilon_i > 0$. Thus, the coefficient at $\varepsilon_i$ in the expression for the change in the (linear) objective function cannot be positive – because then, we would be able to further increase this objective function. So, this coefficient must be non-positive, i.e.,

$$c_i - c_j \, \frac{\Delta x_i^{p-1}}{\Delta x_j^{p-1}} \, \frac{\sigma_j^p}{\sigma_i^p} \leq 0,$$

or, equivalently,

$$\frac{\Delta x_i^{p-1}}{c_i \sigma_i^p} \leq \frac{\Delta x_j^{p-1}}{c_j \sigma_j^p}.$$

Since $\Delta x_i < \Delta_i$, for $i$, we have $\dfrac{\Delta x_i^{p-1}}{c_i \sigma_i^p} = r_0$. Thus, we conclude that

$$\frac{\Delta x_j^{p-1}}{c_j \sigma_j^p} \leq r_0,$$

i.e., $\Delta x_j = \Delta_j \leq \alpha \, c_j^{1/(p-1)} \sigma_j^{p/(p-1)}$.

Hence,

- when $\Delta x_i < \Delta_i$, we get $\Delta x_i = \alpha\, c_i^{1/(p-1)}\sigma_i^{p/(p-1)}$;
- when $\Delta x_j = \Delta_i$, we get $\Delta x_j = \Delta_j \leq \alpha\, c_j^{1/(p-1)}\sigma_j^{p/(p-1)}$.

To complete the proof of our proposition, let us consider two cases.

If $\Delta_i \leq \alpha\, c_i^{1/(p-1)}\sigma_i^{p/(p-1)}$, then we cannot have $\Delta x_i < \Delta_i$ – because then we would have $\Delta x_i = \alpha\, c_i^{1/(p-1)}\sigma_i^{p/(p-1)}$ and thus, $\Delta_i > \Delta x_i = \alpha\, c_i^{1/(p-1)}\sigma_i^{p/(p-1)}$ and $\Delta_i > \alpha\, c_i^{1/(p-1)}\sigma_i^{p/(p-1)}$ – which contradicts our assumption. Thus, the only remaining case here is $\Delta x_i = \Delta_i$.

On the other hand, if $\Delta_j > \alpha\, c_j^{1/(p-1)}\sigma_j^{p/(p-1)}$, then we cannot have $\Delta x_j = \Delta_j$ – because otherwise, we would have $\Delta_j \leq \alpha\, c_j^{1/(p-1)}\sigma_j^{p/(p-1)}$, which also contradicts our assumption. Thus, in this case, we must have $\Delta x_j < \Delta_j$, and we already know that in this case, $\Delta x_j = \alpha\, c_j^{1/(p-1)}\sigma_i^{p/(p-1)}$. So:

- if $\Delta_i \leq \alpha\, c_i^{1/(p-1)}\sigma_i^{p/(p-1)}$ then $\Delta x_i = \Delta_i$;
- if $\Delta_j > \alpha\, c_j^{1/(p-1)}\sigma_j^{p/(p-1)}$ then $\Delta x_j = \alpha\, c_j^{1/(p-1)}\sigma_i^{p/(p-1)}$.

In both cases, we have

$$\Delta x_i = \min(\Delta_i, \alpha\, c_i^{1/(p-1)}\sigma_i^{p/(p-1)}).$$

The proposition is proven.

# Beyond Interval Uncertainty: Case of Discontinuous Processes (Phase Transitions)

One of the main tasks of science and engineering is to use the current values of the physical quantities for predicting the future values of the desired quantities. Due to the (inevitable) measurement inaccuracy, we usually know the current values of the physical quantities with interval uncertainty. Traditionally, it is assumed that all the processes are continuous; as a result, the range of possible values of the future quantities is also known with interval uncertainty. However, in many practical situations (such as phase transitions), the dependence of the future values on the current ones becomes discontinuous. We show that in such cases, initial interval uncertainties can lead to arbitrary bounded closed ranges of possible values of the future quantities. We also show that the possibility of such a discontinuity may drastically increase the computational complexity of the corresponding range prediction problem.

## Formulation and Analysis of the Problem

*Objectives of science and engineering.* One of the main tasks of science and engineering is to use the current values of the physical quantities $x_1, \ldots, x_n$ to predict the future values $y$ of the desired quantities.

To be able to perform this prediction, we must know how $y$ depends on $x_i$, i.e., we must know the algorithm $y = f(x_1, \ldots, x_n)$ which transforms the current values $x_1, \ldots, x_n$ into the desired prediction $y$. Once we know this algorithm, and we know the values of the physical quantities $x_1, \ldots, x_n$, we can then predict $y$ as $y = f(x_1, \ldots, x_n)$.

*Comment.* In reality, often, the algorithm $f$ represents the actual physical dependence only approximately. For example, in quantum physics, only probabilistic predictions are possible, so any deterministic prediction algorithm is approximate.

In many practical situations, however, the real-life dynamics is known reasonably accurately. In such situations, we can safely assume that the

H.T. Nguyen et al.: Computing Stat. under Interval & Fuzzy Uncertain., SCI 393, pp. 349–355.
springerlink.com                                    © Springer-Verlag Berlin Heidelberg 2012

algorithm $f$ describes the exact dependence. This is the assumption that we make in this paper.

*Measurement inaccuracy.* In the above description, we assumed that we know the exact current values of the quantities $x_1, \ldots, x_n$. In practice, however, these values usually come from measurements, and measurements are never 100% accurate. As a result, the measured value $\widetilde{x}_i$ of the $i$-th quantity is, in general, different from its (unknown) actual value $x_i$.

Usually, the manufacturer of the corresponding measuring instrument provides us with a guaranteed upper bound $\Delta_i$ on the (absolute value) of the measurement error $\Delta x_i \overset{\text{def}}{=} \widetilde{x}_i - x_i$ of the $i$-th quantity.

In this case, after we measure $x_i$ and get the measurement result $\widetilde{x}_i$, we can conclude that the actual value of $x_i$ belongs to the interval

$$\boldsymbol{x}_i = [\widetilde{x}_i - \Delta_i, \widetilde{x}_i + \Delta_i].$$

In other words, due to the (inevitable) measurement inaccuracy, we usually know the current values of the physical quantities with interval uncertainty.

*Comment.* Often, in addition to the range $\boldsymbol{x}_i$ of possible values of $x_i$, we also know the probabilities of different values $x_i \in \boldsymbol{x}_i$; see, e.g., [283]. In this chapter, however, we only consider the range information.

*The effect of measurement inaccuracy on prediction.* In this chapter, we assume that we know the exact algorithm $f(x_1, \ldots, x_n)$ which transforms the current values $x_1, \ldots, x_n$ into the desired future value $y$. Under this assumption, in the idealized situation in which we know the exact values $x_i$ of the current quantities, we could compute the exact value $y = f(x_1, \ldots, x_n)$ of the desired future quantity.

In practice, for each $i$, we only know the interval $\boldsymbol{x}_i$ of possible values of $x_i$. In this case, the only thing that we can conclude about the quantity $y$ is that $y$ belongs to the set

$$\boldsymbol{y} = f(\boldsymbol{x}_1, \ldots, \boldsymbol{x}_n) \overset{\text{def}}{=} \{f(x_1, \ldots, x_n) : x_1 \in \boldsymbol{x}_1, \ldots, x_n \in \boldsymbol{x}_n\}.$$

*Traditional assumption: all physical dependencies are continuous.* Traditionally, it is assumed that all the processes are continuous; in particular, that the function $y = f(x_1, \ldots, x_n)$ computed by the algorithm $f$ is continuous. It is well known that the range of a continuous function on a bounded connected set, e.g., on the box $\boldsymbol{x}_1 \times \ldots \times \boldsymbol{x}_n$, is an interval. Thus, for continuous functions $f$, the range $\boldsymbol{y}$ of possible values of the future quantity $y$ is an interval.

Thus, due to inevitable measurement inaccuracy, we can only make predictions with interval uncertainty. Computing such intervals is one of the main tasks of *interval computations*; see, e.g., [142].

*Discontinuous dependencies: a physical possibility.* Some physical processes are discontinuous: e.g., phase transitions. When a water is heated and boils, its density abruptly changes from the density of water to the (orders of magnitude) smaller density of steam.

Of course, all the molecules in water move continuously. So, strictly speaking, the density cannot change abruptly: theoretically, it does continuously change from the density of water to the density of steam. However, for all practical purposes, this transition is so fast that from the prediction viewpoint, we can safely assume that:

- the future density can be equal to the density of water,
- the future density can be equal to the density of steam, but
- the future density cannot be equal to any intermediate value.

*Formulation of the problem.* How does the possibility of discontinuous dependencies change the class of possible ranges? How does it affect the computational complexity of computing these ranges?

These are the questions that we will handle in this chapter.

*How discontinuities affect the class of possible ranges?* Let us first describe how the possibility of discontinuous dependencies changes the class $\mathcal{S}$ of possible ranges $S$. Before we describe this problem in precise terms, let us make some preliminary comments.

*Comment.* From the mathematical viewpoint, the following result is similar to the results from [262] and [263].

*It is sufficient to consider closed ranges.* Let $S \in \mathcal{S}$ be a possible range, i.e., a possible set of values of some physically relevant quantity $y$. Let us also assume that for this range $S$, the values $s_1, s_2, \ldots, s_k, \ldots$ are all possible (i.e., $s_k \in S$), and that the sequence $s_k$ converges to a certain number $s$. In this case, no matter how accurately we compute $s$, we will always find a number $s_k$ that is indistinguishable from $s$ (and possible). Therefore, it is natural to assume that this limit value $s$ is also possible.

In other words, it is natural to assume that every set $S \in \mathcal{S}$ contains all its limit points, i.e., that it is a *closed* set.

*It is sufficient to consider closed classes of sets.* A similar requirement can be formulated for different sets $S \in \mathcal{S}$.

Indeed, on the class of all bounded closed sets, there is a natural metric – Hausdorff distance $d_H(S, S')$. This distance is defined as the smallest $\varepsilon > 0$ for which $S$ is contained in the $\varepsilon$-neighborhood of $S'$ and $S'$ is contained in the $\varepsilon$-neighborhood of $S$. In more precise terms, the Hausdorff distance is the smallest number $\varepsilon$ for which

$$\forall s \in S \, \exists s' \in S' \, (d(s, s') \leq \varepsilon)$$

and

$$\forall s' \in S' \, \exists s \in S \, (d(s, s') \leq \varepsilon),$$

where $d(s, s') = |s - s'|$ is the standard distance between the points on the real line.

Informally, it means that if $d_H(S, S') \leq \varepsilon$, and we only know the values $s \in S$ and $s' \in S'$ with accuracy $\varepsilon$, then we cannot distinguish between the sets $S$ and $S'$.

So, if the sets $S_1, S_2, \ldots, S_k, \ldots$ are all possible (i.e., $S_i \in \mathcal{S}$), and the sequence of sets $S_k$ converges to a certain set $S$ (i.e., $d_H(S_k, S) \to 0$), then no matter how accurately we compute the values, we will always find a set $S_k$ that is indistinguishable from the set $S$ (and possible). Therefore, it is natural to assume that this limit set $S$ is also possible.

In other words, it is natural to assume that the class $\mathcal{S}$ contains all its limit points, i.e., that it is a *closed* class under the Hausdorff metric.

*Towards formalization of the problem.* We know that continuous dynamics functions are physically possible.

We assume that at least one function describing physical dynamics is discontinuous. In the simplest case, we have a monotonic function of one variable that has a "jump": it continuously grows until some threshold value, then makes a jump, and then again continuously grows. In this case, the range of this variable is not a single interval, it is a union of two intervals.

Thus, we arrive at the following definition.

**Definition 42.1.** *A class $\mathcal{S}$ of closed bounded non-empty subsets of the real line is called a* class of ranges *if it satisfies the following conditions:*

*(i) the class $\mathcal{S}$ contains an interval;*
*(ii) the class $\mathcal{S}$ is closed under arbitrary continuous transformations, i.e., if $S \in \mathcal{S}$ and $f(x)$ is a continuous function, then $f(S) \in \mathcal{S}$;*
*(iii) there exist a value $x_0$ and a function $f_0(x)$ such that:*
   *– the function $f_0(x)$ is continuously increasing for $x < x_0$,*
   *– the function $f_0(x)$ is continuously decreasing for $x > x_0$,*
   *– the function $f_0(x)$ has a "jump" at $x_0$, i.e., $f_0(x_0-) < f_0(x_0+)$, and*
   *– the class $\mathcal{S}$ is closed under $f_0$, i.e.,*
$$\text{if } S \in \mathcal{S} \text{ then } \overline{f_0(S)} \in \mathcal{S};$$
*(iv) the class $\mathcal{S}$ is closed under Hausdorff metric.*

**Theorem 42.1.** *The class of ranges coincides with the class of all bounded closed sets.*

*Computational complexity of the prediction problem: interval uncertainty, linear functions.* Before we discuss how discontinuities affect the computational complexity of the prediction problem, let us recall the computational complexity of the prediction problem in the continuous cases, i.e., under interval uncertainty. In Chapter 8, we have mentioned that in the simplest case of a linear function

$$y = f(x_1, \ldots, x_n) = a_0 + \sum_{i=1}^{n} a_i \cdot x_i$$

under interval uncertainty $x_i \in [\widetilde{x}_i - \Delta_i, \widetilde{x}_i + \Delta_i]$, we have explicit formulas for computing the range $[\underline{y}, \overline{y}]$: $\underline{y} = \widetilde{y} - \Delta$ and $\overline{y} = \widetilde{y} + \Delta$, where

$$\widetilde{y} \overset{\text{def}}{=} f(\widetilde{x}_1, \ldots, \widetilde{x}_n) = a_0 + \sum_{i=1}^{n} a_i \cdot \widetilde{x}_i$$

and

$$\Delta = \sum_{i=1}^{n} |a_i| \cdot \Delta_i.$$

The corresponding range can be computed in linear time, i.e., efficiently.

In contrast, for quadratic functions $f(x_1, \ldots, x_n)$, the problem of range computation is, in general, NP-hard.

*Computational complexity of the prediction problem: general uncertainty, linear functions.* We have already mentioned that due to possible discontinuities, the range of possible values of each input $x_i$ is, in general, different from the interval; there may be gaps – specifically, it can be equal to an arbitrary bounded closed set. In particular, when each gap is the largest possible, this range can be equal to the 2-point set $\{\underline{x}_i, \overline{x}_i\}$.

**Theorem 42.2.** *For 2-point inputs, the problem of computing the range becomes NP-hard already for linear functions $f(x_1, \ldots, x_n)$.*

*Conclusions.* One of the main tasks of science and engineering is to use the current values of the physical quantities for predicting the future values of the desired quantities. Due to the measurement inaccuracy, we usually know the current values of the physical quantities with interval uncertainty. Traditionally, it is assumed that all the processes are continuous; as a result, the range of possible values of the future quantities is also known with interval uncertainty.

However, in many practical situations (such as phase transitions), the dependence of the future values on the current ones becomes discontinuous. In this paper, we have shown that in such cases, initial interval uncertainties can lead to arbitrary bounded closed range of possible values of the future quantities. We have also shown that the possibility of such a discontinuity may drastically increase the computational complexity of the corresponding range prediction problem: e.g., for linear functions, the complexity increases from linear time to NP-hard.

## Proofs

*Proof of Theorem 42.1.* First, every two intervals can be obtained from each other by a continuous transformation – e.g., by a linear function. Since the class $\mathcal{S}$ contains an interval and it is closed under arbitrary continuous transformations, we can thus conclude that this class contains all possible intervals.

Let us take an arbitrary interval $I = [a_1, a_2]$ that contains a point $x_0$ inside. We have already shown that this interval belongs to the class $\mathcal{S}$. By the construction of a discontinuous function $f_0(x)$ as monotonic, for this interval, the image $\overline{f_0(I)}$ is the union of two disjoint intervals: $[f_0(a_1), f(x_0-)] \cup [f(x_0+), f_0(a_2)]$.

Now, let us consider unions of two disjoint intervals, i.e., sets of the type $[a_1, a_2] \cup [a_3, a_4]$ with $a_2 < a_3$. Every two sets $[a_1, a_2] \cup [a_3, a_4]$ and $[a_1', a_2'] \cup [a_3', a_4']$ of this type can be obtained from each other by a continuous transformation – e.g., we can take a piece-wise linear transformation $f(x)$ which maps:

- $[a_1, a_2]$ into $[a_1', a_2']$,
- $[a_2, a_3]$ into $[a_2', a_3']$, and
- $[a_3, a_4]$ into $[a_3', a_4']$.

Since the class $\mathcal{S}$ contains one such set and it is closed under arbitrary continuous transformations, we can thus conclude that this class contains all possible two-interval sets.

Let us take an arbitrary two-interval set

$$S = [a_1, a_2] \cup [a_3, a_4]$$

for which the second interval $[a_3, a_4]$ contains a point $x_0$ inside. We have already shown that this two-interval set belongs to the class $\mathcal{S}$. By the construction of a discontinuous function $f_0(x)$ as monotonic, for this two-interval set, the image $\overline{f_0(S)}$ is the union of three disjoint intervals:

$$[f_0(a_1), f_0(a_2)] \cup [f_0(a_3), f_0(x_0-)] \cup [f_0(x_0+), f_0(a_4)].$$

Now, let us consider unions of three disjoint intervals, i.e., sets of the type

$$[a_1, a_2] \cup [a_3, a_4] \cup [a_5, a_6]$$

with $a_2 < a_3$ and $a_4 < a_5$. Every two sets

$$[a_1, a_2] \cup [a_3, a_4] \cup [a_5, a_6]$$

and

$$[a_1', a_2'] \cup [a_3', a_4'] \cup [a_5', a_6']$$

of this type can be obtained from each other by a continuous transformation – e.g., we can take a piece-wise linear transformation $f(x)$ which maps:

- $[a_1, a_2]$ into $[a_1', a_2']$,
- $[a_2, a_3]$ into $[a_2', a_3']$,
- $[a_3, a_4]$ into $[a_3', a_4']$,
- $[a_4, a_5]$ into $[a_4', a_5']$, and
- $[a_5, a_6]$ into $[a_5', a_6']$.

Since the class $\mathcal{S}$ contains one such set and it is closed under arbitrary continuous transformations, we can thus conclude that this class contains all possible three-interval sets.

By applying $f_0$, we can now conclude that the class $\mathcal{S}$ contains all 4-interval sets, etc., and any finite unions of intervals.

Let us now prove that the class $\mathcal{S}$ contains an arbitrary bounded closed set $S$.

Indeed, for every $\varepsilon$, we can consider an interval-based approximation $S_\varepsilon$ to the set $S$, by taking the union $S_\varepsilon$ of all the grid intervals $[k \cdot \varepsilon, (k+1) \cdot \varepsilon]$ (with integer $k$) for which $[k \cdot \varepsilon, (k+1) \cdot \varepsilon] \cap S \neq \emptyset$. One can easily check that in the limit $\varepsilon \to 0$, we have $S_\varepsilon \to S$. Thus, from the fact that the class $\mathcal{S}$ contains all finite unions of intervals $S_\varepsilon$, we conclude that the class $\mathcal{S}$ must also contain their limit $S$.

The theorem is proven.

*Proof of Theorem 42.2.* The proof is typical proof of NP-hardness: we reduce a known NP-hard problem to our problem. Specifically, we take the partition problem [274]. In this problem, we are given $n$ positive integers $s_1, \ldots, s_n$, and we must check whether there exist values $\varepsilon_i \in \{-1, 1\}$ for which $\sum_{i=1}^{n} \varepsilon_i \cdot s_i = 0$. We will reduce each particular case of this problem to the following particular case of our problem: $a_0 = 0$, $a_i = s_i$, $\underline{x}_i = -1$, and $\overline{x}_i = 1$ for all $i$. For the resulting linear function

$$y = f(x_1, \ldots, x_n) = \sum_{i=1}^{n} s_i \cdot x_i,$$

0 belongs to the range

$$f(\{\underline{x}_1, \overline{x}_1\}, \ldots, \{\underline{x}_n, \overline{x}_n\}) =$$

$$\{f(x_1, \ldots, x_n) : x_1 \in \{\underline{x}_1, \overline{x}_1\}, \ldots, x_n \in \{\underline{x}_n, \overline{x}_n\}\}$$

if and only if the original problem has a solution. The reduction is proven, hence our problem is indeed NP-hard.

*Comment.* This result was, in effect, proven in [182, 257]. The difference is that in [182, 257], this NP-hardness was proven to justify the use of intervals, while we already know that we have to go beyond intervals, so our NP-hardness is the (inevitable) complexity of an important practical problem.

# 43

# Beyond Interval Uncertainty in Describing Statistical Characteristics: Case of Smooth Distributions and Info-Gap Decision Theory

In the traditional statistical approach, we assume that we know the exact cumulative distribution function (CDF) $F(x)$. In practice, we often only know the envelopes $[\underline{F}(x), \overline{F}(x)]$ bounding this CDF, i.e., we know the interval-valued "p-box" which contains $F(x)$. P-boxes have been successfully applied to many practical applications. In the p-box approach, we assume that the actual CDF can be any CDF $F(x) \in [\underline{F}(x), \overline{F}(x)]$. In many practical situations, however, we know that the actual distribution is smooth. In such situations, we may wish our model to further restrict the set of CDFs by requiring them to share smoothness (and similar) properties with the bounding envelopes $\underline{F}(x)$ and $\overline{F}(x)$. In previous work, ideas from Info-Gap Decision Theory were used to propose heuristic methods for selecting such distributions. In this chapter, we provide justifications for this heuristic approach.

The main results of this chapter first appeared in [38].

## Formulation and Analysis of the Problem

*Traditional approach: a brief reminder.* In the traditional statistical techniques typically used in science and engineering applications, we assume that we know the exact probability distributions of measurement errors, of different population quantities, etc. (e.g. [305] and [337]).

For each quantity $\xi$, this distribution can be described, e.g., by its cumulative distribution function (CDF)

$$F(x) \stackrel{\text{def}}{=} \text{Prob}(\xi \leq x).$$

Computationally, the CDF is often represented by its *quantiles*, i.e., by the values $x(\alpha)$ for which $F(x(\alpha)) = \alpha$ for some pre-selected values $\alpha$: e.g., $\alpha = 0, 0.1, 0.2, \ldots, 1.0$.

In mathematical terms, the representation means, crudely speaking, that instead of discussing the original CDF function $F(x)$ directly, we discuss

H.T. Nguyen et al.: Computing Stat. under Interval & Fuzzy Uncertain., SCI 393, pp. 357–366.
springerlink.com

the inverse function $x(\alpha)$. (This is exactly true when the CDF is strictly monotonic.)

*P-box approach: main idea.* In practice, we rarely know the exact values of the probabilities of different events. In particular, for real-life quantities, we rarely know the exact values $F(x)$ of the probability $\text{Prob}(\xi \leq x)$. Instead, we have approximate knowledge of $F(x)$.

In some situations, we know the bounding envelopes $[\underline{F}(x), \overline{F}(x)]$ for the unknown CDF $F(x)$. In other situations, we have expert estimates for values $F(x)$ – which can be naturally described as fuzzy numbers.

*Comment.* It is well known that a fuzzy number can be represented as a nested family of its $\alpha$-cuts (intervals), and that processing these fuzzy numbers can be reduced to processing the corresponding $\alpha$-cuts. Moreover, this is usually how fuzzy numbers are processed (see e.g. [90, 156, 230, 246, 252]).

In view of this fact, in the following text we will concentrate on interval uncertainty.

*P-boxes: mathematical representation.* For each dimension $\xi$, uncertainty about probability

$$F(x) = \text{Prob}(\xi \leq x)$$

can be described by an interval

$$\boldsymbol{F}(x) = [\underline{F}(x), \overline{F}(x)]$$

that is guaranteed to contain the unknown actual value of $F(x)$.

The function $\boldsymbol{F}$ that maps each real number $x$ into the interval $\boldsymbol{F}(x)$ is called a *probability box*, or, for short, *p-box* [97].

*P-boxes: computer representation.* As we have mentioned, in the traditional statistical approach a probability distribution is usually represented in the computer by its quantiles. In the p-box case, the fact that we do not know the exact values of $F(x)$ means that we do not know the exact values of the quantiles either.

Instead of the actual value of a quantile $x(\alpha)$, we only know the bounds on the quantile. Namely, from the fact that

$$\underline{F}(x) \leq F(x) \leq \overline{F}(x),$$

we can conclude that

$$\underline{x}(\alpha) \leq x(\alpha) \leq \overline{x}(\alpha),$$

where $\underline{x}(\alpha)$ are the quantiles corresponding to $\overline{F}(x)$, and $\overline{x}(\alpha)$ are the quantiles corresponding to $\underline{F}(x)$.

As a result, in the computer, a p-box is represented by its *interval-valued quantiles*, i.e., by the intervals $[\underline{x}(\alpha), \overline{x}(\alpha)]$ which are guaranteed to contain the actual (unknown) values $x(\alpha)$. These interval-valued quantiles are given for some pre-selected values $\alpha$: e.g., $\alpha = 0, 0.1, 0.2, \ldots, 1.0$ (see e.g. [97]).

*The meaning of p-boxes: reminder.* In short, a p-box expresses the information that for every $x$, the actual (unknown) value $F(x)$ of the CDF is contained in the interval $[\underline{F}(x), \overline{F}(x)]$.

*Limitations of a p-box interpretation.* In many practical situations, the extreme-case bounds $\underline{F}(x)$ and $\overline{F}(x)$ correspond to smooth distributions such as Gaussian, uniform, etc. In such situations, it is often reasonable to expect that the actual distribution $F(x)$ is also smooth. However, in the p-box approach, the only limitation on $F(x)$ is that $F(x) \in [\underline{F}(x), \overline{F}(x)]$. This limitation permits, in addition to smooth functions, very non-smooth – and thus (for many problems) unrealistic – CDF functions $F(x)$.

To take such situations into account, it is desirable to be able to limit ourselves to smooth bounds $\underline{F}(x)$ and $\overline{F}(x)$ and to distributions $F(x)$ which share the same smoothness characteristics as the bounds.

*An approach motivated by Info-Gap Decision Theory: main idea.* To solve this problem, in [35] a new approach is described which is motivated by Info-Gap Decision Theory (see e.g. [29]). Specifically, once we have two possible distributions $F_1(x)$ and $F_2(x)$ which can be described using a set of quantiles $x_1(\alpha)$ and $x_2(\alpha)$ for various values of $\alpha$, we then assume that for every value $\beta \in [0, 1]$, the distribution corresponding to the quantiles

$$x(\alpha) = \beta \cdot x_1(\alpha) + (1 - \beta) \cdot x_2(\alpha)$$

is also possible. Once we fix $F_1(x)$ and $F_2(x)$, we get a 1-parameter class which is much sparser than the p-box of all distributions between $F_1(x)$ and $F_2(x)$.

Let us show that this idea indeed allows us to avoid non-smooth combinations of smooth distributions.

*An approach motivated by Info-Gap Decision Theory avoids non-smooth distributions.* In many practical situations, the uncertainty is in the values of the parameters: we know the shape of the distribution, but we do not know the exact values of the corresponding parameters. For example, we may know that the distribution is Gaussian (or uniform), but not the exact values of the corresponding parameters.

How can we describe this situation in precise terms? For example, let $F_0(x)$ be the CDF of the "standard" normal distribution, with 0 mean and standard deviation 1. Then, the CDF of a general normal distribution, with mean $a$ and standard deviation $\sigma$, can be described as

$$F(x) = F_0\left(\frac{x - a}{\sigma}\right).$$

A similar expression describes a general uniform distribution, etc.

For such distributions $F(x)$, as one can easily check, the corresponding quantiles $x(\alpha)$ are linearly related to the quantiles $x_0(\alpha)$ of the just-mentioned standard normal distribution $F_0(x)$:

$$x(\alpha) = a + \sigma \cdot x_0(\alpha).$$

If we have two distributions with the same property, i.e., if we have

$$x_1(\alpha) = a_1 + \sigma_1 \cdot x_0(\alpha)$$

and

$$x_2(\alpha) = a_2 + \sigma_2 \cdot x_0(\alpha),$$

then their convex combination

$$x(\alpha) = \beta \cdot x_1(\alpha) + (1 - \beta) \cdot x_2(\alpha)$$

also has the same form:

$$x(\alpha) = a + \sigma \cdot x_0(\alpha),$$

with $a = \beta \cdot a_1 + (1 - \beta) \cdot a_2$ and $\sigma = \beta \cdot \sigma_1 + (1 - \beta) \cdot \sigma_2$.

Thus, when applied to two distribution of the same shape, the above procedure leads to the distribution of this same shape: a combination of Gaussian distributions is Gaussian, a combination of uniform distributions is uniform, etc.

*Remaining open problem.* The above procedure seems to work well, but is too *ad hoc*, requiring more justification.

In this chapter, we provide a justification for this procedure.

Specifically, we want to describe an operation $I(x_1, x_2)$ that for every $\alpha$, given two values $x_1(\alpha)$ and $x_2(\alpha)$, returns a suitable intermediate value

$$x(\alpha) = I(x_1(\alpha), x_2(\alpha)).$$

We will call such an operation an *intermediate value operation*.

*Comment.* The main ideas behind our justification are based on the natural notions of symmetry. Similar ideas have been used, e.g., in [189] and in [251].

*Relevant types of invariance.* The three types of invariance described below provide background for the following definitions.

*Scale invariance.* The values $x$ often come from measurements. In this case, if we change the unit of measurement (e.g. from centimeters to meters), numerical values will be multiplied by a constant $\lambda > 0$. It is natural to require that the result of the intermediate value operation not depend on the choice of unit.

How does replacing a unit change the intermediate value operation function $I(x_1, x_2)$? If we replace a unit by a one that is $\lambda$ times smaller, then the quantity that was initially described by the value $x_1$ will be described by a new value $x_1' = \lambda \cdot x_1$, and the quantity that was initially described by the value $x_2$ will be described by a new value $x_2' = \lambda \cdot x_2$. When we combine these values by using the intermediate value operation $I$, we get the resulting value

$$x' = I(x'_1, x'_2) = I(\lambda \cdot x_1, \lambda \cdot x_2).$$

This is the expression of the combined quantile in the new units. In the old units, its expression is

$$x = \lambda^{-1} \cdot x' = \lambda^{-1} \cdot I(\lambda \cdot x_1, \lambda \cdot x_2).$$

We will denote the resulting "re-scaled" intermediate value operation

$$x_1, x_2 \to \lambda^{-1} \cdot I(\lambda \cdot x_1, \lambda \cdot x_2)$$

by $S_\lambda(I)$.

In these terms, the intermediate value operation $I$ is scale invariant if and only if $S_\lambda(I) = I$ for all $\lambda$.

*Reverse invariance.* In addition to changing the units, there can also be changes in sign. For example, when measuring a spatial coordinate, we can change the direction and that will change the sign, or when measuring an electric charge, we usually follow the convention that an electron's charge is negative, but we can also consider electron charges as positive numbers. This possibility is equivalent to a re-scaling with $\lambda = -1$. Therefore we wish to consider not only positive values $\lambda$, but in fact arbitrary non-zero values $\lambda$.

*Shift invariance.* When measuring quantities like time or location, we can also change the starting point. In this case, a constant will be added to all numerical values: $x \to x + a$.

Then the quantity that was initially described by the value $x_1$ will be described by a new value $x'_1 = x_1 + a$, and the quantity that was initially described by the value $x_2$ will be described by a new value $x'_2 = x_2 + a$. When we combine these values by using the intermediate value operation $I$, we get the resulting value

$$x' = I(x'_1, x'_2) = I(x_1 + a, x_2 + a).$$

This is the expression of the combined quantile in the new units. In the old units, its expression is

$$x = x' - a = I(x_1 + a, x_2 + a) - a.$$

We will denote the resulting "shifted" intermediate value operation

$$x_1, x_2 \to I(x_1 + a, x_2 + a) - a$$

by $T_a(I)$.

It is natural to require that the intermediate value operation is invariant w.r.t. these symmetries as well, i.e., that $T_a(I) = I$ for all possible real values $a$.

*Invariance: definitions and the main result.*

### Definition 43.1

- *By an* intermediate value operation, *we mean a function* $I : R^2 \to R$ *from the set of pairs of real numbers into real numbers for which the value* $I(x_1, x_2)$ *is always located in between* $x_1$ *and* $x_2$:

$$\min(x_1, x_2) \leq I(x_1, x_2) \leq \max(x_1, x_2).$$

  *The set of all possible intermediate value operations will be denoted by* $A$.
- *For every intermediate value operation* $I$, *and for every* $\lambda \neq 0$, *by a re-scaled intermediate value operation* $S_\lambda(I)$, *we mean an intermediate value operation*

$$x_1, x_2 \to \lambda^{-1} \cdot I(\lambda \cdot x_1, \lambda \cdot x_2).$$

- *For every intermediate value operation* $I$, *and for every* $a$, *by a* shifted *intermediate value operation* $T_a(I)$, *we mean an operation*

$$x_1, x_2 \to I(x_1 + a, x_2 + a) - a.$$

- *We say that the intermediate value operation* $I$ *is* scale-invariant *if for all* $\lambda$, *we have* $S_\lambda(I) = I$.
- *We say that the intermediate value operation* $I$ *is* shift-invariant *if for all* $a$, *we have* $T_a(I) = I$.

**Proposition 43.1.** *For an intermediate value operation* $I$, *the following two conditions are equivalent to each other:*

- $I$ *is scale-invariant and shift-invariant;*
- $I$ *is described by the expression*

$$I(x_1, x_2) = \beta \cdot x_1 + (1 - \beta) \cdot x_2$$

*for some* $\beta \in [0, 1]$.

*Comment.* As a consequence of Proposition 1, the naturalness of scale- and shift-invariance implies the naturalness of the equivalent intermediate value operation.

*Towards an alternative justification based on optimality: main idea.* Instead of requiring that the intermediate value operation be invariant, it is reasonable to look for *optimal* operations, i.e., operations which are the best in the sense of some optimality criterion.

*What is an "optimality criterion"?* When we say that some *optimality criterion* is given, we mean that, given two different intermediate value operations, we can decide whether the first or the second one is better, or if these operations are equivalent w.r.t. the given criterion. In mathematical terms, this means that we have a *pre-ordering relation* $\preceq$ on the set of all possible intermediate value operations.

*The need to enumerate optimal intermediate value operations.* One way to approach the problem of choosing the "best" intermediate value operation function is to select *one* optimality criterion, and to find an intermediate value operation which is the best with respect to this criterion. The main drawback of this approach is that there can be different optimality criteria, and they can lead to different optimal solutions. It is, therefore, desirable not only to describe an intermediate value operation that is optimal relative to some criterion, but to describe *all* intermediate value operations that are optimal relative to any member of a set of natural criteria.

*Comment.* The word "natural" is used informally. We merely want to say that from the purely mathematical viewpoint, there can be weird ("unnatural") optimality criteria. We will only consider criteria that satisfy some requirements that we would, from a common sense viewpoint, consider reasonable and natural.

*Examples of optimality criteria.* Pre-ordering is the general formulation of optimization problems in general, not just of the problem of choosing an intermediate value operation. In general optimization theory, in which we are comparing arbitrary *alternatives* $a$, $b$, ..., from a given set $A$, the most frequent case of a pre-ordering is when a *numerical criterion* is used, i.e., when a function $J : A \to R$ is given for which $a \preceq b$ if and only if $J(a) \leq J(b)$.

Various natural numerical criteria can be proposed for choosing the intermediate value operations. For example, we could consider cases in which we are given the class of all distributions classified as possible for the given problem, and we have "weights" assigned to different distributions from this class, so that these weights add up to 1. In this case, for some pairs of distributions from this class – characterized by their quantiles $x_1(\alpha)$ and $x_2(\alpha)$ – the distribution corresponding to the quantiles $x(\alpha) = I(x_1(\alpha), x_2(\alpha))$ also belongs to the given class. For some other pairs of distributions $x_1(\alpha)$ and $x_2(\alpha)$, the distribution corresponding to the $x(\alpha)$ does not belong to the given class. We can then take, as $J(I)$, the "ratio" (total weight) of such pairs of distributions for which $x(\alpha)$ also belongs to the given class.

Many other criteria can be proposed. What should be done if there are several different alternatives that perform equally well? In this case, it makes sense to choose the alternative for which the computations are the fastest. This natural idea leads to an optimality criterion that is not describable by a single numerical optimality criterion $J(a)$: in this case, we need *two* functions: $J_1(a)$ describes the "ratio", $J_2(a)$ describes the computation time, and $a \preceq b$ if and only if either $J_1(a) < J_1(b)$, or $J_1(a) = J_1(b)$ and $J_2(a) \geq J_2(b)$.

We could further specify the described optimality criterion so that it disambiguates cases where both $J_1(a) = J_1(b)$ and $J_2(a) = J_2(b)$ with another function $J_3$, etc. However, as we have already mentioned, the goal of this paper is not to find a single intermediate value operation that is optimal relative to some criterion, but to describe *all* intermediate value operations that are optimal relative to any of a set of natural optimality criteria. In view of this goal, in the following, we will not specify the criterion, but rather describe a general class of natural optimality criteria.

So, let us formulate what "natural" means.

*Which optimality criteria are natural?* We have already mentioned that the values $x$ often come from measurements, and that for such values, changing the unit of measurement (e.g. from meters to centimeters) multiplies the measured values by a constant $\lambda$. It is natural to require the relative quality of two intermediate value operations not depend on the choice of units. In other words, we require that if $I$ is better than $I'$, then the "re-scaled" $I$ (i.e., $S_\lambda(I)$) should be better than the "re-scaled" $I'$ (i.e., $S_\lambda(I')$).

It is also natural to require the optimality criterion to be invariant w.r.t. shift transformations. In other words, if $I$ is better than $I'$, then $T_a(I)$ should be better than $T_a(I')$.

There is one more reasonable requirement for a criterion, based on the following idea. If the criterion does not select a single optimal intermediate value operation, i.e., if it considers more than one intermediate value operations equally good, then we can always use some other criterion to help select among them, thus designing a two-step criterion. If this new criterion still does not select a unique intermediate value operation, we can continue this process as many steps as necessary to get only one optimal intermediate value operation. Such a multi-step criterion can always be *final* in this sense.

*An optimization approach: definitions and the main result.*

**Definition 43.2.** *By an* optimality criterion, *we mean a* pre-ordering *(i.e., a transitive, reflexive relation)* $\preceq$ *on the set $A$.*

- *An optimality criterion $\preceq$ is called* scale-invariant *if for all $I$, $I'$, and $\lambda \neq 0$, $I \preceq I'$ implies $S_\lambda(I) \preceq S_\lambda(I')$.*
- *An optimality criterion $\preceq$ is called* shift-invariant *if for all $I$, $I'$, and $a$, $I \preceq I'$ implies $T_a(I) \preceq T_a(I')$.*
- *An optimality criterion $\preceq$ is called* final *if there exists one and only one intermediate value operation $I$ that is preferable to all the others, i.e., for which $I' \preceq I$ for all $I' \neq I$.*

**Proposition 43.2**

- *If an intermediate value operation $I$ is optimal w.r.t. some scale-invariant, shift-invariant, and final optimality criterion, then for some $\beta \in [0, 1]$, the operation $I$ is described by a formula*

$$I(x_1, x_2) = \beta \cdot x_1 + (1 - \beta) \cdot x_2.$$

- *For every $\beta \in [0, 1]$, there exists a scale-invariant, shift-invariant, and final optimality criterion for which the only optimal intermediate value operation is the operation*

$$I(x_1, x_2) = \beta \cdot x_1 + (1 - \beta) \cdot x_2.$$

*Comment.* In other words, if the optimality criterion satisfies the above-described natural properties, then the optimal intermediate value operation coincides with one of $\beta$-operations.

## Proofs

*Proof of Proposition 43.1.* It is easy to check that for every $\beta \in [0,1]$, the formula $I(x_1, x_2) = \beta \cdot x_1 + (1 - \beta) \cdot x_2$ indeed describes a scale- and shift-invariant intermediate value operation.

Let us therefore move on to showing that every scale- and shift-invariant intermediate value operation $I(x_1, x_2)$ has the above form. Indeed, let $I$ be such an operation, and let us define $\beta \stackrel{\text{def}}{=} I(1, 0)$.

For arbitrary $x_1 \neq x_2$, we can apply shift-invariance with $a = -x_2$, and conclude that

$$I(x_1, x_2) = I(x_1 - x_2, 0) + x_2.$$

Now, scale-invariance with $\lambda = 1/(x_1 - x_2)$ implies that

$$I(x_1 - x_2, 0) = (x_1 - x_2) \cdot I(1, 0).$$

By definition of $\beta$, we conclude that

$$I(x_1 - x_2, 0) = (x_1 - x_2) \cdot \beta$$

and, because $I$ was given as shift-invariant, that

$$I(x_1, x_2) = (x_1 - x_2) \cdot \beta + x_2.$$

One can easily see that this expression is exactly equal to $\beta \cdot x_1 + (1 - \beta) \cdot x_2$. So, we have proven that

$$I(x_1, x_2) = \beta \cdot x_1 + (1 - \beta) \cdot x_2$$

for all $x_1 \neq x_2$.

For $x_1 = x_2$, this equality follows from the fact that $I$ is an intermediate value operation and thus, $T(x_1, x_1) = x_1$, just like the convex combination $\beta \cdot x_1 + (1 - \beta) \cdot x_2$ is equal to $x_1$. The proposition is proven.

*Proof of Proposition 43.2.* 1. To prove the first part of Proposition 43.2, we will show that the optimal intermediate value operation $I_{\text{opt}}$ is scale-invariant and shift-invariant, i.e., that $S_\lambda(I_{\text{opt}}) = T_a(I_{\text{opt}}) = I_{\text{opt}}$ for all $\lambda \neq 0$ and $a$. Then, the result will follow from Proposition 43.1.

Indeed, let $X$ be either a scale or a shift transformation. Let us first determine the invertibility of these transformations. Indeed:

- if $X = S_\lambda$, then $X^{-1} = S_{1/\lambda}$;
- if $X = T_a$, then $X^{-1} = T_{-a}$.

Now, from the optimality of $I_{\mathrm{opt}}$, we conclude that for every $I' \in A$, $X^{-1}(I') \preceq I_{\mathrm{opt}}$. From the invariance of the optimality criterion provided as a given in the proposition statement, we next conclude that $I' \preceq X(I_{\mathrm{opt}})$. This is true for all $I' \in A$ and, therefore, the intermediate value operation $X(I_{\mathrm{opt}})$ is optimal. But since the criterion is final (as given in the proposition statement), there is only one optimal intermediate value operation; hence, $X(I_{\mathrm{opt}}) = I_{\mathrm{opt}}$. So, the optimal intermediate value operation is indeed invariant and hence, due to Proposition 1, it coincides with one of the $\beta$-expressions. The first part is proven.

2. Let us now prove the second part of Proposition 2. Let $\beta \in [0, 1]$ be fixed, and let $I_\beta$ be the corresponding intermediate value operation. We will then define the optimality criterion as follows: $I \preceq I'$ if and only if $I' = I_\beta$.

Since the intermediate value operation $I_\beta$ is scale-invariant and shift-invariant, the just-defined optimality criterion is also scale-invariant and shift-invariant. It is also by definition final. The intermediate value operation $I_\beta$ is clearly optimal w.r.t. this scale-invariant, shift-invariant, and final optimality criterion. The proposition is proven.

# Beyond Traditional Interval Uncertainty in Describing Statistical Characteristics: Case of Interval Bounds on the Probability Density Function

In the previous chapter, we considered the case when, in addition to knowing the interval bounds on the cumulative distribution function $F(x)$ (= p-box), we also know that the function $F(x)$ is smooth. In addition to knowing that $F(x)$ is smooth – i.e., that its derivative $F'(x)$ (= a probability density function) is bounded – we sometimes also know the bounds on $F'(x)$. Such a situation is analyzed in this chapter.

We show that in this situations, the exact range of some statistical characteristics can be efficiently computed. Surprisingly, for some other characteristics, similar statistical problems which are efficiently solvable for interval-valued cdf become computationally difficult (NP-hard) for interval-valued pdf.

The results of this chapter have previously appeared in [354].

## Formulation and Analysis of the Problem

*Uncertainty in probability.* In the traditional statistics, we usually assume that we know the exact probability distributions.

In general, a probability distribution can be described by a cumulative probability distribution (cdf) $F(x) = \text{Prob}(t \le x)$.

A continuous probability distribution can also be described by a probability density function (pdf) $\rho(x)$. A discrete distribution can be similarly described by the probabilities $p(x)$ of individual values; however, there are distributions which cannot be described in this way.

In practice, we usually know the probabilities only with some uncertainty. It is therefore desirable to take this uncertainty into account when we compute the values of the statistical characteristics.

*p-boxes: the most computationally developed approach to handling uncertainty with which we know the probabilities.* Since the cdf corresponds to the most general case of a probability distribution, most algorithmic efforts in taking uncertainty into account have been directed towards the case when our knowledge about the probability distribution is represented by a cdf.

H.T. Nguyen et al.: Computing Stat. under Interval & Fuzzy Uncertain., SCI 393, pp. 367–378.
springerlink.com 

In the case of a cdf, uncertainty means that for every $x$, instead of the exact value of $F(x)$, we only know the interval of possible values $[\underline{F}(x), \overline{F}(x)]$.

Situation when we only know the cdf with interval uncertainty $[\underline{F}(x), \overline{F}(x)]$ is known as a *p-box*; see, e.g., [97]. For p-boxes, there are efficient algorithms that compute statistical characteristics such as ranges of moments, ranges of the cdf for the sum $x' + x''$ of two independent random variables in the situation when we know the p-boxes for $x'$ and $x''$, etc.

*Practical situation: bounds on probabilities.* In many practical situations, e.g., in climate modeling, we have bounds on the probability density or, in the discrete case, bounds on the probabilities of individual values; see, e.g., [127] and [128]. How can we process this uncertainty?

*In principle, we can use p-boxes.* One possible approach to dealing with bounds $[\underline{\rho}(x), \overline{\rho}(x)]$ on the (unknown) probability density $\rho(x)$ is to find corresponding range of the cdf $F(x)$, i.e., to find the (smallest possible) p-box which contains all probability distributions for which $\rho(x) \in [\underline{\rho}(x), \overline{\rho}(x)]$. Once we have this p-box, we can use known methods to estimate the ranges of different statistical characteristics.

*Limitations of using p-boxes: general description.* The problem with this approach is that the p-box estimates are based on the assumption that all $F(x) \in [\underline{F}(x), \overline{F}(x)]$ are possible, while we are only interested in cumulative distribution functions $F(x)$ for which $\rho(x) = F'(x)$ is bounded by the given bounds $[\underline{\rho}(x), \overline{\rho}(x)]$. As a result, we often only get an *enclosure* for the desired range, an enclosure which has excess width.

*Limitations of using p-boxes: example.* Let us describe a simple example where the use of p-boxes leads to excess width. Let us have a discrete random variable $x$ which can take 3 possible values 1, 2, and 3. We do not know the exact probabilities $p_1$, $p_2$, and $p_3$ of accessing these values; instead, we only know the intervals of possible values of these probabilities

$$\boldsymbol{p_1} = \boldsymbol{p_2} = \left[\frac{1}{3} - \beta, \frac{1}{3} + \beta\right]$$

for some small positive value $\beta \leq \dfrac{1}{6}$. For $p_3$, we do not have separate interval bounds, only bounds which can be inferred from the bounds on $p_1$ and $p_2$ and the fact that $p_1 + p_2 + p_3 = 1$.

In this simple example, we are interested in the probability that $x = 2$. Of course, based on the known information, we can easily find the interval of possible value of this probability: it is

$$\boldsymbol{p_2} = \left[\frac{1}{3} - \beta, \frac{1}{3} + \beta\right].$$

Let us show that if we convert to p-boxes, we will instead get an enclosure with excess width.

In this discrete situation, a cdf-style description means that we need to describe two numbers:

$$F(1) = p_1 = \mathrm{Prob}(x \leq 1)$$

and

$$F(2) = \mathrm{Prob}(x \leq 2) = p_1 + p_2.$$

These two values uniquely determine the resulting cdf:

- for $x < 1$, we have $F(x) = 0$;
- for $1 < x < 2$, we have $F(x) = F(1)$;
- for $2 < x < 3$, we have $F(x) = F(2)$;
- finally, for $x \geq 3$, we have $F(x) = 1$.

Based on the known intervals $\boldsymbol{x}_1$ and $\boldsymbol{p}_2$ of possible values of $p_1$ and $p_2$, we conclude that the resulting bound on $F(1)$ is $\boldsymbol{F}(1) = \boldsymbol{p}_1 = \left[\dfrac{1}{3} - \beta, \dfrac{1}{3} + \beta\right]$ and that the resulting bound on $F(2) = p_1 + p_2$ is

$$\boldsymbol{F}(2) = \boldsymbol{p}_1 + \boldsymbol{p}_2 = \left[\frac{1}{3} - \beta, \frac{1}{3} + \beta\right] + \left[\frac{1}{3} - \beta, \frac{1}{3} + \beta\right] =$$

$$\left[\frac{2}{3} - 2\beta, \frac{2}{3} + 2\beta\right]$$

(see, e.g., [142]). Thus, the resulting p-box $\boldsymbol{F}(x) = [\underline{F}(x), \overline{F}(x)]$ has the following form:

- for $x < 1$, we have $\boldsymbol{F}(x) = [0, 0]$;
- for $1 \leq x < 2$, we have

$$\boldsymbol{F}(x) = \left[\frac{1}{3} - \beta, \frac{1}{3} + \beta\right];$$

- for $2 \leq x < 3$, we have

$$\boldsymbol{F}(x) = \left[\frac{2}{3} - 2\beta, \frac{2}{3} + 2\beta\right];$$

- finally, for $x \geq 3$, we have $\boldsymbol{F}(x) = [1, 1]$.

Based on this p-box information, the probability that $x = 2$ can be found as $F(2) - F(2 - 0)$, where $F(2 - 0) \overset{\text{def}}{=} \lim F(2 - \delta)$, where $\delta > 0$ and $\delta \to 0$. From the p-box information, we conclude that $F(2)$ can take any values from the interval $\left[\dfrac{2}{3} - 2\beta, \dfrac{2}{3} + 2\beta\right]$ and that $F(2 - 0)$ can take any value from the interval $\left[\dfrac{1}{3} - \beta, \dfrac{1}{3} + \beta\right]$. According to interval computations [142, 147, 148]:

- the largest possible value of the difference

$$F(2) - F(2 - 0)$$

  is when we subtract the smallest possible value of
  $F(2 - 0)$ from the largest possible value of $F(2)$, and
- the smallest possible value of the difference

$$F(2) - F(2 - 0)$$

  is when we subtract the largest possible value of $F(2-0)$ from the smallest
  possible value of $F(2)$.

Thus, we conclude that the resulting interval of possible values of $\mathrm{Prob}(x = 2)$
is

$$\left[\frac{2}{3} - 2\beta, \frac{2}{3} + 2\beta\right] - \left[\frac{1}{3} - \beta, \frac{1}{3} + \beta\right] = \left[\frac{1}{3} - 3\beta, \frac{1}{3} + 3\beta\right].$$

This interval has the width of $6\beta$ – three times wider than the actual interval
of possible values of $p_2$.

This example shows that if we only use p-boxes, we can get estimates with
excess width.

*How can we compute exact ranges: formulation of the problem.* It is desirable to compute the exact ranges for such characteristics as mean, central
moments, convolution of several distributions (corresponding to the distribution of the sum of two independent variables), etc.

*Efficient algorithms for computing moments: formulation of the problem.* In
the discrete case, we know the values $x_1 < x_2 < \ldots < x_n$, we know the
bounds $[\underline{p}_i, \overline{p}_i]$ on the (unknown) actual probabilities $p_i$, and we are given an
integer $m > 0$. Our objective is to find the range for $\sum_{i=1}^{n} p_i \cdot x_i^m$ under the
constraints $p_i \in [\underline{p}_i, \overline{p}_i]$ and $\sum_{i=1}^{n} p_i = 1$.

*This problem is a particular case of a more general problem for which efficient
algorithms are known.* To compute these ranges, we can use the fact that a
linear-time algorithm (i.e., an algorithm with an $O(n)$ running time) is known
for a more general problem. Namely, such an algorithm is known for a general
case of computing the range $[\underline{a}, \overline{a}]$ of the expected value $a = \sum_{i=1}^{n} p_i \cdot a_i$ of a
known variable $(a_1, \ldots, a_n)$ under the constraints $p_i \in [\underline{p}_i, \overline{p}_i]$ and $\sum_{i=1}^{n} p_i = 1$;
see, e.g., [52] and [132]. The case of moments correspond to $a_i = x_i^m$.

*Computing the upper endpoint $\overline{a}$: analysis.* Let us first consider the problem of computing the maximum $\overline{a}$. One can easily show that if for the maximizing vector $(p_1, \ldots, p_n)$, we have two values $p_i > \underline{p}_i$ and $p_j < \overline{p}_j$ for which $a_i < a_j$, then, by adding a small value $\Delta$ to $p_j$ and subtracting this value from $p_i$, we can get a new vector $p'_i$ for which still $\sum p'_i = 1$ but the value of $a = \sum\limits_{i=1}^{n} p_i \cdot a'_i$ is larger. Thus, for $a_i < a_j$, we cannot have $p_i > \underline{p}_i$ and $p_j < \overline{p}_j$ in the maximizing vector. So, we conclude that we can only have one value $k$ for which $p_k \in (\underline{p}_k, \overline{p}_k)$.

- For all values $a_i$ for which $a_i < a_k$, we have $p_i = \underline{p}_i$.
- For all values $a_j$ for which $a_j > a_k$, we have $p_j = \overline{p}_j$.

So, if we sort the values $a_i$ in increasing order, then we conclude that for some $k$, the maximum is attained for the vector $(\underline{p}_1, \ldots, \underline{p}_{k-1}, p_i, \overline{p}_{k+1}, \ldots, \overline{p}_n)$, where $p_k$ can be determined, from the condition that $\sum\limits_{i=1}^{n} p_i = 1$, as

$$p_k = 1 - \underline{p}_1 - \cdots - \underline{p}_{k-1} - \overline{p}_{k+1} - \cdots - \overline{p}_n.$$

The condition that $\underline{p}_k \leq p_k \leq \overline{p}_k$ leads to

$$\sum_{i=1}^{k-1} \underline{p}_i + \sum_{j=k}^{n} \overline{p}_j \leq 1 \leq \sum_{i=1}^{k} \underline{p}_i + \sum_{j=k+1}^{n} \overline{p}_j.$$

This condition uniquely determines the desired value $k$.

The above analysis leads to the following algorithm for computing $\overline{a}$.

*Computing the upper endpoint $\overline{a}$: first algorithm.*

- First, we sort the values $a_i$ in increasing order; sorting can be done in time $O(n \cdot \log(n))$; see, e.g., [73].
- Next, we compute the sums corresponding to $k = 0$; this computation takes linear time.
- Then, for each $k$, we need to change two terms to compute the new sums, so we need linear time for check all possible values of $k$ and find the right one.
- After this, we can compute the maximizing vector $p_i$ and the resulting upper endpoint $\overline{a} = \sum\limits_{i=1}^{n} p_i \cdot a_i$ in linear time.

In general, this algorithm takes $O(n \cdot \log(n)) + O(n) = O(n \cdot \log(n))$ time.

*Comment.* If the values $a_i$ are already sorted, then we only need linear time to compute $\overline{a}$. It turns out that we can have a linear-time algorithm in the general case, when the values $a_i$ are not pre-sorted.

*Computing the upper endpoint $\overline{a}$: linear-time algorithm.* This algorithm is based on the known fact that we can compute the median of a set of $n$ elements in linear time (see, e.g., [73]).

The algorithm is iterative. At each iteration of this algorithm we have three sets:

- the set $I^-$ of all the indices $i$ from 1 to $n$ for which we already know that for the maximizing vector $p$, we have $p_i = \underline{p}_i$;
- the set $I^+$ of all the indices $j$ for which we already know that for the maximizing vector $p$, we have $p_j = \overline{p}_j$;
- the set $I = \{1, \ldots, n\} \setminus (I^- \cup I^+)$ of the indices $i$ for which we are still undecided.

In the beginning, $I^- = I^+ = \emptyset$ and $I = \{1, \ldots, n\}$. At each iteration we also update the values of two auxiliary quantities $E^- \overset{\text{def}}{=} \sum\limits_{i \in I^-} \underline{p}_i$ and $E^+ \overset{\text{def}}{=} \sum\limits_{j \in I^+} \overline{p}_j$. In principle, we could compute these values by computing these sums. However, to speed up computations on each iteration, we update these two auxiliary values in a way that is faster than re-computing the corresponding two sums. Initially, since $I^- = I^+ = \emptyset$, we take $E^- = E^+ = 0$.

At each iteration we do the following:

- first, we compute the median $m$ of the set $I$ (median in terms of sorting by $a_i$);
- then, by analyzing the elements of the undecided set $I$ one by one, we divide them into two subsets $P^- = \{i : a_i \leq a_m\}$ and $P^+ = \{j : a_j > a_m\}$;
- we compute $e^- = E^- + \sum\limits_{i \in P^-} \underline{p}_i$ and $e^+ = E^+ + \sum\limits_{j \in P^+} \overline{p}_j$;
- If $e^- + e^+ > 1$, then we replace $I^-$ with $I^- \cup P^-$, $E^-$ with $e^-$, and $I$ with $P^+$.
- If $e^- + e^+ + 2\Delta_m < 1$, then we replace $I^+$ with $I^+ \cup P^+$, $E^+$ with $e^+$, and $I$ with $P^-$.
- Finally, if $e^- + e^+ \leq 1 \leq e^- + e^+ + 2\Delta_m$, then we replace $I^-$ with $I^- \cup (P^- - \{m\})$, $I^+$ with $I^+ \cup P^+$, $I$ with $\{m\}$, $E^-$ with $e^- - \underline{p}_m$, and $E^+$ with $e^+$.

At each iteration the set of undecided indices is divided in half. Iterations continue until we have only one undecided index $I = \{k\}$. After this we return, as $\overline{a}$, the value of the linear combination $\sum\limits_{i=1}^{n} p_i \cdot a_i$ for the vector $p$ for which $p_i = \underline{p}_i$ for $i \in I^-$, $x_j = \overline{p}_j$ for $j \in I^+$, and $p_k = 1 - e^- - e^+$ for the remaining value $k$.

*Proof that the second algorithm for computing $\overline{a}$ takes linear time.* At each iteration, computing median takes linear time, and all other operations with $I$ take time $t$ linear in the number of elements $|I|$ of $I$: $t \leq C \cdot |I|$ for some $C$. We start with the set $I$ of size $n$. On the next iteration, we have a set of size

$n/2$, then $n/4$, etc. Thus, the overall computation time is $\leq C \cdot (n + n/2 + n/4 + \ldots) \leq C \cdot 2n$, i.e. linear in $n$.

*How to compute $\underline{a}$.* It is known that the smallest possible value $\underline{a}$ of the linear form $\sum_{i=1}^{n} p_i \cdot a_i$ under given constraints is equal to $-\overline{b}$, where $\overline{b}$ is the largest possible value of the form $\sum_{i=1}^{n} p_i \cdot b_i$, with $b_i = -a_i$. Thus, by using the above algorithm, we can compute the lower endpoint as well.

*Computing convolution: a practically important problem.* If we know the distributions $\rho'(x)$ and $\rho''(x)$ of two independent random variables $x'$ and $x''$, then the probability density function $\rho(x)$ for their sum $x = x' + x''$ is described by the convolution $\rho(x) = \int \rho'(z) \cdot \rho''(x - z)\,dz$.

*There exist efficient algorithms for computing convolution of p-boxes.* Researchers have analyzed the problem of computing the convolution in situations when instead of knowing the exact cumulative distribution functions $F'(x)$ and $F''(x)$, we only know p-boxes $[\underline{F}'(x), \overline{F}'(x)]$ and $[\underline{F}''(x), \overline{F}''(x)]$.

In these situations, we can efficiently compute a p-box for $x = x' + x''$; see, e.g., [97]. This possibility comes from the fact that for every $x$, the value $F(x)$ corresponding to the convolution is a (non-strictly) increasing function of the values $F'(x')$ and $F''(x'')$. Thus:

- to compute the lower endpoint $\underline{F}(x)$ for the resulting cdf $F(x)$, it is sufficient to compute the convolution of the distributions corresponding to $\underline{F}'(x)$ and $\underline{F}''(x)$;
- similarly, to compute the upper endpoint $\overline{F}(x)$ for the resulting cdf $F(x)$, it is sufficient to compute the convolution of the distributions corresponding to $\overline{F}'(x)$ and $\overline{F}''(x)$.

*Convolution of interval-valued probabilities: general case.* In line with our previous discussions, let us now consider the situation in which, instead of the p-boxes (i.e., bounds on the cumulative distribution functions), we know the interval bounds $[\underline{\rho}'(x), \overline{\rho}'(x)]$ and $[\underline{\rho}''(x), \overline{\rho}''(x)]$ of the corresponding probability distribution functions.

In this case, for every $x$, we would like to compute the exact range $[\underline{\rho}(x), \overline{\rho}(x)]$ of the convolution

$$\rho(x) = \int \rho'(z) \cdot \rho''(x - z)\,dz$$

when $\rho'(x) \in [\underline{\rho}'(x), \overline{\rho}'(x)]$ and $\rho''(x) \in [\underline{\rho}''(x), \overline{\rho}''(x)]$.

Let us prove that this problem is computationally difficult even in the discrete case, when each of the two variables $x'$ and $x''$ can only takes finitely many values.

*Convolution of interval-valued probabilities: discrete case.* Let us assume that we have two independent discrete random variables $x'$ and $x''$.

- For the variable $x'$, we know its possible values $x'_1, \ldots, x'_{n'}$ and the bounds $\left[\underline{p}'_i, \overline{p}'_i\right]$ on the corresponding (unknown) probabilities $p'_i$.
- Similarly, for the variable $x''$, we know its possible values $x''_1, \ldots, x''_{n''}$ and the bounds $\left[\underline{p}''_j, \overline{p}''_j\right]$ on the corresponding (unknown) probabilities $p''_j$.

For the sum $x = x' + x''$, we have possible values $x_{ij} = x'_i + x''_j$.

The question is to find the ranges $\underline{p}_{ij}$ and $\overline{p}_{ij}$ of possible values of the corresponding probabilities

$$p_{ij} = \sum_{i',j' : x'_{i'} + x''_{j'} = x'_i + x''_j} p'_{i'} \cdot p''_{j'},$$

where the values $p'_i$ and $p''_i$ satisfy the conditions

$$p'_i \in \left[\underline{p}'_i, \overline{p}'_i\right], \quad p''_i \in \left[\underline{p}''_i, \overline{p}''_i\right],$$

$$\sum_{i=1}^{n'} p'_i = 1, \quad \sum_{i=1}^{n''} p''_i = 1.$$

*For interval-valued probabilities, computing convolution is NP-hard.* In this chapter, we prove, that, in contrast to the case of p-boxes, computing the (endpoints of) the exact range $\left[\underline{p}_{ij}, \overline{p}_{ij}\right]$ of the convolution probabilities $p_{ij}$ is computationally difficult (namely, NP-hard).

To be more precise, we prove two results:

- that the problem of computing the upper endpoint $\overline{p}_{ij}$ of the convolution if NP-hard, and
- that the problem of computing the lower endpoint $\underline{p}_{ij}$ is also NP-hard.


## Proofs

*Comment.* Our proofs are somewhat similar to the proof of NP-hardness from [13].

$1°$. Our proof is based on reducing, to this problem, a known NP-hard *subset* problem, where we are given $n$ positive integers $s_1, \ldots, s_n$, and we must find the values $\varepsilon_i \in \{-1, 1\}$ for which $\sum_{i=1}^{n} \varepsilon_i \cdot s_i = 0$.

$2°$. To each instance $s_1, \ldots, s_n$ of the subset problem, we assign the following two interval-valued probability distributions $x'$ and $x''$.

2.1°. The variable $x'$ can only take $n' = n$ values $x'_1 = 1, \ldots, x'_i = i, \ldots,$ $x'_n = n$. For each $i$ from 1 to $n$, the corresponding probability $p'_i$ can take any value from the interval

$$\left[\underline{p}'_i, \overline{p}'_i\right] = \left[\frac{1}{n} - \beta \cdot s_i, \frac{1}{n} + \beta \cdot s_i\right].$$

2.2°. Similarly, the variable $x''$ can only take $n'' = n$ values $x''_1 = -1, \ldots,$ $x''_i = -i, \ldots, x''_n = -n$. For each $i$ from 1 to $n$, the corresponding probability $p''_i$ can take any value from the interval

$$\left[\underline{p}''_i, \overline{p}''_i\right] = \left[\frac{1}{n} - \beta \cdot s_i, \frac{1}{n} + \beta \cdot s_i\right].$$

3°. The value $\beta$ should be selected in such a way as to guarantee that the resulting probabilities are always non-negative, i.e., that $\frac{1}{n} - \beta \cdot s_i \geq 0$ for all $i$. This requirement is equivalent to $\beta \cdot s_i \leq \frac{1}{n}$, i.e., to $\beta \leq \frac{1}{n \cdot s_i}$. This must hold for all $i$, so we must make sure that $\beta$ does not exceed the smallest of these values – i.e., the value corresponding to the largest $s_i$. Thus, we can take

$$\beta = \frac{1}{n \cdot \max\limits_{i} s_i}.$$

4.1°. In this case, for every $i$, the (unknown) actual probability $p'_i$ can be described as

$$p'_i = \frac{1}{n} + \beta \cdot \Delta'_i,$$

where $\Delta'_i \stackrel{\text{def}}{=} p'_i - \frac{1}{n}$ can take any value from the interval $[-s_i, s_i]$.

4.2°. Similarly, for every $i$, the (unknown) actual probability $p''_i$ can be described as

$$p''_i = \frac{1}{n} + \beta \cdot \Delta''_i,$$

where $\Delta''_i \stackrel{\text{def}}{=} p''_i - \frac{1}{n}$ can take any value from the interval $[-s_i, s_i]$.

5.1°. In terms of the auxiliary variables $\Delta'_i$, the requirement that $\sum\limits_{i=1}^{n} p'_i = 1$ means that

$$\sum_{i=1}^{n} \left(\frac{1}{n} + \beta \cdot \Delta'_i\right) = 1,$$

i.e., that $1 + \sum\limits_{i=1}^{n} \Delta'_i = 1$ and $\sum\limits_{i=1}^{n} \Delta'_i = 0$.

5.1°. Similarly, the requirement that $\sum\limits_{i=1}^{n} p_i'' = 1$ means that

$$\sum_{i=1}^{n} \left( \frac{1}{n} + \beta \cdot \Delta_i'' \right) = 1,$$

i.e., that $1 + \sum\limits_{i=1}^{n} \Delta_i'' = 1$ and $\sum\limits_{i=1}^{n} \Delta_i'' = 0$.

6°. Let us now find the range of possible values for the probability that the sum $x = x' + x''$ is equal to 0.

The value 0 can be obtained if $x' = i$ and $x'' = -i$ for the same value $i$. Thus, the desired probability is equal to

$$p_{11} = \sum_{i=1}^{n} p_i' \cdot p_i''.$$

Substituting the expressions $p_i' = \dfrac{1}{n} + \beta \cdot \Delta_i'$ and $p_i'' = \dfrac{1}{n} + \beta \cdot \Delta_i''$ into this formula, we get

$$p_{11} = \sum_{i=1}^{n} \left( \frac{1}{n} + \beta \cdot \Delta_i' \right) \cdot \left( \frac{1}{n} + \beta \cdot \Delta_i'' \right).$$

Here,

$$\left( \frac{1}{n} + \beta \cdot \Delta_i' \right) \cdot \left( \frac{1}{n} + \beta \cdot \Delta_i'' \right) =$$

$$\left( \frac{1}{n} \right)^2 + \frac{1}{n} \cdot \beta \cdot \Delta_i' + \frac{1}{n} \cdot \beta \cdot \Delta_i'' + \beta^2 \cdot \Delta_i' \cdot \Delta_i''.$$

Therefore, we conclude that

$$p_{11} = \sum_{i=1}^{n} \left( \frac{1}{n} \right)^2 + \sum_{i=1}^{n} \frac{1}{n} \cdot \beta \cdot \Delta_i' + \sum_{i=1}^{n} \frac{1}{n} \cdot \beta \cdot \Delta_i'' +$$

$$\sum_{i=1}^{n} \beta^2 \cdot \Delta_i' \cdot \Delta_i''.$$

By moving constant factors outside the sum, we get:

$$p_{11} = \left( \frac{1}{n} \right)^2 \cdot \sum_{i=1}^{n} 1 + \frac{1}{n} \cdot \beta \cdot \sum_{i=1}^{n} \Delta_i' + \frac{1}{n} \cdot \beta \cdot \sum_{i=1}^{n} \Delta_i'' +$$

$$\beta^2 \cdot \sum_{i=1}^{n} \Delta_i' \cdot \Delta_i''.$$

The first sum is equal to $\left(\dfrac{1}{n^2}\right) \cdot n = \dfrac{1}{n}$. The second and the third sums are equal to 0 since $\sum\limits_{i=1}^{n} \Delta'_i = 0$ and $\sum\limits_{i=1}^{n} \Delta''_i = 0$. Thus, we conclude that

$$p_{11} = \frac{1}{n} + \beta^2 \cdot \sum_{i=1}^{n} \Delta'_i \cdot \Delta''_i.$$

$7°$. Let us prove that the number $\dfrac{1}{n} + \beta^2 \cdot \sum\limits_{i=1}^{n} s_i^2$ is a possible value of $p_{11}$ if and only if the original instance of a subset problem has a solution.

This will prove that the problem of computing the upper endpoint $\overline{p}_{11}$ of the range of $p_{11}$ is NP-hard.

$7.1°$. Indeed, if the original instance has a solution $\varepsilon$ for which $\sum\limits_{i=1}^{n} \varepsilon_i \cdot s_i = 0$, then we can take $\Delta'_i = \Delta''_i = \varepsilon_i \cdot s_i$ and get

$$p_{11} = \frac{1}{n} + \beta^2 \cdot \sum_{i=1}^{n} s_i^2.$$

$7.2°$. Vice versa, let us assume that the number $\dfrac{1}{n} + \beta^2 \cdot \sum\limits_{i=1}^{n} s_i^2$ is a possible value of $p_{11}$. Let us prove that in this case, the original instance of the subset problem has a solution.

Indeed, since $|\Delta'_i| \le s_i$ and $|\Delta''_i| \le s_i$, we always have $|\Delta'_i \cdot \Delta''_i| \le s_i^2$ and hence $\Delta'_i \cdot \Delta''_i \le s_i^2$.

So, the only possibility to have

$$p_{11} = \frac{1}{n} + \beta^2 \cdot \sum_{i=1}^{n} \Delta'_i \cdot \Delta''_i = \frac{1}{n} + \beta^2 \cdot \sum_{i=1}^{n} s_i^2$$

is to have $\Delta'_i \cdot \Delta''_i = s_i^2$ for all $i$ – otherwise, we would have

$$p_{11} = \frac{1}{n} + \beta^2 \cdot \sum_{i=1}^{n} \Delta'_i \cdot \Delta''_i < \frac{1}{n} + \beta^2 \cdot \sum_{i=1}^{n} s_i^2.$$

If $|\Delta'_i| < s_i$ or $|\Delta''_i| < s_i$, then we have $|\Delta'_i \cdot \Delta''_i| < s_i^2$ and hence $\Delta'_i \cdot \Delta''_i < s_i^2$. So, the only way to have $\Delta'_i \cdot \Delta''_i = s_i^2$ is to have $|\Delta'_i| = s_i$ and $|\Delta''_i| = s_i$. So, we have $\Delta'_i = \pm s_i$, i.e., $\Delta'_i = \varepsilon_i \cdot s_i$ for some value $\varepsilon_i \in \{-1, 1\}$.

The fact that $\sum\limits_{i=1}^{n} \Delta'_i = 0$ implies that $\sum\limits_{i=1}^{n} \varepsilon_i \cdot s_i = 0$. So, the values $\varepsilon_i$ form a solution to the original instance of the subset problem.

7.3°. The reduction is proven, and so the problem of computing the upper endpoint $\overline{p}_{ij}$ of the convolution is indeed NP-hard in case of interval uncertainty.

8°. Let us prove that the number $\dfrac{1}{n} - \beta^2 \cdot \sum\limits_{i=1}^{n} s_i^2$ is a possible value of $p_{11}$ if and only if the original instance of a subset problem has a solution.

This will prove that the problem of computing the lower endpoint $\underline{p}_{11}$ of the range of $p_{11}$ is also NP-hard.

8.1°. Indeed, if the original instance has a solution $\varepsilon$ for which

$$\sum_{i=1}^{n} \varepsilon_i \cdot s_i = 0,$$

then we can take $\Delta_i' = \varepsilon_i \cdot s_i$ and $\Delta_i'' = -\varepsilon_i \cdot s_i$, and get

$$p_{11} = \frac{1}{n} - \beta^2 \cdot \sum_{i=1}^{n} s_i^2.$$

8.2°. Vice versa, let us assume that the number $\dfrac{1}{n} - \beta^2 \cdot \sum\limits_{i=1}^{n} s_i^2$ is a possible value of $p_{11}$. Let us prove that in this case, the original instance of the subset problem has a solution.

Indeed, since $|\Delta_i'| \le s_i$ and $|\Delta_i''| \le s_i$, we always have $|\Delta_i' \cdot \Delta_i''| \le s_i^2$ and hence $\Delta_i' \cdot \Delta_i'' \ge -s_i^2$.

So, the only possibility to have

$$p_{11} = \frac{1}{n} + \beta^2 \cdot \sum_{i=1}^{n} \Delta_i' \cdot \Delta_i'' = \frac{1}{n} - \beta^2 \cdot \sum_{i=1}^{n} s_i^2$$

is to have $\Delta_i' \cdot \Delta_i'' = -s_i^2$ for all $i$ – otherwise, we would have

$$p_{11} = \frac{1}{n} + \beta^2 \cdot \sum_{i=1}^{n} \Delta_i' \cdot \Delta_i'' > \frac{1}{n} + \beta^2 \cdot \sum_{i=1}^{n} s_i^2.$$

If $|\Delta_i'| < s_i$ or $|\Delta_i''| < s_i$, then we have $|\Delta_i' \cdot \Delta_i''| < s_i^2$ and hence $\Delta_i' \cdot \Delta_i'' > -s_i^2$. So, the only way to have $\Delta_i' \cdot \Delta_i'' = -s_i^2$ is to have $|\Delta_i'| = s_i$ and $|\Delta_i''| = s_i$. So, we have $\Delta_i' = \pm s_i$, i.e., $\Delta_i' = \varepsilon_i \cdot s_i$ for some value $\varepsilon_i \in \{-1, 1\}$.

The fact that $\sum\limits_{i=1}^{n} \Delta_i' = 0$ implies that $\sum\limits_{i=1}^{n} \varepsilon_i \cdot s_i = 0$. So, the values $\varepsilon_i$ form a solution to the original instance of the subset problem.

# Beyond Interval Uncertainty in Describing Statistical Characteristics: Case of Normal Distributions

In the previous two chapters, we considered the case when, in addition to the bounds on the cumulative distribution function $F(x)$, we also have additional information about $F(x)$ – e.g., we know that $F(x)$ is smooth (differentiable), and we know the bounds on the derivative of $F(x)$. Sometimes, we have even more information about $F(x)$; we may even know the analytical expression for $F(x)$ – with the parameters which are only known with uncertainty. For example, in practice, when the observed signal is caused by a joint effect of many small components, it is reasonable to assume that the distribution is normal – but the parameters of this normal distribution are only known with uncertainty. Such a situation is analyzed in this chapter.

In traditional statistical analysis, if we know that the distribution is normal, then the most popular way to estimate its mean $a$ and standard deviation $\sigma$ from the data sample $x_1, \ldots, x_n$ is to equate $a$ and $\sigma$ to the arithmetic mean and sample standard deviation of this sample. After this equation, we get the cumulative distribution function $F(x) = \Phi\left(\dfrac{x-a}{\sigma}\right)$ of the desired distribution.

In many practical situations, we only know intervals $[\underline{x}_i, \overline{x}_i]$ that contain the actual (unknown) values of $x_i$ or, more generally, a fuzzy number that describes $x_i$. Different values of $x_i$ lead, in general, to different values of $F(x)$. In this chapter, we show how to compute, for every $x$, the resulting interval $[\underline{F}(x), \overline{F}(x)]$ of possible values of $F(x)$.

The results of this chapter first appeared in [357].

## Formulation and Analysis of the Problem

*Formulation of the problem.* In many real-life situations, the actual distribution is normal (Gaussian). It is known that a normal distribution is uniquely determined by its mean $a$ and its standard deviation $\sigma$. Usually, a cumulative distribution function corresponding to the distribution (cdf) with 0 mean and

H.T. Nguyen et al.: Computing Stat. under Interval & Fuzzy Uncertain., SCI 393, pp. 379–389.
springerlink.com      © Springer-Verlag Berlin Heidelberg 2012

standard deviation 1 is denoted by $\Phi(x)$. In terms of this function $\Phi(x)$, the cdf $F(x)$ of a general normal distribution has the form

$$F(x) = \Phi\left(\frac{x-a}{\sigma}\right). \tag{45.1}$$

To find the cdf, we must therefore estimate the (unknown) parameters $a$ and $\sigma$ from the (known) sample values $x_1, \ldots, x_n$. In traditional statistical data processing, one of the most widely used methods for estimating $a$ and $\sigma$ is the *method of moments*, when we find the mean and variance of the data, i.e., the values

$$a = \frac{1}{n} \cdot \sum_{i=1}^{n} x_i, \quad \sigma^2 = \frac{1}{n} \cdot \sum_{i=1}^{n}(x_i - a)^2 = \frac{1}{n} \cdot \sum_{i=1}^{n} x_i^2 - a^2,$$

and consider the normal distribution with these values $a$ and $\sigma$ as "fitted" to the data $x_1, \ldots, x_n$; see, e.g., [305, 337].

*Case of interval uncertainty.* In practice, instead of the exact values $x_i$ of the sample, we often only know the intervals $\boldsymbol{x}_i = [\underline{x}_i, \overline{x}_i]$ of possible values of $x_i$. Different values of $x_i \in \boldsymbol{x}_i$ lead, in general, to different values of $a$ and $\sigma$ and thus, for every $x$, to different values of the cdf $F(x) = \Phi\left(\dfrac{x-a}{\sigma}\right)$. It is therefore desirable to find the interval $[\underline{F}(x), \overline{F}(x)]$ of possible values of the cdf, i.e., in terms of [97], to find a *p-box* that contains all possible cumulative distribution functions.

*What is known.* Since the value of $F(x)$ is determined by the values of the mean $a$ and of the standard deviation $\sigma$, it is reasonable to first analyze the intervals of possible values for $a$ and for $\sigma$. For $a$, the interval of possible values is easy to describe: since the average is an increasing function of all its variables, its minimum is attained when all $x_i$ takes their smallest values $\underline{x}_i$, and the maximum is attained when all its variables take their largest values $\overline{x}_i$; as a result, we get the interval $[\underline{a}, \overline{a}]$, where

$$\underline{a} = \frac{1}{n} \cdot \sum_{i=1}^{n} \underline{x}_i, \quad \overline{a} = \frac{1}{n} \cdot \sum_{i=1}^{n} \overline{x}_i.$$

For standard deviation, the problem of computing the corresponding interval $[\underline{\sigma}, \overline{\sigma}]$ is, as we have mentioned, NP-hard. Crudely speaking, this means that unless it turns out that P=NP (which few computer scientists believe), every algorithm that computes this interval exactly in all cases takes time which grows at least exponentially with $n$. Actually, exponential time is sufficient: we can compute the upper endpoint $\overline{\sigma}$ if we consider all $2^n$ possible combinations of the values $\underline{x}_i$ and $\overline{x}_i$, i.e., all the corners of the $n$-dimensional box $[\underline{x}_1, \overline{x}_1] \times \ldots \times [\underline{x}_n, \overline{x}_n]$.

In some practically important cases, there exist efficient algorithms whose running time grows only polynomially with $n$. For example, such algorithms are possible in the "no-nesting" case when no two intervals $[\underline{x}_i, \overline{x}_i]$ and $[\underline{x}_j, \overline{x}_j]$ $(i \neq j)$ are proper subset of one another in the sense that $[\underline{x}_i, \overline{x}_i] \not\subseteq (\underline{x}_j, \overline{x}_j)$.

In principle, we can use the resulting bounds $[\underline{a}, \overline{a}]$ on $a$ and $[\underline{\sigma}, \overline{\sigma}]$ on $\sigma$ to produce bounds on the ratio $\dfrac{x - a}{\sigma}$ and thus, on the desired cumulative distribution function (cdf) $F(x)$. However, the values $a$ and $\sigma$ are dependent in the sense that not all combinations of $a \in [\underline{a}, \overline{a}]$ and $\sigma \in [\underline{\sigma}, \overline{\sigma}]$ are possible; as a result, these bounds contain excess width – a typical situation when computations with intervals ignore dependence (see, e.g., [142]).

How can we compute the exact bounds on $F(x)$? The closest to this are the algorithms from [187] which produce bounds for the absolute value $\dfrac{|x - a|}{\sigma}$ of the desired ratio.

*What we plan to do.* In this chapter, we show how to compute the desired p-box, i.e., the exact bounds for the normal $F(x)$ under interval data.

*Reducing the problem to computing the ratio.* The above cdf $\Phi(x)$ of a "standard" normal distribution is a strictly increasing function. Thus, for every $x$, the interval $[\underline{F}(x), \overline{F}(x)]$ of possible values of the the quantity $F(x) = \Phi\left(\dfrac{x - a}{\sigma}\right)$ can be computed as $[\Phi(\underline{r}(x)), \Phi(\overline{r}(x))]$, where $[\underline{r}(x), \overline{r}(x)]$ is the interval of possible values of the ratio $r \stackrel{\text{def}}{=} \dfrac{x - a}{\sigma}$. Thus, to compute the desired p-box, it is sufficient to compute this interval $[\underline{r}(x), \overline{r}(x)]$.

*Comment.* To make the following text easier to read, we will write $\underline{r}$ instead of $\underline{r}(x)$ and $\overline{r}$ instead of $\overline{r}(x)$. A reader should keep in mind, however, that for different $x$, generally, we get different bounds $\underline{r}$ and $\overline{r}$.

*The need to consider the signs: informal explanation.* We have already mentioned that we know how to compute the bounds on the absolute value $|r|$ of the ratio $r$; see, e.g., [187]. The absolute value can be equal either to the ratio itself or to $-r$. Here:

- If $r \geq 0$ and $|r| = r$, then, e.g., the maximum of $|r|$ is the same as the maximum of $r$.
- On the other hand, if $r < 0$ and $|r| = -r$, then the maximum of the absolute value may correspond to the minimum of $r$.

So, to apply the results and techniques from [187] to our problem, we must first analyze the signs of the corresponding extreme values $\underline{r}$ and $\overline{r}$.

*Signs of the bounds $\underline{r}$ and $\overline{r}$.* In view of the above, it is reasonable to first find out the signs of the bounds $\underline{r}$ and $\overline{r}$ of the desired interval.

**Proposition 45.1**

- *For $x \leq \underline{a}$, we have $\underline{r} \leq \overline{r} \leq 0$.*
- *For $\underline{a} < x < \overline{a}$, we have $\underline{r} < 0 < \overline{r}$.*
- *For $x \geq \overline{a}$, we have $0 \leq \underline{r} \leq \overline{r}$.*

*General idea: using basic facts about derivatives.* Let us fix the value $x$. For this $x$, each tuple $(x_1, \ldots, x_n)$ from the box $B \stackrel{\text{def}}{=} \boldsymbol{x}_1 \times \ldots \ldots \boldsymbol{x}_n$ leads, in general, to a different value of the ratio $r$. The ratio is a continuous function of $(x_1, \ldots, x_n)$; thus, both its smallest and its largest values are attained at some tuple. (To be more precise, we first have to add $-\infty$ and $+\infty$ to the set of possible values of $r$ to take care of the possibility that $\sigma = 0$.)

Let $(x_1^M, \ldots, x_n^M)$ be a tuple at which the ratio $r$ attains its largest possible value. If we fix all the values except one $x_i$, then we conclude that the corresponding function $r(x_1^M, \ldots, x_{i-1}^M, x_i, x_{i+1}^M, \ldots, x_n^M)$ also attains its maximum for $x_i = x_i^M$. There are three possible cases:

- If the ratio $r$ attains its maximum at $x_i \in (\underline{x}_i, \overline{x}_i)$, then, according to calculus, we should have $\dfrac{\partial r}{\partial x_i} = 0$ at this point.

- If $r$ attains its maximum at $x_i = \underline{x}_i$, then the derivative $\dfrac{\partial r}{\partial x_i}$ at this point cannot be positive – else we would have even larger values for $x_i > \underline{x}_i$; thus, we should have $\dfrac{\partial r}{\partial x_i} \leq 0$.

- Similarly, if $r$ attains its maximum at $x_i = \overline{x}_i$, then at this point, $\dfrac{\partial r}{\partial x_i} \geq 0$.

For the point $(x_1^m, \ldots, x_n^m)$ at which the ratio $r$ attains its *minimum*, we similarly have three cases for each $i$:

- If the ratio $r$ attains its minimum for $x_i \in (\underline{x}_i, \overline{x}_i)$, then $\dfrac{\partial r}{\partial x_i} = 0$.

- If $r$ attains its minimum at $x_i = \underline{x}_i$, then $\dfrac{\partial r}{\partial x_i} \geq 0$.

- Finally, if $r$ attains its minimum at $x_i = \overline{x}_i$, then at this point, $\dfrac{\partial r}{\partial x_i} \leq 0$.

The corresponding analysis leads to the following algorithm. In this algorithm, we assumed that the value $x$ is given. If we need to find the range $[\underline{F}(x), \overline{F}(x)]$ for several different values $x$, we need to repeat this algorithm for each of these values $x$.

*Algorithm $A_1$.* In this algorithm, we consider all possible partitions of the set of indices $\{1, \ldots, n\}$ into three disjoint subsets $I^-$, $I^+$, and $I_0$. For each subdivision we set $x_i = \underline{x}_i$ for $i \in I^-$ and $x_i = \overline{x}_i$ for $i \in I^+$. When $I_0 \neq \emptyset$, we set the values $x_i$ for $i \in I_0$ as follows:

- We compute the values

$$\widetilde{a} = \sum_{i \in I^-} \underline{x}_i + \sum_{j \in I^+} \overline{x}_j, \quad \widetilde{m} = \sum_{i \in I^-} (\underline{x}_i)^2 + \sum_{j \in I^+} (\overline{x}_j)^2.$$

- We find the value $a$ from the quadratic equation

$$\widetilde{m} + \frac{1}{N_0} \cdot (n \cdot a - \widetilde{a})^2 = n \cdot \left(a - \frac{n \cdot a - \widetilde{a}}{N_0}\right) \cdot (x - a) + n \cdot a^2,$$

where $N_0$ is the number of elements in the set $I_0$, and then compute $a_0 = \dfrac{n \cdot a - \tilde{a}}{N_0}$.

- If this quadratic equation does not have any real solutions, or if it does but the corresponding value $a_0$ does not belong to all intervals $\boldsymbol{x}_i$ with $i \in I_0$, then we skip this partition and go to the next one.
- For each solution $a_0$ that belongs to all the intervals $\boldsymbol{x}_i$, $i \in I_0$, we set $x_i = a_0$ for $i \in I_0$ and compute $\sigma = \sqrt{(a - a_0) \cdot (x - a)}$ and $r = \dfrac{x - a}{\sigma}$.

The smallest of these values $r$ is returned as $\underline{r}$, and the largest is returned as $\overline{r}$. Then, we compute the desired p-box as $[\Phi(\underline{r}), \Phi(\overline{r})]$.

**Proposition 45.2.** *The above algorithm always computes the exact range*

$$[\underline{F}(x), \overline{F}(x)] \tag{45.1}$$

*of the normal cdf under interval uncertainty.*

*Comments.*

- For each of $n$ indices $i$, we have 3 choices: we can assign this index to $I^-$, to $I^+$, and to $I_0$. For a single index, we have 3 possible assignments; for two indices, we have $3 \cdot 3 = 2^2$ possible assignments; in general, for $n$ indices, we have $3^n$ possible assignments. Thus, this algorithm takes an exponential number of computational steps which grows with $n$ as $3^n$.
- In this algorithm, the values $x_i$ at which the minimum and the maximum of $r$ are assigned depend, in general, on the value $x$ at which we estimate $F(x)$. So, in general, to find the range $[\underline{F}(x), \overline{F}(x)]$ at $N_p$ points $x$, we have to repeat this algorithm $N_p$ times.

*Some bounds can be computed faster.* It turns out that some of the bounds can be computed in polynomial time, namely, the upper bound $\overline{r}$ for $x \geq \underline{a}$ and the lower bound $\underline{r}$ for $x \leq \overline{a}$.

*Algorithm $A_2$.* To find $\overline{r}$ for $x \geq \underline{a}$, we do the following:

- First, we sort all $2n$ values $\underline{x}_i$ and $\overline{x}_i$ into a non-decreasing sequence $x_{(1)} \leq x_{(2)} \leq \ldots \leq x_{(2n)}$. Thus, we subdivide the real line into $2n + 1$ zones $[x_{(0)}, x_{(1)}]$, $[x_{(1)}, x_{(2)}]$, $\ldots$, $[x_{(2n-1)}, x_{(2n)}]$, $[x_{(2n)}, x_{(2n+1)}]$, where we denoted $x_0 \overset{\text{def}}{=} -\infty$ and $x_{(2n+1)} \overset{\text{def}}{=} +\infty$.
- For each zone $[x_{(k)}, x_{(k+1)}]$, we partition indices $i = 1, \ldots, n$ into three sets

$$I^- = \{i : x_{(k+1)} \leq \underline{x}_i\}, \quad I^+ = \{i \notin I^- : x_{(k)} \geq \overline{x}_i\},$$

$$I_0 = \{1, \ldots, n\} - I^- - I^+.$$

Based on this partition, we compute $\tilde{a}$, $\tilde{m}$, $a$, and $a_0$ as in Algorithm $A_1$. For each value $a_0$ which is within the zone, we compute $\sigma = \sqrt{(a - a_0) \cdot (x - a)}$ and $r = \dfrac{x - a}{\sigma}$.

- The largest of the resulting values $r$ is the desired $\overline{r}$.

*Comment.* To find $\underline{r}$ for $x \leq \overline{a}$, we perform the same computations, with the only difference that at the end, instead of finding the *largest* of the resulting values $r$, we find the *smallest* of these values.

**Proposition 45.3.** *The above algorithms always computes the exact bound $\overline{r}$ for $x \geq \underline{a}$ and $\underline{r}$ for $x \leq \overline{a}$, and they take quadratic time $O(n^2)$.*

*Efficient algorithm for the no-nesting case.* Let us show that in a no-nesting case, when no two intervals are nested, i.e., when $[\underline{x}_i, \overline{x}_i] \not\subseteq (\underline{x}_j, \overline{x}_j)$ for all $i \neq j$. In this case, we can compute the remaining bounds $\underline{r}$ for $x > \overline{a}$ and $\overline{r}$ for $x < \underline{a}$ also in polynomial time.

It is known that intervals which satisfy the no-nesting property can be ordered in "lexicographic" order, i.e., the order in which $[\underline{x}_i, \overline{x}_i] < [\underline{x}_j, \overline{x}_j]$ if and only if either $\underline{x}_i < \underline{x}_j$ or ($\underline{x}_i = \underline{x}_j$ and $\overline{x}_i \leq \overline{x}_j$); see, e.g., [188, 197]. With respect to this order, both sequences $\underline{x}_i$ and $\overline{x}_i$ become monotonic: $\underline{x}_1 \leq \underline{x}_2 \leq \ldots \leq \underline{x}_n$ and $\overline{x}_1 \leq \overline{x}_2 \leq \ldots \leq \overline{x}_n$. We have used this order in our previous algorithms [188, 197], and we will use it here as well.

*Algorithm $A_3$.* To find $\underline{r}$ for $x > \overline{a}$, we do the following:

- First, we sort all $n$ intervals $[\underline{x}_i, \overline{x}_i]$ in the lexicographic order. As a result, we get two monotonic sequences

$$\underline{x}_1 \leq \underline{x}_2 \leq \ldots \leq \underline{x}_n, \quad \overline{x}_1 \leq \overline{x}_2 \leq \ldots \leq \overline{x}_n.$$

- For every $n^-$ from 1 to $n$, we consequently compute $\sum_{i=1}^{n^-} \underline{x}_i$ and $\widetilde{m} = \sum_{i=1}^{n^-} (\underline{x}_i)^2$: we start with 0 and we consequently add, correspondingly, $\underline{x}_i$ or $(\underline{x}_i)^2$.

- For every $n^+$ from 1 to $n+1$, we consequently compute $\sum_{i=n^++1}^{n} \overline{x}_i$ and $\widetilde{m} = \sum_{i=n^++1}^{n} (\overline{x}_i)^2$: we start with 0 for $n^+ = n+1$ and then we take $n^+ = n, n-1, \ldots, 1$ by consequently adding, correspondingly, $\overline{x}_i$ or $(\overline{x}_i)^2$.

- For every two natural numbers $n^-$ and $n^+$ for which $0 \leq n^- < n^+ \leq n+1$, we do the following:
  - We compute the values $N_0 = n - n^- - (n+1-n^+)$ and

$$\widetilde{a} = \sum_{i=1}^{n^-} \underline{x}_i + \sum_{j=n^+}^{n} \overline{x}_j, \quad \widetilde{m} = \sum_{i=1}^{n^-} (\underline{x}_i)^2 + \sum_{j=n^+}^{n} (\overline{x}_j)^2.$$

- We find the value $a$ from the same quadratic equation

$$\widetilde{m} + \frac{1}{N_0} \cdot (n \cdot a - \widetilde{a})^2 =$$

$$n \cdot \left( a - \frac{n \cdot a - \widetilde{a}}{N_0} \right) \cdot (x - a) + n \cdot a^2,$$

as in Algorithm $A_1$ (with $N_0 = n^+ - n^- - 1$), and then compute $a_0 = \dfrac{n \cdot a - \widetilde{a}}{N_0}$.

- If this quadratic equation does not have any real solutions, or if it does but the corresponding value $a_0$ does not belong to the interval $[\underline{x}_{n^+ - 1}, \overline{x}_{n^- + 1}]$, then we skip this partition and go to the next one.
- For each solution $a_0$ which belongs to the interval $[\underline{x}_{n^+ - 1}, \overline{x}_{n^- + 1}]$, we compute $\sigma = \sqrt{(a - a_0) \cdot (x - a)}$ and $r = \dfrac{x - a}{\sigma}$.
- The smallest of the resulting values $r$ is the desired $\overline{r}$.

*Comments.* To find $\overline{r}$ for $x < \underline{a}$, we perform the same computations, with the only difference that at the end, instead of finding the *smallest* of the resulting values $r$, we find the *largest* of these values.

**Proposition 45.4.** *The above algorithms always computes the exact bound $\underline{r}$ for $x > \overline{a}$ and $\overline{r}$ for $x < \underline{a}$, and they take quadratic time $O(n^2)$.*


## Proofs


*Proof of Proposition 45.1.* We know that the mean $a$ can take any values from the interval $[\underline{a}, \overline{a}]$. When $x \leq \underline{a}$, this means that the value $x - a$ is always non-positive. Since the standard deviation $\sigma$ is always non-negative, the ratio $\dfrac{x - a}{\sigma}$ is also non-positive. Therefore, both the smallest and the largest values of this ratio are non-positive: $\underline{r} \leq 0$ and $\overline{r} \geq 0$.

Similarly, when $x \geq \overline{a}$, we have $x - a \geq 0$, hence the ratio $r$ is non-negative and its bounds are also non-negative.

When $\underline{a} < x < \overline{a}$, the difference $x - a$ can take both positive values (e.g., when $x = \overline{a}$) and negative values (e.g., when $x = \underline{a}$). Thus, the ratio $r$ can also be both positive and negative. Hence, the largest possible value of this ratio is positive, and the smallest possible value of this ratio is negative.

*Proof of Proposition 45.2.*

1°. Within each interval $[\underline{x}_i, \overline{x}_i]$, the value $x_i$ corresponding to the optimal tuple can be either at the left endpoint $\underline{x}_i$ or at the right endpoint $\overline{x}_i$, or inside the interval $(\underline{x}_i, \overline{x}_i)$. Let us denote the set of all indices for which $x_i = \underline{x}_i$ by $I^-$, the set of all indices for which $x_i = \overline{x}_i$ by $I^+$, and the set of all remaining indices by $I_0$.

$2°$. According to the arguments described before the formulation of this proposition, for every $i$, either $x_i = \underline{x}_i$ or $x_i = \overline{x}_i$, or $\dfrac{\partial r}{\partial x_i} = 0$. Let us therefore describe an explicit formula for this derivative.

$2.1°$. Since $r$ is defined in terms of $a$ and $\sigma$, let us first find the formulas for the derivatives of $a$ and $\sigma$.

Since $a = \dfrac{1}{n} \cdot \sum\limits_{i=1}^{n} x_i$, we have $\dfrac{\partial a}{\partial x_i} = \dfrac{1}{n}$. Since $\sigma = \sqrt{V}$, where

$$V \stackrel{\text{def}}{=} \frac{1}{n} \cdot \sum_{i=1}^{n} (x_i - a)^2 = \frac{1}{n} \cdot \sum_{i=1}^{n} x_i^2 - a^2,$$

we have

$$\frac{\partial \sigma}{\partial x_i} = \frac{1}{2\sigma} \cdot \frac{\partial V}{\partial x_i}.$$

Here,

$$\frac{\partial V}{\partial x_i} = \frac{2}{n} \cdot x_i - 2a \cdot \frac{\partial a}{\partial x_i} = \frac{2}{n} \cdot (x_i - a).$$

Therefore, we have

$$\frac{\partial \sigma}{\partial x_i} = \frac{1}{n \cdot \sigma} \cdot (x_i - a).$$

$2.2°$. Now, we are ready to compute the desired derivative. Here,

$$\frac{\partial r}{\partial x_i} = \frac{\partial}{\partial x_i} \left( \frac{x - a}{\sigma} \right) = \frac{-\dfrac{\partial a}{\partial x_i} \cdot \sigma - (x - a) \cdot \dfrac{\partial \sigma}{\partial x_i}}{\sigma^2}.$$

In view of the analysis that preceded the formulation of this proposition, we are only interested in the sign of the derivative $\dfrac{\partial r}{\partial x_i}$. Since the denominator $\sigma^2$ of the expression describing this derivative is always non-negative, this sign coincides with the sign of the numerator

$$N_i \stackrel{\text{def}}{=} -\frac{\partial a}{\partial x_i} \cdot \sigma - (x - a) \cdot \frac{\partial \sigma}{\partial x_i}.$$

Substituting the above expressions for $\dfrac{\partial a}{\partial x_i}$ and $\dfrac{\partial \sigma}{\partial x_i}$ into this formula, we conclude that

$$N_i = -\frac{1}{n} \cdot \sigma - (x - a) \cdot \frac{1}{n \cdot \sigma} \cdot (x_i - a).$$

We can simplify this expression even further if we multiply it by $n \cdot \sigma$ – which also does not change the signs. Thus, the sign of the desired derivative $\dfrac{\partial r}{\partial x_i}$ coincides with the sign of the product $p_i \stackrel{\text{def}}{=} n \cdot \sigma \cdot N_i$, which is equal to

$$p_i = -(x_i - a) \cdot (x - a) - \sigma^2.$$

$3°$. The only possibility for $x_i$ to be inside the interval $(\underline{x}_i, \overline{x}_i)$ is to have $p_i = 0$. Dividing both sides by $x - a$, we conclude that $x_i = a_0$, where we denoted

$$a_0 \overset{\text{def}}{=} a - \frac{\sigma^2}{x - a}.$$

Thus, all the values $x_i$ with $i \in I_0$ have exactly the same value $a_0$.

Once we know the partition into the sets $I^-$, $I^+$, and $I_0$, we also know the values $x_i$ for $i \in I^-$ and $i \in I^+$. To find the values $x_i$ for $i \in I_0$, we need to find the value $a_0$.

$4°$. By definition of the sample mean $a$, the sum of all $n$ values $x_i$ is equal to $n \cdot a$, i.e.,

$$\sum_{i \in I^-} \underline{x}_i + \sum_{i \in I^+} \overline{x}_i + N_0 \cdot a_0 = n \cdot a.$$

The sum of the first two sums is what we denoted by $\widetilde{a}$; so, we conclude that $\widetilde{a} + N_0 \cdot a_0 = n \cdot a$ and hence, that

$$a_0 = \frac{n \cdot a - \widetilde{a}}{N_0}. \tag{45.2}$$

Since $a_0 = a - \dfrac{\sigma^2}{x - a}$, we conclude that

$$\sigma^2 = (a - a_0) \cdot (x - a) = \left( a - \frac{n \cdot a - \widetilde{a}}{N_0} \right) \cdot (x - a). \tag{45.3}$$

By definition of the sample variance, we have $\sum_{i=1}^{n} x_i^2 = n \cdot a^2 + n \cdot \sigma^2$, i.e.,

$$\sum_{i \in I^-} (\underline{x}_i)^2 + \sum_{i \in I^+} (\overline{x}_i)^2 + N_0 \cdot a_0^2 = n \cdot a^2 + n \cdot \sigma^2.$$

The sum of the first two sums is what we denoted by $\widetilde{m}$; so, we conclude that $\widetilde{m} + N_0 \cdot a_0^2 = n \cdot a^2 + n \cdot \sigma^2$. Substituting the expressions (45.2) and (45.3) for $a_0$ and $\sigma^2$ into this formula, we get the quadratic equation given in the algorithm.

So, the optimal solution is indeed among those processed by the algorithm. The proposition is proven.

*Proof of Proposition 45.3.*

$1°$. We have already proven that the sign of the desired derivative $\dfrac{\partial r}{\partial x_i}$ coincides with the sign of the product $p_i = -(x_i - a) \cdot (x - a) - \sigma^2$. According to Proposition 45.1, when we are looking for $\overline{r}$ and $x \geq \overline{a}$, then $x - a \geq 0$. In this case, the sign of $p_i$ coincides with the sign of the ratio

$$r_i \overset{\text{def}}{=} \frac{p_i}{x - a} = a_0 - x_i.$$

So we make the following conclusions:

(i) If the maximum $\overline{r}$ is attained for $x_i \in (\underline{x}_i, \overline{x}_i)$, then (the derivative is 0 hence) $x_i = a_0$.

(ii) If the maximum is attained for $x_i = \underline{x}_i$, then (the derivative is non-positive hence) $\underline{x}_i \geq a_0$.

(iii) Finally, if the maximum is attained for $x_i = \overline{x}_i$, then (the derivative is non-negative hence) $\overline{x}_i \leq a_0$.

Therefore, if $a_0 < \underline{x}_i$, then we cannot have the case (i) when $\underline{x}_i \leq x_i = a_0$ and we cannot have the case (iii) when $\underline{x}_i \leq \overline{x}_i \leq a_0$, so we must have case (ii), i.e., we must have $x_i = \underline{x}_i$.

Similarly, if $a_0 > \overline{x}_i$, then our only option is $x_i = \overline{x}_i$, and if $\underline{x}_i \leq a_0 \leq \overline{x}_i$, then our only option is $x_i = a_0$. Thus, as soon as we know the location of the value $a_0$ in comparison to the bounds $\underline{x}_i$ and $\overline{x}_i$ – i.e., as soon as we know the zone which contains $a_i$ – we can (almost) uniquely determine the values $x_i$ for all $x_i$ – with the only additional problem that we still need to determine the value $a_0$. We already described, in Algorithm $A_1$, how we can find $a_0$.

The case of $\underline{r}$ for $x \leq \underline{a}$ is handled similarly.

$2°$. To complete the proof, it is sufficient to show that these algorithms take quadratic time. Indeed, in addition to sorting (time $O(n \cdot \log(n))$), this algorithm takes linear time for each of $2n + 1$ zones. So, overall, we perform $O(n^2)$ computational steps. The proposition is proven.

*Proof of Proposition 45.4.*

$1°$. We have already proven that in the optimal tuple, each value of $x_i$ is either equal to $\underline{x}_i$ or to $\overline{x}_i$, or to a common value $a_0$. Let us now prove that in the no-nesting case, we can also assume that the optimal sequence $x_i$ is monotonic, i.e., $x_1 \leq x_2 \leq \ldots \leq x_n$.

Indeed, let us assume that in the optimal sequence, we have $x_i > x_j$ for some $i < j$. Here, we have $\underline{x}_i \leq x_i \leq \overline{x}_i$, $\underline{x}_j \leq x_j \leq \overline{x}_j$, and, since $i < j$, we also have $\underline{x}_i \leq \underline{x}_j$ and $\overline{x}_i \leq \overline{x}_j$. Let us show that in this case, $x_i \in \boldsymbol{x}_j$ and $x_j \in \boldsymbol{x}_i$.

Indeed, from $x_i \leq \overline{x}_i$ and $\overline{x}_i \leq \overline{x}_j$, we conclude that $x_i \leq \overline{x}_j$. Similarly, from $\underline{x}_j \leq x_j$ and $x_j < x_i$, we conclude that $\underline{x}_j < x_i$. Thus, indeed $x_i \in [\underline{x}_j, \overline{x}_j]$. Similarly, we have $x_j \in [\underline{x}_i, \overline{x}_i]$.

Because of these two inclusions, we can "swap" the values $x_i$ and $x_j$, i.e., produce a new tuple in which $x_i^{\text{new}} = x_j$ and $x_j^{\text{new}} = x_i$. The values of sample mean $a$ and sample standard deviation $\sigma$ do not change if we simply swap two values. So, for this new tuple, we can the exact same values of $a$, $\sigma$ and therefore, the same value of the ratio $r$. Since the original tuple maximized $r$, the new tuple is also maximizing $r$.

By repeating this swapping sufficiently many times, we will get a monotonic optimizing tuple.

$2^\circ$. Let us now prove that if in the optimal solution, we have $x_i > \underline{x}_i$ and $x_j < \overline{x}_j$, then we should have $x_i \geq x_j$.

Indeed, in this case, for sufficiently small $h > 0$, we can simultaneously do the following:

- decrease $x_i$ by $h$, i.e., replace it with $x_i^{\text{new}} = x_i - h$, and
- increase $x_j$ by $h$, i.e., replace it with $x_j^{\text{new}} = x_j + h$,

and still keep both values $x_i$ and $x_j$ within the corresponding intervals $[\underline{x}_i, \overline{x}_i]$ and $[\underline{x}_j, \overline{x}_j]$.

Since the value of $r$ was originally the smallest, it cannot decrease under this replacement. After the replacement, the sum $\sum x_i$ does not change hence the average $a$ does not change, and the value of the numerator $x - a > 0$ does not change either.

The value of $\sigma^2 = \dfrac{1}{n} \cdot \sum x_i^2 - a^2$ changes since two terms change: $x_i^2$ to $(x_i - h)^2 = x_i^2 - 2h \cdot x_i + o(h)$ and $x_j^2$ to $(x_j + h)^2 = x_j^2 + 2h \cdot x_j + o(h)$. Thus, overall, $\sigma^2$ is replaced with $\sigma^2 + \dfrac{2h}{n} \cdot (x_j - x_i) + o(h)$. We cannot have an increase in $\sigma$ – that would lead to an impossible decrease of $r$ below it smallest value. Thus, the new value of $\sigma^2$ cannot be larger than its original value. In other words, we must have

$$\sigma^2 + \frac{2h}{n} \cdot (x_j - x_i) + o(h) \geq \sigma^2.$$

Subtracting $\sigma^2$ from both sides and dividing both sides by $h \geq 0$, we conclude that $x_j - x_i + o(1) \leq 0$. In the limit $h \to 0$, we get the desired inequality $x_j \leq x_i$.

$3^\circ$. So, in the optimal tuple, every value $x_i = \underline{x}_i$ must precede every value $x_j = \overline{x}_j$, and all the values $x_i = a_0 \in (\underline{x}_i, \overline{x}_i)$ must be in between. Due to monotonicity, we therefore conclude that first we have a sequence of several values $\underline{x}_i$, then several values equal to $a_0$, and after that, several values equal to $\overline{x}_j$. This is exactly the type of solution we analyze in the algorithm.

For each selection of $n^-$ and $n^+$, we need to check whether the value $a_0$ is indeed contained in all the corresponding intermediate intervals $[\underline{x}_i, \overline{x}_i]$ for $i - n^- + 1, \ldots, n^+ - 1$. Since the sequence $\underline{x}_i$ is increasing, it is sufficient to check the inequality $a_0 \geq \underline{x}_i$ only for the largest of these bounds, i.e., for the bound $\underline{x}_{n^+ - 1}$. Similarly, since the sequence $\overline{x}_i$ is increasing, it is sufficient to check the inequality $a_0 \leq \overline{x}_i$ only for the smallest of these bounds, i.e., for the bound $\overline{x}_{n^- + 1}$. Thus, the algorithm is justified.

$4^\circ$. To complete the proof, it is sufficient to show that the algorithm $A_3$ takes quadratic time. Indeed, in addition to sorting (time $O(n \cdot \log(n))$) and linear time for computing the sums $\sum \underline{x}_i, \sum (\underline{x}_i)^2, \sum \overline{x}_i, \sum (\overline{x}_i)^2$, we need a constant time for each of $n^2$ pairs of indices – i.e., $O(n^2)$ computational steps overall. The proposition is proven.

# Beyond Traditional Fuzzy Uncertainty: Interval-Valued Fuzzy Techniques

For *fuzzy* information, we assumed that we have exact numerical degrees describing expert uncertainty. This is, of course, a simplifying assumption. In practice, an expert can, at best, provide bounds (i.e., an interval) or his or her degree of certainty – or even produce a *fuzzy* degree of certainty (such as "about 0.6"). Situations with interval-valued fuzzy degrees are analyzed in this chapter, and the situations with more general fuzzy-valued degrees (called *type 2*) are analyzed in the next chapter.

*Intervals are necessary to describe degrees of belief.* In the previous text, we described an idealized situation, in which we can describe degrees of belief by exact real numbers. In practice, the situation is more complicated, because experts cannot describe their degrees of belief precisely; see, e.g., [251] and references therein.

Indeed, let us start by reviewing the above-described methods of eliciting degrees of belief. If an expert describes his or her degree of belief by selecting, e.g., 8 on a scale from 0 to 10, this does not mean that his or her degree of belief is exactly 0.8: if instead, we ask him or her to select on a scale from 0 to 9, then whatever he or she chooses, after dividing it by 9, we will never get 0.8. If an expert chooses a value 8 on a 0 to 10 scale, then the only thing that we know about the expert's degree of belief is that it is closer to 8 than to 7 or to 9, i.e., that this degree of belief belongs to the *interval* $[0.75, 0.85]$.

Another possible source of interval uncertainty is when we have *several* experts, and their estimates differ. If, e.g., two equally good experts point to 7 and 8, then, if we are cautious, we would rather describe the resulting degree of belief as the interval $[0.7, 0.8]$ (or, in view of the above remark, as the interval $[0, 65, 0.85]$).

If we determine the degree of belief by polling, then the same argument shows that the resulting numbers are not precise: e.g., if 8 out of 10 experts voted for $A$, then we cannot say that the actual degree of belief is exactly 0.8, because, if we repeated this procedure with 9 experts, we will never get exactly 0.8. In this case, there are two other sources of uncertainty: First, picking experts is sort of a random procedure, so, the result of voting is a

H.T. Nguyen et al.: Computing Stat. under Interval & Fuzzy Uncertain., SCI 393, pp. 391–393.

springerlink.com                         © Springer-Verlag Berlin Heidelberg 2012

statistical estimate that is not precise (just like a statistical frequency estimate of probability). A better description will be to give an *interval* of possible values of $d(A)$.

The polling method of estimating the degree of belief is based on the assumption that an expert can always tell whether he believes in a given statement $S$ or not. Then, we take the ratio $d(S) = N(S)/N$ of the number $N(S)$ of experts who believe in $S$ to the total number $N$ of experts as the desired estimate. For $\neg S$, we thus have $N(\neg S) = N - N(S)$, so $d(\neg S) = N(\neg S)/N = 1 - d(S)$. In reality, an expert is often unsure about $S$. In this case, instead of dividing the experts into two categories: those who believe in $S$ and those who do not, we must divide them into *three* categories: those who believe that $S$ is true (we will denote their number by $N(S)$), those who believe that $S$ is false (we will denote their number by $N(\neg S)$), and those who do not have the definite opinion about $S$; there are $N - N(S) - N(\neg S)$ of them. In this situation, one number is not sufficient to describe the experts' degree of belief in $S$, we need at least two. There are two ways to describe it: We can describe the degree of belief in $S$ as $d(S) = N(S)/N$ and the degree of belief in $\neg S$ as $d(\neg S) = N(\neg S)/N$. These two numbers must satisfy the condition $d(S) + d(\neg S) \leq 1$. This description is known as *intuitionistic fuzzy logic*. (The reason for the word "intuitionistic" is that this logic is close to the original intuitionistic idea that the law of excluded middle is not always true.)

Alternatively, we can describe the degree of belief $d(S)$ in $S$ and the degree of *plausibility* of $S$ estimating as the fraction of experts who do not consider $S$ impossible, i.e., as $pl(S) = 1 - d(\neg S)$, i.e., as an interval $[d(S), pl(S)]$. This representation corresponds to the *Dempster-Shafer formalism*.

So, to describe degrees of belief adequately, we must use *intervals* instead of real numbers.

*Interval computations for processing interval-valued degrees of belief: general idea.* For an expert system with interval-valued degrees of belief, the following problem arises: suppose that we have an expert system whose knowledge base consists of statements $S_1, \ldots, S_N$, and we have an algorithm $f(Q, d_1, \ldots, d_N)$ (called *inference engine*) that for any given query $Q$, transforms the degrees of belief $d(S_1), \ldots, d(S_N)$ in the statements from the knowledge base into a degree of belief $d(Q) = f(Q, d(S_1), \ldots, d(S_N))$ in $Q$ (for example, if $Q = S_1 \& S_2$, then $f(d_1, \ldots, d_N) = f_\&(d_1, d_2)$). Suppose now that we know only the intervals $\mathbf{d}(S_1), \ldots, \mathbf{d}(S_N)$ that contain the desired degree of belief. Then, the degree of belief in $Q$ can take any value from the set

$$f(Q, \mathbf{d}(S_1), \ldots, \mathbf{d}(S_N)) = \{f(Q, d_1, \ldots, d_N) \mid d_i \in \mathbf{d}(S_i)\}.$$

Computing such an interval is a typical problem of *interval computations*.

In particular, since the functions $f_\&$ and $f_\vee$ are increasing in both arguments, we have

$$f_\&([\underline{x}, \overline{x}], [\underline{y}, \overline{y}]) = [f_\&(\underline{x}, \underline{y}), f_\&(\overline{x}, \overline{y})]$$

and
$$f_\vee([\underline{x}, \overline{x}], [\underline{y}, \overline{y}]) = [f_\vee(\underline{x}, \underline{y}), f_\vee(\overline{x}, \overline{y})].$$

For example,
$$\min([\underline{x}, \overline{x}], [\underline{y}, \overline{y}]) = [\min(\underline{x}, \underline{y}), \min(\overline{x}, \overline{y})]$$

and
$$\max([\underline{x}, \overline{x}], [\underline{y}, \overline{y}]) = [\max(\underline{x}, \underline{y}), \max(\overline{x}, \overline{y})].$$

# Beyond Traditional Fuzzy Uncertainty: Type-2 Fuzzy Techniques

For *fuzzy* information, we assumed that we have exact numerical degrees describing expert uncertainty. As we have mentioned in the previous chapter, in practice, an expert can, at best, provide bounds (i.e., an interval) or his or her degree of certainty – or even produce a "type-2" *fuzzy* degree of certainty (such as "about 0.6").

As we have shown in Part I, processing of data under general type-1 fuzzy uncertainty can be reduced to the simplest case – of interval uncertainty: namely, Zadeh's extension principle is equivalent to level-by-level interval computations applied to $\alpha$-cuts of the corresponding fuzzy numbers.

The more general inter-valued and type-2 fuzzy numbers more adequately represent the expert's opinions, but their practical use is limited by the seeming computational complexity of their use. In his recent research, J. Mendel has shown that for the practically important case of interval-valued fuzzy sets, processing such sets can also be reduced to interval computations. In this chapter, we show that Mendel's idea can be naturally extended to arbitrary type-2 fuzzy numbers.

The results of this chapter first appeared in [194].

*Need for type-2 fuzzy sets.* The main objective of fuzzy logic is to describe uncertain ("fuzzy") knowledge, when an expert cannot describe his or her knowledge by an exact value or by a precise set of possible values. Instead, the expert describe this knowledge by using words from natural language. Fuzzy logic provides a procedure for formalizing these words into a computer-understandable form – as fuzzy sets.

In the traditional approach to fuzzy logic, the expert's degree of certainty in a statement – such as the value $m_A(x)$ describing that the value $x$ satisfies the property $A$ (e.g., "small") – is described by a number from the interval $[0, 1]$. However, we are considering situations in which an expert is unable to describe his or her knowledge in precise terms. It is not very reasonable to expect that in this situation, the same expert will be able to meaningfully express his or her degree of certainty by a precise number. It is much more

H.T. Nguyen et al.: Computing Stat. under Interval & Fuzzy Uncertain., SCI 393, pp. 395–399.
springerlink.com                                              © Springer-Verlag Berlin Heidelberg 2012

reasonable to assume that the expert will describe these degrees also by words from natural language.

Thus, for every $x$, a natural representation of the degree $m(x)$ is not a number, but rather a new fuzzy set. Such situations, in which to every value $x$ we assign a fuzzy number $m(x)$, are called *type-2* fuzzy sets.

*Successes of type-2 fuzzy sets.* Type-2 fuzzy sets are actively used in practice; see, e.g., [217] and [218]. Since type-2 fuzzy sets provide a more adequate representation of expert knowledge, it is not surprising that such sets lead to a higher quality control, higher quality clustering, etc., in comparison with the more traditional type-1 sets.

*The main obstacle to using type-2 fuzzy sets.* If type-2 fuzzy sets are more adequate, why are not they used more? The main reason why their use is limited is that the transition from type-1 to type-1 fuzzy sets leads to an increase in computation time. Indeed, to describe a traditional (type-1) membership function function, it is sufficient to describe, for each value $x$, a single number $m(x)$. In contrast, to describe a type-2 set, for each value $x$, we must describe the entire membership function – which needs several parameters to describe. Since we need more numbers just to store such information, we need more computational time to process all the numbers representing these sets.

*Interval-valued fuzzy sets.* In line with this reasoning, the most widely used type-2 fuzzy sets are the ones which take the smallest number of parameters to store. We are talking about *interval-valued* fuzzy numbers, in which for each $x$, the degree of certainty $m(x)$ is an interval $\boldsymbol{m}(x) = [\underline{m}(x), \overline{m}(x)]$. To store each interval, we need exactly two numbers – the smallest possible increase over the single number needed to store the type-1 value $m(x)$.

*Mendel's 2007 algorithm for processing interval-valued fuzzy data.* In his plenary talk [219], J. M. Mendel provided a new algorithm which drastically reduced the computational complexity of processing interval-valued fuzzy data. Specifically, he showed that processing interval-valued fuzzy data can be efficiently reduced to interval computations. Since there exist many efficient algorithms and software packages for solving interval computation problems, Mendel's reduction means that we can use these packages to also process interval-valued fuzzy data – and thus, that processing interval-valued fuzzy data is (almost) as efficient as processing the traditional (type-1) fuzzy data.

Mendel's algorithm can be explained as follows. In the case of interval-valued fuzzy data, we do not know the exact numerical values $m_i(x_i)$ of the membership functions, we only know the interval $\boldsymbol{m}_i(x_i) = [\underline{m}_i(x), \overline{m}_i(x)]$ of possible values of $m_i(x_i)$, By applying Zadeh's extension principle to different combinations of values $m_i(x_i) \in [\underline{m}_i(x), \overline{m}_i(x)]$, we can get, in general, different values of

$$m(y) = \sup\{\min(m_1(x_1), m_2(x_2), \ldots) : y = f(x_1, \ldots, x_n)\}.$$

The result of processing interval-valued fuzzy numbers can be thus described if for each $y$, we describe the set of possible values of $m(y)$.

When the values $m_i(x_i)$ continuously change, the value $m(y)$ also continuously change. So, for every $y$, the set $\boldsymbol{m}(y)$ of all possible values of $m(y)$ is an interval: $\boldsymbol{m}(y) = [\underline{m}(y), \overline{m}(y)]$. Thus, to describe this set, it is sufficient, for each $y$, to provide the lower endpoint $\underline{m}(y)$ and the upper endpoint $\overline{m}(y)$ of this interval.

This computation is a particular case of the general problem of interval computations. Indeed, in general, we start with the intervals of possible values of the input, and we want to compute the interval of possible values of the output. In our case, we start with the intervals of possible values of $m_i(x_i)$, and we want to find the set of possible values of $m(y)$.

It is worth mentioning that the corresponding interval computation problem is easier than the general problem because the expression described by Zadeh's extension principle is monotonic – to be more precise, (non-strictly) increasing. Namely, if we increase one of the values $m_i(x_i)$, then the resulting value $m(y)$ can only increase (or stay the same). For monotonic functions, the range of possible values is easy to compute:

- the function attains its smallest value when all the inputs are the smallest, and
- the function attains its largest value when all the inputs are the largest.

In our case, for each input $m_i(x_i)$, the smallest possible value of $\underline{m}_i(x_i)$, and the largest possible value is $\overline{m}_i(x_i)$. Thus, we conclude that:

$$\underline{m}(y) = \sup\{\min(\underline{m}_1(x_1), \underline{m}_2(x_2), \ldots) : y = f(x_1, \ldots, x_n)\};$$

$$\overline{m}(y) = \sup\{\min(\overline{m}_1(x_1), \overline{m}_2(x_2), \ldots) : y = f(x_1, \ldots, x_n)\}.$$

In other words,

- to compute the lower membership function $\underline{m}(y)$, it is sufficient to apply the standard Zadeh's extension principle to the lower membership functions $\underline{m}_i(x_i)$, and
- to compute the upper membership function $\overline{m}(y)$, it is sufficient to apply the standard Zadeh's extension principle to the upper membership functions $\overline{m}_i(x_i)$.

We already know that for type 1 fuzzy sets, Zadeh's extension principle can be reduced to interval computations. Thus, we conclude that for every level $\alpha \in (0, 1]$, we have

$$\underline{\boldsymbol{x}}(\alpha) = f(\underline{\boldsymbol{x}}_1(\alpha), \ldots, \underline{\boldsymbol{x}}_n(\alpha))$$

and

$$\overline{\boldsymbol{x}}(\alpha) = f(\overline{\boldsymbol{x}}_1(\alpha), \ldots, \overline{\boldsymbol{x}}_n(\alpha)),$$

where

$$\underline{\boldsymbol{x}}_i \stackrel{\mathrm{def}}{=} \{x_i : \underline{m}_i(x_i) \geq \alpha\} \text{ and } \overline{\boldsymbol{x}}_i \stackrel{\mathrm{def}}{=} \{x_i : \overline{m}_i(x_i) \geq \alpha\}.$$

These are, in effect, the formulas proposed by Mendel in [219].

*New result: extension of Mendel's formulas to general type-2 fuzzy numbers.*
Let us show that Mendel's idea can be extended beyond interval-valued fuzzy
numbers, to arbitrary type-2 fuzzy numbers. Indeed, for arbitrary type-2
fuzzy numbers, for each $x_i$, the value $m_i(x_i)$ is also a fuzzy number. The relation between the input fuzzy numbers $m_i(x_i)$ and the desired fuzzy number
$m(y)$ can be expressed by the same Zadeh's principle:

$$m(y) = \sup\{\min(m_1(x_1), m_2(x_2), \ldots) : y = f(x_1, \ldots, x_n)\},$$

but this time, all the values $m_i(x_i)$ and $m(y)$ are fuzzy numbers. How can
we describe this relation between fuzzy numbers?

Let us first describe the fuzzy numbers themselves. By definition, a fuzzy
number is a function that maps every possible value to a degree from the
interval $[0, 1]$ describing to what extend this value is possible. Thus, e.g., for
each $y$, the corresponding fuzzy number $m(y)$ is a mapping which maps all
possible values $t \in [0, 1]$ into a degree (from the interval $[0, 1]$) with which $t$
is a possible value of $m(y)$. Let us denote this degree by $m(y, t)$.

Similarly, for each $i$ and for each real number $x_i$, the fuzzy number $m_i(x_i)$
is a mapping which maps all possible values $t \in [0, 1]$ into a degree (from the
interval $[0, 1]$) with which $t$ is a possible value of $m_i(x_i)$. Let us denote this
degree by $m_i(x_i, t)$.

As we have already mentioned, processing fuzzy numbers can be reduced
to processing the corresponding $\alpha$-cuts. In this case, all the values $m_i(x_i)$
and $m(y)$ are fuzzy numbers, we conclude that, for every $\alpha \in (0, 1]$, the
$\alpha$-cut $(m(y))(\alpha)$ for the fuzzy number $m(y)$ can be obtained by processing
the corresponding $\alpha$-cuts $(m(y))(\alpha)$ for $m_i(x_i)$. To avoid confusion between
standard $\alpha$-cuts, let us denote the corresponding threshold not as $\alpha$ but as
$\beta$. As a result, we conclude that

$$m(y)(\beta) = \sup\{\min(m_1(x_1)(\beta), m_2(x_2)(\beta), \ldots) : y = f(x_1, \ldots, x_n)\}.$$

For fuzzy numbers, the corresponding $\beta$-cuts are intervals:

$$m(y)(\beta) = [\underline{m(y)(\beta)}, \overline{m(y)(\beta)}] \text{ and } m_i(x_i)(\beta) = [\underline{m_i(x_i)(\beta)}, \overline{m_i(x_i)(\beta)}].$$

From our description of Mendel's result, we already know that in the interval case, since the expression corresponding to Zadeh's extension principle
is monotonic,

- the lower endpoints of the output can be obtained form the lower endpoints
  of the inputs, and
- the upper endpoint of the output can be obtained from the upper endpoints
  of the inputs,

hence, that

$$\underline{m(y)(\beta)} = \sup\{\min(\underline{m_1(x_1)(\beta)}, \underline{m_2(x_2)(\beta)}, \ldots) : y = f(x_1, \ldots, x_n)\};$$

$$\overline{m(y)(\beta)} = \sup\{\min(\overline{m_1(x_1)(\beta)}, \overline{m_2(x_2)(\beta)}, \ldots) : y = f(x_1, \ldots, x_n)\}.$$

For the corresponding functions $\underline{m(y)(\beta)}$, $\underline{m_i(x_i)(\beta)}$, $\overline{m(y)(\beta)}$, and $\overline{m_i(x_i)(\beta)}$, we get the standard Zadeh's extension principle relation between membership functions. We already know that this relation can be described in terms of interval computations. Thus, we conclude that

$$\underline{\boldsymbol{y}}(\alpha, \beta) = f(\underline{\boldsymbol{x}}_1(\alpha, \beta), \ldots, \underline{\boldsymbol{x}}_n(\alpha, \beta))$$

and

$$\overline{\boldsymbol{y}}(\alpha, \beta) = f(\overline{\boldsymbol{x}}_1(\alpha, \beta), \ldots, \overline{\boldsymbol{x}}_n(\alpha, \beta)),$$

where

$$\underline{\boldsymbol{y}}(\alpha, \beta) = \{x : \underline{y(\beta)} \geq \alpha\} \text{ and } \overline{\boldsymbol{y}}(\alpha, \beta) = \{x : \overline{y(\beta)} \geq \alpha\}$$

are the $\alpha$-cuts of the corresponding membership functions $\underline{m(y)(\beta)}$, and $\overline{m(y)(\beta)}$, and similarly,

$$\underline{\boldsymbol{x}}_i(\alpha, \beta) = \{x : \underline{x_i(\beta)} \geq \alpha\} \text{ and } \overline{\boldsymbol{x}}_i(\alpha, \beta) = \{x : \overline{x_i(\beta)} \geq \alpha\}$$

are the $\alpha$-cuts of the corresponding membership functions $\underline{m_i(x_i)(\beta)}$, and $\overline{m_i(x_i)(\beta)}$.

Thus, from the computational viewpoint, the problem of processing data under type-2 fuzzy uncertainty can be reduced to several problems of data processing under interval uncertainty – as many problems as there are $(\alpha, \beta)$-levels.

*Conclusion.* Type-2 fuzzy sets more adequately describe expert's opinion than the more traditional type-1 fuzzy sets. Because of this, in many practical applications, the use of type-2 fuzzy sets has led to better quality control, better quality clustering, etc. The main reason why they are not universally used is that when we go from type-1 sets to type-2 sets, the computational time of data processing increases. In his 2007 paper, J. Mendel has shown that for the practically important case of interval-valued fuzzy numbers, processing of such such data can be reduced to processing interval data – and is, thus, (almost) as fast as processing type-1 fuzzy data. In this chapter, we show that Mendel's idea can be extended to arbitrary type-2 fuzzy numbers – and thus, that processing general type-2 fuzzy numbers is also (almost) as fast as processing type-1 fuzzy data. This result will hopefully lead to more practical applications of type-2 fuzzy sets – which more adequately describe expert knowledge.

# References


1. Abellan, J., Moral, S.: A non-specificity measure for convex sets of probability distributions. Intern. J. of Uncertainty, Fuzziness and Knowledge-Based Systems 8(3), 357–367 (2000)
2. Abellan, J., Moral, S.: Maximum of entropy for credal sets. Intern. J. of Uncertainty, Fuzziness, and Knowledge-Based Systems 11(5), 587–597 (2003)
3. Abellan, J., Moral, S.: Range of entropy for credal sets. In: López-Diaz, M., et al. (eds.) Soft Methodology and Random Information Systems, pp. 157–164. Springer, Heidelberg (2004)
4. Abellan, J., Moral, S.: Difference of entropies as a nonspecificity function on credal sets. Intern. J. of General Systems 34(3), 201–214 (2005)
5. Abellan, J., Moral, S.: An algorithm to compute the upper entropy for order-2 capacities. Intern. J. of Uncertainty, Fuzziness, and Knowledge-Based Systems 14(2), 141–154 (2006)
6. Aida, K., Takefusa, A., Nakada, H., Matsuoka, S., Nagashima, U.: A performance evaluation model for effective job scheduling in global computing systems. In: Proceedings of the Seventh International IEEE Symposium on High Performance Distributed Computing, July 28-31, pp. 352–353 (1998)
7. Ainsworth, M., Senior, B.: Aspects of an hp-Adaptive Finite Element Method: Adaptive Strategy, Conforming Approximation and Efficient Solvers. Technical Report 1997/2. University of Leicester, England, U.K., Department of Mathematics and Computer Science (1997)
8. Akpan, U.O., Koko, T.S., Orisamolu, I.R., Gallant, B.K.: Practical fuzzy finite element analysis of structures. Finite Elem. Anal. Des. 38, 93–111 (2001)
9. Aldouri, A., Keller, G.R., Gates, A., Rasillo, J., Salayandia, L., Kreinovich, V., Seeley, J., Taylor, P., Holloway, S.: GEON: Geophysical data add the 3rd dimension in geospatial studies. In: Proceedings of the ESRI International User Conference 2004, San Diego, California, August 9-13, Paper 1898 (2005)
10. Aló, R., Beheshti, M., Xiang, G.: Computing variance under interval uncertainty: a new algorithm and its potential application to privacy in statistical databases. In: Proceedings of the International Conference on Information Processing and Management of Uncertainty in Knowledge-Based Systems IPMU 2006, Paris, France, July 2-7, pp. 810–816 (2006)
11. Annan, A.P.: Ground Penetrating Radar Workshop Notes. Sensors and Software Inc., Mississauga (1992)

12. Araiza, R., Taufer, M., Leung, M.Y.: Towards optimal scheduling for global computing under probabilistic, interval, and fuzzy uncertainty, with potential applications to bioinformatics. In: Reformat, R., Berthold, M.R. (eds.) Proceedings of the 26th International Conference of the North American Fuzzy Information Processing Society NAFIPS 2007, San Diego, California, June 24-27, pp. 520–525 (2007)

13. Araiza, R., Xiang, G., Kosheleva, O., Skulj, D.: Under interval and fuzzy uncertainty, symmetric markov chains are more difficult to predict. In: Reformat, M., Berthold, M.R. (eds.) Proceedings of the 26th International Conference of the North American Fuzzy Information Processing Society NAFIPS 2007, San Diego, California, June 24-27, pp. 526–531 (2007)

14. Archimedes: On the measurement of the circle. In: Heath, T.L. (ed.) The Works of Archimedes. Cambridge University Press, Cambridge (1897); reprinted by Dover (1953)

15. Atlantis Scientific, Inc., Theory of Synthetic Aperture Radar (2005), http://www.geo.unizh.ch/fpaul/sar-theory.html

16. Ausiello, G., Crescenzi, P., Kann, V., Marchetti-Spaccamela, A., Protasi, M.: Complexity and Approximation: Combinatorial Optimization Problems and Their Approximability Properties. Springer, Heidelberg (1999)

17. Au-Yeung, S.W.M.: Finding Probability Distributions from Moments. Master's Thesis. Imperial College, London (2003)

18. Averill, M.G., Miller, K.C., Keller, G.R., Kreinovich, V., Araiza, R., Starks, S.A.: Using expert knowledge in solving the seismic inverse problem. International Journal of Approximate Reasoning 45(3), 564–587 (2006)

19. Averill, M.G., Xiang, G., Kreinovich, V., Keller, G.R., Starks, S.A., Debroux, P.S., Boehm, J.: How to reconstruct the original shape of a radar signal? In: Proceedings of the 24th International Conference of the North American Fuzzy Information Processing Society NAFIPS 2005, Ann Arbor, Michigan, June 22-25, pp. 717–721 (2005)

20. Babuška, I., Griebel, M., Pitkäranta, J.: The problem of selecting the shape functions for $p$-type elements. Int. J. Numer. Meth. Engrg. 28, 1891–1908 (1999)

21. Babuška, I., Gui, W.: The $h$, $p$, and $hp$-versions of the finite element method in one dimension. Numer. Math. 49(I–III), 577–683 (1986)

22. Babuška, I., Oden, T.J.: Verification and validation in computational engineering and science: basic concepts. Computer Methods in Applied Mechanics and Engineering 193, 4057–4066 (2004)

23. Bardossy, G., Fodor, J.: Evaluation of Uncertainties and Risks in Geology. Springer, Berlin (2004)

24. Barmish, B.R.: New Tools for Robustness of Linear Systems. McMillan, New York (1994)

25. Beck, J.B., Kreinovich, V., Wu, B.: Interval-valued and fuzzy-valued random variables: from computing sample variances to computing sample covariances. In: Lopez, M., Gil, M.A., Grzegorzewski, P., Hrynewicz, O., Lawry, J. (eds.) Soft Methodology and Random Information Systems, pp. 85–92. Springer, Heidelberg (2004)

26. Beheshti, M., Han, J., Longpré, L., Starks, S.A., Vargas, J.I., Xiang, G.: Interval and fuzzy techniques in business-related computer security: intrusion detection, privacy protection. In: Proceedings of the Second International Conference on Fuzzy Sets and Soft Computing in Economics and Finance FSSCEF 2006, St. Petersburg, Russia, June 28-July 1, pp. 23–30 (2006)

27. Belforte, G., Bona, B.: An improved parameter identification algorithm for signal with unknown-but-bounded errors. In: Proceeding of the 7th IFAC Symposium on Identification and Parameter Estimation, York, U.K (1985)
28. Belytschko, T., Moës, N., Usui, S., Parimi, C.: Arbitrary discontinuities in finite elements. International Journal of Numerical Methods in Engineering 50(4), 993–1013 (2001)
29. Ben-Haim, Y.: Info-Gap Decision Theory: decisions under severe uncertainty. Academic Press, New York (2006)
30. Ben-Haim, Y., Elishakoff, I.: Convex Models of Uncertainty in Applied Mechanics. Elsevier Science, Amsterdam (1990)
31. Berberian, S.: Lectures in Functional Analysis and Operator Theory. Springer, Heidelberg (1974)
32. Berleant, D.: Automatically verified arithmetic with both intervals and probability density functions. Interval Computations 2, 48–70 (1993)
33. Berleant, D.: Automatically verified arithmetic on probability distributions and intervals. In: Kearfott, R.B., Kreinovich, V. (eds.) Applications of Interval Computations. Kluwer, Dordrecht (1996)
34. Berleant, D., Anderson, G., Goodman-Strauss, C.: Arithmetic on bounded families of distributions: a DEnv algorithm tutorial. In: Hu, C., Kearfott, R.B., de Korvin, A., Kreinovich, V. (eds.) Knowledge Processing with Interval and Soft Computing, pp. 183–210. Springer, London (2008)
35. Berleant, D., Andrieu, L., Argaud, J.P., Barjon, F., Cheong, M.P., Dancre, M., Sheble, G., Teoh, C.C.: Portfolio management under epistemic uncertainty using stochastic dominance and Information-Gap Theory. International Journal of Approximate Reasoning 49(1), 101–116 (2008)
36. Berleant, D., Cheong, M.-P., Chu, C., Guan, Y., Kamal, A., Sheblé, G., Ferson, S., Peters, J.F.: Dependable handling of uncertainty. Reliable Computing 9(6), 407–418 (2003)
37. Berleant, D., Goodman-Strauss, C.: Bounding the results of arithmetic operations on random variables of unknown dependency using intervals. Reliable Computing 4(2), 147–165 (1998)
38. Berleant, D., Villaverde, K., Kosheleva, O.: Towards a more realistic representation of uncertainty: an approach motivated by Info-Gap Decision Theory. In: Proceedings of the 27th International Conference of the North American Fuzzy Information Processing Society NAFIPS 2008, New York, May 19-22 (2008)
39. Berleant, D., Xie, L., Zhang, J.: Statool: a tool for Distribution Envelope Determination (DEnv), an interval-based algorithm for arithmetic on random variables. Reliable Computing 9(2), 91–108 (2003)
40. Berleant, D., Zhang, J.: Using Pearson correlation to improve envelopes around the distributions of functions. Reliable Computing 10(2), 139–161 (2004)
41. Berleant, D., Zhang, J.: Representation and problem solving with the Distribution Envelope Determination (DEnv) method. Reliability Engineering and System Safety 85(1-3), 153–168 (2004)
42. Berz, M., Hoffstätter, G.: Computation and Application of Taylor Polynomials with Interval Remainder Bounds. Reliable Computing 4(1), 83–97 (1998)
43. Bhatia, K., Stearn, B., Taufer, M., Zamudio, R., Catarino, D.: Extending grid protocols onto the desktop using the Mozilla framework. In: Proceedings of the 2nd International Workshop on Grid Computing Environments (GCE 2006), A Workshop in Conjunction with SuperComputing 2006 Conference, Tampa, Florida (November 2006)

44. Bhatia, K., Stearn, B., Taufer, M., Zamudio, R., Catarino, D.: Integrate GridFTP into Firefox – Build grid protocols into Mozilla-based tools. IBM DeveloperWorks (October 10, 2006a),
http://www-128.ibm.com/developerworks/grid/library/gr-firefoxftp/

45. Bhattacharyya, S.P., Chapellat, H., Keel, L.: Robust Control: The Parametric Approach. Prentice-Hall, Englewood Cliffs (1995)

46. Blelloch, G.E.: Prefix sums and their applications. In: Reif, J.H. (ed.) Synthesis of Parallel Algorithms, pp. 35–60. Morgan Kaufmann, San Mateo (1993)

47. Bojadziev, G., Bojadziev, M.: Fuzzy Sets, Fuzzy Logic, Applications. World Scientific, Singapore (1995)

48. Boning, D., Nassif, S.: Models of process variations in device and interconnect. In: Chandrakasan, A. (ed.) Design of High-Performance Microcomputer Circuits (2000)

49. Boyd, S., Vandenberghe, L.: Convex Optimization. Cambridge University Press (2004)

50. Bozkurt-Unal, G., Cetin, A.E.: Restoration of error-diffused images using projection onto convex sets. IEEE Trans. Image Processing 10(12), 1836–1841 (2001)

51. Brito, A.E., Kosheleva, O.: Interval + image = wavelet: for image processing under interval uncertainty, wavelets are optimal. Reliable Computing 4(3), 291–301 (1998)

52. van den Broek, P., Noppen, J.: Fuzzy weighted average: alternative approach. In: Proceedings of the 25th International Conference of the North American Fuzzy Information Processing Society NAFIPS 2006, Montreal, Quebec, Canada, June 3-6 (2006)

53. Cabrera, S.D., Iyer, K., Xiang, G., Kreinovich, V.: On inverse halftoning: computational complexity and interval computations. In: Proceedings of the 39th Conference on Information Sciences and Systems CISS 2005, March 16-18, Paper 164. John Hopkins University (2005)

54. Ceberio, M., Ferson, S., Kreinovich, V., Chopra, S., Xiang, G., Murguia, A., Santillan, J.: How to take into account dependence between the inputs: from interval computations to constraint-related set computations, with potential applications to nuclear safety, bio- and geosciences. In: Proceedings of the Second International Workshop on Reliable Engineering Computing, Savannah, Georgia, February 22-24, pp. 127–154 (2006)

55. Ceberio, M., Ferson, S., Kreinovich, V., Chopra, S., Xiang, G., Murguia, A., Santillan, J.: How to take into account dependence between the inputs: from interval computations to constraint-related set computations, with potential applications to nuclear safety, bio- and geosciences. Journal of Uncertain Systems 1(1), 11–34 (2007)

56. Ceberio, M., Kreinovich, V., Chopra, S., Longpré, L., Nguyen, H.T., Ludäscher, B., Baral, C.: Interval-type and affine arithmetic-type techniques for handling uncertainty in expert systems. Journal of Computational and Applied Mathematics 199(2), 403–410 (2007)

57. Ceberio, M., Kreinovich, V., Pownuk, A., Bede, B.: From interval computations to constraint-related set computations: towards faster estimation of statistics and odes under interval, p-box, and fuzzy uncertainty. In: Melin, P., Castillo, O., Aguilar, L.T., Kacprzyk, J., Pedrycz, W. (eds.) IFSA 2007. LNCS (LNAI), vol. 4529, pp. 33–42. Springer, Heidelberg (2007)

58. Ceberio, M., Xiang, G., Longpré, L., Kreinovich, V., Nguyen, H.T., Berleant, D.: Two Etudes on Combining Probabilistic and Interval Uncertainty: Processing Correlations and Measuring Loss of Privacy. In: Proceedings of the 7th International Conference on Intelligent Technologies InTech 2006, Taipei, Taiwan, December 13-15, pp. 8–17 (2006)
59. Chen, H., Belytschko, T.: An enriched finite element method for elastodynamic crack propagation. International Journal of Numerical Methods in Engineering 58(12), 1873–1905 (2003)
60. Chen, S.H., Lian, H.D., Yang, X.W.: Interval static displacement analysis for structures with interval parameters. Int. J. Numer. Methods Engrg. 53, 393–407 (2002)
61. Chernousko, F.L.: Estimation of the Phase Space of Dynamic Systems. Nauka publ., Moscow (1988) (in Russian)
62. Chernousko, F.L.: State Estimation for Dynamic Systems. CRC Press, Boca Raton (1994)
63. Chessa, J., Belytschko, T.: An extended finite element method for two-phase fluids. J. Appl. Mech. 70(1), 10–17 (2003)
64. Chessa, J., Belytschko, T.: A local space-time discontinuous finite element method. Computer Methods in Applied Mechanics and Engineering 195, 1325–1343 (2006)
65. Chessa, J., Smolinski, P., Belytschko, T.: The extended finite element method (X-FEM) for Solidification problems. Int. J. Numer. Methods Engrg. 53, 1959–1977 (2002)
66. Chessa, J., Wang, H., Belytschko, T.: On the construction of blending elements for local partition of unity enriched finite elements. Int. J. Numer. Methods Engrg. 57(7), 1015–1038 (2003)
67. Chinnery, D., Keutzer, K.: Closing the gap between ASICs and custom. In: Proc. of the Design Automation Conference DAC 2000 (2000)
68. Chinnery, D., Keutzer, K. (eds.): Closing the Gap Between ASICs and Custom. Kluwer, Dordrecht (2002)
69. Chokr, B., Kreinovich, V.: How far are we from the complete knowledge: complexity of knowledge acquisition in Dempster-Shafer approach. In: Yager, R.R., Kacprzyk, J., Pedrizzi, M. (eds.) Advances in the Dempster-Shafer Theory of Evidence, pp. 555–576. Wiley, New York (1994)
70. Ciarlet, P.G.: Discrete maximum principle for finite difference operators. Aequationes Math. 4, 338–352 (1970)
71. Ciarlet, P.G., Raviart, P.A.: Maximum principle and uniform convergence for the finite element method. Computer Methods in Applied Mechanics and Engineering 2, 17–31 (1973)
72. Corliss, G., Foley, C., Kearfott, R.B.: Formulation for reliable analysis of structural frames. In: Muhanna, R.L., Mullen, R.L. (eds.) Proc. NSF Workshop on Reliable Engineering Computing, Savannah, Georgia (2004), http://www.gtsav.gatech.edu/rec/recworkshop/index.html
73. Cormen, T.H., Leiserson, C.E., Rivest, R.L., Stein, C.: Introduction to Algorithms. MIT Press, Cambridge (2009)
74. Damera-Venkata, N., Kite, T.D., Geisler, W.S., Evans, B.L., Bovik, A.C.: Image quality assessment based on a degradation model. IEEE Transactions on Image Processing 9(4), 636–650 (2000)
75. Damera-Venkata, N., Kite, T.D., Venkataraman, M., Evans, B.L.: Fast blind inverse halftoning. In: Proc. IEEE Int. Conf. on Image Processing, Chicago, IL, October 4-7, vol. 2, pp. 64–68 (1998)

76. Daniels, J.J.: Fundamentals of ground penetrating radar. In: Proceedings of the Symposium on the Application of Geophysics to Engineering and Environmental Problems, pp. 62–142. Colorado School of Mines, Golden (1989)

77. Daniels, J.J., Bower, J., Baumgartner, F.: High resolution Gpr at Brookhaven National Lab to delineate complex subsurface structure. Journal Environmental and Engineering Geophysics 3(1), 1–6 (1989)

78. Daniels, J.J., Grumman, D.L., Vendl, M.: Coincident antenna three dimensional GPR. Journal of Environmental and Engineering Geophysics 2 (1997)

79. Dantsin, E., Kreinovich, V., Wolpert, A., Xiang, G.: Population variance under interval uncertainty: a new algorithm. Reliable Computing 12(4), 273–280 (2006)

80. Dantsin, E., Wolpert, A., Ceberio, M., Xiang, G., Kreinovich, V.: Detecting outliers under interval uncertainty: a new algorithm based on constraint satisfaction. In: Proceedings of the International Conference on Information Processing and Management of Uncertainty in Knowledge-Based Systems IPMU 2006, Paris, France, July 2-7, pp. 802–809 (2006)

81. Davis, J.L., Annan, A.P.: Ground penetrating radar for soil and rock stratigraphy. Geophysical Prospecting 37, 531–551 (1997)

82. Debroux, P., Boehm, J., Modave, F., Kreinovich, V., Xiang, G., Beck, J., Tupelly, K., Kandathi, R., Longpré, L., Villaverde, K.: Using 1-D radar observations to detect a space explosion core among the explosion fragments: sequential and distributed algorithms. In: Proceedings of the 11th IEEE Digital Signal Processing Workshop, Taos, New, Mexico, August 1-4, pp. 273–277 (2004)

83. Demicco, R., Klir, J. (eds.): Fuzzy Logic in Geology. Academic Press (2003)

84. Demkowicz, L., Rachowicz, W., Devloo, P.: A Fully Automatic hp-Adaptivity. TICAM Report 01-28, The University of Texas at Austin (2001)

85. Dempster, A.P.: Upper and lower probabilities induced by a multi-valued mapping. Ann. Mat. Stat. 38, 325–339 (1967)

86. Denning, D.: Cryptography and Data Security. Addison-Wesley, Reading (1982)

87. Dessombz, O., Thouverez, F., Laîné, J.P., Jézéquel, L.: Analysis of mechanical systems using interval computations applied to finite elements methods. J. Sound. Vib. 238(5), 949–968 (2001)

88. Devore, J., Peck, R.: Statistics: the Exploration and Analysis of Data, Duxbury, Pacific Grove, California (1999)

89. Doser, D.I., Crain, K.D., Baker, M.R., Kreinovich, V., Gerstenberger, M.C.: Estimating uncertainties for geophysical tomography. Reliable Computing 4(3), 241–268 (1998)

90. Dubois, D., Prade, H.: Operations on fuzzy numbers. International Journal of Systems Science 9, 613–626 (1978)

91. Dubois, D., Fargier, H., Fortin, J.: The empirical variance of a set of fuzzy intervals. In: Proceedings of the 2005 IEEE International Conference on Fuzzy Systems FUZZ-IEEE 2005, Reno, Nevada, May 22-25, pp. 885–890 (2005)

92. Dwyer, P.S.: Linear Computations. J. Wiley, New York (1951)

93. Edwards, R.E.: Functional Analysis: Theory and Applications. Dover Publ., NY (1995)

94. Elishakoff, I., Ren, Y.: The bird's eye View on finite element method for stochastic structures. Computer Methods in Applied Mechanics and Engineering 168, 51–61 (1999)

95. Ferregut, C., Osegueda, R.A., Nuñez, A. (eds.): Proceedings of the International Workshop on Intelligent NDE Sciences for Aging and Futuristic Aircraft, El Paso, TX, September 30-October 2 (1997)

96. Ferson, S.: RAMAS Risk Calc: Risk Assessment with Uncertain Numbers. Applied Biomathematics (1999)

97. Ferson, S.: RAMAS Risk Calc 4.0: Risk Assessment with Uncertain Numbers. CRC Press, Boca Raton (2002)

98. Ferson, S., Ginzburg, L.R.: Different methods are needed to propagate ignorance and variability. Reliab. Engng. Syst. Saf. 54, 133–144 (1996)

99. Ferson, S., Ginzburg, L., Akcakaya, R.: Whereof One Cannot Speak: When Input Distributions Are Unknown. Technical Report, Applied Biomathematics (2001)

100. Ferson, S., Ginzburg, L., Kreinovich, V., Aviles, M.: Exact bounds on sample variance of interval data. In: Extended Abstracts of the 2002 SIAM Workshop on Validated Computing, Toronto, Canada, May 23-25, pp. 67–69 (2002)

101. Ferson, S., Ginzburg, L., Kreinovich, V., Longpré, L., Aviles, M.: Computing variance for interval data is NP-hard. ACM SIGACT News 33(2), 108–118 (2002)

102. Ferson, S., Ginzburg, L., Kreinovich, V., Longpré, L., Aviles, M.: Exact bounds on finite populations of interval data. Reliable Computing 11(3), 207–233 (2005)

103. Ferson, S., Ginzburg, L., Kreinovich, V., Lopez, J.: Absolute bounds on the mean of sum, product, etc.: a probabilistic extension of interval arithmetic. In: Extended Abstracts of the 2002 SIAM Workshop on Validated Computing, Toronto, Canada, May 23-25, pp. 70–72 (2002)

104. Ferson, S., Hajagos, J., Berleant, D., Zhang, J., Tucker, W.T., Ginzburg, L., Oberkampf, W.: Dependence in Dempster-Shafer Theory and Probability Bounds Analysis. Technical Report SAND2004-3072, Sandia National Laboratory (2004)

105. Ferson, S., Kreinovich, V., Ginzburg, L., Myers, D.S., Sentz, K.: Constructing Probability Boxes and Dempster-Shafer Structures. Technical Report SAND2002-4015, Sandia National Laboratories (2003)

106. Ferson, S., Kreinovich, V., Hajagos, J., Oberkampf, W., Ginzburg, L.: Experimental Uncertainty Estimation and Statistics for Data Having Interval Uncertainty. Technical Report SAND2007-0939, Sandia National Laboratories (2007)

107. Ferson, S., Myers, D., Berleant, D.: Distribution-free Risk Analysis: I. Range, Mean, and Variance. Technical Report, Applied Biomathematics (2001)

108. Ferson, S., Oberkampf, W.L., Ginzburg, L.: Validation of imprecise probability models. In: Muhanna, R.L., Mullen, R.L. (eds.) Proceedings of the International Workshop on Reliable Engineering Computing REC 2008, Savannah, Georgia, February 20-22, pp. 23–44 (2008)

109. Ferson, S., Oberkampf, W.L., Ginzburg, L.: Model validation and predictive capability for the thermal challenge problem. Computer Methods in Applied Mechanics and Engineering 197(29-32), 2408–2430 (2008)

110. Filippov, A.F.: Ellipsoidal estimates for a solution of a system of differential equations. Interval Computations 2(2), 6–17 (1992)

111. Finkelstein, A., Kosheleva, O., Kreinovich, V.: Astrogeometry, error estimation, and other applications of set-valued analysis. ACM SIGNUM Newsletter 31(4), 3–25 (1996)

408    References

112. Fisher, E., McMechan, G.A., Annan, A.P.: Acquisition and processing of wide-aperture ground-penetrating radar data. Geophysics 57(3), 495–504 (1992)
113. Fogel, E., Huang, Y.F.: On the value of information in system identification. Bounded noise case. Automatica 18(2), 229–238 (1982)
114. Fuller, W.A.: Measurement Error Models. J. Wiley & Sons, New York (1987)
115. Galambos, J.: The Asymptotic Theory of Extreme Order Statistics. Wiley, New York (1987)
116. Ganzerli, S., Pantelides, C.P.: Load and resistance convex models for optimum design. Struct. Optim. 17, 259–268 (1999)
117. Garey, M.E., Johnson, D.S.: Computers and Intractability: A Guide to the Theory of NP-Completeness. Freeman, San Francisco (1979)
118. Garling, D.J.H.: A book review. American Mathematical Monthly 112(6), 575–579 (2005)
119. Gates, A., Kreinovich, V., Longpré, L., Pinheiro da Silva, P., Keller, G.R.: Towards secure cyberinfrastructure for sharing border information. In: Proceedings of the Lineae Terrarum: International Border Conference, El Paso, Las Cruces, and Cd. Juárez, March 27-30 (2006)
120. Goldsztejn, A.: Private communication (2007)
121. Goodchild, M., Gopal, S.: Accuracy of Spatial Databases. Taylor & Francis, London (1989)
122. Granvilliers, L., Kreinovich, V., Müller, N.: Novel Approaches to Numerical Software with Result Verification. In: Alt, R., Frommer, A., Kearfott, R.B., Luther, W. (eds.) Dagstuhl Seminar 2003. LNCS, vol. 2991, pp. 274–305. Springer, Heidelberg (2004)
123. Gros, X.E.: NDT Data Fusion. J. Wiley, London (1997)
124. Guy, E.D., Daniels, J.J., Radzevicius, S.J., Vendl, M.A.: Demonstration of using crossed dipole Gpr antenna for site characterization. Geophysical Research Letters 26(22), 3421–3424 (1999)
125. Haeni, F.P.: Use of ground-penetrating radar and continuous seismic-reflection on surface-water bodies in environmental and engineering studies. Journal of Environmental & Engineering Geophysics 1(1), 27–36 (1996)
126. Haldar, A., Mahadevan, S.: Reliability Assessment Using Stochastic Finite Element Analysis. John Wiley & Sons, New York (2000)
127. Hall, J.M.: Soft methods in Earth science engineering. In: Lawry, J., et al. (eds.) Soft Methods for Integrated Uncertainty Modeling, pp. 7–10. Springer, Heidelberg (2006)
128. Hall, K.H., Fu, G., Lawry, J.: Imprecise probabilities of climate change: aggregation of fuzzy scenarios and model uncertainties. Climatic Change 81(3-4), 265–281 (2007)
129. Han, J., Park, D.: Scheduling proxy: enabling adaptive-grained scheduling for global computing system. In: Proceedings of the Fifth IEEE/ACM International Workshop on Grid Computing, pp. 415–420 (November 8, 2004)
130. Hansen, E.: Global Optimization Using Interval Analysis. Marcel Dekker, Inc., New York (1992)
131. Hansen, E.: Sharpness in interval computations. Reliable Computing 3, 7–29 (1997)
132. Hansen, P., de Aragao, M.V.P., Ribeiro, C.C.: Hyperbolic 0-1 programming and optimization in information retrieval. Math. Programming 52, 255–263 (1991)

133. Hardy, G.H., Littlewood, J.E., Pólya, G.: Inequalities. Cambridge University Press (1988)
134. Hein, S., Zakhor, A.: Halftone to continuous tone conversion of error-diffusion coded images. IEEE Transactions on Image Processing 4(2), 208–216 (1995)
135. Höhn, W., Mittelmann, H.D.: Some remarks on the discrete maximum principle for finite elements of higher-order. Computing 27, 145–154 (1981)
136. Hole, J.A.: Nonlinear high-resolution three-dimensional seismic travel time tomography. J. Geophysical Research 97, 6553–6562 (1992)
137. Holt, J., Vendl, M., Baumgartner, F., Radziviscius, S., Daniels, J.J.: Brownfields site-characterization using geophysics: a case history from east Chicago. In: Transactions of the Symposium on the Application of Geophysics to Engineering and Environmental Problems: Soc. Eng. and Min. Expl. Geophys., Chicago, IL, April 26-29 (1998)
138. Huber, P.J.: Robust Statistics. Wiley, New York (2004)
139. Interval computations website, http://www.cs.utep.edu/interval-comp
140. Jájá, J.: An Introduction to Parallel Algorithms. Addison-Wesley, Reading (1992)
141. Jaksurat, P., Freudenthal, E., Ceberio, M., Kreinovich, V.: Probabilistic approach to trust: ideas, algorithms, and simulations. In: Proceedings of the Fifth International Conference on Intelligent Technologies InTech 2004, Houston, Texas, December 2-4 (2004)
142. Jaulin, L., Kieffer, M., Didrit, O., Walter, E.: Applied interval analysis: with examples in parameter and state estimation, robust control and robotics. Springer, London (2001)
143. Jaynes, E.T.: Where do we Stand on Maximum Entropy? In: Levine, R.D., Tribus, M. (eds.) The Maximum Entropy Formalism, pp. 15–118. M.I.T. Press, Cambridge (1979)
144. Jaynes, E.T.: Probability Theory: The Logic of Science. Cambridge University Press, Massachusetts (2003)
145. Karátson, J., Korotov, S.: Discrete maximum principles for finite element solutions of nonlinear elliptic problems with mixed boundary conditions. Numer. Math. 99, 669–698 (2005)
146. Karniadakis, G.E., Sherwin, S.J.: Spectral/$hp$ Element Methods for CFD. Oxford University Press, Oxford (1999)
147. Kearfott, R.B.: Rigorous Global Search: Continuous Problems. Kluwer, Dordrecht (1996)
148. Kearfott, R.B., Kreinovich, V. (eds.): Applications of Interval Computations. Kluwer, Dordrecht (1996)
149. Kearfott, R.B., Kreinovich, V.: Beyond convex? Global optimization is feasible only for convex objective functions: a theorem. Journal of Global Optimization 33(4), 617–624 (2005)
150. Keeney, R.L., Raiffa, H.: Decisions with Multiple Objectives. John Wiley and Sons, New York (1976)
151. Keller, G.R., Starks, S.A., Velasco, A., Averill, M., Araiza, R., Xiang, G., Kreinovich, V.: Towards combining probabilistic, interval, fuzzy uncertainty, and constraints: on the example of inverse problem in geophysics. In: Proceedings of the Second International Conference on Fuzzy Sets and Soft Computing in Economics and Finance FSSCEF 2006, St. Petersburg, Russia, June 28-July 1, pp. 47–54 (2006)

152. Keller, G.R., Hildenbrand, T.G., Kucks, R., Webring, M., Briesacher, A., Rujawitz, K., Hittleman, A.M., Roman, D.R., Winester, D., Aldouri, R., Seeley, J., Rasillo, J., Torres, R., Hinze, W.J., Gates, A., Kreinovich, V., Salayandia, S.: A community effort to construct a gravity database for the United States and an associated Web portal. In: Sinha, A.K. (ed.) Geoinformatics: Data to Knowledge, pp. 21–34. Geological Society of America Publ., Boulder (2006)

153. Kelley, J.L.: General Topology. Springer, Heidelberg (1975)

154. Kendall, D.G.: Foundations of a theory of random sets. In: Harding, E., Kendall, D. (eds.) Stochastic Geometry, New York, pp. 322–376 (1974)

155. Kite, T.D., Damera-Venkata, N., Evans, B.L., Bovik, A.C.: A fast, high-quality inverse halftoning algorithm for error diffused halftones. IEEE Transactions on Image Processing 9(9), 1583–1592 (2000)

156. Klir, G., Yuan, B.: Fuzzy Sets and Fuzzy Logic. Prentice Hall, New Jersey (1995)

157. Klir, G.J., Wierman, M.J.: Uncertainty-based information: elements of Generalized Information Theory. Springer, Heidelberg (1999)

158. Klir, G.J.: Uncertainty and Information: Foundations of Generalized Information Theory. J. Wiley, Hoboken (2005)

159. Kohli, R., Krishnamurthi, R., Mirchandani, P.: The Minimum Satisfiability Problem. SIAM Journal on Discrete Mathematics 7(2), 275–283 (1994)

160. Kondo, D., Casanova, H., Wing, E., Berman, F.: Models and scheduling mechanisms for global computing applications. In: Proceedings of the International IEEE Symposium on Parallel and Distributed Processing IPDPS 2002, April 15-19, pp. 79–86 (2002)

161. Korotov, S., Křížek, M., Neittaanmäki, P.: Křížek M, Neittaanmäki P Weakened acute type condition for tetrahedral triangulations and the discrete maximum principle. Math. Comp. 70, 107–119 (2000)

162. Kosheleva, O.: Symmetry-group justification of maximum entropy method and generalized maximum entropy methods in image processing. In: Erickson, G.J., Rychert, J.T., Smith, C.R. (eds.) Maximum Entropy and Bayesian Methods, pp. 101–113. Kluwer, Dordrecht (1998)

163. Kosheleva, O.: On the optimal choice of quality metric in image compression: a soft computing approach. Soft Computing 8(4), 268–273 (2004)

164. Kosheleva, O., Cabrera, S., Osegueda, R., Nazarian, S., George, D.L., George, M.J., Kreinovich, V., Worden, K.: Case study of non-linear inverse problems: mammography and non-destructive evaluation. In: Mohamad-Djafari, A. (ed.) Bayesian Inference for Inverse Problems, Proceedings of the SPIE/International Society for Optical Engineering, San Diego, California, vol. 3459, pp. 128–135 (1998)

165. Kosheleva, O., Cabrera, S., Usevitch, B., Vidal Jr., E.: Compressing 3D Measurement Data Under Interval Uncertainty. In: Dongarra, J., Madsen, K., Waśniewski, J. (eds.) PARA 2004. LNCS, vol. 3732, pp. 142–150. Springer, Heidelberg (2006)

166. Kotz, S., Nadarajah, S.: Extreme Value Distributions: Theory and Applications. Imperial College Press, London (2000)

167. Koyluoglu, U., Cakmak, S., Ahmet, N., Soren, R.K.: Interval algebra to deal with pattern loading and structural uncertainty. J. Engrg. Mech. 121(11), 1149–1157 (1995)

168. Kreinovich, V.: Maximum entropy and interval computations. Reliable Computing 2(1), 63–79 (1996)

169. Kreinovich, V.: Probabilities, Intervals, What Next? Optimization Problems Related to Extension of Interval Computations to Situations with Partial Information about Probabilities. Journal of Global Optimization 29(3), 265–280 (2004)

170. Kreinovich, V.: Optimal finite characterization of linear problems with inexact data. Reliable Computing 11(6), 479–489 (2005)

171. Kreinovich, V.: Interval computations and interval-related statistical techniques: tools for estimating uncertainty of the results of data processing and indirect measurements. In: Pavese, F., Forbes, A.B. (eds.) Advances in Data Modeling for Measurements in the Metrology and Testing Fields, pp. 119–147. Birkhauser-Springer, Boston (2007)

172. Kreinovich, V.: Application-motivated combinations of fuzzy, interval, and probability approaches, with application to geoinformatics, bioinformatics, and engineering. In: Proceedings of the International Conference on Information Technology InTech 2007, Sydney, Australia, December 12-14, pp. 11–20 (2007)

173. Kreinovich, V.: Application-motivated combinations of fuzzy, interval, and probability approaches, and their use in geoinformatics, bioinformatics, and engineering. International Journal of Automation and Control (IJAAC) 2(2/3), 317–339 (2008)

174. Kreinovich, V.: Interval computations as an important part of granular computing: an introduction. In: Pedrycz, W., Skowron, A., Kreinovich, V. (eds.) Handbook on Granular Computing, pp. 3–32. Wiley, Chichester (2008)

175. Kreinovich, V.: Relation between interval computing and soft computing. In: Hu, C., Kearfott, R.B., de Korvin, A., Kreinovich, V. (eds.) Knowledge Processing with Interval and Soft Computing, pp. 73–94. Springer, Heidelberg (2008)

176. Kreinovich, V.: From interval computations to constraint-related set computations: towards faster estimation of statistics and odes under interval and p-box uncertainty. In: Bauer, A., Dillhage, R., Hertling, P., Ko, K.-I., Rettinger, R. (eds.) Proceedings of the Sixth International Conference on Computability and Complexity in Analysis CCA 2009, Ljubljana, Slovenia, August 18-22, pp. 4–15 (2009)

177. Kreinovich, V., Beck, J., Ferregut, C., Sanchez, A., Keller, G.R., Averill, M., Starks, S.A.: Monte-Carlo-type techniques for processing interval uncertainty, and their potential engineering applications. Reliable Computing 13(1), 25–69 (2007)

178. Kreinovich, V., Ferson, S.: A new Cauchy-Based black-box technique for uncertainty in risk analysis. Reliability Engineering and Systems Safety 85(1-3), 267–279 (2004)

179. Kreinovich, V., Ferson, S.: Computing Best-Possible Bounds for the Distribution of a Sum of Several Variables is NP-Hard. International Journal of Approximate Reasoning 41, 331–342 (2006)

180. Kreinovich, V., Ferson, S., Ginzburg, L.: Exact upper bound on the mean of the product of many random variables with known expectations. Reliable Computing 9(6), 441–463 (2003)

181. Kreinovich, V., Ferson, S., Ginzburg, L.: Exact upper bound on the mean of the product of many random variables with known expectations. Reliable Computing 9(6), 441–463 (2004)

182. Kreinovich, V., Lakeyev, A., Rohn, J., Kahl, P.: Computational Complexity and Feasibility of Data Processing and Interval Computations. Kluwer, Dordrecht (1997)

183. Kreinovich, V., Longpré, L.: Computational complexity and feasibility Of data processing and interval computations, with extension to cases When we have partial information about probabilities. In: Brattka, V., Schroeder, M., Weihrauch, K., Zhong, N. (eds.) Proceedings of the Conference on Computability and Complexity in Analysis CCA 2003, Cincinnati, Ohio, USA, August 28-30, pp. 19–54 (2003)

184. Kreinovich, V., Longpré, L.: Fast quantum algorithms for handling probabilistic and interval uncertainty. Mathematical Logic Quarterly 50(4/5), 507–518 (2004)

185. Kreinovich, V., Longpré, L., Ferson, S., Ginzburg, L.: Computing Higher Central Moments for Interval Data. Technical Report UTEP-CS-03-14b. University of Texas at El Paso, Department of Computer Science (2004), http://www.cs.utep.edu/vladik/2003/tr03-14b.pdf

186. Kreinovich, V., Longpré, L., Patangay, P., Ferson, S., Ginzburg, L.: Outlier detection under interval uncertainty: Algorithmic solvability and computational complexity. In: Lirkov, I., Margenov, S., Waśniewski, J., Yalamov, P. (eds.) LSSC 2003. LNCS, vol. 2907, pp. 238–245. Springer, Heidelberg (2004)

187. Kreinovich, V., Longpré, L., Patangay, P., Ferson, S., Ginzburg, L.: Outlier detection under interval uncertainty: algorithmic solvability and computational complexity. Reliable Computing 11(1), 59–76 (2005)

188. Kreinovich, V., Longpré, L., Starks, S.A., Xiang, G., Beck, J., Kandathi, R., Nayak, A., Ferson, S., Hajagos, J.: Interval versions of statistical techniques, with applications to environmental analysis, bioinformatics, and privacy in statistical databases. Journal of Computational and Applied Mathematics 199(2), 418–423 (2007)

189. Kreinovich, V., Mayer, G., Starks, S.: On a theoretical justification of the choice of epsilon-inflation in PASCAL-XSC. Reliable Computing 3(4), 437–452 (1997)

190. Kreinovich, V., Nguyen, H.T., Wu, B.: On-line algorithms for computing mean and variance of interval data, and their use in intelligent systems. Information Sciences 177(16), 3228–3238 (2007)

191. Kreinovich, V., Patangay, P., Longpré, L., Starks, S.A., Campos, C., Ferson, S., Ginzburg, L.: Outlier detection under interval and fuzzy uncertainty: algorithmic solvability and computational complexity. In: Proceedings of the 22nd International Conference of the North American Fuzzy Information Processing Society NAFIPS 2003, Chicago, Illinois, July 24-26, pp. 401–406 (2003)

192. Kreinovich, V., Solopchenko, G.N., Ferson, S., Ginzburg, L., Aló, R.: Probabilities, intervals, what next? Extension of interval computations to situations with partial information about probabilities. In: Proceedings of the 10th IMEKO TC7 International Symposium on Advances of Measurement Science, St. Petersburg, Russia, June 30-July 2, vol. 1, pp. 137–142 (2004)

193. Kreinovich, V., Xiang, G.: Fast algorithms for computing statistics under interval uncertainty: an overview. In: Huynh, V.N., Nakamori, Y., Ono, H., Lawry, J., Kreinovich, V., Nguyen, H.T. (eds.) Interval/Probabilistic Uncertainty and Non-Classical Logics, pp. 19–31. Springer, Heidelberg (2008)

194. Kreinovich, V., Xiang, G.: Towards fast algorithms for processing type-2 fuzzy data: extending Mendel's algorithms from interval-valued to a more general case. In: Proceedings of the 27th International Conference of the North American Fuzzy Information Processing Society NAFIPS 2008, New York, May 19-22 (2008)

195. Kreinovich, V., Xiang, G., Ferson, S.: How the concept of information as average number of "yes"-"no" questions (bits) can be extended to intervals, P-boxes, and more general uncertainty. In: Proceedings of the 24th International Conference of the North American Fuzzy Information Processing Society NAFIPS 2005, Ann Arbor, Michigan, June 22-25, pp. 80–85 (2005)

196. Kreinovich, V., Xiang, G., Ferson, S.: Efficient algorithms for computing mean and variance under Dempster-Shafer uncertainty. International Journal of Approximate Reasoning 42, 212–227 (2006)

197. Kreinovich, V., Xiang, G., Starks, S.A., Longpré, L., Ceberio, M., Araiza, R., Beck, J., Kandathi, R., Nayak, A., Torres, R., Hajagos, J.: Towards combining probabilistic and interval uncertainty in engineering calculations: algorithms for computing statistics under interval uncertainty, and their computational complexity. Reliable Computing 12(6), 471–501 (2006)

198. Křížek, M., Liu, L.: On the maximum and comparison principles for a steady-state nonlinear heat conduction problem. ZAMM Z. Angew. Math. Mech. 83, 559–563 (2003)

199. Kuznetsov, V.P.: Interval Statistical Models. Radio i Svyaz, Moscow (1991) (in Russian)

200. Langewisch, A.T., Choobineh, F.F.: Mean and variance bounds and propagation for ill-specified random variables. IEEE Transactions on Systems, Man, and Cybernetics, Part A 34(4), 494–506 (2004)

201. Li, S., Ogura, Y., Kreinovich, V.: Limit Theorems and Applications of Set Valued and Fuzzy Valued Random Variables. Kluwer Academic Publishers, Dordrecht (2002)

202. Liu, W.K., Belytschko, T., Mani, A.: Probabilistic finite elements for nonlinear structural dynamics. Computer Methods in Applied Mechanics and Engineering 56, 61–81 (1986)

203. Lodwick, W.A., Jamison, K.D.: Special issue: interface between fuzzy set theory and interval analysis. Fuzzy Sets and Systems 135(1-3) (2002)

204. Lodwick, W.A., Jamison, K.D.: Estimating and validating the cumulative distribution of a function of random variables: toward the development of distribution arithmetic. Reliable Computing 9(2), 127–141 (2003)

205. Lodwick, W.A., Neumaier, A., Newman, F.: Optimization under uncertainty: methods and applications in radiation therapy. In: Proc. 10th IEEE Int. Conf. Fuzzy Systems, Melbourne, Australia, December 2-5 (2001)

206. Longpé, L., Ferson, S., Tucker, W.T.: How to measure a degree of mismatch between probability models, p-boxes, etc.: a decision-theory-motivated utility-based approach. In: Proceedings of the 27th International Conference of the North American Fuzzy Information Processing Society NAFIPS 2008, New York, May 19-22 (2008)

207. Longpé, L., Kreinovich, V.: How to efficiently process uncertainty within a cyberinfrastructure without sacrificing privacy and confidentiality. In: Nedjah, N., Abraham, A., de Macedo Mourelle, L. (eds.) Computational Intelligence in Information Assurance and Security, pp. 155–173. Springer, Heidelberg (2007)

208. Longpré, L., Servin, S.: Quantum computations techniques for gauging reliability of interval and fuzzy data. In: Proceedings of the 27th International Conference of the North American Fuzzy Information Processing Society NAFIP 2008, New York, May 19-22 (2008)

209. Longpré, L., Xiang, G., Kreinovich, V., Freudenthal, E.: Interval approach to preserving privacy in statistical databases: related challenges and algorithms of computational statistics. In: Gorodetsky, V., Kotenko, I., Skormin, V.A. (eds.) Proceeedings of the International Conference "Mathematical Methods, Models and Architectures for Computer Networks Security" MMM-ACNS-2007, St. Petersburg, Russia, September 13-15. LNCS, vol. CCIS-1, pp. 346–361 (2007)

210. Luce, R.D., Raiffa, H.: Games and Decisions: Introduction and Critical Survey. Dover, New York (1989)

211. Maceira, M., Taylor, S.R., Ammon, C.J., Yang, X., Velasco, A.A.: High-resolution rayleigh wave slowness tomography of Central Asia. Journal of Geophysical Research 110, paper B06304 (2005)

212. Manski, C.: Partial Identification of Probability Distributions. Springer, New York (2003)

213. Martinez, M., Longpré, L., Kreinovich, V., Starks, S.A., Nguyen, H.T.: Fast quantum algorithms for handling probabilistic, interval, and fuzzy uncertainty. In: Proceedings of the 22nd International Conference of the North American Fuzzy Information Processing Society NAFIPS 2003, Chicago, Illinois, July 24-26, pp. 395–400 (2003)

214. McCain, M., William, C.: Integrating quality assurance into the GIS project life cycle. In: Proceedings of the 1998 ESRI Users Conference (1998)

215. McWilliam, S.: Anti-optimisation of uncertain structures using interval analysis. Comput. Struct. 79, 421–430 (2000)

216. Melchers, R.E.: Structural Reliability Analysis and Prediction, 2nd edn. John Wiley & Sons, West Sussex (1999)

217. Mendel, J.M.: Uncertain Rule-Based Fuzzy Logic Systems: Introduction and New Directions. Prentice-Hall (2001)

218. Mendel, J.M.: Type-2 fuzzy sets and systems: an overview. IEEE Computational Intelligence Magazine 2, 20–29 (2007)

219. Mendel, J.M.: Novel weighted averages as a computing with words engine. In: Plenary talk at IFSA World Congress IFSA 2007, Cancun, Mexico, June 18-21 (2007)

220. Mese, M., Vaidyanathan, P.P.: Optimized halftoning using dot diffusion and methods for inverse halftoning. IEEE Transactions on Image Processing 9(4), 691–709 (2000)

221. Modave, F., Ceberio, M., Wang, X., Xiang, G., Garay, O., Ramirez, R., Tejeda, R.: Comparison of computer attacks: an application of interval-based fuzzy integration. In: Proceedings of the 24th International Conference of the North American Fuzzy Information Processing Society NAFIPS 2005, Ann Arbor, Michigan, June 22-25, pp. 676–681 (2005)

222. Moës, N., Dolbow, J., Belytschko, T.: A finite element method for crack growth without remeshing. Int. J. Numer. Methods Engrg. 46, 131–150 (1999)

223. Möller, B., Graf, W., Beer, M.: Fuzzy structural analysis using level-optimization. Comput. Mech. 26(6), 547–565 (2000)

224. Moore, R.E.: Automatic Error Analysis in Digital Computation. Technical Report Space Div. Report LMSD84821, Lockheed Missiles and Space Co. (1959)

225. Moore, R.E.: Interval Analysis. Prentice-Hall, Inc., Englewood Cliffs (1966)

226. Moore, R.E.: Methods and Applications of Interval Analysis. SIAM, Philadelphia (1979)
227. Moore, R.E.: Risk analysis without Monte Carlo methods. Freiburger Intervall-Berichte, Inst. F. Angew. Math., Universitaet Freiburg I. Br. 84(1), 1–48 (1984)
228. Moore, A.S.: Interval risk analysis of real estate investment: a non-Monte-Carlo approach. Freiburger Intervall-Berichte, Inst. F. Angew. Math., Universitaet Freiburg I. Br. 85(3), 23–49 (1985)
229. Moore, R.E., Kearfott, R.B., Cloud, M.J.: Introduction to Interval Analysis. SIAM Press, Philadelphia (2009)
230. Moore, R.E., Lodwick, W.A.: Interval analysis and fuzzy set theory. Fuzzy Sets and Systems 135(1), 5–9 (2003)
231. Morgenstein, D., Kreinovich, V.: Which algorithms are feasible and which are not depends on the geometry of space-time. Geombinatorics 4(3), 80–97 (1995)
232. Muhanna, R., Kreinovich, V., Solin, P., Chessa, J., Araiza, R., Xiang, G.: Interval finite element methods: new directions. In: Proceedings of the Second International Workshop on Reliable Engineering Computing, Savannah, Georgia, February 22-24, pp. 229–243 (2006)
233. Muhanna, R.L., Mullen, R.L.: Development of interval based methods for fuzziness in continuum mechanics. In: Proc. ISUMA-NAFIPS 1995, pp. 23–45 (1995)
234. Muhanna, R.L., Mullen, R.L.: Formulation of fuzzy finite element methods for mechanics problems. Compu.-Aided Civ. Infrastruct. Engrg. 14, 107–117 (1999)
235. Muhanna, R.L., Mullen, R.L.: Uncertainty in mechanics problems-interval-based approach. J. Engrg. Mech. 127(6), 557–566 (2001)
236. Muhanna, R., Mullen, R. (eds.): Proceedings of the NSF Workshop on Reliable Engineering Computing, Savannah, Georgia, September 15-17 (2004)
237. Muhanna, R.L., Mullen, R.L., Zhang, H.: Penalty-based solution for the interval finite-element methods. ASCE, Engineering Mechanics 131(10), 1102–1111 (2005)
238. Mullen, R.L., Muhanna, R.L.: Structural analysis with fuzzy-based load uncertainty. In: Proc. 7th ASCE EMD/STD Joint Spec. Conf. on Probabilistic Mech. and Struct. Reliability, Mass., pp. 310–313 (1996)
239. Mullen, R.L., Muhanna, R.L.: Bounds of structural response for all possible loadings. J. Struct. Engrg., ASCE 125(1), 98–106 (1999)
240. Neumaier, A.: Interval Methods for Systems of Equations. Cambridge University Press, Cambridge (1990)
241. Neumaier, A.: Introduction to Numerical Analysis. Cambridge Univ. Press, Cambridge (2001)
242. Neumaier, A.: Taylor forms. Reliable Computing 9, 43–79 (2002)
243. Neumaier, A.: Fuzzy modeling in terms of surprise. Fuzzy Sets and Systems 135, 21–38 (2003)
244. Neumaier, A., Pownuk, A.: Linear systems with large uncertainties, with applications to truss structures. Reliable Computing 13(2), 149–172 (2004)
245. Nguyen, H.T.: A note on the extension principle for fuzzy sets. J. Math. Anal. and Appl. 64, 369–380 (1978)
246. Nguyen, H.T., Kreinovich, V.: Nested intervals and sets: concepts, relations to fuzzy sets, and applications. In: Kearfott, R.B., Kreinovich, V. (eds.) Applications of Interval Computations, pp. 245–290. Kluwer, Dordrecht (1996)

247. Nguyen, H.T., Kreinovich, V.: Applications of Continuous Mathematics to Computer Science. Kluwer, Dordrecht (1997)

248. Nguyen, H.T., Kreinovich, V.: Methodology of fuzzy control: an introduction. In: Nguyen, H.T., Sugeno, M. (eds.) Fuzzy Systems: Modeling and Control, pp. 19–62. Kluwer, Dordrecht (1998)

249. Nguyen, H.T., Kreinovich, V., Gorodetski, V.I., et al.: Applications of interval-valued degrees of belief. In: Information Technologies and Intellectual Methods, St. Petersburg, Russia. Inst. for Information and Automation, vol. 3, pp. 6–61 (1987) (1999) (in Russian)

250. Nguyen, H.T., Kreinovich, V., Xiang, G.: Foundations of statistical processing of set-valued data: towards efficient algorithms. In: Proceedings of the Fifth International Conference on Intelligent Technologies InTech 2004, Houston, Texas, December 2-4 (2004)

251. Nguyen, H.T., Kreinovich, V., Zuo, G.: Interval-valued degrees of belief: applications of interval computations to expert systems and intelligent control. International Journal of Uncertainty, Fuzziness, and Knowledge-Based Systems (IJUFKS) 5(3), 317–358 (1997)

252. Nguyen, H.T., Walker, E.A.: First Course in Fuzzy Logic. CRC Press, Boca Raton (2006)

253. Nguyen, H.T., Wang, T., Kreinovich, V.: Towards foundations of processing imprecise data: from traditional statistical techniques of processing crisp data to statistical processing of fuzzy data. In: Liu, Y., Chen, D., Ying, M., Cai, K.Y. (eds.) Proceedings of the International Conference on Fuzzy Information Processing: Theories and Applications FIP 2003, Beijing, China, March 1-4, vol. 2, pp. 895–900 (2003)

254. Nguyen, H.T., Wu, B., Kreinovich, V.: Shadows of fuzzy sets – a natural approach towards describing 2-D and multi-D fuzzy uncertainty in linguistic terms. In: Proc. 9th IEEE Int'l Conference on Fuzzy Systems FUZZ-IEEE 2000, San Antonio, Texas, May 7-10, vol. 1, pp. 340–345 (2000)

255. Nivlet, P., Fournier, F., Royer, J.: A new methodology to account for uncertainties in 4-D seismic interpretation. In: Proc. 71st Annual Int'l Meeting of Soc. of Exploratory Geophysics SEG 2001, San Antonio, TX, September 9-14, pp. 1644–1647 (2001)

256. Nivlet, P., Fournier, F., Royer, J.: Propagating interval uncertainties in supervised pattern recognition for reservoir characterization. In: Proc. 2001 Society of Petroleum Engineers Annual Conf. SPE 2001, New Orleans, LA, September 30-October 3, paper SPE-71327 (2001)

257. Nogueira, M., Nandigam, A.: Why intervals? Because if we allow other sets, tractable problems become intractable. Reliable Computing 4(4), 389–394 (1998)

258. Norton, J.P.: Identification and application of bounded parameter models. In: Proceeding of the 7th IFAC Symposium on Identification and Parameter Estimation, York, U.K (1985)

259. Novitskii, P.V., Zograph, I.A.: Estimating the Measurement Errors. Energoatomizdat, Leningrad (1991) (in Russian)

260. Olhoeft, G.R.: Electrical, magnetic, and geometric properties that determine ground penetrating performance. In: Proc. Seventh International Conference on Ground Penetrating Radar, May 27-30, The University of Kansas, Lawrence (1998)

261. Orlov, A.I.: How often are the observations normal? Industrial Laboratory 57(7), 770–772 (1991)

262. Ornelas, G.: Set-Valued Extensions of Fuzzy Logic: Classification Theorems. Master's Thesis, University of Texas at El Paso, Department of Computer Science (2007)
263. Ornelas, G., Kreinovich, V.: Set-valued extensions of fuzzy logic: classification theorems. In: Reformat, M., Berthold, M.R. (eds.) Proceedings of the 26th International Conference of the North American Fuzzy Information Processing Society NAFIPS 2007, San Diego, California, June 24-27, pp. 549–553 (2007)
264. Orshansky, M.: Increasing circuit performance through statistical design techniques. In: Chinnery, D., Keutzer, K. (eds.) Closing the Gap Between ASICs and Custom. Kluwer (2002)
265. Orshansky, M., Bandyopadhyay, A.: Fast statistical timing analysis handling arbitrary delay correlations. In: Proc. of the Design Automation Conference DAC 2004, San Diego, California, June 7-11, pp. 337–342 (2004)
266. Orshansky, M., Keutzer, K.: A general probabilistic framework for worst case timing analysis. In: Proceedings of the Design Automation Conference DAC 2002, pp. 556–561 (June 2002)
267. Orshansky, M., Keutzer, K.: From blind certainty to informed uncertainty. In: Proceedings of the TAU Workshop (December 2002)
268. Orshansky, M., Nassif, S., Boning, D.: Design for Manufacturability and Statistical Design. Springer, New York (2008)
269. Orshansky, M., Wang, W., Ceberio, M., Xiang, G.: Interval-based robust statistical techniques for non-negative convex functions, with application to timing analysis of computer chips. In: Proceedings of the ACM Symposium on Applied Computing SAC 2006, Dijon, France, April 23-27, pp. 1629–1633 (2006)
270. Orshansky, M., Wang, W., Xiang, C., Kreinovich, V.: Interval-based robust statistical techniques for non-negative convex functions with application to timing analysis of computer chips. In: Proceedings of the Second International Workshop on Reliable Engineering Computing, Savannah, Georgia, February 22-24, pp. 197–212 (2006)
271. Osegueda, R., Kreinovich, V., Potluri, L., Aló, R.: Non-destructive testing of aerospace structures: granularity and data mining approach. In: Proc. FUZZ-IEEE 2002, Honolulu, HI, May 12-17, vol. 1, pp. 685–689 (2002)
272. Osegueda, R.A., Seelam, S.R., Holguin, A.C., Kreinovich, V., Tao, C.W.: Statistical and Dempster-Shafer techniques in testing structural integrity of aerospace structures. International Journal of Uncertainty, Fuzziness, Knowledge-Based Systems (IJUFKS) 9(6), 749–758 (2001)
273. Pantelides, C.P., Ganzerli, S.: Comparison of fuzzy set and convex model theories in structural design. Mech. Systems Signal Process. 15(3), 499–511 (2001)
274. Papadimitriou, C.H.: Computational Complexity. Addison Wesley, San Diego (1994)
275. Papadimitriou, C.H., Steiglitz, K.: Combinatorial Optimization: Algorithms and Complexity. Dover Publications, Inc., Mineola (1998)
276. Papoulis, A.: Probability, Random Variables, and Stochastic Processes. McGraw-Hill, New York (2002)
277. Parker, R.L.: Geophysical Inverse Theory. Princeton University Press, Princeton (1994)
278. Paszynski, M., Kurtz, J., Demkowicz, L.: Parallel, Fully Automatic hp-Adaptive 2D Finite Element Package. TICAM Report 04-07, The University of Texas at Austin (2004)

279. Pawlak, Z.: Rough Sets. Kluwer Academic Publishers, Dodrecht (1991)
280. Polkowski, L.: Rough sets: Mathematical foundations. Physica Verlag, A Springer-Verlag Company, Heidelberg, New York (2002)
281. Popova, E.D., Datcheva, M., Iankov, R., Schanz, T.: Mechanical models with interval parameters. In: Gürlebeck, G., Hempel, L., Könke, C. (eds.) IKM 2003: Digital Proceedings of 16th International Conference on the Applications of Computer Science and Mathematics in Architecture and Civil Engineering. Weimar, Germany (2003)
282. Proceedings of the 10th International Conference on Ground Penetrating Radar GPR 2004, Delft, The Netherlands, June 21-24 (2004)
283. Rabinovich, S.: Measurement Errors and Uncertainties: Theory and Practice. Springer, New York (2005)
284. Rachowicz, W., Pardo, D., Demkowicz, L.: Fully Automatic hp-Adaptivity in Three Dimensions. ICES Report 04-22, The University of Texas at Austin (2004)
285. Raiffa, H.: Decision Analysis. Addison-Wesley, Reading (1970)
286. Ramer, A., Kreinovich, V.: Information complexity and fuzzy control. In: Kandel, A., Langholtz, G. (eds.) Fuzzy Control Systems, pp. 75–97. CRC Press, Boca Raton (1994)
287. Ramer, A., Kreinovich, V.: Maximum entropy approach to fuzzy control. Information Sciences 81(3-4), 235–260 (1994)
288. Rajasekaran, S., Pardalos, P., Reif, J., Rolim, J. (eds.): Handbook on Randomized Computing. Kluwer (2001)
289. Rao, S.S., Berke, L.: Analysis of uncertain structural systems using interval analysis. AIAA J. 35(4), 727–735 (1997)
290. Rao, S.S., Chen, L.: Numerical solution of fuzzy linear equations in engineering analysis. Int. J. Numer. Meth. Engng. 43, 391–408 (1998)
291. Regan, H., Ferson, S., Berleant, D.: Equivalence of five methods for bounding uncertainty. Journal of Approximate Reasoning 36(1), 1–30 (2004)
292. Roberts, A.W., Varberg, D.E.: Convex Functions. Academic Press, New York (1973)
293. Rowe, N.C.: Absolute bounds on the mean and standard deviation of transformed data for constant-sign-derivative transformations. SIAM Journal of Scientific Statistical Computing 9, 1098–1113 (1988)
294. Rump, S.M.: Solving algebraic problems with high accuracy. In: Kulisch, U., Miranker, W. (eds.) A New Approach to Scientific Computation. Academic Press, New York (1983)
295. Scott, L.: Identification of GIS attribute error using exploratory data analysis. Professional Geographer 46(3), 378–386 (1994)
296. Schröcker, H., Wallner, J.: Geometric constructions with discretized random variables. Reliable Computing 12(3), 203–223 (2006)
297. Schuëller, C.: Computational stochastic mechanics – recent advances. Computers and Structures 79, 2225–2234 (2001)
298. Schweppe, F.C.: Recursive state estimation: unknown but bounded errors and system inputs. IEEE Transactions on Automatic Control 13, 22 (1968)
299. Schweppe, F.C.: Uncertain Dynamic Systems. Prentice Hall, Englewood Cliffs (1973)
300. Sentz, K., Ferson, S.: Combination of Evidence in Dempster-Shafer Theory. Technical Report SAND2002-0835, Sandia National Laboratories (2002)
301. Sethian, J.A.: Level Set Methods and Fast Marching Methods. Cambridge University Press (1999)

302. Shafer, G.: A Mathematical Theory of Evidence. Princeton University Press, Princeton (1976)
303. Shary, S.P.: Parameter partitioning scheme for interval linear systems with constraints. In: Proceedings of the International Workshop on Interval Mathematics and Constraint Propagation Methods (ICMP 2003), Novosibirsk, Akademgorodok, Russia, July 8-9, pp. 1–12 (2003) (in Russian)
304. Shary, S.P.: Solving tied interval linear systems. Siberian Journal of Numerical Mathematics 7(4), 363–376 (2004) (in Russian)
305. Sheskin, D.J.: Handbook of Parametric and Nonparametric Statistical Procedures. Chapman & Hall/CRC, Boca Raton, Florida (2004)
306. Shmulevich, I., Zhang, W.: Binary analysis and optimization-based normalization of gene expression data. Bioinformatics 18(4), 555–565 (2002)
307. Sinha, A.K. (ed.): Geoinformatics: Data to Knowledge. Geological Society of America Publ., Boulder (2006)
308. Šolín, P.: Partial Differential Equations and the Finite Element Methods. J. Wiley & Sons, Hoboken (2005)
309. Šolín, P., Demkowicz, L.: Goal-oriented $hp$-adaptivity for elliptic problems. Computer Methods in Applied Mechanics and Engineering 193, 449–468 (2004)
310. Šolín, P., Segeth, K., Doležel, I.: Higher-Order Finite Element Methods. Chapman & Hall/CRC Press, Boca Raton (2003)
311. Šolín, P., Vejchodský, T.: On the Discrete Maximum Principle for the $hp$-FEM. Technical Report, University of Texas at El Paso, Department of Mathematical Science (2005),
http://www.math.utep.edu/Faculty/solin/new_papers/dmp.pdf; see also
http://www.math.utep.edu/Faculty/solin/new_papers/dmp-coll.pdf
312. Šolín, P., Vejchodský, T., Araiza, R.: Discrete Conservation of Nonnegativity or Elliptic Problems Solved by the $hp$-FEM. Technical Report UTEP-CS-05-29, University of Texas at El Paso, Department of Computer Science (2005),
http://www.cs.utep.edu/vladik/2005/tr05-29.pdf
313. Srikrishnan, S., Araiza, R., Kreinovich, V., Starks, S.A., Xie, H.: Automatic referencing of satellite and radar images. In: Proceedings of the 2001 IEEE Systems, Man, and Cybernetics Conference, Tucson, Arizona, October 7-10, pp. 2176–2181 (2001)
314. Starks, S.A., Kreinovich, V., Longpré, L., Ceberio, M., Xiang, G., Araiza, R., Beck, J., Kandathi, R., Nayak, A., Torres, R.: Towards combining probabilistic and interval uncertainty in engineering calculations. In: Proceedings of the Workshop on Reliable Engineering Computing, Savannah, Georgia, September 15-17, pp. 193–213 (2004)
315. Stazi, F., Budyn, E., Chessa, J., Belytschko, T.: An extended finite element method with higher-order elements for crack problems with curvature. Computational Mechanics 31(1-2), 38–48 (2003)
316. Sukumar, N., Chopp, D.L., Moës, N., Belytschko, T.: Modeling holes and inclusions by level sets in the extended finite element method. Computer Methods in Applied Mechanics and Engineering 190(46-47), 6183–6200 (2000)
317. Sun, G., Fan, B., Chen, G., Zhou, Y.: Study on scheduling strategy for global computing application. In: Proceedings of the Seventh IEEE International Conference on Parallel and Distributed Computing, Applications and Technologies PDCAT 2006, pp. 368–372 (December 2006)

318. Sun, G., Shan, J., Chen, G.: Job scheduling for campus-scale global computing with machine availability constraints. In: Proceedings of the First International IEEE Multi-Symposiums on Computer and Computational Sciences IMSCCS 2006, June 20-24, vol. 1, pp. 385–388 (2006)

319. Sunaga, T.: Theory of interval algebra and its application to numerical analysis. RAAG Memoirs, Ggujutsu Bunken Fukuy-kai, Tokyo 2, 29–46, 547–564 (1958)

320. Suvorov, P.Y.: On the recognition of the tautological nature of propositional formulas. J. Sov. Math. 14, 1556–1562 (1980)

321. Sweeney, L.: Weaving technology and policy together to maintain confidentiality. Journal of Law, Medicine and Ethics 25, 98–110 (1997)

322. Sweeney, L.: k-anonymity: a model for protecting privacy. International Journal on Uncertainty, Fuzziness and Knowledge-based Systems 10(5), 557–570 (2002)

323. Takefusa, A., Matsuoka, S., Nakada, H., Aida, K., Nagashima, U.: Overview of a performance evaluation system for global computing scheduling algorithms. In: Proceedings of the Eighth International IEEE Symposium on High Performance Distributed Computing, August 3-6, pp. 97–104 (1999)

324. Tarantola, A.: Inverse Problem Theory: Methods for Data Fitting and Model Parameter Estimation. Elsevier, Amsterdam (1987)

325. Taubman, D.S., Marcellin, M.W.: JPEG2000 Image Compression Fundamentals, Standards and Practice. Kluwer, Boston (2002)

326. Taufer, M., An, C., Kerstens, A., Brooks III, C.L.: Predictor@Home: a protein structure prediction supercomputer based on global computing. IEEE Transactions on Parallel and Distributed Systems 17(8), 786–796 (2006)

327. Taufer, M., Kerstens, A., Estrada, T., Flores, D.A., Teller, P.J.: SimBA: a discrete event simulator for performance prediction of volunteer computing projects. In: Proceedings of the International Workshop on Principles of Advanced and Distributed Simulation 2007 PADS 2007, San Diego, California (June 2007)

328. Taufer, M., Kerstens, A., Estrada, T., Flores, D.A., Zamudio, R., Teller, P.J., Armen, R., Brooks III, C.L.: Moving volunteer computing towards knowledge-constructed, dynamically-adaptive modeling and scheduling. In: Proceedings of the First Workshop on Large-Scale, Volatile Desktop Grids PCGrid 2007, in Conjunction with IPDPS 2007, Long Beach, California (March 2007)

329. Taufer, M., Leung, M.Y., Johnson, K.L., Licon, A.: RNAVLab: A unified environment for computational RNA structure analysis based on grid computing technology. In: Proceedings of the 6th IEEE International Workshop on High Performance Computational Biology HiCOMB 2007, in Conjunction with IPDPS 2007, Long Beach, California (March 2007)

330. Tikhonov, A.N., Arsenin, V.Y.: Solutions of Ill-Posed Problems. W. H. Whinston & Sons, Washington, D.C (1977)

331. Trejo, R., Kreinovich, V.: Error estimations for indirect measurements, randomized vs. deterministic algorithms for 'black-box' programs. In: Rajasekaran, S., Pardalos, P., Reif, J., Rolim, J. (eds.) Handbook on Randomized Computing, pp. 673–729. Kluwer (2001)

332. Tucker, W.T., Ferson, S.: Probability Bounds Analysis in Environmental Risk Assessments. Technical Report, Applied Biomathematics (2003)

333. Ullman, J.D., Widom, J.: A First Course in Database Systems. Prentice Hall, Upper Saddle River (2002)

334. Vavasis, S.A.: Nonlinear Optimization: Complexity Issues. Oxford University Press, New York (1991)
335. Vernon, J.: Introduction to Engineering Materials. Palgrave Macmillan, New York (2003)
336. Villaverde, K., Xiang, G.: Estimating variance under interval and fuzzy uncertainty: parallel algorithms. In: Proceedings of the IEEE World Congress on Computational Intelligence WCCI 2008, Hong Kong, China, June 1-6, pp. 1030–1033 (2008)
337. Wadsworth Jr., H.M. (ed.): Handbook of Statistical Methods for Engineers and Scientists. McGraw-Hill Publishing Co., NY (1990)
338. Walley, P.: Statistical Reasoning with Imprecise Probabilities. Chapman & Hall, NY (1991)
339. Walster, G.W.: Philosophy and practicalities of interval arithmetic. In: Reliability in Computing, pp. 309–323. Academic Press, N.Y (1988)
340. Walster, G.W., Kreinovich, V.: For unknown-but-bounded errors, interval estimates are often better than averaging. ACM SIGNUM Newsletter 31(2), 6–19 (1996)
341. Wang, W.-S., Kreinovich, V., Orshansky, M.: Static timing analysis based on partial probabilistic description of delay uncertainty. In: Proceedings of the 43rd Design Automation Conference, San Francisco, California, July 24-28 (2006)
342. Warmus, M.: Calculus of approximations. Bulletin de l'Academie Polonaise de Sciences 4(5), 253–257 (1956)
343. Webster, R.: Convexity. Oxford University Press, Oxford (1994)
344. Wen, Q., Gates, A.Q., Beck, J., Kreinovich, V., Keller, G.R.: Towards automatic detection of erroneous measurement results in a gravity database. In: Proceedings of the 2001 IEEE Systems, Man, and Cybernetics Conference, Tucson, Arizona, October 7-10, pp. 2170–2175 (2001)
345. Williamson, R., Downs, T.: Probabilistic arithmetic I: numerical methods for calculating convolutions and dependency bounds. International Journal of Approximate Reasoning 4, 89–158 (1990)
346. Willenborg, L., De Waal, T.: Statistical Disclosure Control in Practice. Springer, New York (1996)
347. Wong, P.W.: Image quantization, halftoning, and printing. In: Bovik, A. (ed.) Handbook of Image and Video Processing, pp. 657–667. Academic Press, New York (2000)
348. Worden, K., Osegueda, R., Ferregut, C., Nazarian, S., George, D.L., George, M.J., Kreinovich, V., Kosheleva, O., Cabrera, C.: Interval methods in non destructive testing of aerospace structures and in mammography. In: Abstracts of the International Conference on Interval Methods and their Application in Global Optimization Interval 1998, Nanjing, China, April 20-23, pp. 152–154 (1998)
349. Worden, K., Osegueda, R., Ferregut, C., Nazarian, S., Rodriguez, E., George, D.L., George, M.J., Kreinovich, V., Kosheleva, O., Cabrera, S.: Interval approach to non-destructive testing of aerospace structures and to mammography. In: Alefeld, G., Trejo, R.A. (eds.) Workshop on Interval Computations and its Applications to Reasoning Under Uncertainty, Knowledge Representation, and Control Theory. Proceedings of MEXICON 1998, 4th World Congress on Expert Systems, México City, México (1998)

350. Worden, K., Osegueda, R., Ferregut, C., Nazarian, S., George, D.L., George, M.J., Kreinovich, V., Kosheleva, O., Cabrera, S.: Interval methods in non-destructive testing of material structures. Reliable Computing 7(4), 341–352 (2001)

351. Wu, B., Nguyen, H.T., Kreinovich, V.: Real-time algorithms for statistical analysis of interval data. In: Proceedings of the International Conference on Information Technology InTech 2003, Chiang Mai, Thailand, December 17-19, pp. 483–490 (2003)

352. Xiang, G.: Fast algorithm for computing the upper endpoint of sample variance for interval data: case of sufficiently accurate measurements. Reliable Computing 12(1), 59–64 (2006)

353. Xiang, G., Ceberio, M., Kreinovich, V.: Computing population variance and entropy under interval uncertainty: linear-time algorithms. Reliable Computing 13(6), 467–488 (2007)

354. Xiang, G., Hall, J.H.: Computing statistical characteristics when we know probabilities with interval or fuzzy uncertainty: computational complexity. In: Reformat, M., Berthold, M.R. (eds.) Proceedings of the 26th International Conference of the North American Fuzzy Information Processing Society NAFIPS 2007, San Diego, California, June 24-27, pp. 576–581 (2007)

355. Xiang, G., Kosheleva, O., Klir, G.J.: Estimating information amount under interval uncertainty: algorithmic solvability and computational complexity. In: Proceedings of the International Conference on Information Processing and Management of Uncertainty in Knowledge-Based Systems IPMU 2006, Paris, France, July 2-7, pp. 840–847 (2006)

356. Xiang, G., Kreinovich, V.: Estimating Variance Under Interval and Fuzzy Uncertainty: Case of Hierarchical Estimation. In: Melin, P., Castillo, O., Aguilar, L.T., Kacprzyk, J., Pedrycz, W. (eds.) IFSA 2007. LNCS (LNAI), vol. 4529, pp. 3–12. Springer, Heidelberg (2007)

357. Xiang, G., Kreinovich, V., Ferson, S.: Fitting a normal distribution to interval and fuzzy data. In: Reformat, M., Berthold, M.R. (eds.) Proceedings of the 26th International Conference of the North American Fuzzy Information Processing Society NAFIPS 2007, San Diego, California, June 24-27, pp. 560–565 (2007)

358. Xiang, G., Pownuk, A., Kosheleva, O., Starks, S.A.: Von Mises failure criterion in mechanics of materials: how to efficiently use it under interval and fuzzy uncertainty. In: Reformat, M., Berthold, M.R. (eds.) Proceedings of the 26th International Conference of the North American Fuzzy Information Processing Society NAFIPS 2007, San Diego, California, June 24-27, pp. 570–575 (2008)

359. Xiang, G., Starks, S.A., Kreinovich, V., Longpré, L.: New Algorithms for Statistical Analysis of Interval Data. In: Dongarra, J., Madsen, K., Waśniewski, J. (eds.) PARA 2004. LNCS, vol. 3732, pp. 189–196. Springer, Heidelberg (2006)

360. Xiang, G., Starks, S.A., Kreinovich, V., Longpré, L.: New Algorithms for Statistical Analysis of Interval Data. In: Dongarra, J., Madsen, K., Waśniewski, J. (eds.) PARA 2004. LNCS, vol. 3732, pp. 189–196. Springer, Heidelberg (2006)

361. Yager, R.R., Kreinovich, V.: Decision making under interval probabilities. International Journal of Approximate Reasoning 22(3), 195–215 (1999)

362. Young, W.H.: Sull due funzioni a piu valori constituite dai limiti d'una funzione di variable reale a destra ed a sinistra di ciascun punto. Rendiconti Academia di Lincei, Classes di Scienza Fiziche 17(5), 582–587 (1908)

363. Young, R.C.: The algebra of multi-valued quantities. Mathematische Annalen 104, 260–290 (1931)

364. Zadeh, L.A.: Fuzzy sets as a basis for a theory of possibility. Fuzzy Sets and Systems 1, 3–28 (1978)
365. Zamudio, R., Catarino, D., Taufer, M., Bhatia, K., Stearn, B.: Topaz: Extending Firefox to accommodate the GridFTP protocol. In: Proceedings of the Fourth High-Performance Grid Computing Workshop HPGC 2007, in Conjunction with IPDPS 2007, Long Beach, California (March 2007)
366. Zelt, C.A., Barton, P.J.: Three-dimensional seismic refraction tomography: A comparison of two methods applied to data from the Faeroe Basin. J. Geophysical Research 103, 7187–7210 (1998)
367. Zhang, J., Berleant, D.: Envelopes around cumulative distribution functions from interval parameters of standard continuous distributions. In: Proceedings, North American Fuzzy Information Processing Society (NAFIPS 2003), Chicago, pp. 407–412 (2003)
368. Zhang, W., Shmulevich, I., Astola, J.: Microarray Quality Control. Wiley, Hoboken (2004)

# Index

CPSIA information can be obtained at www.ICGtesting.com
Printed in the USA
LVOW01s1333020214

371966LV00004B/189/P